T0256950

Dodge Pick-ups Automotive Repair Manual

by Tim Imhoff
and John H Haynes
Member of the Guild of Motoring Writers

Models covered:
Dodge Full-size Pick-ups
2009 through 2018
2WD & 4WD, V6 and V8 gasoline engines
and Cummins turbo-diesel engine

Does not include information specific to the Ram 4500 or 5500, 2009 fleet models equipped with the 5.9L diesel engine or models with the 3.0L V6 diesel engine

(30043-8Y10)

Haynes Group Limited
Haynes North America, Inc.
www.haynes.com

Acknowledgements

Wiring diagrams originated exclusively for Haynes North America, Inc. by Valley Forge Technical Information Services.

© **Haynes North America, Inc. 2012, 2015, 2016, 2018**
With permission from Haynes Group Limited

A book in the Haynes Automotive Repair Manual Series

ISBN-13: 978-1-62092-342-9
ISBN-10: 1-62092-342-4

Library of Congress Control Number: 2018958423

While every attempt is made to ensure that the information in this manual is correct, no liability can be accepted by the authors or publishers for loss, damage or injury caused by any errors in, or omissions from, the information given.

Contents

Haynes mechanic and photographer with a 2012 Dodge Ram pickup

About this manual

Its purpose

The purpose of this manual is to help you get the best value from your vehicle. It can do so in several ways. It can help you decide what work must be done, even if you choose to have it done by a dealer service department or a repair shop; it provides information and procedures for routine maintenance and servicing; and it offers diagnostic and repair procedures to follow when trouble occurs.

We hope you use the manual to tackle the work yourself. For many simpler jobs, doing it yourself may be quicker than arranging an appointment to get the vehicle into a shop and making the trips to leave it and pick it up. More importantly, a lot of money can be saved by avoiding the expense the shop must pass on to you to cover its labor and overhead costs. An added benefit is the sense of satisfaction and accomplishment that you feel after doing the job yourself.

Using the manual

The manual is divided into Chapters. Each Chapter is divided into numbered Sections, which are headed in bold type between horizontal lines. Each Section consists of consecutively numbered paragraphs.

At the beginning of each numbered Section you will be referred to any illustrations which apply to the procedures in that Section. The reference numbers used in illustration captions pinpoint the pertinent Section and the Step within that Section. That is, illustration 3.2 means the illustration refers to Section 3 and Step (or paragraph) 2 within that Section.

Procedures, once described in the text, are not normally repeated. When it's necessary to refer to another Chapter, the reference will be given as Chapter and Section number. Cross references given without use of the word "Chapter" apply to Sections and/or paragraphs in the same Chapter. For example, "see Section 8" means in the same Chapter.

References to the left or right side of the vehicle assume you are sitting in the driver's seat, facing forward.

Even though we have prepared this manual with extreme care, neither the publisher nor the author can accept responsibility for any errors in, or omissions from, the information given.

NOTE

A **Note** provides information necessary to properly complete a procedure or information which will make the procedure easier to understand.

CAUTION

A **Caution** provides a special procedure or special steps which must be taken while completing the procedure where the Caution is found. Not heeding a Caution can result in damage to the assembly being worked on.

WARNING

A **Warning** provides a special procedure or special steps which must be taken while completing the procedure where the Warning is found. Not heeding a Warning can result in personal injury.

Introduction

Dodge Ram pick-ups are available in regular, mega-cab, quad-cab and four door crew-cab body styles. All cabs are single welded unit construction and bolted to the frame. They are available in short-bed and long-bed models. All models are available in two-wheel drive (2WD) and four-wheel drive (4WD) versions.

Powertrain options include a 6.7L inline six-cylinder diesel engine and 3.6L V6, 3.7L V6, 4.7L V8 and 5.7L/6.4L Hemi V8 gasoline engines. Transmissions used are either a four-speed or five-speed automatic, or five-speed, six-speed or eight-speed manual.

Chassis layout is conventional, with the engine mounted at the front and the power being transmitted through either the manual or automatic transmission to a driveshaft and solid rear axle. On 4WD models, a transfer case also transmits power to the front axle by way of a driveshaft.

The front suspension on 2WD models features an independent coil spring and shock absorber upper and lower A-arm type front suspension, while heavy duty 4WD models use coil springs and a solid axle located by four links. 4WD models with independent front suspension use coil-over shock absorbers and upper and lower control arms. 2500 and 3500 models have a solid axle and leaf springs at the rear. 1500 models have a coil spring rear suspension.

All models are equipped with power assisted disc front and rear brakes. Rear Wheel Anti-Lock (RWAL) brakes are standard with a four-wheel Anti-lock Braking System (ABS) used on some models.

Vehicle identification numbers

Modifications are a continuing and unpublicized process in vehicle manufacturing. Since spare parts manuals and lists are compiled on a numerical basis, the individual vehicle numbers are essential to correctly identify the component required.

Vehicle Identification Number (VIN)

This very important identification number is stamped on a plate attached to the left side of the dashboard just inside the windshield on the driver's side of the vehicle (see illustration). The VIN also appears on the Vehicle Certificate of Title and Registration. It contains information such as where and when the vehicle was manufactured, the model year and the body style.

VIN year and engine codes

Two particularly important pieces of information located in the VIN are the model year and engine codes. Counting from the left, the engine code is the eighth digit and the model year code is the 10th digit.

On the models covered by this manual the engine codes are:

G	3.6L V6
K	3.7L V6
P	4.7L V8
T	5.7L V8
2	5.7L V8 CNG
J	6.4L V8
L	6.7L diesel

On the models covered by this manual the model year codes are:

9	2009
A	2010
B	2011
C	2012
D	2013
E	2014
F	2015
G	2016
H	2017
J	2018

Equipment identification plate

This plate is located on the inside of the hood. It contains valuable information concerning the production of the vehicle as well as information on all production or special equipment.

Safety Certification label

The Safety Certification label is affixed to the left front door (see illustration). The label contains the name of the manufacturer, the month and year of production, the Gross Vehicle Weight Rating (GVWR) and the safety certification statement. This label also contains the paint code. It is especially useful for matching the color and type of paint during repair work.

Engine identification number

The engine ID number on gasoline engines is located at the right side of the engine block (see illustration). The engine ID number on diesel engines is on the data plate which is on the valve cover; it's also stamped on the right side of the block.

The VIN is visible from the outside of the vehicle, through the driver's side of the windshield

Typical Safety Certification label

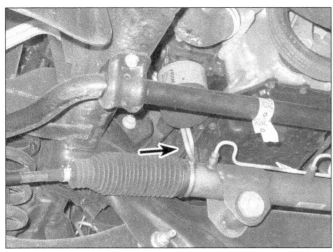

On Hemi engines, the identification number is located on a pad near the right side engine mount

Automatic transmission identification number location

Transmission identification number

The ID number on manual transmissions is located on the left side of the case. On automatic transmissions, the number is stamped on the left side of the transmission case above the oil pan flange **(see illustration)**.

Transfer case identification number

The transfer case identification plate is attached to the rear side of the case **(see illustration)**.

Typical transfer case identification tag location

Buying parts

Replacement parts are available from many sources, which generally fall into one of two categories - authorized dealer parts departments and independent retail auto parts stores. Our advice concerning these parts is as follows:

Retail auto parts stores: Good auto parts stores will stock frequently needed components which wear out relatively fast, such as clutch components, exhaust systems, brake parts, tune-up parts, etc. These stores often supply new or reconditioned parts on

an exchange basis, which can save a considerable amount of money. Discount auto parts stores are often very good places to buy materials and parts needed for general vehicle maintenance such as oil, grease, filters, spark plugs, belts, touch-up paint, bulbs, etc. They also usually sell tools and general accessories, have convenient hours, charge lower prices and can often be found not far from home.

Authorized dealer parts department: This is the best source for parts which are

unique to the vehicle and not generally available elsewhere (such as major engine parts, transmission parts, trim pieces, etc.).

Warranty information: If the vehicle is still covered under warranty, be sure that any replacement parts purchased - regardless of the source - do not invalidate the warranty!

To be sure of obtaining the correct parts, have engine and chassis numbers available and, if possible, take the old parts along for positive identification.

Recall information

Vehicle recalls are carried out by the manufacturer in the rare event of a possible safety-related defect. The vehicle's registered owner is contacted at the address on file at the Department of Motor Vehicles and given the details of the recall. Remedial work is carried out free of charge at a dealer service department.

If you are the new owner of a used vehicle which was subject to a recall and you want to be sure that the work has been carried out, it's best to contact a dealer service department and ask about your individual vehicle - you'll need to furnish them your Vehicle Identification Number (VIN).

The table below is based on information provided by the National Highway Traffic Safety Administration (NHTSA), the body which oversees vehicle recalls in the United States. The recall database is updated constantly. **Note:** *This a partial list containing only dealer recalls. There are additional aftermarket recalls available.* For the latest information on vehicle recalls, check the NHTSA website at www.nhtsa.gov, www.safercar.gov, or call the NHTSA hotline at 1-888-327-4236.

Recall date	Recall campaign number	Model(s) affected	Concern
JAN 06, 2009	09V005000	2009 Ram 2500 2009 Ram 3500	On some models, the steering linkage drag link inner tie rod-to-Pitman arm ballstud may fracture. Also, the steering linkage damper attaching the bracket may yield and shift on the linkage. This could result in a loss of steering control and the restricted ability to turn the vehicle in one direction, increasing the risk of a crash.
MAR 09, 2009	09E009000	2009 Ram 1500	On some models, the windshield wiper module may be susceptible to water intrusion that could result in partial or complete loss of windshield wiper capability. When the wiper system fails, the operator will have reduced visibility which could result in a crash.
MAR 09, 2009	09V078000	2009 Ram 1500	On some models, the clutch pedal connecting rod to the clutch master cylinder may separate from the master cylinder. This may not allow disengagement of the clutch when the pedal is depressed, which could result in unintended vehicle movement. Increased stopping distance and engine stalling may occur, increasing the risk of a crash.
MAY 05, 2009	09V158000	2009 Ram 1500	On some models, the software programmed into the heating, ventilation and air conditioning module may cause the windshield defrosting and defogging functions to become inoperative. This can decrease the driver's visibility under certain driving conditions and result in a crash.
JAN 07, 2010	10V009000	2009 Ram 1500 2010 Ram 1500	Some models may have been built with an improperly formed or missing brake booster input rod retaining clip. This could result in brake failure without warning which could cause a crash.
MAY 06, 2010	10V200000	2010 Ram 1500	Some models may have been built with a Wireless Ignition Node (WIN) module exhibiting a binding condition of the solenoid latch. The result of the defect could lead to a condition where the key may be removed from the ignition switch (WIN module) prior to placing the shifter in Park. This could result in the potential for unintended vehicle movement and could increase the risk of a crash.

Recall date	Recall campaign number	Model(s) affected	Concern
JUL 07, 2010	10V315000	2010 Ram 1500	Some models may have been built with an improperly formed master cylinder-to-Hydraulic Control Unit (HCU) brake tube assembly end flare. This could lead to loss of brake fluid and reduced braking performance increasing the risk of a crash.
OCT 07, 2010	10V474000	2011 Ram 1500	Some models have been built with certification and/or supplemental tire pressure information labels printed with incorrect weight or seating capacity information. The misinformation on these labels could lead to, among other things, failure to follow proper vehicle loading specifications that could increase the risk of a crash.
OCT 07, 2010	10V475000	2011 Ram 1500	Some models may experience a separation at the crimped end of the power steering pressure hose assembly. Leaked power steering fluid onto hot engine components could cause a fire.
DEC 07, 2010	10V616000	2010 Ram 1500 2011 Ram 1500	Some models equipped with a diesel engine and a hydroboost brake system may be equipped with a power steering reservoir cap with excessive vent pressure levels that may result in brake pedals that are slow to return. Brake lights that are slow to extinguish could increase the risk of a crash.
DEC 23, 2010	10V657000	2009 Ram 1500 2010 Ram 1500 2011 Ram 1500	Some models may experience a weakening and fracture of the left ballstud on the tie-rod resulting in the potential loss of steering, which could result in a crash.
JUL 06, 2011	11V350000	2009 Ram 2500, 3500 2010 Ram 2500, 3500 2011 Ram 2500, 3500	On some models, the left tie-rod ballstud may fracture. This condition tends to occur during low speed parking lot type maneuvers when the driver is making a tight turn. This condition could result in the potential loss of directional stability in the left-hand front wheel, increasing the risk of a crash.
MAY 01, 2012	12V192000	2012 Ram 1500	Some models have a spare tire that does not match the specifications listed on the tire placard. These vehicles fail to comply with the requirements of Federal Motor Vehicle Safety Standard no. 110, "Tire selection and rims." If installed, the incorrectly sized spare tire may unexpectedly activate the Electronic Stability Control (ESC) system, altering the effectiveness of the ESC system, increasing the risk of a crash.
APR 3, 2013	13V128000	2013 Ram 1500	On some models, the parking brake cable equalizer was set incorrectly and may not hold the vehicle on a 20% grade. If the parking brake fails to operate as designed, the vehicle could roll when parked, increasing the risk of a crash.

Recall date	Recall campaign number	Model(s) affected	Concern
MAY 7, 2013	13V177000	2013 Ram 1500	On some models, the coolant bypass valve may stick in a position that does not allow coolant to flow into the heater core. Without a properly working windshield defrosting system, a buildup of moisture or ice could limit the driver's ability to see, increasing the risk of a crash.
JUN 2013	13V237000	2013 Ram 1500 2013 Ram 2500 2013 Ram 3500	On some models, the Body Control Module (BCM) may not give a double-flash notification of the turn signal malfunction. Loss of turn signal function could result in an accident
JUL 2013	13V240000	2013 Ram 2500 2013 Ram 3500	On some models, the premium headlamps were not properly configured for high beam aim and intensity, in the Body Control Module (BCM). The loss of high beam intensity could result in an accident.
AUG 2013	13V22000	2013 Ram 2500 2013 Ram 3500 6.7L diesel	On some vehicles, the engine cover may experience a condition where it does not insulate the area on the passenger side of the engine compartment. This could cause engine components under the engine cover to overheat and possibly cause an underhood fire.
OCT 17, 2013	13V486000	2014, Ram 1500, 2500 and 3500	On some models, the warning lights in the instrument cluster may fail to illuminate as designed. Inoperative warning lights may not properly warn the driver of the vehicle problems, increasing the risk of a crash
APR 01, 2014	14V161000	2014 Ram 1500	On some models, the transmission case may have been improperly machined which can result in the park pawl not properly engaging when the transmission is in the 'Park' position. If the park pawl does not properly engage, the vehicle may roll away, increasing the risk of a crash and occupant or pedestrian injuries.
JUL 2, 2014	14V392000	2014 Ram 1500	On some models, due to an insufficient weld, the rear shocks may detach from the vehicle at one end and possibly damage other chassis components, the tire or result in reduced braking.
OCT 8, 2014	14V635000	2010 - 2014 Ram 2500, 3500	On some models, the electrical connectors of the diesel fuel heater may overheat. If the connectors overheat, the fuel heater may leak fuel, increasing the risk of a fire.
JUL 22, 2015	15V459000	2012 – 2014 Ram 1500, 2500 and 3500	On some models equipped with the Electronic Vehicle Information Center option, a steering wheel wiring harness may rub against the driver's airbag module retainer spring. This abrasion may result in an electrical short that could cause the driver's frontal airbag to unexpectedly deploy. Inadvertant deployment of the airbag may increase the risk of injury and the possibility of a crash.

Recall date	Recall campaign number	Model(s) affected	Concern
JUL 22, 2015	15V460000	2013-2015 Ram 1500, 2500 and 3500	On some models, the side impact sensor calibrations may be overly sensitive, and as a result, the side air bag inflatable curtains and seat air bags may unexpectedly deploy and the seat belt pre-tensioners may activate, which could result in an accident or injury
JUL 23, 2015	15V461000	2013-2015 Ram 1500, 2500 and 3500	Some models are equipped with radios that have software vulnerabilities that can allow third-party access to certain networked vehicle control systems. Exploitation of the software vulnerability may result in unauthorized remote modification and control of certain vehicle systems, increasing the risk of a crash.
AUG 14, 2015	15V517000	2014-2015 Ram 1500 (diesel, 2WD)	Some models have a battery harness that may rub against the frame bracket for the right engine mount, resulting in an electrical short. If the harness shorts, the vehicle may stall and there is an increased risk of a fire.
AUG 21, 2015	15V534000	2014-2015 Ram 1500 Quad Cab	On some models, the side-impact airbag(s) may not inflate as intended (to fully overlap the C-pillar), increasing the risk of injury to a rear-seat occupant in an accident.
SEP 23, 2015	15V592000	2016 Ram 2500, 3500	On some models, the Engine Control Module (ECM) may cause the engine to stall without warning due to a short, increasing the risk of an accident
OCT 15, 2015	15V661000	2015-2016 Ram 1500	On some models, the rear axles were incorrectly heat treated. If the rear axle shaft was not properly heated treated, it may fracture and a wheel separation could occur, increasing the risk of a crash.
OCT 16, 2015	15V663000	2016 Ram 1500	On some models, the dual-exhaust heat shield may be missing, resulting in heat damage to the spare tire. If used, the tire tread may separate, increasing the risk of a crash.
MAR 23, 2016	16V167000	2015-2016 Ram 1500	On some models, the Electric Power Steering (EPS) control circuit board may cause a short and result in the loss of power steering assist, increasing the risk of a crash.
MAY 16, 2016	16V301000	2016 Ram 1500	On some models, the front driveaxles may fracture and cause damage to the undercarriage, which could result in road debris and/or cause an accident.
MAY 16, 2016	16E041000	2012-2016 RAM 1500, 2500, and 3500	On some models, the aftermarket Mopar canvas seat covers can impede the deployment of the seat thorax airbags, increasing the risk of a injury in a crash.
MAR 23, 2017	17V198000	2016 – 2017 Ram 1500	On the affected vehicles, the differential pin retaining screw may come loose while driving, potentially causing the differential to break or lock up. A broken or locked-up differential may result in a loss of motive power and possibly a loss of vehicle control. Either scenario increases the risk of a crash.

Recall date	Recall campaign number	Model(s) affected	Concern
MAY 9, 2017	17V302000	2013 – 2016 Ram 1500, 2500 and 2014 – 2016 Ram 2500	On the affected vehicles, certain driving conditions, such as driving off-road or debris striking the vehicle may cause the roll rate sensor to trigger a fault within the Occupant Restraint Controller (ORC). If this fault occurs, the rollover side curtain airbag and the seat belt pretensioner will be disabled from deploying. If the rolloever side curtain airbags and seat belt pretensioners are disabled, there is an increased risk of injury to the vehicle occupants in the event of a crash that necessistates activation of these safety systems.
JUL 7, 2017	17V434000	2017 Ram 1500	On the affected vehicles, the fuel tank may have a broken fuel tank control valve which might leak fuel if the vehicle were to become inverted. In the event of a rollover crash, leaking fuel in the presence of an ignition source can increase the risk of a fire.
AUG 10, 2017	17V499000	2013 – 2017, 2500 and 3500	On trucks, modified by AEV to be equipped with cast aluminum Katla 8.5" road wheels. The outer surface of the wheel may fracture, resulting in rapid air loss. If the wheel fractures, a loss of vehicle control can occur, increasing the risk of a crash.
SEP 12, 2017	17V562000	2013 – 2017 Ram 2500 and 3500	On models equipped with Cummins 6.7L Turbo Diesel engines with a Concentric-brand water pump without a vent hole, the water pump may leak coolant. The leaking water pump can increase the risk of an engine compartment fire.
DEC 20, 2017	17V821000	2009 – 2017 Ram 1500, 2500 and 3500	In the affected vehicles, all equipped with a column shifter, pushing the brake pedal for prolonged periods when the vehicle is running and in PARK may cause the Brake Transmission Shift Interlock (BTSI) pin to stick in the open position. With the pin in the open position, the transmission can be shifted out of PARK into any gear without pushing the brake pedal or having the key in the ignition. Being able to shift the transmission without pushing the brake pedal and/or without a key in the ignition. This can cause the risk of an unitended vehicle rollaway that may result in personal injury or a crash.
DEC 21, 2017	17V824000	2017 – 2018 Ram 1500, 2500 and 3500	Dodge is recalling certain vehicles equipped with Kidde Plastic-Handle or Push Button 'Pindicator' Fire Extinguishers. These extinguishers may become clogged, preventing the extinguisher from discharging as expected or requiring excessive force to activate the extinguisher. Additionally, in certain models, the nozzle may detach from the valve assembly with enough force that it could cause injury and also render the product inoperable. If the fire extinguisher does not function properly, it can increase the risk of injury in the event of a fire.

Recall date	Recall campaign number	Model(s) affected	Concern
FEB 8, 2018	18V100000	2017 – 2018 Ram 1500, 2500 and 3500	Dodge is recalling certain vehicles equipped with Kidde Plastic-Handle or Push Button 'Pindicator' Fire Extinguishers. These extinguishers may become clogged, preventing the extinguisher from discharging as expected or requiring excessive force to activate the extinguisher. Additionally, in certain models, the nozzle may detach from the valve assembly with enough force that it could cause injury and also render the product inoperable. If the fire extinguisher does not function properly, it can increase the risk of injury in the event of a fire.
FEB 8, 2018	18V100000	2017 – 2018 Ram 1500, 2500 and 3500	On models equipped with a column shifter, pushing the brake pedal for prolonged periods when the vehicle is running and in PARK may cause the Brake Transmission Shift Interlock (BTSI) pin to stick in the open position. With the pin in the open position, the transmission can be shifted out of PARK into any gear without pushing the brake pedal or having the key in the ignition. Being able to shift the transmission without pushing the brake pedal and/or without a key in the ignition can increase the risk of an unintended vehicle rollaway that may result in personal injury or a crash.
MAY 1, 2018	18V280000	2018 Ram 1500	An incorrect transmission park lock rod may have been installed in the transmission. If the incorrect park lock rod is installed, the transmission may not shift into "PARK" and keep the vehicle from moving, increasing the risk of unitended vehicle movement and the risk of a crash.
MAY 17, 2018	18V332000	2014 - 2019 Ram 1500, 2016 - 2018 Ram 3500	These vehicles are being recalled to address a defect that could prevent the cruise control system from disengaging. If, when using cruise control, there is a short circuit within the vehicle's wiring, the driver may not be able to shut off the cruise control either by depressing the brake pedal or manually turning the system off once it has been engaged, resulting in either the vehicle maintaining its current speed or possibly accelerating. If the vehicle maitains its speed or accelerates despite attempts tp deactivate the cruise control, there would be an increased risk of a crash.
JUN 14, 2018	18V398000	2018 Ram 1500, 2500 and 3500	On these models, the backup camera may experience a loss of image display while backing up. A loss of image in the rearview camera while backing up can increse the risk of a crash.

Maintenance techniques, tools and working facilities

Maintenance techniques

There are a number of techniques involved in maintenance and repair that will be referred to throughout this manual. Application of these techniques will enable the home mechanic to be more efficient, better organized and capable of performing the various tasks properly, which will ensure that the repair job is thorough and complete.

Fasteners

Fasteners are nuts, bolts, studs and screws used to hold two or more parts together. There are a few things to keep in mind when working with fasteners. Almost all of them use a locking device of some type,

either a lockwasher, locknut, locking tab or thread adhesive. All threaded fasteners should be clean and straight, with undamaged threads and undamaged corners on the hex head where the wrench fits. Develop the habit of replacing all damaged nuts and bolts with new ones. Special locknuts with nylon or fiber inserts can only be used once. If they are removed, they lose their locking ability and must be replaced with new ones.

Rusted nuts and bolts should be treated with a penetrating fluid to ease removal and prevent breakage. Some mechanics use turpentine in a spout-type oil can, which works quite well. After applying the rust penetrant, let it work for a few minutes before trying to loosen the nut or bolt. Badly rusted fasten-

ers may have to be chiseled or sawed off or removed with a special nut breaker, available at tool stores.

If a bolt or stud breaks off in an assembly, it can be drilled and removed with a special tool commonly available for this purpose. Most automotive machine shops can perform this task, as well as other repair procedures, such as the repair of threaded holes that have been stripped out.

Flat washers and lockwashers, when removed from an assembly, should always be replaced exactly as removed. Replace any damaged washers with new ones. Never use a lockwasher on any soft metal surface (such as aluminum), thin sheet metal or plastic.

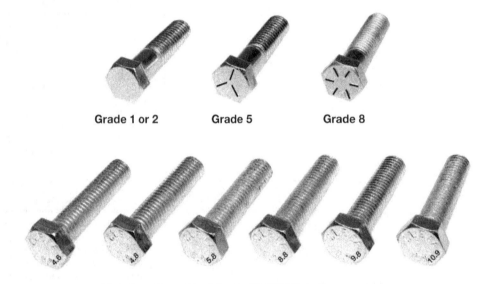

Grade 1 or 2   Grade 5   Grade 8

Bolt strength marking (standard/SAE/USS; bottom - metric)

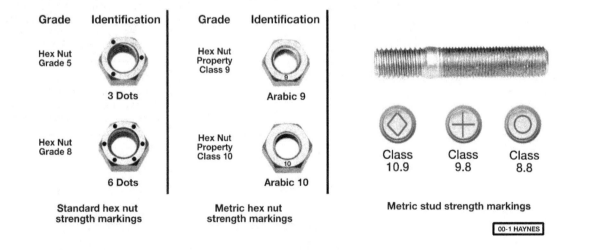

Grade	Identification	Grade	Identification
Hex Nut Grade 5	3 Dots	Hex Nut Property Class 9	Arabic 9
Hex Nut Grade 8	6 Dots	Hex Nut Property Class 10	Arabic 10

Standard hex nut strength markings

Metric hex nut strength markings

Class 10.9   Class 9.8   Class 8.8

Metric stud strength markings

00-1 HAYNES

Fastener sizes

For a number of reasons, automobile manufacturers are making wider and wider use of metric fasteners. Therefore, it is important to be able to tell the difference between standard (sometimes called U.S. or SAE) and metric hardware, since they cannot be interchanged.

All bolts, whether standard or metric, are sized according to diameter, thread pitch and length. For example, a standard 1/2 - 13 x 1 bolt is 1/2 inch in diameter, has 13 threads per inch and is 1 inch long. An M12 - 1.75 x 25 metric bolt is 12 mm in diameter, has a thread pitch of 1.75 mm (the distance between threads) and is 25 mm long. The two bolts are nearly identical, and easily confused, but they are not interchangeable.

In addition to the differences in diameter, thread pitch and length, metric and standard bolts can also be distinguished by examining the bolt heads. To begin with, the distance across the flats on a standard bolt head is measured in inches, while the same dimension on a metric bolt is sized in millimeters (the same is true for nuts). As a result, a standard wrench should not be used on a metric bolt and a metric wrench should not be used on a standard bolt. Also, most standard bolts have slashes radiating out from the center of the head to denote the grade or strength of the bolt, which is an indication of the amount of torque that can be applied to it. The greater the number of slashes, the greater the strength of the bolt. Grades 0 through 5 are commonly used on automobiles. Metric bolts have a property class (grade) number, rather than a slash, molded into their heads to indicate bolt strength. In this case, the higher the number, the stronger the bolt. Property class numbers 8.8, 9.8 and 10.9 are commonly used on automobiles.

Strength markings can also be used to distinguish standard hex nuts from metric hex nuts. Many standard nuts have dots stamped into one side, while metric nuts are marked with a number. The greater the number of dots, or the higher the number, the greater the strength of the nut.

Metric studs are also marked on their ends according to property class (grade). Larger studs are numbered (the same as metric bolts), while smaller studs carry a geometric code to denote grade.

It should be noted that many fasteners, especially Grades 0 through 2, have no distinguishing marks on them. When such is the case, the only way to determine whether it is standard or metric is to measure the thread pitch or compare it to a known fastener of the same size.

Standard fasteners are often referred to as SAE, as opposed to metric. However, it should be noted that SAE technically refers to a non-metric fine thread fastener only. Coarse thread non-metric fasteners are referred to as USS sizes.

Since fasteners of the same size (both standard and metric) may have different

Metric thread sizes	Ft-lbs	Nm
M-6	6 to 9	9 to 12
M-8	14 to 21	19 to 28
M-10	28 to 40	38 to 54
M-12	50 to 71	68 to 96
M-14	80 to 140	109 to 154

Pipe thread sizes		
1/8	5 to 8	7 to 10
1/4	12 to 18	17 to 24
3/8	22 to 33	30 to 44
1/2	25 to 35	34 to 47

U.S. thread sizes		
1/4 - 20	6 to 9	9 to 12
5/16 - 18	12 to 18	17 to 24
5/16 - 24	14 to 20	19 to 27
3/8 - 16	22 to 32	30 to 43
3/8 - 24	27 to 38	37 to 51
7/16 - 14	40 to 55	55 to 74
7/16 - 20	40 to 60	55 to 81
1/2 - 13	55 to 80	75 to 108

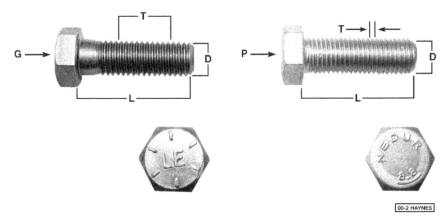

Standard (SAE and USS) bolt dimensions/ grade marks

G   Grade marks (bolt strength)
L   Length (in inches)
T   Thread pitch (number of threads per inch)
D   Nominal diameter (in inches)

Metric bolt dimensions/grade marks

P   Property class (bolt strength)
L   Length (in millimeters)
T   Thread pitch (distance between threads in millimeters)
D   Diameter

strength ratings, be sure to reinstall any bolts, studs or nuts removed from your vehicle in their original locations. Also, when replacing a fastener with a new one, make sure that the new one has a strength rating equal to or greater than the original.

Tightening sequences and procedures

Most threaded fasteners should be tightened to a specific torque value (torque is the twisting force applied to a threaded component such as a nut or bolt). Overtightening the fastener can weaken it and cause it to break, while undertightening can cause it to eventually come loose. Bolts, screws and studs,

depending on the material they are made of and their thread diameters, have specific torque values, many of which are noted in the Specifications at the beginning of each Chapter. Be sure to follow the torque recommendations closely. For fasteners not assigned a specific torque, a general torque value chart is presented here as a guide. These torque values are for dry (unlubricated) fasteners threaded into steel or cast iron (not aluminum). As was previously mentioned, the size and grade of a fastener determine the amount of torque that can safely be applied to it. The figures listed here are approximate for Grade 2 and Grade 3 fasteners. Higher grades can tolerate higher torque values.

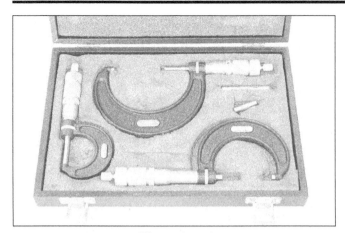

Micrometer set

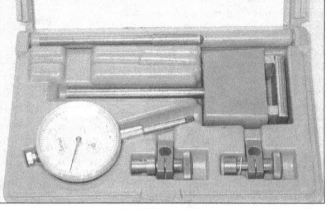

Dial indicator set

Fasteners laid out in a pattern, such as cylinder head bolts, oil pan bolts, differential cover bolts, etc., must be loosened or tightened in sequence to avoid warping the component. This sequence will normally be shown in the appropriate Chapter. If a specific pattern is not given, the following procedures can be used to prevent warping.

Initially, the bolts or nuts should be assembled finger-tight only. Next, they should be tightened one full turn each, in a criss-cross or diagonal pattern. After each one has been tightened one full turn, return to the first one and tighten them all one-half turn, following the same pattern. Finally, tighten each of them one-quarter turn at a time until each fastener has been tightened to the proper torque. To loosen and remove the fasteners, the procedure would be reversed.

Component disassembly

Component disassembly should be done with care and purpose to help ensure that the parts go back together properly. Always keep track of the sequence in which parts are removed. Make note of special characteristics or marks on parts that can be installed more than one way, such as a grooved thrust washer on a shaft. It is a good idea to lay the disassembled parts out on a clean surface in the order that they were removed. It may also be helpful to make sketches or take instant photos of components before removal.

When removing fasteners from a component, keep track of their locations. Sometimes threading a bolt back in a part, or putting the washers and nut back on a stud, can prevent mix-ups later. If nuts and bolts cannot be returned to their original locations, they should be kept in a compartmented box or a series of small boxes. A cupcake or muffin tin is ideal for this purpose, since each cavity can hold the bolts and nuts from a particular area (i.e. oil pan bolts, valve cover bolts, engine mount bolts, etc.). A pan of this type is especially helpful when working on assemblies with very small parts, such as the carburetor, alternator, valve train or interior dash and trim pieces. The cavities can be marked with paint or tape to identify the contents.

Whenever wiring looms, harnesses or connectors are separated, it is a good idea to identify the two halves with numbered pieces of masking tape so they can be easily reconnected.

Gasket sealing surfaces

Throughout any vehicle, gaskets are used to seal the mating surfaces between two parts and keep lubricants, fluids, vacuum or pressure contained in an assembly.

Many times these gaskets are coated with a liquid or paste-type gasket sealing compound before assembly. Age, heat and pressure can sometimes cause the two parts to stick together so tightly that they are very difficult to separate. Often, the assembly can be loosened by striking it with a soft-face hammer near the mating surfaces. A regular hammer can be used if a block of wood is placed between the hammer and the part. Do not hammer on cast parts or parts that could be easily damaged. With any particularly stubborn part, always recheck to make sure that every fastener has been removed.

Avoid using a screwdriver or bar to pry apart an assembly, as they can easily mar the gasket sealing surfaces of the parts, which must remain smooth. If prying is absolutely necessary, use an old broom handle, but keep in mind that extra clean up will be necessary if the wood splinters.

After the parts are separated, the old gasket must be carefully scraped off and the gasket surfaces cleaned. Stubborn gasket material can be soaked with rust penetrant or treated with a special chemical to soften it so it can be easily scraped off. **Caution:** *Never use gasket removal solutions or caustic chemicals on plastic or other composite components.* A scraper can be fashioned from a piece of copper tubing by flattening and sharpening one end. Copper is recommended because it is usually softer than the surfaces to be scraped, which reduces the chance of gouging the part. Some gaskets can be removed with a wire brush, but regardless of the method used, the mating surfaces must be left clean and smooth. If for some reason the gasket surface is gouged, then a gasket

sealer thick enough to fill scratches will have to be used during reassembly of the components. For most applications, a non-drying (or semi-drying) gasket sealer should be used.

Hose removal tips

Warning: *If the vehicle is equipped with air conditioning, do not disconnect any of the A/C hoses without first having the system depressurized by a dealer service department or a service station.*

Hose removal precautions closely parallel gasket removal precautions. Avoid scratching or gouging the surface that the hose mates against or the connection may leak. This is especially true for radiator hoses. Because of various chemical reactions, the rubber in hoses can bond itself to the metal spigot that the hose fits over. To remove a hose, first loosen the hose clamps that secure it to the spigot. Then, with slip-joint pliers, grab the hose at the clamp and rotate it around the spigot. Work it back and forth until it is completely free, then pull it off. Silicone or other lubricants will ease removal if they can be applied between the hose and the outside of the spigot. Apply the same lubricant to the inside of the hose and the outside of the spigot to simplify installation.

As a last resort (and if the hose is to be replaced with a new one anyway), the rubber can be slit with a knife and the hose peeled from the spigot. If this must be done, be careful that the metal connection is not damaged.

If a hose clamp is broken or damaged, do not reuse it. Wire-type clamps usually weaken with age, so it is a good idea to replace them with screw-type clamps whenever a hose is removed.

Tools

A selection of good tools is a basic requirement for anyone who plans to maintain and repair his or her own vehicle. For the owner who has few tools, the initial investment might seem high, but when compared to the spiraling costs of professional auto maintenance and repair, it is a wise one.

To help the owner decide which tools are

Dial caliper

Hand-operated vacuum pump

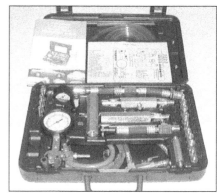

Fuel pressure gauge set

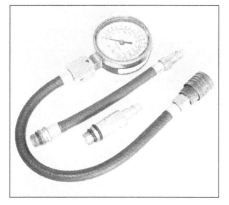

Compression gauge with spark plug
hole adapter

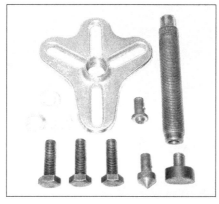

Damper/steering wheel puller

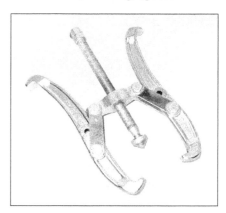

General purpose puller

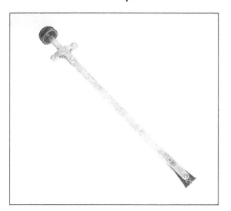

Hydraulic lifter removal tool

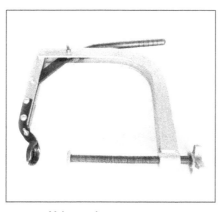

Valve spring compressor

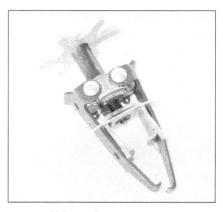

Valve spring compressor

Ridge reamer

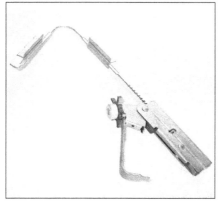

Piston ring groove cleaning tool

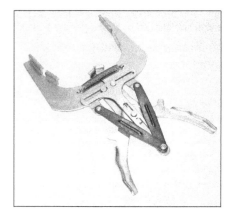

Ring removal/installation tool

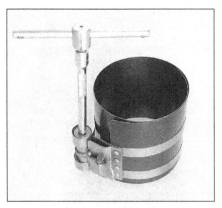

Ring compressor

Cylinder hone

Brake hold-down spring tool

Torque angle gauge

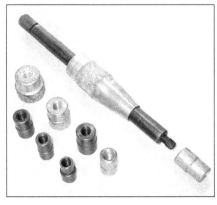

Clutch plate alignment tool

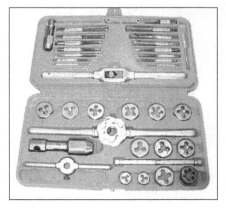

Tap and die set

needed to perform the tasks detailed in this manual, the following tool lists are offered: *Maintenance and minor repair, Repair/overhaul* and *Special*.

The newcomer to practical mechanics should start off with the *maintenance and minor repair* tool kit, which is adequate for the simpler jobs performed on a vehicle. Then, as confidence and experience grow, the owner can tackle more difficult tasks, buying additional tools as they are needed. Eventually the basic kit will be expanded into the *repair and overhaul* tool set. Over a period of time, the experienced do-it-yourselfer will assemble a tool set complete enough for most repair and overhaul procedures and will add tools from the special category when it is felt that the expense is justified by the frequency of use.

Maintenance and minor repair tool kit

The tools in this list should be considered the minimum required for performance of routine maintenance, servicing and minor repair work. We recommend the purchase of combination wrenches (box-end and open-end combined in one wrench). While more expensive than open end wrenches, they offer the advantages of both types of wrench.

> *Combination wrench set (1/4-inch to 1 inch or 6 mm to 19 mm)*
> *Adjustable wrench, 8 inch*
> *Spark plug wrench with rubber insert*

Spark plug gap adjusting tool
Feeler gauge set
Brake bleeder wrench
Standard screwdriver (5/16-inch x 6 inch)
Phillips screwdriver (No. 2 x 6 inch)
Combination pliers - 6 inch
Hacksaw and assortment of blades
Tire pressure gauge
Grease gun
Oil can
Fine emery cloth
Wire brush
Battery post and cable cleaning tool
Oil filter wrench
Funnel (medium size)
Safety goggles
Jackstands (2)
Drain pan

Note: *If basic tune-ups are going to be part of routine maintenance, it will be necessary to purchase a good quality stroboscopic timing light and combination tachometer/dwell meter. Although they are included in the list of special tools, it is mentioned here because they are absolutely necessary for tuning most vehicles properly.*

Repair and overhaul tool set

These tools are essential for anyone who plans to perform major repairs and are in addition to those in the maintenance and minor

repair tool kit. Included is a comprehensive set of sockets which, though expensive, are invaluable because of their versatility, especially when various extensions and drives are available. We recommend the 1/2-inch drive over the 3/8-inch drive. Although the larger drive is bulky and more expensive, it has the capacity of accepting a very wide range of large sockets. Ideally, however, the mechanic should have a 3/8-inch drive set and a 1/2-inch drive set.

> *Socket set(s)*
> *Reversible ratchet*
> *Extension - 10 inch*
> *Universal joint*
> *Torque wrench (same size drive as sockets)*
> *Ball peen hammer - 8 ounce*
> *Soft-face hammer (plastic/rubber)*
> *Standard screwdriver (1/4-inch x 6 inch)*
> *Standard screwdriver (stubby - 5/16-inch)*
> *Phillips screwdriver (No. 3 x 8 inch)*
> *Phillips screwdriver (stubby - No. 2)*
> *Pliers - vise grip*
> *Pliers - lineman's*
> *Pliers - needle nose*
> *Pliers - snap-ring (internal and external)*
> *Cold chisel - 1/2-inch*
> *Scribe*
> *Scraper (made from flattened copper tubing)*
> *Centerpunch*

Pin punches (1/16, 1/8, 3/16-inch)
Steel rule/straightedge - 12 inch
Allen wrench set (1/8 to 3/8-inch or
  4 mm to 10 mm)
A selection of files
Wire brush (large)
Jackstands (second set)
Jack (scissor or hydraulic type)

Note: *Another tool which is often useful is an electric drill with a chuck capacity of 3/8-inch and a set of good quality drill bits.*

Special tools

The tools in this list include those which are not used regularly, are expensive to buy, or which need to be used in accordance with their manufacturer's instructions. Unless these tools will be used frequently, it is not very economical to purchase many of them. A consideration would be to split the cost and use between yourself and a friend or friends. In addition, most of these tools can be obtained from a tool rental shop on a temporary basis.

This list primarily contains only those tools and instruments widely available to the public, and not those special tools produced by the vehicle manufacturer for distribution to dealer service departments. Occasionally, references to the manufacturer's special tools are included in the text of this manual. Generally, an alternative method of doing the job without the special tool is offered. However, sometimes there is no alternative to their use. Where this is the case, and the tool cannot be purchased or borrowed, the work should be turned over to the dealer service department or an automotive repair shop.

Valve spring compressor
Piston ring groove cleaning tool
Piston ring compressor
Piston ring installation tool
Cylinder compression gauge
Cylinder ridge reamer
Cylinder surfacing hone
Cylinder bore gauge
Micrometers and/or dial calipers
Hydraulic lifter removal tool
Balljoint separator
Universal-type puller
Impact screwdriver
Dial indicator set
Stroboscopic timing light (inductive
  pick-up)
Hand operated vacuum/pressure pump
Tachometer/dwell meter
Universal electrical multimeter
Cable hoist
Brake spring removal and installation
  tools
Floor jack

Buying tools

For the do-it-yourselfer who is just starting to get involved in vehicle maintenance and repair, there are a number of options available when purchasing tools. If maintenance and minor repair is the extent of the work to be done, the purchase of individual tools is satisfactory. If, on the other hand, extensive work is planned, it would be a good idea to purchase

a modest tool set from one of the large retail chain stores. A set can usually be bought at a substantial savings over the individual tool prices, and they often come with a tool box. As additional tools are needed, add-on sets, individual tools and a larger tool box can be purchased to expand the tool selection. Building a tool set gradually allows the cost of the tools to be spread over a longer period of time and gives the mechanic the freedom to choose only those tools that will actually be used.

Tool stores will often be the only source of some of the special tools that are needed, but regardless of where tools are bought, try to avoid cheap ones, especially when buying screwdrivers and sockets, because they won't last very long. The expense involved in replacing cheap tools will eventually be greater than the initial cost of quality tools.

Care and maintenance of tools

Good tools are expensive, so it makes sense to treat them with respect. Keep them clean and in usable condition and store them properly when not in use. Always wipe off any dirt, grease or metal chips before putting them away. Never leave tools lying around in the work area. Upon completion of a job, always check closely under the hood for tools that may have been left there so they won't get lost during a test drive.

Some tools, such as screwdrivers, pliers, wrenches and sockets, can be hung on a panel mounted on the garage or workshop wall, while others should be kept in a tool box or tray. Measuring instruments, gauges, meters, etc. must be carefully stored where they cannot be damaged by weather or impact from other tools.

When tools are used with care and stored properly, they will last a very long time. Even with the best of care, though, tools will wear out if used frequently. When a tool is damaged or worn out, replace it. Subsequent jobs will be safer and more enjoyable if you do.

How to repair damaged threads

Sometimes, the internal threads of a nut or bolt hole can become stripped, usually from overtightening. Stripping threads is an all-too-common occurrence, especially when working with aluminum parts, because aluminum is so soft that it easily strips out.

Usually, external or internal threads are only partially stripped. After they've been cleaned up with a tap or die, they'll still work. Sometimes, however, threads are badly damaged. When this happens, you've got three choices:

1) *Drill and tap the hole to the next suitable oversize and install a larger diameter bolt, screw or stud.*
2) *Drill and tap the hole to accept a threaded plug, then drill and tap the plug to the original screw size. You can also buy a plug already threaded to the origi-*

nal size. Then you simply drill a hole to the specified size, then run the threaded plug into the hole with a bolt and jam nut. Once the plug is fully seated, remove the jam nut and bolt.
3) *The third method uses a patented thread repair kit like Heli-Coil or Slimsert. These easy-to-use kits are designed to repair damaged threads in straight-through holes and blind holes. Both are available as kits which can handle a variety of sizes and thread patterns. Drill the hole, then tap it with the special included tap. Install the Heli-Coil and the hole is back to its original diameter and thread pitch.*

Regardless of which method you use, be sure to proceed calmly and carefully. A little impatience or carelessness during one of these relatively simple procedures can ruin your whole day's work and cost you a bundle if you wreck an expensive part.

Working facilities

Not to be overlooked when discussing tools is the workshop. If anything more than routine maintenance is to be carried out, some sort of suitable work area is essential.

It is understood, and appreciated, that many home mechanics do not have a good workshop or garage available, and end up removing an engine or doing major repairs outside. It is recommended, however, that the overhaul or repair be completed under the cover of a roof.

A clean, flat workbench or table of comfortable working height is an absolute necessity. The workbench should be equipped with a vise that has a jaw opening of at least four inches.

As mentioned previously, some clean, dry storage space is also required for tools, as well as the lubricants, fluids, cleaning solvents, etc. which soon become necessary.

Sometimes waste oil and fluids, drained from the engine or cooling system during normal maintenance or repairs, present a disposal problem. To avoid pouring them on the ground or into a sewage system, pour the used fluids into large containers, seal them with caps and take them to an authorized disposal site or recycling center. Plastic jugs, such as old antifreeze containers, are ideal for this purpose.

Always keep a supply of old newspapers and clean rags available. Old towels are excellent for mopping up spills. Many mechanics use rolls of paper towels for most work because they are readily available and disposable. To help keep the area under the vehicle clean, a large cardboard box can be cut open and flattened to protect the garage or shop floor.

Whenever working over a painted surface, such as when leaning over a fender to service something under the hood, always cover it with an old blanket or bedspread to protect the finish. Vinyl covered pads, made especially for this purpose, are available at auto parts stores.

Jacking and towing

Jacking

The jack supplied with the vehicle should only be used for raising the vehicle when changing a tire or placing jackstands under the frame. NEVER work under the vehicle or start the engine when the vehicle supported only by a jack.

The vehicle should be parked on level ground with the wheels blocked, the parking brake applied and the transmission in Park (automatic) or Reverse (manual). If the vehicle is parked alongside the roadway, or in any other hazardous situation, turn on the emergency hazard flashers. If a tire is to be changed, loosen the lug nuts one-half turn before raising off the ground.

Place the jack under the vehicle in the

indicated positions (see illustrations). Operate the jack with a slow, smooth motion until the wheel is raised off the ground. Remove the lug nuts, pull off the wheel, install the spare and thread the lug nuts back on with the beveled or flanged side facing in. Tighten the lug nuts snugly, lower the vehicle until some weight is on the wheel, tighten them completely in a criss-cross pattern and remove the jack. On models with dual rear wheels, it is important that the lug nuts are clean and properly lubricated. Put two drops of oil at the point where the flange portion of the nut assembly attaches to the nut itself. This will ensure consistent torque readings. Note that some spare tires are designed for temporary use only - don't exceed the recommended

speed, mileage or other restriction instructions accompanying the spare.

Towing

Equipment specifically designed for towing should be used and attached to the main structural members of the vehicle. Optional tow hooks may be attached to the frame at both ends of the vehicle; they are intended for emergency use only, such as rescuing a stranded vehicle. Do not use the tow hooks for highway towing. Stand clear when using tow straps or chains, as they could break and cause serious injury.

Safety is a major consideration when towing and all applicable state and local laws must be obeyed. In addition to a tow bar, a safety chain must be used for all towing.

Two-wheel drive vehicles with automatic transmission may be towed with four wheels on the ground for a distance of 15 miles or less, as long as the speed doesn't exceed 30 mph. If the vehicle has to be towed more than 15 miles, place the rear wheels on a towing dolly.

Four-wheel drive vehicles should be towed on a flatbed or with all four wheels off the ground to avoid damage to the transfer case.

If any vehicle is to be towed with the front wheels on the ground and the rear wheels raised, the ignition key must be turned to the OFF position to unlock the steering column and a steering wheel clamping device designed for towing must be used or damage to the steering column lock may occur.

Front jacking point (all 1500 2WD and 4WD models)

Front jacking point, 2500/3500 2WD models - on 2500/3500 4WD models, the jack head must be placed under the front axle tube, nearest the wheel to be changed

Rear jacking point (all models)

Booster battery (jump) starting

Observe these precautions when using a booster battery to start a vehicle:

a) *Before connecting the booster battery, make sure the ignition switch is in the Off position.*
b) *Turn off the lights, heater and other electrical loads.*
c) *Your eyes should be shielded. Safety goggles are a good idea.*
d) *Make sure the booster battery is the same voltage as the dead one in the vehicle.*
e) *The two vehicles MUST NOT TOUCH each other!*
f) *Make sure the transmission is in Neutral (manual) or Park (automatic).*
g) *If the booster battery is not a maintenance-free type, remove the vent caps and lay a cloth over the vent holes.*

Connect the red jumper cable to the positive (+) terminals of each battery **(see illustration)**. **Note:** *On diesel models, the connections must be made to the left (driver's side) battery.*

Connect one end of the black jumper cable to the negative (-) terminal of the booster battery. The other end of this cable should be connected to a good ground on the vehicle to be started, such as a bolt or bracket on the body.

Start the engine using the booster battery, then, with the engine running at idle speed, disconnect the jumper cables in the reverse order of connection.

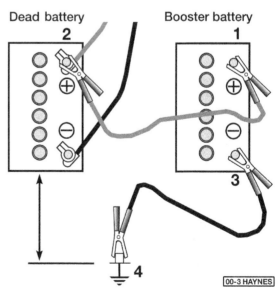

Make the booster battery cable connections in the numerical order shown (note that the negative cable of the booster battery is NOT attached to the negative terminal of the dead battery)

Automotive chemicals and lubricants

A number of automotive chemicals and lubricants are available for use during vehicle maintenance and repair. They include a wide variety of products ranging from cleaning solvents and degreasers to lubricants and protective sprays for rubber, plastic and vinyl.

Cleaners

Carburetor cleaner and choke cleaner is a strong solvent for gum, varnish and carbon. Most carburetor cleaners leave a dry-type lubricant film which will not harden or gum up. Because of this film it is not recommended for use on electrical components.

Brake system cleaner is used to remove brake dust, grease and brake fluid from the brake system, where clean surfaces are absolutely necessary. It leaves no residue and often eliminates brake squeal caused by contaminants.

Electrical cleaner removes oxidation, corrosion and carbon deposits from electrical contacts, restoring full current flow. It can also be used to clean spark plugs, carburetor jets, voltage regulators and other parts where an oil-free surface is desired.

Demoisturants remove water and moisture from electrical components such as alternators, voltage regulators, electrical connectors and fuse blocks. They are non-conductive and non-corrosive.

Degreasers are heavy-duty solvents used to remove grease from the outside of the engine and from chassis components. They can be sprayed or brushed on and, depending on the type, are rinsed off either with water or solvent.

Lubricants

Motor oil is the lubricant formulated for use in engines. It normally contains a wide variety of additives to prevent corrosion and reduce foaming and wear. Motor oil comes in various weights (viscosity ratings) from 0 to 50. The recommended weight of the oil depends on the season, temperature and the demands on the engine. Light oil is used in cold climates and under light load conditions. Heavy oil is used in hot climates and where high loads are encountered. Multiviscosity oils are designed to have characteristics of both light and heavy oils and are available in a number of weights from 0W-20 to 20W-50.

Gear oil is designed to be used in differentials, manual transmissions and other areas where high-temperature lubrication is required.

Chassis and wheel bearing grease is a heavy grease used where increased loads and friction are encountered, such as for wheel bearings, balljoints, tie-rod ends and universal joints.

High-temperature wheel bearing grease is designed to withstand the extreme temperatures encountered by wheel bearings in disc brake equipped vehicles. It usually contains molybdenum disulfide (moly), which is a dry-type lubricant.

White grease is a heavy grease for metal-to-metal applications where water is a problem. White grease stays soft under both low and high temperatures (usually from -100 to +190-degrees F), and will not wash off or dilute in the presence of water.

Assembly lube is a special extreme pressure lubricant, usually containing moly, used to lubricate high-load parts (such as main and rod bearings and cam lobes) for initial start-up of a new engine. The assembly lube lubricates the parts without being squeezed out or washed away until the engine oiling system begins to function.

Silicone lubricants are used to protect rubber, plastic, vinyl and nylon parts.

Graphite lubricants are used where oils cannot be used due to contamination problems, such as in locks. The dry graphite will lubricate metal parts while remaining uncontaminated by dirt, water, oil or acids. It is electrically conductive and will not foul electrical contacts in locks such as the ignition switch.

Moly penetrants loosen and lubricate frozen, rusted and corroded fasteners and prevent future rusting or freezing.

Heat-sink grease is a special electrically non-conductive grease that is used for mounting electronic ignition modules where it is essential that heat is transferred away from the module.

Sealants

RTV sealant is one of the most widely used gasket compounds. Made from silicone, RTV is air curing, it seals, bonds, waterproofs, fills surface irregularities, remains flexible, doesn't shrink, is relatively easy to remove, and is used as a supplementary sealer with almost all low and medium temperature gaskets.

Anaerobic sealant is much like RTV in that it can be used either to seal gaskets or to form gaskets by itself. It remains flexible, is solvent resistant and fills surface imperfections. The difference between an anaerobic sealant and an RTV-type sealant is in the curing. RTV cures when exposed to air, while an anaerobic sealant cures only in the absence of air. This means that an anaerobic sealant cures only after the assembly of parts, sealing them together.

Thread and pipe sealant is used for sealing hydraulic and pneumatic fittings and vacuum lines. It is usually made from a Teflon compound, and comes in a spray, a paint-on liquid and as a wrap-around tape.

Chemicals

Anti-seize compound prevents seizing, galling, cold welding, rust and corrosion in fasteners. High-temperature anti-seize, usually made with copper and graphite lubricants, is used for exhaust system and exhaust manifold bolts.

Anaerobic locking compounds are used to keep fasteners from vibrating or working loose and cure only after installation, in the absence of air. Medium strength locking compound is used for small nuts, bolts and screws that may be removed later. High-strength locking compound is for large nuts, bolts and studs which aren't removed on a regular basis.

Oil additives range from viscosity index improvers to chemical treatments that claim to reduce internal engine friction. It should be noted that most oil manufacturers caution against using additives with their oils.

Gas additives perform several functions, depending on their chemical makeup. They usually contain solvents that help dissolve gum and varnish that build up on carburetor, fuel injection and intake parts. They also serve to break down carbon deposits that form on the inside surfaces of the combustion chambers. Some additives contain upper cylinder lubricants for valves and piston rings, and others contain chemicals to remove condensation from the gas tank.

Miscellaneous

Brake fluid is specially formulated hydraulic fluid that can withstand the heat and pressure encountered in brake systems. Care must be taken so this fluid does not come in contact with painted surfaces or plastics. An opened container should always be resealed to prevent contamination by water or dirt.

Weatherstrip adhesive is used to bond weatherstripping around doors, windows and trunk lids. It is sometimes used to attach trim pieces.

Undercoating is a petroleum-based, tar-like substance that is designed to protect metal surfaces on the underside of the vehicle from corrosion. It also acts as a sound-deadening agent by insulating the bottom of the vehicle.

Waxes and polishes are used to help protect painted and plated surfaces from the weather. Different types of paint may require the use of different types of wax and polish. Some polishes utilize a chemical or abrasive cleaner to help remove the top layer of oxidized (dull) paint on older vehicles. In recent years many non-wax polishes that contain a wide variety of chemicals such as polymers and silicones have been introduced. These non-wax polishes are usually easier to apply and last longer than conventional waxes and polishes.

Conversion factors

Length (distance)

Inches (in)	X	25.4	= Millimeters (mm)	X	0.0394	= Inches (in)
Feet (ft)	X	0.305	= Meters (m)	X	3.281	= Feet (ft)
Miles	X	1.609	= Kilometers (km)	X	0.621	= Miles

Length (distance)

Inches (in)	X 25.4	= Millimeters (mm)	X 0.0394	= Inches (in)
Feet (ft)	X 0.305	= Meters (m)	X 3.281	= Feet (ft)
Miles	X 1.609	= Kilometers (km)	X 0.621	= Miles

Volume (capacity)

Cubic inches (cu in; in^3)	X 16.387	= Cubic centimeters (cc; cm^3)	X 0.061	= Cubic inches (cu in; in^3)
Imperial pints (Imp pt)	X 0.568	= Liters (l)	X 1.76	= Imperial pints (Imp pt)
Imperial quarts (Imp qt)	X 1.137	= Liters (l)	X 0.88	= Imperial quarts (Imp qt)
Imperial quarts (Imp qt)	X 1.201	= US quarts (US qt)	X 0.833	= Imperial quarts (Imp qt)
US quarts (US qt)	X 0.946	= Liters (l)	X 1.057	= US quarts (US qt)
Imperial gallons (Imp gal)	X 4.546	= Liters (l)	X 0.22	= Imperial gallons (Imp gal)
Imperial gallons (Imp gal)	X 1.201	= US gallons (US gal)	X 0.833	= Imperial gallons (Imp gal)
US gallons (US gal)	X 3.785	= Liters (l)	X 0.264	= US gallons (US gal)

Mass (weight)

Ounces (oz)	X 28.35	= Grams (g)	X 0.035	= Ounces (oz)
Pounds (lb)	X 0.454	= Kilograms (kg)	X 2.205	= Pounds (lb)

Force

Ounces-force (ozf; oz)	X 0.278	= Newtons (N)	X 3.6	= Ounces-force (ozf; oz)
Pounds-force (lbf; lb)	X 4.448	= Newtons (N)	X 0.225	= Pounds-force (lbf; lb)
Newtons (N)	X 0.1	= Kilograms-force (kgf; kg)	X 9.81	= Newtons (N)

Pressure

Pounds-force per square inch (psi; lbf/in^2; lb/in^2)	X 0.070	= Kilograms-force per square centimeter (kgf/cm^2; kg/cm^2)	X 14.223	= Pounds-force per square inch (psi; lbf/in^2; lb/in^2)
Pounds-force per square inch (psi; lbf/in^2; lb/in^2)	X 0.068	= Atmospheres (atm)	X 14.696	= Pounds-force per square inch (psi; lbf/in^2; lb/in^2)
Pounds-force per square inch (psi; lbf/in^2; lb/in^2)	X 0.069	= Bars	X 14.5	= Pounds-force per square inch (psi; lbf/in^2; lb/in^2)
Pounds-force per square inch (psi; lbf/in^2; lb/in^2)	X 6.895	= Kilopascals (kPa)	X 0.145	= Pounds-force per square inch (psi; lbf/in^2; lb/in^2)
Kilopascals (kPa)	X 0.01	= Kilograms-force per square centimeter (kgf/cm^2; kg/cm^2)	X 98.1	= Kilopascals (kPa)

Torque (moment of force)

Pounds-force inches (lbf in; lb in)	X 1.152	= Kilograms-force centimeter (kgf cm; kg cm)	X 0.868	= Pounds-force inches (lbf in; lb in)
Pounds-force inches (lbf in; lb in)	X 0.113	= Newton meters (Nm)	X 8.85	= Pounds-force inches (lbf in; lb in)
Pounds-force inches (lbf in; lb in)	X 0.083	= Pounds-force feet (lbf ft; lb ft)	X 12	= Pounds-force inches (lbf in; lb in)
Pounds-force feet (lbf ft; lb ft)	X 0.138	= Kilograms-force meters (kgf m; kg m)	X 7.233	= Pounds-force feet (lbf ft; lb ft)
Pounds-force feet (lbf ft; lb ft)	X 1.356	= Newton meters (Nm)	X 0.738	= Pounds-force feet (lbf ft; lb ft)
Newton meters (Nm)	X 0.102	= Kilograms-force meters (kgf m; kg m)	X 9.804	= Newton meters (Nm)

Vacuum

Inches mercury (in. Hg)	X 3.377	= Kilopascals (kPa)	X 0.2961	= Inches mercury
Inches mercury (in. Hg)	X 25.4	= Millimeters mercury (mm Hg)	X 0.0394	= Inches mercury

Power

Horsepower (hp)	X 745.7	= Watts (W)	X 0.0013	= Horsepower (hp)

Velocity (speed)

Miles per hour (miles/hr; mph)	X 1.609	= Kilometers per hour (km/hr; kph)	X 0.621	= Miles per hour (miles/hr; mph)

Fuel consumption*

Miles per gallon, Imperial (mpg)	X 0.354	= Kilometers per liter (km/l)	X 2.825	= Miles per gallon, Imperial (mpg)
Miles per gallon, US (mpg)	X 0.425	= Kilometers per liter (km/l)	X 2.352	= Miles per gallon, US (mpg)

Temperature

Degrees Fahrenheit = (°C x 1.8) + 32

Degrees Celsius (Degrees Centigrade; °C) = (°F - 32) x 0.56

*It is common practice to convert from miles per gallon (mpg) to liters/100 kilometers (l/100km), where mpg (Imperial) x l/100 km = 282 and mpg (US) x l/100 km = 235

DECIMALS to MILLIMETERS

Decimal	mm	Decimal	mm
0.001	0.0254	0.500	12.7000
0.002	0.0508	0.510	12.9540
0.003	0.0762	0.520	13.2080
0.004	0.1016	0.530	13.4620
0.005	0.1270	0.540	13.7160
0.006	0.1524	0.550	13.9700
0.007	0.1778	0.560	14.2240
0.008	0.2032	0.570	14.4780
0.009	0.2286	0.580	14.7320
		0.590	14.9860
0.010	0.2540		
0.020	0.5080		
0.030	0.7620		
0.040	1.0160	0.600	15.2400
0.050	1.2700	0.610	15.4940
0.060	1.5240	0.620	15.7480
0.070	1.7780	0.630	16.0020
0.080	2.0320	0.640	16.2560
0.090	2.2860	0.650	16.5100
		0.660	16.7640
0.100	2.5400	0.670	17.0180
0.110	2.7940	0.680	17.2720
0.120	3.0480	0.690	17.5260
0.130	3.3020		
0.140	3.5560		
0.150	3.8100		
0.160	4.0640	0.700	17.7800
0.170	4.3180	0.710	18.0340
0.180	4.5720	0.720	18.2880
0.190	4.8260	0.730	18.5420
		0.740	18.7960
0.200	5.0800	0.750	19.0500
0.210	5.3340	0.760	19.3040
0.220	5.5880	0.770	19.5580
0.230	5.8420	0.780	19.8120
0.240	6.0960	0.790	20.0660
0.250	6.3500		
0.260	6.6040		
0.270	6.8580	0.800	20.3200
0.280	7.1120	0.810	20.5740
0.290	7.3660	0.820	21.8280
		0.830	21.0820
0.300	7.6200	0.840	21.3360
0.310	7.8740	0.850	21.5900
0.320	8.1280	0.860	21.8440
0.330	8.3820	0.870	22.0980
0.340	8.6360	0.880	22.3520
0.350	8.8900	0.890	22.6060
0.360	9.1440		
0.370	9.3980		
0.380	9.6520		
0.390	9.9060		
		0.900	22.8600
0.400	10.1600	0.910	23.1140
0.410	10.4140	0.920	23.3680
0.420	10.6680	0.930	23.6220
0.430	10.9220	0.940	23.8760
0.440	11.1760	0.950	24.1300
0.450	11.4300	0.960	24.3840
0.460	11.6840	0.970	24.6380
0.470	11.9380	0.980	24.8920
0.480	12.1920	0.990	25.1460
0.490	12.4460	1.000	25.4000

FRACTIONS to DECIMALS to MILLIMETERS

Fraction	Decimal	mm	Fraction	Decimal	mm
1/64	0.0156	0.3969	33/64	0.5156	13.0969
1/32	0.0312	0.7938	17/32	0.5312	13.4938
3/64	0.0469	1.1906	35/64	0.5469	13.8906
1/16	0.0625	1.5875	9/16	0.5625	14.2875
5/64	0.0781	1.9844	37/64	0.5781	14.6844
3/32	0.0938	2.3812	19/32	0.5938	15.0812
7/64	0.1094	2.7781	39/64	0.6094	15.4781
1/8	0.1250	3.1750	5/8	0.6250	15.8750
9/64	0.1406	3.5719	41/64	0.6406	16.2719
5/32	0.1562	3.9688	21/32	0.6562	16.6688
11/64	0.1719	4.3656	43/64	0.6719	17.0656
3/16	0.1875	4.7625	11/16	0.6875	17.4625
13/64	0.2031	5.1594	45/64	0.7031	17.8594
7/32	0.2188	5.5562	23/32	0.7188	18.2562
15/64	0.2344	5.9531	47/64	0.7344	18.6531
1/4	0.2500	6.3500	3/4	0.7500	19.0500
17/64	0.2656	6.7469	49/64	0.7656	19.4469
9/32	0.2812	7.1438	25/32	0.7812	19.8438
19/64	0.2969	7.5406	51/64	0.7969	20.2406
5/16	0.3125	7.9375	13/16	0.8125	20.6375
21/64	0.3281	8.3344	53/64	0.8281	21.0344
11/32	0.3438	8.7312	27/32	0.8438	21.4312
23/64	0.3594	9.1281	55/64	0.8594	21.8281
3/8	0.3750	9.5250	7/8	0.8750	22.2250
25/64	0.3906	9.9219	57/64	0.8906	22.6219
13/32	0.4062	10.3188	29/32	0.9062	23.0188
27/64	0.4219	10.7156	59/64	0.9219	23.4156
7/16	0.4375	11.1125	15/16	0.9375	23.8125
29/64	0.4531	11.5094	61/64	0.9531	24.2094
15/32	0.4688	11.9062	31/32	0.9688	24.6062
31/64	0.4844	12.3031	63/64	0.9844	25.0031
1/2	0.5000	12.7000	1	1.0000	25.4000

Safety first!

Regardless of how enthusiastic you may be about getting on with the job at hand, take the time to ensure that your safety is not jeopardized. A moment's lack of attention can result in an accident, as can failure to observe certain simple safety precautions. The possibility of an accident will always exist, and the following points should not be considered a comprehensive list of all dangers. Rather, they are intended to make you aware of the risks and to encourage a safety conscious approach to all work you carry out on your vehicle.

Essential DOs and DON'Ts

DON'T rely on a jack when working under the vehicle. Always use approved jackstands to support the weight of the vehicle and place them under the recommended lift or support points.

DON'T attempt to loosen extremely tight fasteners (i.e. wheel lug nuts) while the vehicle is on a jack - it may fall.

DON'T start the engine without first making sure that the transmission is in Neutral (or Park where applicable) and the parking brake is set.

DON'T remove the radiator cap from a hot cooling system - let it cool or cover it with a cloth and release the pressure gradually.

DON'T attempt to drain the engine oil until you are sure it has cooled to the point that it will not burn you.

DON'T touch any part of the engine or exhaust system until it has cooled sufficiently to avoid burns.

DON'T siphon toxic liquids such as gasoline, antifreeze and brake fluid by mouth, or allow them to remain on your skin.

DON'T inhale brake lining dust - it is potentially hazardous (see *Asbestos* below).

DON'T allow spilled oil or grease to remain on the floor - wipe it up before someone slips on it.

DON'T use loose fitting wrenches or other tools which may slip and cause injury.

DON'T push on wrenches when loosening or tightening nuts or bolts. Always try to pull the wrench toward you. If the situation calls for pushing the wrench away, push with an open hand to avoid scraped knuckles if the wrench should slip.

DON'T attempt to lift a heavy component alone - get someone to help you.

DON'T *rush or take unsafe shortcuts to finish a job.*

DON'T allow children or animals in or around the vehicle while you are working on it.

DO wear eye protection when using power tools such as a drill, sander, bench grinder, etc. and when working under a vehicle.

DO keep loose clothing and long hair well out of the way of moving parts.

DO make sure that any hoist used has a safe working load rating adequate for the job.

DO get someone to check on you periodically when working alone on a vehicle.

DO carry out work in a logical sequence and make sure that everything is correctly assembled and tightened.

DO keep chemicals and fluids tightly capped and out of the reach of children and pets.

DO remember that your vehicle's safety affects that of yourself and others. If in doubt on any point, get professional advice.

Steering, suspension and brakes

These systems are essential to driving safety, so make sure you have a qualified shop or individual check your work. Also, compressed suspension springs can cause injury if released suddenly - be sure to use a spring compressor.

Airbags

Airbags are explosive devices that can **CAUSE** injury if they deploy while you're working on the vehicle. Follow the manufacturer's instructions to disable the airbag whenever you're working in the vicinity of airbag components.

Asbestos

Certain friction, insulating, sealing, and other products - such as brake linings, brake bands, clutch linings, torque converters, gaskets, etc. - may contain asbestos or other hazardous friction material. Extreme care must be taken to avoid inhalation of dust from such products, since it is hazardous to health. If in doubt, assume that they do contain asbestos.

Fire

Remember at all times that gasoline is highly flammable. Never smoke or have any kind of open flame around when working on a vehicle. But the risk does not end there. A spark caused by an electrical short circuit, by two metal surfaces contacting each other, or even by static electricity built up in your body under certain conditions, can ignite gasoline vapors, which in a confined space are highly explosive. Do not, under any circumstances, use gasoline for cleaning parts. Use an approved safety solvent.

Always disconnect the battery ground (-) cable at the battery before working on any part of the fuel system or electrical system. Never risk spilling fuel on a hot engine or exhaust component. It is strongly recommended that a fire extinguisher suitable for use on fuel and electrical fires be kept handy in the garage or workshop at all times. Never try to extinguish a fuel or electrical fire with water.

Fumes

Certain fumes are highly toxic and can quickly cause unconsciousness and even death if inhaled to any extent. Gasoline vapor falls into this category, as do the vapors from some cleaning solvents. Any draining or pouring of such volatile fluids should be done in a well ventilated area.

When using cleaning fluids and solvents, read the instructions on the container carefully. Never use materials from unmarked containers.

Never run the engine in an enclosed space, such as a garage. Exhaust fumes contain carbon monoxide, which is extremely poisonous. If you need to run the engine, always do so in the open air, or at least have the rear of the vehicle outside the work area.

The battery

Never create a spark or allow a bare light bulb near a battery. They normally give off a certain amount of hydrogen gas, which is highly explosive.

Always disconnect the battery ground (-) cable at the battery before working on the fuel or electrical systems.

If possible, loosen the filler caps or cover when charging the battery from an external source (this does not apply to sealed or maintenance-free batteries). Do not charge at an excessive rate or the battery may burst.

Take care when adding water to a non maintenance-free battery and when carrying a battery. The electrolyte, even when diluted, is very corrosive and should not be allowed to contact clothing or skin.

Always wear eye protection when cleaning the battery to prevent the caustic deposits from entering your eyes.

Household current

When using an electric power tool, inspection light, etc., which operates on household current, always make sure that the tool is correctly connected to its plug and that, where necessary, it is properly grounded. Do not use such items in damp conditions and, again, do not create a spark or apply excessive heat in the vicinity of fuel or fuel vapor.

Secondary ignition system voltage

A severe electric shock can result from touching certain parts of the ignition system (such as the spark plug wires) when the engine is running or being cranked, particularly if components are damp or the insulation is defective. In the case of an electronic ignition system, the secondary system voltage is much higher and could prove fatal.

Hydrofluoric acid

This extremely corrosive acid is formed when certain types of synthetic rubber, found in some O-rings, oil seals, fuel hoses, etc. are exposed to temperatures above 750-degrees F (400-degrees C). The rubber changes into a charred or sticky substance containing the acid. *Once formed, the acid remains dangerous for years. If it gets onto the skin, it may be necessary to amputate the limb concerned.*

When dealing with a vehicle which has suffered a fire, or with components salvaged from such a vehicle, wear protective gloves and discard them after use.

Troubleshooting

Contents

This section provides an easy reference guide to the more common problems which may occur during the operation of your vehicle. These problems and possible causes are grouped under various components or systems (Engine, Cooling System, etc.), and also refer to the Chapter and/or Section which deals with the problem.

Remember that successful troubleshooting is not a mysterious black art practiced only by professional mechanics. It's simply the result of a bit of knowledge combined with an intelligent, systematic approach to the problem. Always work by a process of elimination, starting with the simplest solution and working through to the most complex - and never overlook the obvious. Anyone can forget to fill the gas tank or leave the lights on overnight, so don't assume that you are above such oversights.

Finally, always get clear in your mind why a problem has occurred and take steps to ensure that it doesn't happen again. If the electrical system fails because of a poor connection, check all other connections in the system to make sure that they don't fail as well. If a particular fuse continues to blow, find out why - don't just go on replacing fuses. Remember, failure of a small component can often be indicative of potential failure or incorrect functioning of a more important component or system.

Engine

1 Engine will not rotate when attempting to start

1 Battery terminal connections loose or corroded. Check the cable terminals at the battery. Tighten the cable or remove corrosion as necessary.
2 Battery discharged or faulty. If the cable connections are clean and tight on the battery posts, turn the key to the On position and switch on the headlights and/or windshield wipers. If they fail to function, the battery is discharged.
3 Automatic transmission not completely engaged in Park or Neutral, or clutch pedal not completely depressed.
4 Broken, loose or disconnected wiring in the starting circuit. Inspect all wiring and connectors at the battery, starter solenoid and ignition switch.
5 Starter motor pinion jammed in flywheel ring gear. If manual transmission, place transmission in gear and rock the vehicle to manually turn the engine. Remove starter and inspect pinion and flywheel (Chapter 5).
6 Starter solenoid faulty (Chapter 5).
7 Starter motor faulty (Chapter 5).
8 Ignition switch faulty (Chapter 12).

2 Engine rotates but will not start

1 Fuel tank empty, fuel filter plugged or fuel line restricted.
2 Fault in the fuel injection system (Chapter 4).
3 Battery discharged (engine rotates slowly). Check the operation of electrical components (see previous Section).
4 Battery terminal connections loose or corroded (see previous Section).
5 Fuel pump faulty (Chapter 4).
6 Excessive moisture on, or damage to, ignition components (see Chapter 5).
7 Worn, faulty or incorrectly gapped spark plugs (Chapter 1).
8 Broken, loose or disconnected wiring in the starting circuit (see previous Section).
9 Broken, loose or disconnected wires at the ignition coil (Chapter 5).
10 Broken, loose or disconnected wires at the fuel shutdown solenoid (diesel) (Chapter 4).
11 Air in the fuel system or defective fuel injection pump or injector (diesel) (Chapter 4).
12 Contaminated fuel.

3 Starter motor operates without rotating engine

1 Starter pinion sticking. Remove the starter (Chapter 5) and inspect.
2 Starter pinion or flywheel teeth worn or broken. Remove the flywheel/driveplate access cover and inspect.

4 Engine hard to start when cold

1 Battery discharged or low. Check as described in Section 1.
2 Fault in the fuel or electrical systems (Chapters 4 and 5).
3 Fault in the intake manifold heater or fuel heater systems (diesel) (Chapter 4).
4 Air in the fuel system or defective fuel injection pump or injector (diesel) (Chapter 4).

5 Engine hard to start when hot

1 Air filter clogged (Chapter 1).
2 Fault in the fuel or electrical systems (Chapters 4 and 5).
3 Fuel not reaching the injectors (see Chapter 4).
4 Air in the fuel system or defective fuel injection pump or injector (diesel) (Chapter 4).
5 Low cylinder compression (Chapter 2).

6 Starter motor noisy or excessively rough in engagement

1 Pinion or flywheel gear teeth worn or broken. Remove the cover at the rear of the engine (if equipped) and inspect.
2 Starter motor mounting bolts loose or missing.

7 Engine starts but stops immediately

1 Loose or faulty electrical connections at distributor, coil or alternator.
2 Fault in the fuel or electrical systems (Chapters 4 and 5).
3 Vacuum leak at the gasket surfaces of the intake manifold or throttle body. Make sure all mounting bolts/nuts are tightened securely and all vacuum hoses connected to the manifold are positioned properly and in good condition.
4 Restricted intake or exhaust systems (Chapter 4)
5 Fault in the fuel heater system (diesel) (Chapter 4).
6 Air in the fuel system or defective fuel injection pump or injector (diesel) (Chapter 4).
7 Contaminated fuel.

8 Engine lopes while idling or idles erratically

1 Vacuum leakage. Check the mounting bolts/nuts at the throttle body and intake manifold for tightness. Make sure all vacuum hoses are connected and in good condition. Use a stethoscope or a length of fuel hose held against your ear to listen for vacuum leaks while the engine is running. A hissing sound will be heard. A soapy water solution will also detect leaks.
2 Fault in the fuel or electrical systems (Chapters 4 and 5).
3 Plugged PCV valve or hose (Chapters 1 and 6).
4 Air filter clogged (Chapter 1).
5 Fuel pump not delivering sufficient fuel to the fuel injectors (see Chapter 4).
6 Leaking head gasket. Perform a compression check (Chapter 2).
7 Camshaft lobes worn (Chapter 2).
8 Air in the fuel system or defective fuel injection pump or injector (diesel) (Chapter 4).

9 Engine misses at idle speed

1 Spark plugs worn, fouled or not gapped properly (Chapter 1).
2 Fault in the fuel or electrical systems (Chapters 4 and 5).
3 Faulty spark plug wires (Chapter 1).
4 Vacuum leaks at intake or hose connections. Check as described in Section 8.
5 Uneven or low cylinder compression. Check compression (Chapter 2).
6 Air in the fuel system or defective fuel injection pump or injector (diesel) (Chapter 4).

10 Engine misses throughout driving speed range

1   Fuel filter clogged and/or impurities in the fuel system (Chapter 1).
2   Faulty or incorrectly gapped spark plugs (Chapter 1).
3   Fault in the fuel or electrical systems (Chapters 4 and 5).
4   Defective spark plug wires (Chapter 1).
5   Faulty emissions system components (Chapter 6).
6   Low or uneven cylinder compression pressures. Remove the spark plugs and test the compression with a gauge (Chapter 2).
7   Weak or faulty ignition system (Chapter 5).
8   Vacuum leaks at the throttle body, intake manifold or vacuum hoses (see Section 8).
9   Air in the fuel system or defective fuel injection pump or injector (diesel) (Chapter 4).

11 Engine stalls

1   Idle speed incorrect. Refer to the VECI label.
2   Fuel filter clogged and/or water and impurities in the fuel system (Chapter 1).
3   Fault in the fuel system or sensors (Chapters 4 and 6).
4   Faulty emissions system components (Chapter 6).
5   Faulty or incorrectly gapped spark plugs (Chapter 1). Also check the spark plug wires (Chapter 1).
6   Vacuum leak at the throttle body, intake manifold or vacuum hoses. Check as described in Section 8.
7   Air in the fuel system or defective fuel injection pump or injector (diesel) (Chapter 4).

12 Engine lacks power

1   Fault in the fuel or electrical systems (Chapters 4 and 5).
2   Faulty or incorrectly gapped spark plugs (Chapter 1).
3   Faulty coil (Chapter 5).
4   Brakes binding (Chapter 1).
5   Automatic transmission fluid level incorrect (Chapter 1).
6   Clutch slipping (Chapter 8).
7   Fuel filter clogged and/or impurities in the fuel system (Chapter 1).
8   Emissions control system not functioning properly (Chapter 6).
9   Use of substandard fuel. Fill the tank with the proper fuel.
10   Low or uneven cylinder compression pressures. Test with a compression tester, which will detect leaking valves and/or a blown head gasket (Chapter 2).
11   Air in the fuel system or defective fuel injection pump or injector (diesel) (Chapter 4).

12   Defective turbocharger or wastegate (diesel) (Chapter 4).
13   Restriction in the intake or exhaust system (Chapter 4).

13 Engine backfires

1   Emissions system not functioning properly (Chapter 6).
2   Fault in the fuel or electrical systems (Chapters 4 and 5).
3   Faulty secondary ignition system (cracked spark plug insulator or faulty plug wires) (Chapters 1 and 5).
4   Vacuum leak at the throttle body, intake manifold or vacuum hoses. Check as described in Section 8.
5   Valves sticking (Chapter 2).
6   Crossed plug wires (Chapter 1).

14 Pinging or knocking engine sounds during acceleration or uphill

1   Incorrect grade of fuel. Fill the tank with fuel of the proper octane rating.
2   Fault in the fuel or electrical systems (Chapters 4 and 5).
3   Improper spark plugs. Check the plug type against the VECI label located in the engine compartment. Also check the plugs and wires for damage (Chapter 1).
4   Faulty emissions system (Chapter 6).
5   Vacuum leak. Check as described in Section 9.

15 Engine continues to run after switching off

1   Defective ignition switch (Chapter 12).
2   Faulty Powertrain Control Module (Chapter 6).
3   Faulty Body Control Module.
4   Leaking fuel injector(s) (Chapter 4).
5   Defective fuel control actuator (diesel).

Engine electrical system

16 Battery will not hold a charge

1   Alternator drivebelt defective or not adjusted properly (Chapter 1).
2   Electrolyte level low or battery discharged (Chapter 1).
3   Battery terminals loose or corroded (Chapter 1).
4   Alternator not charging properly (Chapter 5).
5   Loose, broken or faulty wiring in the charging circuit (Chapter 5).
6   Short in the vehicle wiring causing a con-

tinuous drain on the battery (refer to Chapter 12 and the Wiring Diagrams).
7   Battery defective internally.

17 Ignition light fails to go out

1   Fault in the alternator or charging circuit (Chapter 5).
2   Alternator drivebelt defective or not properly adjusted (Chapter 1).

18 Ignition light fails to come on when key is turned on

1   Instrument cluster warning light bulb defective (Chapter 12).
2   Alternator faulty (Chapter 5).
3   Fault in the instrument cluster printed circuit, dashboard wiring or bulb holder (Chapter 12).

Fuel system

19 Excessive fuel consumption

1   Dirty or clogged air filter element (Chapter 1).
2   Emissions system not functioning properly (Chapter 6).
3   Fault in the fuel or electrical systems (Chapters 4 and 5).
4   Low tire pressure or incorrect tire size (Chapter 1).
5   Restricted exhaust system (Chapter 4).

20 Fuel leakage and/or fuel odor

1   Leak in a fuel feed line (Chapter 4).
2   Tank overfilled.
3   Evaporative emissions system canister clogged (Chapter 6).
4   Vapor leaks from system lines (Chapter 4).

Cooling system

21 Overheating

1   Insufficient coolant in the system (Chapter 1).
2   Water pump drivebelt defective or not adjusted properly (Chapter 1).
3   Radiator core blocked or radiator grille dirty and restricted (see Chapter 3).
4   Thermostat faulty (Chapter 3).
5   Fan blades broken or cracked (Chapter 3).
6   Cooling system pressure cap not maintaining proper pressure (Chapter 3).

22 Overcooling

1   Thermostat faulty (Chapter 3).
2   Inaccurate temperature gauge (Chapter 12).

23 External coolant leakage

1   Deteriorated or damaged hoses or loose clamps. Replace hoses and/or tighten the clamps at the hose connections (Chapter 1).
2   Water pump seals defective. If this is the case, water will drip from the weep hole in the water pump body (Chapter 3).
3   Leakage from the radiator core or side tank(s). This will require the radiator to be professionally repaired (see Chapter 3 for removal procedures).
4   Engine drain plug(s) leaking (Chapter 1) or water jacket core plugs leaking (Chapter 2).

24 Internal coolant leakage

Note: *Internal coolant leaks can usually be detected by examining the oil. Check the dipstick and inside of the valve cover for water deposits and an oil consistency like that of a milkshake.*
1   Leaking cylinder head gasket. Have the cooling system pressure tested.
2   Cracked cylinder bore or cylinder head. Dismantle the engine and inspect (Chapter 2).
3   Leaking intake manifold gasket (gasoline engines).

25 Coolant loss

1   Too much coolant in the system (Chapter 1).
2   Coolant boiling away due to overheating (see Section 15).
3   External or internal leakage (see Sections 23 and 24).
4   Faulty pressure cap (Chapter 3).

26 Poor coolant circulation

1   Inoperative water pump. A quick test is to pinch the top radiator hose closed with your hand while the engine is idling, then let it loose. You should feel the surge of coolant if the pump is working properly (Chapter 1).
2   Restriction in the cooling system. Drain, flush and refill the system (Chapter 1). If necessary, remove the radiator (Chapter 3) and have it reverse flushed.
3   Water pump drivebelt defective or not adjusted properly (Chapter 1).
4   Thermostat sticking (Chapter 3).

5   Drivebelt incorrectly routed, causing the pump to turn backwards (Chapter 1).

Clutch

27 Fails to release (pedal pressed to the floor - shift lever does not move freely in and out of Reverse)

1   Leak in the clutch hydraulic system. Check the master cylinder, release cylinder and lines (Chapters 1 and 8).
2   Clutch plate warped or damaged (Chapter 8).

28 Clutch slips (engine speed increases with no increase in vehicle speed)

1   Clutch plate oil soaked or lining worn. Remove clutch and inspect (Chapter 8).
2   Clutch plate not seated (Chapter 8).
3   Pressure plate worn (Chapter 8).
4   Weak diaphragm springs (Chapter 8).
5   Clutch plate overheated. Allow to cool.

29 Grabbing (chattering) as clutch is engaged

1   Oil on clutch plate lining. Remove and inspect (Chapter 8). Correct any leakage source.
2   Worn or loose engine or transmission mounts. These units move slightly when the clutch is released. Inspect the mounts and bolts (Chapter 2).
3   Worn splines on clutch plate hub. Remove the clutch components and inspect (Chapter 8).
4   Warped pressure plate or flywheel. Remove the clutch components and inspect.

30 Squeal or rumble with clutch fully engaged (pedal released)

Release bearing binding on transmission bearing retainer. Remove clutch components (Chapter 8) and check bearing. Remove any burrs or nicks; clean and re-lubricate bearing retainer before installing.

31 Squeal or rumble with clutch fully disengaged (pedal depressed)

1   Worn, defective or broken release bearing (Chapter 8).
2   Worn or broken pressure plate springs (or diaphragm fingers) (Chapter 8).

32 Clutch pedal stays on floor when disengaged

1   Linkage or release bearing binding. Inspect the linkage or remove the clutch components as necessary.
2   Make sure proper pedal stop (bumper) is installed.

Manual transmission
Note: *All the following references are in Chapter 7A, unless noted.*

33 Noisy in Neutral with engine running

1   Input shaft bearing worn.
2   Damaged main drive gear bearing.
3   Worn countershaft bearings.
4   Worn or damaged countershaft endplay shims.

34 Noisy in all gears

1   Any of the above causes, and/or:
2   Insufficient lubricant (see the checking procedures in Chapter 1).

35 Noisy in one particular gear

1   Worn, damaged or chipped gear teeth for that particular gear.
2   Worn or damaged synchronizer for that particular gear.

36 Slips out of high gear

1   Transmission loose on clutch housing.
2   Internal transmission problem.

37 Difficulty in engaging gears

1   Clutch not releasing completely (see clutch adjustment in Chapter 1).
2   Loose, damaged or out-of-adjustment shift linkage. Make a thorough inspection, replacing parts as necessary.

38 Oil leakage

1   Excessive amount of lubricant in the transmission (see Chapter 1 for correct checking procedures). Drain lubricant as required.
2   Transmission oil seal or speedometer oil seal in need of replacement.

Automatic transmission

Note: *Due to the complexity of the automatic transmission, it's difficult for the home mechanic to properly diagnose and service this component. For problems other than the following, the vehicle should be taken to a dealer service department or a transmission shop.*

39 General shift mechanism problems

1   Chapter 7B deals with checking and adjusting the shift linkage on automatic transmissions. Common problems which may be attributed to poorly adjusted linkage are:

a) *Engine starting in gears other than Park or Neutral.*
b) *Indicator on shifter pointing to a gear other than the one actually being selected.*
c) *Vehicle moves when in Park.*

2   Refer to Chapter 7B to adjust the linkage.

40 Transmission will not downshift with accelerator pedal pressed to the floor

Throttle valve (TV) cable misadjusted (if equipped).

41 Transmission slips, shifts rough, is noisy or has no drive in forward or reverse gears

1   There are many probable causes for the above problems, but the home mechanic should be concerned with only one possibility - fluid level.

2   Before taking the vehicle to a repair shop, check the level and condition of the fluid (Chapter 1). Correct fluid level as necessary or change the fluid and filter if needed. If the problem persists, have a professional diagnose the probable cause.

42 Fluid leakage

1   Automatic transmission fluid is a deep red color. Fluid leaks should not be confused with engine oil, which can easily be blown by air flow to the transmission.

2   To pinpoint a leak, first remove all built-up dirt and grime from around the transmission. Degreasing agents and/or steam cleaning will achieve this. With the underside clean, drive the vehicle at low speeds so air flow will not blow the leak far from its source. Raise the vehicle and determine where the leak is coming from. Common areas of leakage are:

a) *Pan: Tighten the mounting bolts and/ or replace the pan gasket as necessary (Chapter 7).*
b) *Filler pipe: Replace the rubber seal where the pipe enters the transmission case.*
c) *Transmission oil lines: Tighten the connectors where the lines enter the transmission case and/or replace the lines.*
d) *Vent pipe: Transmission overfilled and/ or water in fluid (see checking procedures, Chapter 1).*
e) *Speedometer connector: Replace the O-ring where the speedometer sensor enters the transmission case (Chapter 7).*

Transfer case

43 Transfer case is difficult to shift into the desired range

1   Speed may be too great to permit engagement. Stop the vehicle and shift into the desired range.

2   Shift linkage loose, bent or binding. Check the linkage for damage or wear and replace or lubricate as necessary (Chapter 7C).

3   If the vehicle has been driven on a paved surface for some time, the driveline torque can make shifting difficult. Stop and shift into two-wheel drive on paved or hard surfaces.

4   Insufficient or incorrect grade of lubricant. Drain and refill the transfer case with the specified lubricant (Chapter 1).

5   Worn or damaged internal components. Disassembly and overhaul of the transfer case may be necessary (Chapter 7C).

44 Transfer case noisy in all gears

Insufficient or incorrect grade of lubricant. Drain and refill (Chapter 1).

45 Noisy or jumps out of four-wheel drive Low range

1   Transfer case not fully engaged. Stop the vehicle, shift into Neutral and then engage 4L.

2   Shift linkage loose, worn or binding. Tighten, repair or lubricate linkage as necessary.

3   Shift fork cracked, inserts worn or fork binding on the rail. Disassemble and repair as necessary (Chapter 7C).

46 Lubricant leaks from the vent or output shaft seals

1   Transfer case is overfilled. Drain to the proper level (Chapter 1).

2   Vent is clogged or jammed closed. Clear or replace the vent.

3   Output shaft seal incorrectly installed or damaged. Replace the seal and check contact surfaces for nicks and scoring.

Driveshaft

47 Oil leak at seal end of driveshaft

Defective transmission or transfer case oil seal. See Chapter 7 for replacement procedures. While this is done, check the splined yoke for burrs or a rough condition which may be damaging the seal. Burrs can be removed with crocus cloth or a fine whetstone.

48 Knock or clunk when the transmission is under initial load (just after transmission is put into gear)

1   Loose or disconnected rear suspension components. Check all mounting bolts, nuts and bushings (Chapter 10).

2   Loose driveshaft bolts. Inspect all bolts and nuts and tighten them to the specified torque.

3   Worn or damaged universal joint bearings. Check for wear (Chapter 8).

49 Metallic grinding sound consistent with vehicle speed.

Pronounced wear in the universal joint bearings. Check as described in Chapter 8.

50 Vibration

Note: *Before assuming that the driveshaft is at fault, make sure the tires are perfectly balanced and perform the following test.*

1   Install a tachometer inside the vehicle to monitor engine speed as the vehicle is driven. Drive the vehicle and note the engine speed at which the vibration (roughness) is most pronounced. Now shift the transmission to a different gear and bring the engine speed to the same point.

2   If the vibration occurs at the same engine speed (rpm) regardless of which gear the transmission is in, the driveshaft is NOT at fault since the driveshaft speed varies.

3   If the vibration decreases or is eliminated when the transmission is in a different gear at the same engine speed, refer to the following probable causes.

4   Bent or dented driveshaft. Inspect and replace as necessary (Chapter 8).

5   Undercoating or built-up dirt, etc. on the driveshaft. Clean the shaft thoroughly and recheck.

6   Worn universal joint bearings. Remove and inspect (Chapter 8).
7   Driveshaft and/or companion flange out of balance. Check for missing weights on the shaft. Remove the driveshaft (Chapter 8) and reinstall 180-degrees from original position, then retest. Have the driveshaft professionally balanced if the problem persists.

Axles

51   Noise

1   Road noise. No corrective procedures available.
2   Tire noise. Inspect tires and check tire pressures (Chapter 1).
3   Rear wheel bearings loose, worn or damaged (Chapter 8).

52   Vibration

See probable causes under *Driveshaft*. Proceed under the guidelines listed for the driveshaft. If the problem persists, check the rear wheel bearings by raising the rear of the vehicle and spinning the rear wheels by hand. Listen for evidence of rough (noisy) bearings. Remove and inspect (Chapter 8).

53   Oil leakage

1   Pinion seal damaged (Chapter 8).
2   Axleshaft oil seals damaged (Chapter 8).
3   Differential inspection cover leaking. Tighten the bolts or replace the gasket as required (Chapters 1 and 8).

Brakes

Note: *Before assuming that a brake problem exists, make sure that the tires are in good condition and inflated properly (Chapter 1), that the front end alignment is correct and that the vehicle is not loaded with weight in an unequal manner.*

54   Vehicle pulls to one side during braking

1   Defective, damaged or oil contaminated disc brake pads on one side. Inspect (Chapter 9).
2   Excessive wear of pad material or disc on one side. Inspect and correct as necessary.
3   Loose or disconnected front suspension components. Inspect and tighten all bolts to the specified torque (Chapter 10).
4   Defective caliper assembly. Remove the caliper and inspect for a stuck piston or other

damage (Chapter 9).
5   Inadequate lubrication of front brake caliper slide rails. Remove caliper and lubricate slide rails (Chapter 9).

55   Noise (high-pitched squeal with the brakes applied)

1   Disc brake pads worn out. The noise comes from the wear sensor rubbing against the disc (does not apply to all vehicles) or the actual pad backing plate itself if the material is completely worn away. Replace the pads with new ones immediately (Chapter 9). If the pad material has worn completely away, the brake discs should be inspected for damage (Chapter 9).
2   Missing or damaged brake pad insulators. Replace pad insulators (see Chapter 9).
3   Linings contaminated with dirt or grease. Replace pads.
4   Incorrect linings. Replace with correct linings.

56   Excessive brake pedal travel

1   Partial brake system failure. Inspect the entire system (Chapter 9) and correct as required.
2   Insufficient fluid in the master cylinder. Check (Chapter 1), add fluid and bleed the system if necessary (Chapter 9).

57   Brake pedal feels spongy when depressed

1   Air in the hydraulic lines. Bleed the brake system (Chapter 9).
2   Faulty flexible hoses. Inspect all system hoses and lines. Replace parts as necessary.
3   Master cylinder mounting bolts/nuts loose.
4   Master cylinder defective (Chapter 9).

58   Excessive effort required to stop vehicle

1   Power brake booster or vacuum pump (diesel models) not operating properly (Chapter 9).
2   Excessively worn pads. Inspect and replace if necessary (Chapter 9).
3   One or more caliper pistons seized or sticking. Inspect and replace as required (Chapter 9).
4   Brake pads contaminated with oil or grease. Inspect and replace as required (Chapter 9).
5   New pads installed and not yet seated. It will take a while for the new material to seat against the disc.

59   Pedal travels to the floor with little resistance

1   Little or no fluid in the master cylinder reservoir caused by leaking caliper piston(s), loose, damaged or disconnected brake lines. Inspect the entire system and correct as necessary.
2   Worn master cylinder seals (Chapter 9).

60   Brake pedal pulsates during brake application

1   Caliper improperly installed. Remove and inspect (Chapter 9).
2   Disc defective. Remove (Chapter 9) and check for excessive lateral runout and parallelism. Have the disc resurfaced or replace it with a new one.

Suspension and steering systems

61   Vehicle pulls to one side

1   Tire pressures uneven (Chapter 1).
2   Defective tire (Chapter 1).
3   Excessive wear in suspension or steering components (Chapter 10).
4   Front end in need of alignment.
5   Front brakes dragging. Inspect the brakes (Chapter 9).

62   Shimmy, shake or vibration

1   Tire or wheel out-of-balance or out-of-round. Have professionally balanced.
2   Loose, worn or out-of-adjustment front wheel bearings (Chapter 1).
3   Shock absorbers and/or suspension components worn or damaged (Chapter 10).

63   Excessive pitching and/or rolling around corners or during braking

1   Defective shock absorbers. Replace as a set (Chapter 10).
2   Broken or weak springs and/or suspension components. Inspect (Chapter 10).

64   Excessively stiff steering

1   Lack of fluid in power steering fluid reservoir (Chapter 1).
2   Incorrect tire pressures (Chapter 1).
3   Lack of lubrication at steering joints (Chapter 1).
4   Front end out of alignment.
5   Lack of power assistance (see Section 66).

65 Excessive play in steering

1   Loose front wheel bearings (Chapters 1 and 10).
2   Excessive wear in suspension or steering components (Chapter 10).
3   Steering gearbox damaged or out of adjustment (Chapter 10).

66 Lack of power assistance

1   Steering pump drivebelt faulty or not adjusted properly (Chapter 1).
2   Fluid level low (Chapter 1).
3   Hoses or lines restricted. Inspect and replace parts as necessary.
4   Air in power steering system. Bleed the system (Chapter 10).

67 Excessive tire wear (not specific to one area)

1   Incorrect tire pressures (Chapter 1).
2   Tires out-of-balance. Have professionally balanced.
3   Wheels damaged. Inspect and replace as necessary.
4   Suspension or steering components excessively worn (Chapter 10).

68 Excessive tire wear on outside edge

1   Inflation pressures incorrect (Chapter 1).
2   Excessive speed in turns.
3   Front end alignment incorrect. Have professionally aligned.
4   Suspension arm bent or twisted (Chapter 10).

69 Excessive tire wear on inside edge

1   Inflation pressures incorrect (Chapter 1).
2   Front end alignment incorrect. Have professionally aligned.
3   Loose or damaged steering components (Chapter 10).

70 Tire tread worn in one place

1   Tires out-of-balance.
2   Damaged or buckled wheel. Inspect and replace if necessary.
3   Defective tire (Chapter 1).

Chapter 1
Tune-up and routine maintenance

Contents

Specifications

Recommended lubricants and fluids

Note: *Listed here are manufacturer recommendations at the time this manual was written. Manufacturers occasionally upgrade their fluid and lubricant specifications, so check with your local auto parts store for current recommendations.*

Engine oil
  Type
    Gasoline engines .... API grade "certified for gasoline engines"
    Diesel engine .... API grade CJ-4 multi-grade and low sulfated ash limit engine oil
  Viscosity
    Gasoline
      All except 6.4L models 5W-20
      2013 and later 6.4L models .... 0W-20 fully synthetic
    Diesel engine .... 15W-40 (5W-40 in temperatures below 0-degrees F)
Automatic transmission fluid
  All models except 8HP45/845RE and 8HP70 models .... Mopar® type ATF+4 or equivalent
  8HP45/845RE and 8HP70 models .... Mopar® ZF 8&9 speed Automatic Transmission Fluid
Manual transmission lubricant type .... Mopar® type ATF+4 or equivalent
Transfer case lubricant type
  BW44 series and NV246 .... Mopar® type BW44 transfer case fluid or equivalent
  All others .... Mopar® type ATF+4 or equivalent
Coolant
  2012 and earlier models .... 50/50 mixture of Mopar® 5 year/100,000 mile Formula antifreeze/coolant with HOAT (Hybrid Organic Additive Technology) and water
  2013 and later models .... 50/50 mixture of Mopar® 10 year/150,000 mile Formula antifreeze/coolant with OAT (Organic Additive Technology) and water

Note: *Most models are filled with either 50/50 mixture of Mopar® 5 year/100,000 mile or 50/50 mixture of Mopar® 10 year/150,000 mile coolant that shouldn't be mixed with other coolants. Some early model diesel engines are filled with a 50/50 mixture of ethylene glycol-based antifreeze and water. Refer to the owners manual for your vehicle to determine what type coolant you have. Always refill with the correct coolant.*

Recommended lubricants and fluids (continued)

Note: *Listed here are manufacturer recommendations at the time this manual was written. Manufacturers occasionally upgrade their fluid and lubricant specifications, so check with your local auto parts store for current recommendations.*

Differential lubricant type
    Front axle
        1500 models
            2012 and earlier models.. Synthetic GL-5 SAE 75W-90 gear lubricant
            2013 and later models.. Synthetic GL-5 SAE 75W-85 gear lubricant
        2500/3500 models .... Synthetic GL-5 SAE 75W-90 or equivalent
    Rear axle
        1500 models*
            2010 and earlier models.. Synthetic GL-5 SAE 75W-140 gear lubricant
            2011 and 2012 models
                C235 (9-1/4 inch) axle.. Synthetic GL-5 SAE 75W-140 gear lubricant
                All othersSynthetic .... GL-5 SAE 75W-90 gear lubricant
            2013 and later models.. Synthetic GL-5 SAE 75W-140 gear lubricant
        2500/3500 models .... Synthetic GL-5 SAE 75W-90 or equivalent - use no additives for limited slip

Brake/clutch fluid type .... DOT 3 or DOT 4 brake fluid
Power steering fluid .... Mopar® type ATF+4 or equivalent
Chassis grease type... NLGI no. 2 EP chassis grease

* Limited-slip rear axles add 4 oz. of Mopar® limited slip additive or equivalent, to the specified lubricant.

Capacities*

Cooling system
    3.6L V6 engine .... 14.8 quarts
    3.7L V6 engine .... 14 quarts
    4.7L V8 engine .... 14 quarts
    5.7L V8 engine
        2009
            1500 models.. 16 quarts
            2500 and 3500 cab chassis models.................................... 18.7 quarts
        2010 and later models .... 16 quarts
    6.4L engine .... 16.6 quarts
    6.7L diesel engine .... 23 quarts
Engine oil (with filter change)
    3.6L, 3.7L V6 engine .... 5 quarts
    4.7L V8 engine .... 6 quarts
    5.7L and 6.4L V8 engines .... 7 quarts
    6.7L diesel engine .... 12 quarts
Automatic transmission (drain and refill) .... 4 to 7.2 quarts

Note: *The best way to determine the amount of fluid to add during a routine fluid change is to measure the amount drained. It is important not to overfill the transmission. After draining the transmission, begin the refilling procedure by initially adding 3 quarts, then adding one pint at a time until the level is correct on the dipstick.*

Manual transmission
    G56 .... 6 quarts
    Getrag .... 2.3 quarts
Transfer case.. 1.5 to 2 quarts
Differential
    Front axle .... 2 to 3 quarts
    Rear axle .... 2.5 quarts

*All capacities approximate. Add as necessary to bring to the appropriate levels.

Ignition system

Spark plug type
    3.6L V6 engine .... RER8ZWYCB4 (Champion)
    3.7L V6 engine .... ZFR6F-11G (NGK)
    4.7L V8 engine
        Intake (upper) .... FR9 TE2 (Bosch)
        Exhaust (lower)... FR8 T1332 (Bosch)
    5.7L V8 engine
        2010 and earlier models .... REC14MCC4 (Champion)
        2011 and later models .... LZFR5C-11G (NGK)
    6.4L engine .... LZFR5C-11G (NGK)
Spark plug gap
    3.6L V6 engine .... 0.043 inch
    3.7L V6 engine .... 0.044 inch
        4.7L V8 engine
        Intake (upper) .... 0.039 inch
        Exhaust (lower)... 0.050 inch
        5.7L V8 engine
            2010 and earlier models .... 0.040 inch
            2011 and later models .... 0.043 inch
        6.4L engine .... 0.043 inch

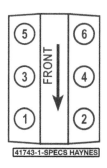

3.6L V6 engine cylinder locations

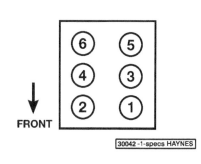

3.7L V6 engine cylinder locations

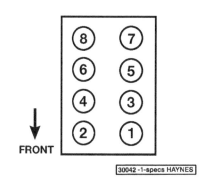

4.7L, 6.4L and 5.7L (Hemi) V8 engine
cylinder locations

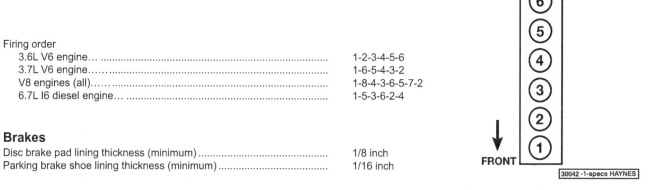

6.7L diesel engine cylinder locations

Firing order
   3.6L V6 engine… 1-2-3-4-5-6
   3.7L V6 engine….. 1-6-5-4-3-2
   V8 engines (all)…... 1-8-4-3-6-5-7-2
   6.7L I6 diesel engine… 1-5-3-6-2-4

Brakes
Disc brake pad lining thickness (minimum) .... 1/8 inch
Parking brake shoe lining thickness (minimum) .... 1/16 inch

Torque specifications   Ft-lbs (unless otherwise specified)

Note: *One foot-pound (ft-lb) of torque is equivalent to 12 inch-pounds (in-lbs) of torque. Torque values below approximately 15 ft-lbs are expressed in inch-pounds, since most foot-pound torque wrenches are not accurate at these smaller values.*

Automatic transmission pan bolts
   545RFE, 68RFE, 66RFE, 65RFE .... 105 in-lbs
   42RLE .... 174 in-lbs
   AS68RC, AS69RC .... 62 in-lbs
   8HP45/845RE and 8HP70 .... 89 in-lbs
Drivebelt tensioner bolt
   3.6L, 3.7L, 4.7L and 5.7L engines................................... 30
   6.7L diesel engine 32
Engine oil filter cap (3.6L V6 engine)..................................... 18
Fuel filter (diesel, 2009 models) .... Tighten until it stops, then 1/2-inch additional turn
Fuel filter cap (diesel, 2010 and later models)........................ 23
Manual transmission drain/fill plug 42
Spark plugs
   3.6L engine 15
   3.7L and 4.7L engines.. 20
   5.7L and 6.4L engines.. 156 in-lbs
Note: *On models with tapered designed spark plugs, do not tighten the spark plugs more than 15 ft-lbs.*
Transmission fill/check plug (8-speed models)........................ 26
Transmission drain plug (8-speed models) .... 80 in-lbs
Wheel lug nuts
   1500 models
      2009 through 2011.. 130
      2012 and later.. 135
   2500 and 3500 models
      2009 through 2011
         Single rear wheels... 140
         Dual rear wheels.. 145
      2012 and later models
         Single rear wheels... 150
         Dual rear wheels.. 145

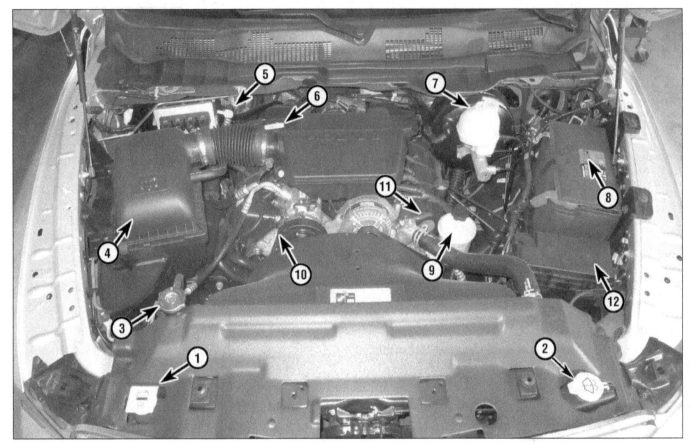

4.7L V8 engine compartment layout

1	Coolant reservoir	5	Automatic transmission fluid dipstick	9	Power steering fluid reservoir
2	Windshield washer fluid reservoir	6	Engine oil dipstick	10	Drivebelt
3	Radiator cap	7	Brake fluid reservoir	11	Engine oil fill cap
4	Air filter housing	8	Battery	12	Fuse and relay center

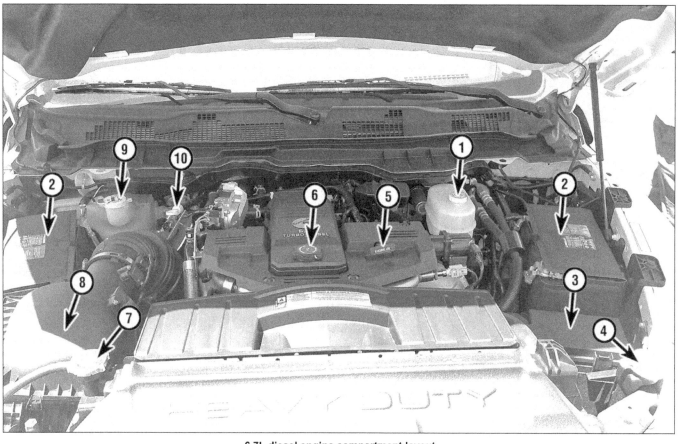

6.7L diesel engine compartment layout

1	Brake fluid reservoir	5	Engine oil dipstick
2	Battery	6	Engine oil filler cap
3	Underhood fuse/relay block	7	Radiator cap
4	Windshield washer fluid reservoir	8	Air filter housing

9   Coolant reservoir
10 Automatic transmission
    fluid dipstick

Typical underside components (1500 model shown)

1 Engine oil drain plug	4 Steering gear boot	7 Engine oil filter
2 Lower radiator hose	5 Drivebelt	(not visible in photo)
3 Tie-rod end	6 Automatic transmission pan	

Typical rear underside components (1500 model)

1	Differential check/fill plug	4	Spring
2	Fuel tank	5	Brake caliper
3	Brake hose		

1 Maintenance schedule

The maintenance intervals in this manual are provided with the assumption that you, not the dealer, will be doing the work. These are the minimum maintenance intervals recommended by the factory for vehicles that are driven daily. If you wish to keep your vehicle in peak condition at all times, you may wish to perform some of these procedures even more often. Because frequent maintenance enhances the efficiency, performance and resale value of your car, we encourage you to do so. If you drive in dusty areas, tow a trailer, idle or drive at low speeds for extended periods or drive for short distances (less than four miles) in below freezing temperatures, shorter intervals are also recommended.

When your vehicle is new, it should be serviced by a factory authorized dealer service department to protect the factory warranty. In many cases, the initial maintenance check is done at no cost to the owner.

Every 250 miles or weekly, whichever comes first

Check the engine oil level (Section 4)
Check the engine coolant level (Section 4)
Check the brake fluid level (Section 4)
Check the power steering fluid level (Section 4)
Check the windshield washer fluid level (Section 4)
Check the automatic transmission fluid level (Section 4)
Check the tires and tire pressures (Section 5)
Check the Water In Fuel (WIF) warning light on the instrument panel. If it's illuminated, drain the fuel filter (Section 21)
Check the operation of all lights
Check the horn operation

Every 6,000 miles or 6 months, whichever comes first

All items listed above, plus:
Change the engine oil and filter (Section 6)
Lubricate the front driveshaft (2500/3500 diesel models) (Section 8)

Lubricate the outer tie-rod ends (diesel models) (Section 9)
Check the wiper blade condition (Section 7)
Check and clean the battery and terminals (see Section 10)
Rotate the tires (Section 11)
Check the seatbelts (Section 12)
Inspect underhood hoses (Section 13)
Check the cooling system hoses and connections for leaks and damage (Section 14)
Check the brake hoses (Section 15)
Check the exhaust pipes and hangers (Section 16)

Every 15,000 miles or 12 months, whichever comes first

All items listed above, plus:
Check the suspension system, steering system and the driveaxle boots (Section 17)
Change the differential lubricant on diesel models or any model operated under the conditions described as "severe" (*) at the end of this schedule (Section 28)
Replace the fuel filter (diesel engine) (Section 21)
Check the brake system (see Section 18)
Check the drivebelts and replace if necessary (Section 19)
Check the fuel system hoses and connections for leaks and damage (Section 20)
Replace the fuel filter (diesel engine) (Section 21)
Check the manual transmission, transfer case (4WD models) and differential lubricant level (Section 22)

Every 30,000 miles or 24 months, whichever comes first

All items listed above, plus:
Change the brake fluid (Section 23)
Replace the air filter element (gasoline engines) (Section 24)
Replace the spark plugs (3.7L and 5.7L engines) (Section 25)

Every 45,000 miles or 36 months, whichever comes first

Replace upper spark plugs (4.7L engine) (Section 25)

Every 60,000 miles or 48 months, whichever comes first

All items listed above, plus:

Replace the spark plug wires (3.7L engine) (Section 26)
Clean the Exhaust Gas Recirculation (EGR) valve and
   cooler (diesel engine) (Section 27)
Change the manual transmission lubricant (Section 28)
Change the transfer case lubricant (Section 28)*
Change the differential lubricant (Section 28)*
Replace the Closed Crankcase Ventilation (CCV) filter (die-
   sel engine) (Section 29)

Every 90,000 miles or 72 months, whichever comes first

Replace the PCV valve (gasoline engines) (Section 30)
Change the automatic transmission fluid and filter
   (Section 31)**
Change the power steering fluid and flush the system
   (Section 33)
Replace the spark plug wires (4.7L engine) (Section 26)
Replace the lower spark plugs (4.7L engine) (Section 25)
Replace the spark plugs (3.6L and 6.4L engines) (Section 25)
Service the cooling system (drain, flush and refill) (Section 32)

Every 150,000 miles

Check and adjust if necessary, the valve clearances (diesel
   engine) (Chapter 2C)

Note: *Valve clearance check and adjustment is not a routine maintenance item on diesel engines. It must be
checked and, if necessary, adjusted only after the valve
train assembly has been removed and installed. And it
might be necessary to check and adjust the valve clearance when troubleshooting a performance, emissions,
fuel economy or noise problem.*

*This item is affected by "severe" operating conditions as
described below. If your vehicle is operated under "severe"
conditions, perform all maintenance indicated with an asterisk
(*) at 15,000 mile/12 month intervals (unless otherwise specified in the schedule). Severe conditions are indicated if you
mainly operate your vehicle under one or more of the following
conditions:

> *Operating in dusty areas*
> *Towing a trailer*
> *Idling for extended periods and/or low speed operation*
> *Operating in extended temperatures below freezing
>   (32-degrees F/0-degrees C)*
> *If more the half of your driving is at high speeds in
>   temperatures above 90-degrees F (32-degrees C)*

**If operated under one or more of the following conditions,
change the automatic transmission fluid every 30,000 miles:

> *In heavy city traffic where the outside temperature regularly
>   reaches 90-degrees F (32-degrees C) or higher*
> *In hilly or mountainous terrain*
> *Frequent towing of a trailer*

2   Introduction

This Chapter is designed to help the home mechanic maintain the Dodge Ram pickup truck with the goals of maximum per-formance, economy, safety and reliability in mind.

Included is a master maintenance sched-ule, followed by procedures dealing specifi-cally with each item on the schedule. Visual checks, adjustments, component replacement and other helpful items are included. Refer to the accompanying illustrations of the engine compartment and the underside of the vehicle for the locations of various components.

Servicing your vehicle in accordance with the mileage/time maintenance schedule and the step-by-step procedures will result in a planned maintenance program that should produce a long and reliable service life. Keep in mind that it's a comprehensive plan, so maintaining some items but not others at the specified intervals will not produce the same results.

As you service your vehicle, you will dis-cover that many of the procedures can - and should - be grouped together because of the nature of the particular procedure you're per-forming or because of the close proximity of two otherwise unrelated components to one another.

For example, if the vehicle is raised for chassis lubrication, you should inspect the exhaust, suspension, steering and fuel sys-tems while you're under the vehicle. When you're rotating the tires, it makes good sense to check the brakes since the wheels are already removed. Finally, let's suppose you have to borrow or rent a torque wrench. Even if you only need it to tighten the spark plugs, you might as well check the torque of as many critical fasteners as time allows.

The first step in this maintenance pro-gram is to prepare yourself before the actual work begins. Read through all the procedures you're planning to do, then gather up all the parts and tools needed. If it looks like you might run into problems during a particular job, seek advice from a mechanic or an expe-rienced do-it-yourselfer.

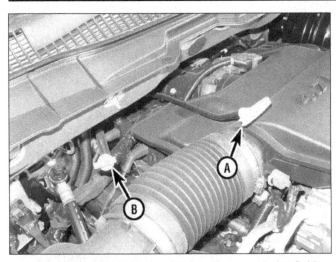

4.2 Engine oil dipstick (A) and automatic transmission fluid dipstick (B) locations (most models)

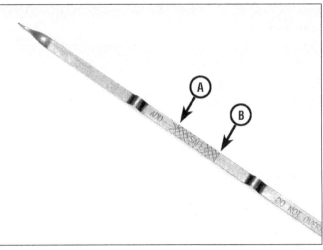

4.4 On gasoline models, it takes one quart of oil to raise the level from ADD mark (A) to the FULL mark (B). On diesel models, it takes two quarts

Owner's Manual and VECI label information

Your vehicle owner's manual was written for your year and model and contains very specific information on component locations, specifications, fuse ratings, part numbers, etc. The Owner's Manual is an important resource for the do-it-yourselfer to have; if one was not supplied with your vehicle, it can generally be ordered from a dealer parts department.

Among other important information, the Vehicle Emissions Control Information (VECI) label contains specifications and procedures for applicable tune-up adjustments and, in some instances, spark plugs. The information on this label is the exact maintenance data recommended by the manufacturer. This data often varies by intended operating altitude, local emissions regulations, month of manufacture, etc.

This Chapter contains procedural details, safety information and more ambitious maintenance intervals than you might find in manufacturer's literature. However, you may also find procedures or specifications in your Owner's Manual or VECI label that differ with what's printed here. In these cases, the Owner's Manual or VECI label can be considered correct, since it is specific to your particular vehicle.

3   Tune-up general information

The term tune-up is used in this manual to represent a combination of individual operations rather than one specific procedure.

If, from the time the vehicle is new, the routine maintenance schedule is followed closely and frequent checks are made of fluid levels and high wear items, as suggested throughout this manual, the engine will be kept in relatively good running condition and the need for additional work will be minimized.

More likely than not, however, there will be times when the engine is running poorly due to lack of regular maintenance. This is even more likely if a used vehicle, which has not received regular and frequent maintenance checks, is purchased. In such cases, an engine tune-up will be needed outside of the regular routine maintenance intervals.

The first step in any tune-up or diagnostic procedure to help correct a poor running engine is a cylinder compression check. A compression check (see Chapter 2D) will help determine the condition of internal engine components and should be used as a guide for tune-up and repair procedures. If, for instance, a compression check indicates serious internal engine wear, a conventional tune-up will not improve the performance of the engine and would be a waste of time and money. Because of its importance, the compression check should be done by someone with the right equipment and the knowledge to use it properly.

The following procedures are those most often needed to bring a generally poor running engine back into a proper state of tune.

Minor tune-up
Check all engine-related fluids (Section 4)
Check the air filter (Section 24)
Clean, inspect and test the battery (Section 10)
Check all underhood hoses (Section 13)
Check the cooling system (Section 14)
Check the drivebelt (Section 19)

Major tune-up
All items listed under Minor tune-up, plus . . .
Replace the air filter (Section 24)
Replace the spark plugs (Section 25)
Check the charging system (Chapter 5)

4   Fluid level checks (every 250 miles or weekly)

1   Fluids are an essential part of the lubrication, cooling, brake and windshield washer systems. Because the fluids gradually become depleted and/or contaminated during normal operation of the vehicle, they must be periodically replenished. See *Recommended lubricants and fluids* in this Chapter's Specifications before adding fluid to any of the following components. **Note:** *The vehicle must be on level ground when fluid levels are checked.*

Engine oil
Refer to illustrations 4.2, 4.4 and 4.6

2   The oil level is checked with a dipstick, which is located on the side of the engine **(see illustration)**. The dipstick extends through a metal tube down into the oil pan.
3   The oil level should be checked before the vehicle has been driven, or about 5 minutes after the engine has been shut off. If the oil is checked immediately after driving the vehicle, some of the oil will remain in the upper part of the engine, resulting in an inaccurate reading on the dipstick.
4   Pull the dipstick out of the tube and wipe all the oil from the end with a clean rag or paper towel. Insert the clean dipstick all the way back into the tube and pull it out again. Note the oil at the end of the dipstick. At its highest point, the level should be between the ADD and FULL marks on the dipstick **(see illustration)**.
5   On gasoline engines, it takes one quart of oil to raise the level from the ADD mark to the FULL mark on the dipstick. Do not allow the level to drop below the ADD mark or oil starvation may cause engine damage. Conversely, overfilling the engine (adding oil above the FULL mark) may cause oil fouled spark plugs or oil foaming. On diesel engines,

4.6 Engine oil filler cap location (4.7L V8 shown, others similar)

4.8 Location of the coolant reservoir/expansion tank
(some gasoline-engine models)

it takes two quarts of oil to raise the level from the ADD to the FULL mark on the dipstick. Maintaining the oil level above the FULL mark can cause excessive oil consumption.

6   To add oil, remove the filler cap from the valve cover **(see illustration)**. After adding oil, wait a few minutes to allow the level to stabilize, then pull out the dipstick and check the level again. Add more oil if required. Install the filler cap and tighten it by hand only.

7   Checking the oil level is an important preventive maintenance step. A consistently low oil level indicates oil leakage through damaged seals, defective gaskets or past worn rings or valve guides. If the oil looks milky in color or has water droplets in it, the cylinder head gasket(s) may be blown or the head(s) or block may be cracked. The engine should be checked immediately. The condition of the oil should also be checked. Whenever you check the oil level, slide your thumb and index finger up the dipstick before wiping off the oil. If you see small dirt or metal particles clinging to the dipstick, the oil should be changed (see Section 6). Most models are equipped with an engine oil change reminder system. This system will alert you by displaying "Oil Change Required" on the Electronic Vehicle Information Center (EVIC) when it is time to change your engine oil. Never exceed 7,500 mile oil change intervals even if the light doesn't come on.

Engine coolant

Refer to illustration 4.8

Warning: *Do not allow antifreeze to come in contact with your skin or painted surfaces of the vehicle. Flush contaminated areas immediately with plenty of water. Don't store new coolant or leave old coolant lying around where it's accessible to children or pets - they're attracted by its sweet smell. Ingestion of even a small amount of coolant can be fatal! Wipe up garage floor and drip pan spills immediately. Keep antifreeze containers cov-*

ered and repair cooling system leaks as soon as they're noticed.

8   All vehicles covered by this manual are equipped with a pressurized coolant recovery system. A plastic coolant reservoir is located at the right-front corner of the engine compartment, just behind the grille **(see illustration)**, mounted to the right side of the radiator fan shroud, or at the right-rear corner of the engine compartment, depending on model.

9   The coolant level in the tank should be checked regularly. **Warning:** *Do not remove the cap to check the coolant level when the engine is warm!* The level in the tank varies with the temperature of the engine.

a) *Start the engine. Once the engine has warmed up, let the engine idle and check the coolant level in the reservoir. The level should be between the marks.*

b) *If it isn't, allow the engine to cool, then remove the cap from the reservoir tank and add a 50/50 mixture of ethylene glycol based antifreeze and water.*

10   Drive the vehicle and recheck the coolant level. If only a small amount of coolant is required to bring the system up to the proper level, water can be used. However, repeated additions of water will dilute the antifreeze and water solution. In order to maintain the proper ratio of antifreeze and water, always top up the coolant level with the correct mixture. Don't use rust inhibitors or additives. An empty plastic milk jug or bleach bottle makes an excellent container for mixing coolant.

11   If the coolant level drops consistently, there may be a leak in the system. Inspect the radiator, hoses, filler cap, drain plugs and water pump (see Section 14). If no leaks are noted, have the pressure cap or expansion tank cap pressure tested by a service station.

12   If you have to remove the pressure cap or expansion tank cap, wait until the engine has cooled completely, then wrap a thick cloth around the cap and turn it to the first stop. If

coolant or steam escapes, or if you hear a hissing noise, let the engine cool down longer, then remove the cap.

13   Check the condition of the coolant as well. It should be relatively clear. If it's brown or rust colored, the system should be drained, flushed and refilled. Even if the coolant appears to be normal, the corrosion inhibitors wear out, so it must be replaced at the specified intervals.

Brake and clutch fluid

Refer to illustration 4.15

14   The brake master cylinder is located in the driver's side of the engine compartment, near the firewall. The hydraulic clutch master cylinder used on manual transmission vehicles is sealed at the factory and requires replacement if leaks develop.

15   To check the fluid level of the brake master cylinder, simply look at the MAX and MIN marks on the reservoir **(see illustration)**. The level should be within the specified distance from the maximum fill line.

4.15 Never let the brake fluid level drop
below the MIN mark

4.23   The power steering fluid reservoir is located on the left side of the engine (5.7L V8 engine shown, others similar)

16   If the level is low, wipe the top of the reservoir cover with a clean rag to prevent contamination of the brake system before lifting the cover.

17   Add only the specified brake fluid to the brake reservoir (refer to *Recommended lubricants and fluids* in this Chapter's Specifications or your owner's manual). Mixing different types of brake fluid can damage the system. Fill the brake master cylinder reservoir only to the MAX line. **Warning:** *Use caution when filling the reservoir - brake fluid can harm your eyes and damage painted surfaces. Do not use brake fluid that is more than one year old or has been left open. Brake fluid absorbs moisture from the air. Excess moisture can cause a dangerous loss of braking.*

18   While the reservoir cap is removed, inspect the master cylinder reservoir for contamination. If deposits, dirt particles or water droplets are present, the system should be drained and refilled.

19   After filling the reservoir to the proper level, make sure the lid is properly seated to prevent fluid leakage and/or system pressure loss.

20   The fluid in the brake master cylinder will drop slightly as the brake pads at each wheel wear down during normal operation. If

the master cylinder requires repeated replenishing to keep it at the proper level, this is an indication of leakage in the brake system, which should be corrected immediately. If the brake system shows an indication of leakage check all brake lines and connections, along with the calipers and booster (see Section 19 for more information). If the hydraulic clutch system shows an indication of leakage, check all clutch lines and connections, along with the clutch release cylinder (see Chapter 8 for more information).

21   If, upon checking the brake master cylinder fluid level, you discover the reservoir empty or nearly empty, the system should be bled (see Chapter 9).

Power steering fluid

Refer to illustration 4.23

Note: *On 2013 and later 1500 models with 3.6L V6 or 5.7L V8 engines, a dry-gear Electric Power Steering (EPS) system is used and there is no power steering fluid.*

22   Check the power steering fluid level periodically to avoid steering system problems, such as damage to the pump. **Caution:** *DO NOT hold the steering wheel against either stop (extreme left or right turn) for more than five seconds. If you do, the power steering pump could be damaged.*

23   The power steering reservoir, located at the right side of the engine compartment **(see illustration)**, has MIN and MAX fluid level marks on the side. The fluid level can be seen without removing the reservoir cap.

24   Park the vehicle on level ground and apply the parking brake.

25   Run the engine until it has reached normal operating temperature.

26   With the engine at idle, turn the steering wheel back and forth about 10 times to get any air out of the steering system. Shut the engine off with the wheels in the straight-ahead position.

27   Note the fluid level on the side of the reservoir. It should be between the two marks.

28   Add small amounts of fluid until the level is correct. **Caution:** *Do not overfill the reservoir. If too much fluid is added, remove the*

excess with a clean syringe or suction pump.

29   Check the power steering hoses and connections for leaks and wear.

Windshield washer fluid

Refer to illustration 4.30

30   Fluid for the windshield washer system is stored in a plastic reservoir located at the left front of the engine compartment **(see illustration)**.

31   In milder climates, plain water can be used in the reservoir, but it should be kept no more than 2/3 full to allow for expansion if the water freezes. In colder climates, use windshield washer system antifreeze, available at any auto parts store, to lower the freezing point of the fluid. Mix the antifreeze with water in accordance with the manufacturer's directions on the container. **Caution:** *Do not use cooling system antifreeze - it will damage the vehicle's paint.*

Automatic transmission fluid

Refer to illustration 4.37

32   The automatic transmission fluid level should be carefully maintained. Low fluid level can lead to slipping or loss of drive, while overfilling can cause foaming and loss of fluid.

33   With the parking brake set, start the engine, then move the shift lever through all the gear ranges, ending in Park. The fluid level must be checked with the vehicle level and the engine running at idle. **Note:** *Incorrect fluid level readings will result if the vehicle has just been driven at high speeds for an extended period, in hot weather in city traffic, or if it has been pulling a trailer. If any of these conditions apply, wait until the fluid has cooled (about 30 minutes).*

All models except (8HP45/845RE and 8HP70) 8-speed transmissions

34   With the transmission at normal operating temperature, remove the dipstick from the filler tube. The dipstick is located at the rear of the engine compartment on the passenger's side **(see illustration 4.2)**. **Note:** *Normal operating temperature is reached after a few minutes of engine operation or after 15 miles of driving.*

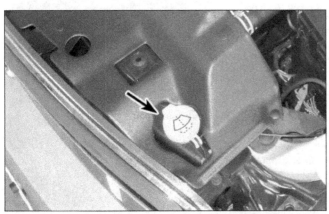

4.30   Windshield washer fluid reservoir location (5.7L V8 shown, others similar)

4.34   The automatic transmission dipstick is located at the rear of the engine compartment

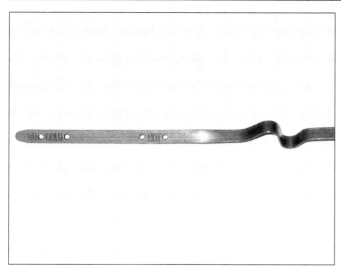

4.37 Check the fluid with the transmission at normal operating temperature - the level should be in the HOT range

5.2 A tire tread depth indicator should be used to monitor tire wear - they are available at auto parts stores and cost very little

35   Wipe the fluid from the dipstick with a clean rag and push it back into the filler tube until the cap seats.

36   Pull the dipstick out again and note the fluid level.

37   At normal operating temperature, the fluid level should be between the two upper reference holes (HOT) **(see illustration)**. On diesel models, if the fluid is warm, the level should be between the two holes. If it's hot, the level should be in the area marked OK (crosshatched area). If additional fluid is required, add it directly into the tube using a funnel. Add the fluid a little at a time and keep checking the level until it's correct. **Note:** *Wait at least two minutes before rechecking the fluid level allowing the fluid to fully drain into the transmission.*

38   The condition of the fluid should also be checked along with the level. If the fluid at the end of the dipstick is a dark reddish-brown color, or if it smells burned, it should be changed. If you are in doubt about the condition of the fluid, purchase some new fluid and compare the two for color and smell.

(8HP45/845RE and 8HP70)
8-speed transmissions

Warning: *Never get underneath the vehicle when it is supported only by a jack. The jack provided with your vehicle is designed solely for raising the vehicle to remove and install a wheel. Always use jackstands to support the vehicle when it becomes necessary to place your body underneath the vehicle.*

Warning: *These transmissions do not have a dipstick to check fluid level. The fluid level is checked at a filler plug on the right rear of the transmission housing. To accurately check the fluid level on this transmission, the fluid temperature must be below 86-degrees F, the engine must be running in Park and the vehi-*

cle has to be raised off the ground. There is a risk of personal injury from hot transmission fluid and also from the nearby exhaust system. Therefore if attempting this procedure, wear heat-proof gloves and position yourself beneath the vehicle so you are not at risk of being splashed by any transmission fluid. Do not remove the fill plug unless the engine is running at idle and in Park, because the transmission fluid drains from the torque converter once the engine is switched off and flows back to the transmission pan. If removed with the engine Off, a large amount of transmission fluid will be expelled from the plug hole.

Note: *The vehicle must be level to accurately check the automatic transmission fluid, so it will be necessary to raise both the front and rear of the vehicle.*

38   With the parking brake set, start the engine, then move the shift lever through all the gear ranges, ending in Park. The fluid level must be checked with the vehicle level and the engine running at idle. It is critical that the automatic transmission fluid temperature be below 86-degrees F to obtain an accurate fluid level reading.

39   Raise the vehicle and support it securely on jackstands.

Note: *On models with adequate clearance, this won't be necessary.*

40   Disable the traction control (ESC) by turning the system off at the instrument panel.

41   Make sure the engine is idling in "Park", then place a drain pan underneath the fill plug to catch any fluid that may come from the plug and remove the plug from the right rear side of the transmission.

Warning: *Position your body far enough away from the filler plug hole as hot transmission fluid can cause burns. Also take care as the exhaust is positioned close to the filler plug.*

42   Add fluid to the fill plug hole, slightly overfilling the transmission and allowing the

excess to drain until fluid only drips out of the oil fill plug hole, then install the plug and tighten it securely.

43   Apply the brake, then place the transmission in Reverse and hold it there for five seconds, then into Drive for five seconds.

44   Release the brakes and slowly accelerate until the transmission shifts into second gear and hold it there for five seconds.

45   Slowly apply the brake and place the transmission in to neutral, hold the brake and raise the engine RPM to 2,000 rpm and hold five seconds.

46   Release the accelerator and allow the transmission to return to idle, then place the transmission in park.

47   Remove the fill plug and allow any excess fluid to drain until the fluid only drips out.

48   Install the fill plug and tighten it to the torque listed in this Chapter's Specifications.

5   Tire and tire pressure checks (every 250 miles or weekly)

Refer to illustrations 5.2, 5.3, 5.4a, 5.4b and 5.8

1   Periodic inspection of the tires may spare you the inconvenience of being stranded with a flat tire. It can also provide you with vital information regarding possible problems in the steering and suspension systems before major damage occurs.

2   The original tires on this vehicle are equipped with 1/2-inch wide bands that will appear when tread depth reaches 1/16-inch, at which point they can be considered worn out. Tread wear can be monitored with a simple, inexpensive device known as a tread depth indicator **(see illustration)**.

UNDERINFLATION

CUPPING

Cupping may be caused by:
- Underinflation and/or mechanical irregularities such as out-of-balance condition of wheel and/or tire, and bent or damaged wheel.
- Loose or worn steering tie-rod or steering idler arm.
- Loose, damaged or worn front suspension parts.

OVERINFLATION

INCORRECT TOE-IN OR EXTREME CAMBER

FEATHERING DUE TO MISALIGNMENT

5.3 This chart will help you determine the condition of your tires, the probable cause(s) of abnormal wear and the corrective action necessary

3   Note any abnormal tread wear **(see illustration)**. Tread pattern irregularities such as cupping, flat spots and more wear on one side than the other are indications of front end alignment and/or balance problems. If any of these conditions are noted, take the vehicle to a tire shop or service station to correct the problem.

4   Look closely for cuts, punctures and embedded nails or tacks. Sometimes a tire will hold air pressure for a short time or leak down very slowly after a nail has embedded itself in the tread. If a slow leak persists, check the valve stem core to make sure it is tight **(see illustration)**. Examine the tread for an object that may have embedded itself in the tire or for a plug that may have begun to leak (radial tire punctures are repaired with a plug that is installed in a puncture). If a puncture is suspected, it can be easily verified by spraying a solution of soapy water onto the puncture area **(see illustration)**. The soapy solution will bubble if there is a leak. Unless the puncture is unusually large, a tire shop or service station can usually repair the tire.

5   Carefully inspect the inner sidewall of each tire for evidence of brake fluid leakage. If you see any, inspect the brakes immediately.

6   Correct air pressure adds miles to the life span of the tires, improves mileage and enhances overall ride quality. Tire pressure

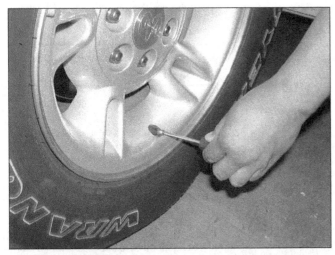

5.4a If a tire loses air on a steady basis, check the valve core first to make sure it's snug (special inexpensive wrenches are commonly available at auto parts stores)

5.4b If the valve core is tight, raise the corner of the vehicle with the low tire and spray a soapy water solution onto the tread as the tire is turned slowly - slow leaks will cause small bubbles to appear

cannot be accurately estimated by looking at a tire, especially if it's a radial. A tire pressure gauge is essential. Keep an accurate gauge in the glove compartment. The pressure gauges attached to the nozzles of air hoses at gas stations are often inaccurate.

7   Always check tire pressure when the tires are cold. Cold, in this case, means the vehicle has not been driven over a mile in the three hours preceding a tire pressure check. A pressure rise of four to eight pounds is not uncommon once the tires are warm.

8   Unscrew the valve cap protruding from the wheel or hubcap and push the gauge firmly onto the valve stem **(see illustration)**. Note the reading on the gauge and compare the figure to the recommended tire pressure shown on the tire placard on the driver's side door. Reinstall the valve cap to keep dirt and moisture out of the valve stem mechanism. Check all four tires and, if necessary, add enough air to bring them up to the recommended pressure.

9   Don't forget to keep the spare tire inflated to the specified pressure (refer to the pressure molded into the tire sidewall).

5.8 To extend the life of your tires, check the air pressure at least once a week with an accurate gauge (don't forget the spare!)

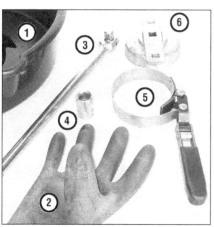

6.2 These tools are required when changing the engine oil and filter

1   **Drain pan** - *It should be fairly shallow in depth, but wide in order to prevent spills*
2   **Rubber gloves** - *When removing the drain plug and filter, it is inevitable that you will get oil on your hands (the gloves will prevent burns)*
3   **Breaker bar** - *Sometimes the oil drain plug is pretty tight and a long breaker bar is needed to loosen it*
4   **Socket** - *To be used with the breaker bar or a ratchet (must be the correct size to fit the drain plug)*
5   **Filter wrench** - *This is a metal band-type wrench, which requires clearance around the filter to be effective*
6   **Filter wrench** - *This type fits on the bottom of the filter and can be turned with a ratchet or beaker bar (different size wrenches are available for different types of filters)*

6   Engine oil and filter change (every 3,000 miles or 3 months)

Refer to illustrations 6.2, 6.7, 6.12 and 6.14

1   Frequent oil changes are the best preventive maintenance the home mechanic can give the engine, because aging oil becomes diluted and contaminated, which leads to premature engine wear.

2   Make sure you have all the necessary tools before you begin this procedure **(see illustration)**. You should also have plenty of rags or newspapers handy for mopping up any spills.

3   Access to the underside of the vehicle is greatly improved if the vehicle can be lifted on a hoist, driven onto ramps or supported by jackstands. **Warning:** *Do not work under a vehicle which is supported only by a bumper, hydraulic or scissors-type jack.*

4   If this is your first oil change, get under the vehicle and familiarize yourself with the locations of the oil drain plug and the oil filter. The engine and exhaust components will be warm during the actual work, so try to anticipate any potential problems before the engine and accessories are hot.

5   Park the vehicle on a level spot. Start the engine and allow it to reach its normal operating temperature. Warm oil and sludge will flow out more easily. Turn off the engine when it's warmed up. Remove the filler cap from the valve cover. **Note:** *On 3.6L engines, the oil filler cap must be removed to allow all the oil to properly drain out.*

6   Raise the vehicle and support it securely on jackstands. **Warning:** *Never get beneath the vehicle when it is supported only by a jack. The jack provided with your vehicle is designed solely for raising the vehicle to remove and replace the wheels. Always use jackstands to support the vehicle when it becomes necessary to place your body underneath the vehicle.*

7   Being careful not to touch the hot exhaust components, place the drain pan under the drain plug in the bottom of the pan and remove the plug **(see illustration)**. You may want to wear gloves while unscrewing the plug the final few turns if the engine is hot.

8   Allow the old oil to drain into the pan. It may be necessary to move the pan farther under the engine as the oil flow slows to a trickle. Inspect the old oil for the presence of metal shavings and chips.

9   After all the oil has drained, wipe off the drain plug with a clean rag. Even minute metal

6.7 Use a proper size box-end wrench or socket to remove the oil drain plug and avoid rounding it off

6.12 Use an oil filter wrench to remove the filter

6.14 Lubricate the oil filter gasket with clean engine oil before installing the filter on the engine

6.17 Using a box-end wrench or socket, turn the filter cap counterclockwise to remove it (3.6L engine)

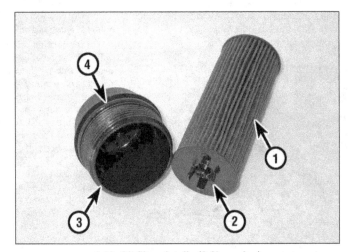

6.20 Oil filter details (3.6L engine)

1	Filter element	3	Filter cap
2	Filter clips	4	O-ring

particles clinging to the plug would immediately contaminate the new oil.

10   Clean the area around the drain plug opening, reinstall the plug and tighten it securely, but do not strip the threads.

11   Move the drain pan into position under the oil filter.

All except 3.6L V6 engines

12   Loosen the oil filter **(see illustration)** by turning it counterclockwise with an oil filter wrench. Once the filter is loose, use your hands to unscrew it from the block. Keep the open end pointing up to prevent the oil inside the filter from spilling out. **Warning:** *The exhaust system may still be hot, so be careful.*

13   With a clean rag, wipe off the mounting surface on the block. If a residue of old oil is allowed to remain, it will smoke when the block is heated up. Also make sure that none of the old gasket remains stuck to the mounting surface. It can be removed with a scraper

if necessary.

14   Compare the old filter with the new one to make sure they are the same type. Smear some clean engine oil on the rubber gasket of the new filter **(see illustration)**.

15   Attach the new filter to the engine, following the tightening directions printed on the filter canister or packing box. Most filter manufacturers recommend against using a filter wrench due to the possibility of overtightening and damaging the seal.

3.6L V6 engines

Refer to illustrations 6.17 and 6.20

16   Lift up on the outer edges of the engine cover to disengage the rubber mounts from the ballstuds, then remove the engine cover.

17   Place a rag at the base of the filter housing, then loosen the filter cap by turning it counterclockwise **(see illustration)**.

18   Remove the cap and filter from the engine, then pull the filter out of the cap.

19   Remove and discard the O-ring from the filter cap.

20   Insert a new filter into the cap, making sure the filter clips lock into the cap **(see illustration)**.

21   Install a new O-ring onto the cap and apply a film of clean engine oil to the O-ring.

22   Place the filter assembly into the filter housing and carefully thread the cap into the housing. Tighten the cap to the torque listed in this Chapter's Specifications.

All models

23   Remove all tools, rags, etc. from under the vehicle, being careful not to spill the oil in the drain pan, then lower the vehicle.

24   Add new oil to the engine through the oil filler cap in the valve cover. Use a funnel, if necessary, to prevent oil from spilling onto the top of the engine. Pour one quart less than the indicated capacity (see this Chapter's Specifications) of fresh oil into the engine. Wait a

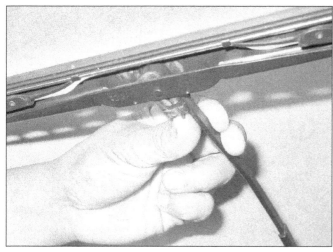

7.5a To release the blade holder, push the release lever . . .

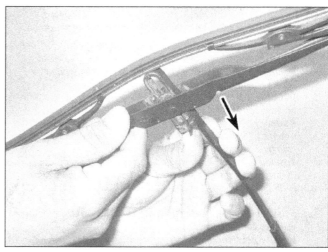

7.5b . . . and pull the wiper blade in the direction of the arrow to separate it from the arm

few minutes to allow the oil to drain into the pan, then check the level on the oil dipstick (see Section 4). If the oil level is at or near the FULL mark on the dipstick, install the filler cap hand tight, start the engine and allow the new oil to circulate.

25   Allow the engine to run for about a minute. While the engine is running, look under the vehicle and check for leaks at the oil pan drain plug and around the oil filter. If either is leaking, stop the engine and tighten the plug or filter.

26   Wait a few minutes to allow the oil to trickle down into the pan, then recheck the level on the dipstick and, if necessary, add enough oil to bring the level to the FULL mark.

27   During the first few trips after an oil change, make it a point to check frequently for leaks and proper oil level.

28   The old oil drained from the engine cannot be reused in its present state and should be disposed of. Check with your local auto parts store, disposal facility or environmental agency to see if they will accept the oil for recycling. After the oil has cooled it can be drained into a container (capped plastic jugs, topped bottles, milk cartons, etc.) for transport to one of these disposal sites. Don't dispose of the oil by pouring it on the ground or down a drain!

Resetting the Oil Change Required light

Note: *It is possible that, driving under the best possible conditions, the oil life monitoring system may not indicate the oil needs to be changed. The manufacturer states that the oil and filter must be changed at least once every year and the oil life monitor reset.*

Note: *If the "Oil Change Required" message comes on when the vehicle is immediately restarted, the oil life monitor was not reset and the reset procedure must be done again.*

Note: *If the message is not reset, it will continue to show up each time you turn the ignition switch On or start the vehicle. It is possible to temporarily turn off the message by*

pressing and releasing the "Menu" button.

29   Some later models have an "Oil Change Required" light. After completing your scheduled oil change, it will be necessary to reset the light.

30   Without starting the engine, turn the key to the On position.

31   Slowly depress the accelerator pedal to the floor three times within 10 seconds.

32   Turn the key to the Off/Lock position.

33   Start the engine and verify on the Electronic Vehicle Information Center (EVIC) that the "Oil Change Required" indicator message is on longer illuminated. Repeat the steps above if the "Oil Change Required" indicator message still appears.

7   Windshield wiper blade inspection and replacement (every 6,000 miles or 6 months)

Refer to illustrations 7.5a and 7.5b

1   The windshield wiper and blade assembly should be inspected periodically for damage, loose components and cracked or worn blade elements.

2   Road film can build up on the wiper blades and affect their efficiency, so they should be washed regularly with a mild detergent solution.

3   The action of the wiping mechanism can loosen bolts, nuts and fasteners, so they should be checked and tightened, as necessary, at the same time the wiper blades are checked.

4   If the wiper blade elements are cracked, worn or warped, or no longer clean adequately, they should be replaced with new ones.

5   Lift the arm assembly away from the glass for clearance, press the release lever, then slide the wiper blade assembly out of the hook at the end of the arm **(see illustrations)**.

6   Attach the new wiper to the arm. Connection can be confirmed by an audible click.

8   Front driveshaft lubrication (2500/3500 4WD) (every 6,000 miles or 6 months)

1   Raise the vehicle and support it securely on jackstands. Clean the grease fitting on the front driveshaft with a rag.

2   Use a conventional grease gun to apply sufficient lubricant to the fitting. Don't apply so much that an excessive amount comes out.

3   Wipe any extra grease and dirt from the area with a rag.

9   Tie-rod end lubrication (diesel models) (every 6,000 miles or 6 months)

1   Raise the vehicle and support it securely on jackstands. Clean the grease fitting on each outer tie-rod end with a rag.

2   Use a conventional grease gun to apply sufficient lubricant to the fittings. Don't apply so much that an excessive amount comes out.

3   Wipe any extra grease and dirt from the areas with a rag.

10   Battery check, maintenance and charging (every 6000 miles or 6 months)

Refer to illustrations 10.1, 10.6a, 10.6b, 10.7a and 10.7b

Warning: *Certain precautions must be followed when checking and servicing the battery. Hydrogen gas, which is highly flammable, is always present in the battery cells, so keep lighted tobacco and all other open flames and*

10.1 Tools and materials required for battery maintenance

1 *Face shield/safety goggles -* When removing corrosion with a brush, the acidic particles can easily fly up into your eyes

2 *Baking soda -* A solution of baking soda and water can be used to neutralize corrosion

3 *Petroleum jelly -* A layer of this on the battery posts will help prevent corrosion

4 *Battery post/cable cleaner -* This wire brush cleaning tool will remove all traces of corrosion from the battery posts and cable clamps

5 *Treated felt washers -* Placing one of these on each post, directly under the cable clamps, will help prevent corrosion

6 *Puller -* Sometimes the cable clamps are very difficult to pull off the posts, even after the nut/bolt has been completely loosened. This tool pulls the clamp straight up and off the post without damage

7 *Battery post/cable cleaner -* Here is another cleaning tool which is a slightly different version of number 4 above, but it does the same thing

8 *Rubber gloves -* Another safety item to consider when servicing the battery; remember that's acid inside the battery!

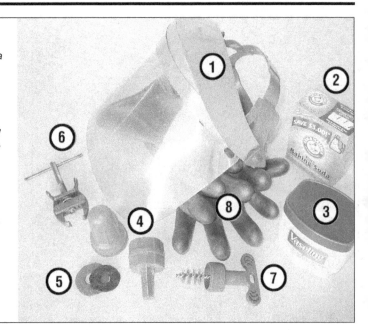

sparks away from the battery. The electrolyte inside the battery is actually diluted sulfuric acid, which will cause injury if splashed on your skin or in your eyes. It will also ruin clothes and painted surfaces. When removing the battery cables, always detach the negative cable first and hook it up last!

1   A routine preventive maintenance program for the battery in your vehicle is the only way to ensure quick and reliable starts. But before performing any battery maintenance, make sure that you have the proper equipment necessary to work safely around the battery **(see illustration).**

2   There are also several precautions that should be taken whenever battery maintenance is performed. Before servicing the battery, always turn the engine and all accessories off and disconnect the cable from the negative

terminal of the battery (see Chapter 5).

3   The battery produces hydrogen gas, which is both flammable and explosive. Never create a spark, smoke or light a match around the battery. Always charge the battery in a ventilated area.

4   Electrolyte contains poisonous and corrosive sulfuric acid. Do not allow it to get in your eyes, on your skin on your clothes. Never ingest it. Wear protective safety glasses when working near the battery. Keep children away from the battery.

5   Note the external condition of the battery. If the positive terminal and cable clamp on your vehicle's battery is equipped with a rubber protector, make sure that it's not torn or damaged. It should completely cover the terminal. Look for any corroded or loose connections, cracks in the case or cover or loose

hold-down clamps. Also check the entire length of each cable for cracks and frayed conductors.

6   If corrosion, which looks like white, fluffy deposits **(see illustration)** is evident, particularly around the terminals, the battery should be removed for cleaning. Loosen the cable clamp bolts with a wrench, being careful to remove the ground cable first, and slide them off the terminals **(see illustration).** Then disconnect the hold-down clamp bolt and nut, remove the clamp and lift the battery from the engine compartment.

7   Clean the cable clamps thoroughly with a battery brush or a terminal cleaner and a solution of warm water and baking soda **(see illustration).** Wash the terminals and the top of the battery case with the same solution but make sure that the solution doesn't get into

10.6a Battery terminal corrosion usually appears as light, fluffy powder

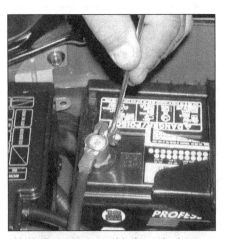

10.6b Removing a cable from the battery post with a wrench - sometimes a pair of special battery pliers are required for this procedure if corrosion has caused deterioration of the nut hex (always remove the ground [-] cable first and hook it up last!)

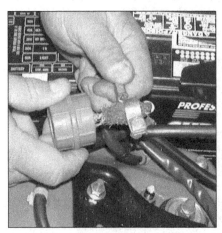

10.7a When cleaning the cable clamps, all corrosion must be removed

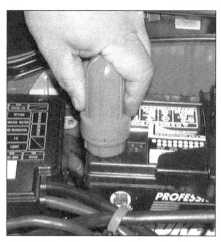

10.7b Regardless of the type of tool used to clean the battery posts, a clean, shiny surface should be the result

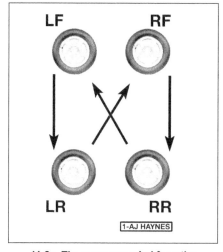

11.2a The recommended four-tire rotation pattern

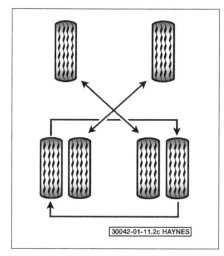

11.2b Six-tire rotation pattern for models with dual rear wheels

the battery. When cleaning the cables, terminals and battery top, wear safety goggles and rubber gloves to prevent any solution from coming in contact with your eyes or hands. Wear old clothes too - even diluted, sulfuric acid splashed onto clothes will burn holes in them. If the terminals have been extensively corroded, clean them up with a terminal cleaner **(see illustration)**. Thoroughly wash all cleaned areas with plain water.

8   Make sure that the battery tray is in good condition and the hold-down clamp fasteners are tight. If the battery is removed from the tray, make sure no parts remain in the bottom of the tray when the battery is reinstalled. When reinstalling the hold-down clamp bolts, do not overtighten them.

9   Information on removing and installing the battery can be found in Chapter 5. If you disconnected the cable(s) from the negative and/or positive battery terminals, see Chapter 5. Information on jump starting can be found at the front of this manual. For more detailed battery checking procedures, refer to the *Haynes Automotive Electrical Manual*.

Cleaning

10   Corrosion on the hold-down components, battery case and surrounding areas can be removed with a solution of water and baking soda. Thoroughly rinse all cleaned areas with plain water.

11   Any metal parts of the vehicle damaged by corrosion should be covered with a zinc-based primer, then painted.

Charging

Warning: *When batteries are being charged, hydrogen gas, which is very explosive and flammable, is produced. Do not smoke or allow open flames near a charging or a recently charged battery. Wear eye protection when near the battery during charging. Also, make sure the charger is unplugged before connecting or disconnecting the battery from the charger.*

12   Slow-rate charging is the best way to restore a battery that's discharged to the point where it will not start the engine. It's also a good way to maintain the battery charge in a vehicle that's only driven a few miles between starts. Maintaining the battery charge is particularly important in the winter when the battery must work harder to start the engine and electrical accessories that drain the battery are in greater use.

13   It's best to use a one or two-amp battery charger (sometimes called a "trickle" charger). They are the safest and put the least strain on the battery. They are also the least expensive. For a faster charge, you can use a higher amperage charger, but don't use one rated more than 1/10th the amp/hour rating of the battery. Rapid boost charges that claim to restore the power of the battery in one to two hours are hardest on the battery and can damage batteries not in good condition. This type of charging should only be used in emergency situations.

14   The average time necessary to charge a battery should be listed in the instructions that come with the charger. As a general rule, a trickle charger will charge a battery in 12 to 16 hours.

11   Tire rotation (every 6000 miles or 6 months)

Refer to illustrations 11.2a and 11.2b

1   The tires should be rotated at the specified intervals and whenever uneven wear is noticed.

2   Refer to the **accompanying illustrations** for the preferred tire rotation pattern. **Note:** *If the front tires are different than the rear tires, rotate the tires from side-to-side only. Additionally, the tires on 3500 models with dual rear wheels are directional; when rotating (or installing the spare) the tires may have to be remounted so they rotate in the proper direction.*

3   Refer to the information in "Jacking and towing" at the front of this manual for the proper procedures to follow when raising the vehicle and changing a tire. If the brakes are to be checked, don't apply the parking brake as stated. Make sure the tires are blocked to prevent the vehicle from rolling as it's raised.

4   Preferably, the entire vehicle should be raised at the same time. This can be done on a hoist or by jacking up each corner and then lowering the vehicle onto jackstands placed under the frame rails. Always use four jackstands and make sure the vehicle is safely supported.

5   After rotation, check and adjust the tire pressures as necessary. Tighten the lug nuts to the torque listed in this Chapter's Specifications.

12   Seat belt check (every 6000 miles or 6 months)

1   Check seat belts, buckles, latch plates and guide loops for obvious damage and signs of wear.

2   Where the seat belt receptacle bolts to the floor of the vehicle, check that the bolts are secure.

3   Check that the seat belt reminder light comes on when the key is turned to the Run or Start position.

13   Underhood hose check and replacement (every 6000 miles or 6 months)

General

Caution: *Replacement of air conditioning hoses must be left to a dealer service department or air conditioning shop that has the equipment to depressurize the system safely and recover the refrigerant. Never remove air* *conditioning components or hoses until the system has been depressurized.*

1 High temperatures in the engine compartment can cause the deterioration of the rubber and plastic hoses used for engine, accessory and emission systems operation. Periodic inspection should be made for cracks, loose clamps, material hardening and leaks. Information specific to the cooling system hoses can be found in Section 14.

2 Some, but not all, hoses are secured to their fittings with clamps. Where clamps are used, check to be sure they haven't lost their tension, allowing the hose to leak. If clamps aren't used, make sure the hose has not expanded and/or hardened where it slips over the fitting, allowing it to leak.

Vacuum hoses

3 It's quite common for vacuum hoses, especially those in the emissions system, to be color-coded or identified by colored stripes molded into them. Various systems require hoses with different wall thickness, collapse resistance and temperature resistance. When replacing hoses, be sure the new ones are made of the same material.

4 Often the only effective way to check a hose is to remove it completely from the vehicle. If more than one hose is removed, label the hoses and fittings to ensure correct installation.

5 When checking vacuum hoses, include any plastic T-fittings in the check. Inspect the fittings for cracks and the hose where it fits over the fitting for distortion, which could cause leakage.

6 A small piece of vacuum hose (1/4-inch inside diameter) can be used as a stethoscope to detect vacuum leaks. Hold one end of the hose to your ear and probe around vacuum hoses and fittings, listening for the hissing sound characteristic of a vacuum leak. **Warning:** *When probing with the vacuum hose stethoscope, be very careful not to come into contact with moving engine components such as the drivebelt, cooling fan, etc.*

Fuel hose

Warning: *There are certain precautions that must be taken when inspecting or servicing fuel system components. Work in a well-ventilated area and do not allow open flames (cigarettes, appliances, etc.) or bare light bulbs near the work area. Mop up any spills immediately and do not store fuel soaked rags where they could ignite. The fuel system is under high pressure, so if any fuel lines are to be disconnected, the pressure in the system must be relieved first (see Chapter 4 for more information).*

7 Check all rubber fuel lines for deterioration and chafing. Check especially for cracks in areas where the hose bends and just before fittings, such as where a hose attaches to the fuel filter.

8 High quality fuel line, made specifically for high-pressure fuel injection systems, must

be used for fuel line replacement. Never, under any circumstances, use un-reinforced vacuum line, clear plastic tubing or water hose for fuel lines.

9 Spring-type clamps are commonly used on fuel lines. These clamps often lose their tension over a period of time, and can be sprung during removal. Replace all spring-type clamps with screw clamps whenever a hose is replaced.

Metal lines

10 Sections of metal line are routed along the frame, between the fuel tank and the engine. Check carefully to be sure the line has not been bent or crimped and that cracks have not started in the line.

11 If a section of metal fuel line must be replaced, only seamless steel tubing should be used, since copper and aluminum tubing don't have the strength necessary to withstand normal engine vibration.

12 Check the metal brake lines where they enter the master cylinder and brake proportioning unit for cracks in the lines or loose fittings. Any sign of brake fluid leakage calls for an immediate and thorough inspection of the brake system.

14 Cooling system check (every 6000 miles or 6 months)

Refer to illustration 14.4

1 Many major engine failures can be attributed to a faulty cooling system. If the vehicle is equipped with an automatic transmission, the cooling system also cools the transmission fluid and thus plays an important role in prolonging transmission life.

2 The cooling system should be checked with the engine cold. Do this before the vehicle is driven for the day or after it has been shut off for at least three hours.

3 Remove the cooling system pressure cap and thoroughly clean the cap, inside and out, with clean water. Also clean the filler neck on the radiator. All traces of corrosion should be removed. The coolant inside the radiator should be relatively transparent. If it is rust-colored, the system should be drained, flushed and refilled (see Section 32). If the coolant level is not up to the top, add additional antifreeze/coolant mixture (see Section 4).

4 Carefully check the large upper and lower radiator hoses along with the smaller diameter heater hoses that run from the engine to the firewall. Inspect each hose along its entire length, replacing any hose that is cracked, swollen or shows signs of deterioration. Cracks may become more apparent if the hose is squeezed **(see illustration)**. Regardless of condition, it's a good idea to replace hoses with new ones every two years.

5 Make sure all hose connections are tight. A leak in the cooling system will usually show up as white or rust-colored deposits on the

areas adjoining the leak. If wire-type clamps are used at the ends of the hoses, it may be a good idea to replace them with more secure screw-type clamps.

6 Use compressed air or a soft brush to remove bugs, leaves, etc. from the front of the radiator or air conditioning condenser. Be careful not to damage the delicate cooling fins or cut yourself on them.

7 Every other inspection, or at the first indication of cooling system problems, have the cap and system pressure tested. If you don't have a pressure tester, most repair shops will do this for a minimal charge.

15 Brake hose check (every 6000 miles or 6 months)

With the vehicle raised and supported securely on jackstands, the rubber hoses which connect the steel brake lines with the front and rear brake assemblies should be

Check for a chafed area that could fail prematurely.

Check for a soft area indicating the hose has deteriorated inside.

Overtightening the clamp on a hardened hose will damage the hose and cause a leak.

Check each hose for swelling and oil-soaked ends. Cracks and breaks can be located by squeezing the hose.

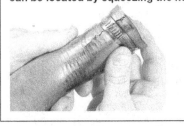

14.4 Hoses, like drivebelts, have a habit of failing at the worst possible time - to prevent the inconvenience of a blown radiator or heater hose, inspect them carefully as shown here

16.2a Inspect the muffler (A) and all hangers (B) for signs of deterioration

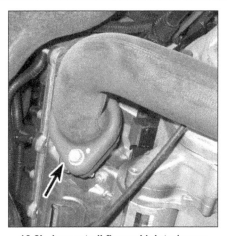

16.2b Inspect all flanged joints (arrow indicates pipe-to-manifold joint) for signs of exhaust gas leakage

17.6 Check for signs of fluid leakage at this point on shock absorbers (rear shock shown)

inspected for cracks, chafing of the outer cover, leaks, blisters and other damage. These are important and vulnerable parts of the brake system and inspection should be complete. A light and mirror will be helpful for a thorough check. If a hose exhibits any of the above conditions, replace it with a new one (see Chapter 9).

16   Exhaust system check (every 6000 miles or 6 months)

Refer to illustrations 16.2a and 16.2b

1   With the engine cold (at least three hours after the vehicle has been driven), check the complete exhaust system from the manifold to the end of the tailpipe. Be careful around the catalytic converter, which may be hot even after three hours. The inspection should be done with the vehicle on a hoist to permit unrestricted access. If a hoist isn't available, raise the vehicle and support it securely on jackstands.
2   Check the exhaust pipes and connections for signs of leakage and/or corrosion indicating a potential failure. Make sure that

all brackets and hangers are in good condition and tight **(see illustrations)**.
3   Inspect the underside of the body for holes, corrosion, open seams, etc. which may allow exhaust gasses to enter the passenger compartment. Seal all body openings with silicone sealant or body putty.
4   Rattles and other noises can often be traced to the exhaust system, especially the hangers, mounts and heat shields. Try to move the pipes, mufflers and catalytic converter. If the components can come in contact with the body or suspension parts, secure the exhaust system with new brackets and hangers.

17   Suspension, steering and driveaxle boot check (every 6000 miles or 6 months)

Note: *The steering linkage and suspension components should be checked periodically. Worn or damaged suspension and steering linkage components can result in excessive and abnormal tire wear, poor ride quality and vehicle handling and reduced fuel economy.*

For detailed illustrations of the steering and suspension components, refer to Chapter 10.

Shock absorber check

Refer to illustration 17.6

1   Park the vehicle on level ground, turn the engine off and set the parking brake. Check the tire pressures.
2   Push down at one corner of the vehicle, then release it while noting the movement of the body. It should stop moving and come to rest in a level position within one or two bounces.
3   If the vehicle continues to move up-and-down or if it fails to return to its original position, a worn or weak shock absorber is probably the reason.
4   Repeat the above check at each of the three remaining corners of the vehicle.
5   Raise the vehicle and support it securely on jackstands.
6   Check the shock absorbers for evidence of fluid leakage **(see illustration)**. A light film of fluid is no cause for concern. Make sure that any fluid noted is from the shocks and not from some other source. If leakage is noted, replace the shocks as a set.
7   Check the shocks to be sure that they are securely mounted and undamaged. Check the upper mounts for damage and wear. If damage or wear is noted, replace the shocks as a set (front or rear).
8   If the shocks must be replaced, refer to Chapter 10 for the procedure.

Steering and suspension check

Refer to illustrations 17.9 and 17.11

9   Visually inspect the steering and suspension components (front and rear) for damage and distortion. Look for damaged seals, boots and bushings and leaks of any kind. Examine the bushings where the control arms meet the chassis **(see illustration)**.

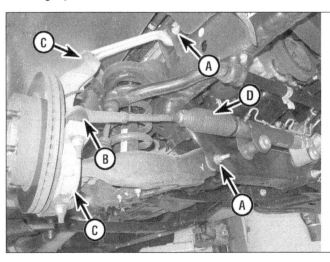

17.9 Examine the mounting points for the upper and lower control arms on the front suspension (A), the tie-rod ends (B), the balljoints (C), and the steering gear boots (D)

17.11 With the steering wheel in the locked position and the vehicle raised, grasp the front tire as shown and try to move it back and forth - if any play is noted, check the steering gear mounts and tie-rod ends for looseness

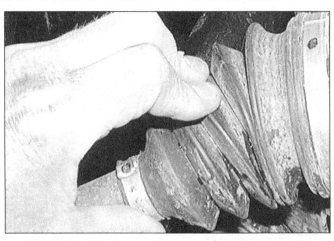

17.14 Inspect the inner and outer driveaxle boots on 4WD models for loose clamps, cracks or signs of leaking lubricant

10   Clean the lower end of the steering knuckle. Have an assistant grasp the lower edge of the tire and move the wheel in-and-out while you look for movement at the steering knuckle-to-control arm balljoint. If there is any movement the suspension balljoint(s) must be replaced.

11   Grasp each front tire at the front and rear edges, push in at the front, pull out at the rear and feel for play in the steering system components. If any freeplay is noted, check the idler arm and the tie-rod ends for looseness **(see illustration)**.

12   Additional steering and suspension system information and illustrations can be found in Chapter 10.

Driveaxle boot check (4WD models)

Refer to illustration 17.14

13   The driveaxle boots are very important because they prevent dirt, water and foreign material from entering and damaging the constant velocity (CV) joints. Oil and grease can cause the boot material to deteriorate prematurely, so it's a good idea to wash the boots with soap and water. Because it constantly pivots back and forth following the steering action of the front hub, the outer CV boot wears out sooner and should be inspected regularly.

14   Inspect the boots for tears and cracks as well as loose clamps **(see illustration)**. If there is any evidence of cracks or leaking lubricant, they must be replaced (see Chapter 8).

---

18   Brake system check (every 15,000 miles or 12 months)

Warning: *The dust created by the brake system is harmful to your health. Never blow it out with compressed air and don't inhale any of it. An approved filtering mask should be worn when working on the brakes. Do not, under* any circumstances, use petroleum-based solvents to clean brake parts. Use brake system cleaner only!

Note: *For detailed photographs of the brake system, refer to Chapter 9.*

1   In addition to the specified intervals, the brakes should be inspected every time the wheels are removed or whenever a defect is suspected.

2   Any of the following symptoms could indicate a potential brake system defect: The vehicle pulls to one side when the brake pedal is depressed; the brakes make squealing or dragging noises when applied; brake pedal travel is excessive; the pedal pulsates; or brake fluid leaks, usually onto the inside of the tire or wheel.

3   Loosen the wheel lug nuts.

4   Raise the vehicle and place it securely on jackstands.

5   Remove the wheels (see "Jacking and towing" at the front of this book, or your owner's manual, if necessary).

Disc brakes

Refer to illustrations 18.7a, 18.7b and 18.9

6   There are two pads (an outer and an inner) in each caliper. The pads are visible with the wheels removed.

7   Check the pad thickness by looking at each end of the caliper and through the inspection window in the caliper body **(see illustrations)**. If the lining material is less than the thickness listed in this Chapter's Specifications, replace the pads. **Note:** *Keep in mind that the lining material is riveted or bonded to a metal backing plate and the metal portion is not included in this measurement.*

8   If it is difficult to determine the exact thickness of the remaining pad material by the above method, or if you are at all concerned about the condition of the pads, remove the caliper(s), then remove the pads from the calipers for further inspection (see Chapter 9).

18.7a With the wheel off, check the thickness of the inner pad through the inspection hole (front disc shown, rear disc caliper similar)

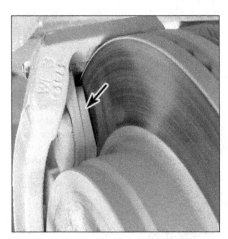

18.7b The outer pad is more easily checked at the edge of the caliper

9   Once the pads are removed from the calipers, clean them with brake cleaner and re-measure them with a ruler or a vernier caliper **(see illustration)**.
10   Measure the disc thickness with a micrometer to make sure that it still has service life remaining. If any disc is thinner than the specified minimum thickness, replace it (see Chapter 9). Even if the disc has service life remaining, check its condition. Look for scoring, gouging and burned spots. If these conditions exist, remove the disc and have it resurfaced (see Chapter 9).
11   Before installing the wheels, check all brake lines and hoses for damage, wear, deformation, cracks, corrosion, leakage, bends and twists, particularly in the vicinity of the rubber hoses at the calipers. Check the clamps for tightness and the connections for leakage. Make sure that all hoses and lines are clear of sharp edges, moving parts and the exhaust system. If any of the above conditions are noted, repair, reroute or replace the lines and/ or fittings as necessary (see Chapter 9).

Brake booster check

12   Sit in the driver's seat and perform the following sequence of tests.
13   With the brake fully depressed, start the engine - the pedal should move down a little when the engine starts.
14   With the engine running, depress the brake pedal several times - the travel distance should not change.
15   Depress the brake, stop the engine and hold the pedal in for about 30 seconds - the pedal should neither sink nor rise.
16   Restart the engine, run it for about a minute and turn it off. Then firmly depress the brake several times - the pedal travel should decrease with each application.

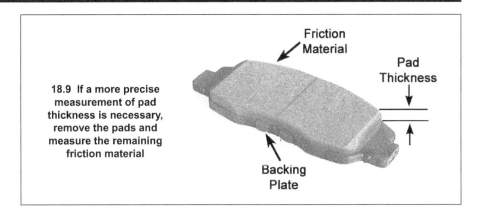

18.9 If a more precise measurement of pad thickness is necessary, remove the pads and measure the remaining friction material

17   If your brakes do not operate as described, the brake booster has failed. Refer to Chapter 9 for the replacement procedure.

Parking brake

18   One method of checking the parking brake is to park the vehicle on a steep hill with the parking brake set and the transmission in Neutral (be sure to stay in the vehicle for this check!). If the parking brake cannot prevent the vehicle from rolling, it's in need of adjustment (see Chapter 9).

19   Drivebelt check and replacement (every 15,000 miles or 12 months)

1   The drivebelt is located at the front of the engine and plays an important role in the overall operation of the vehicle and its components. Due to its function and material make-up, the drivebelt is prone to failure after a period of time and should be inspected and adjusted periodically to prevent major engine damage.
2   The vehicles covered by this manual are equipped with a single self-adjusting serpentine drivebelt, which is used to drive all of the accessory components such as the alternator, power steering pump, water pump and air conditioning compressor.

Inspection

Refer to illustrations 19.4 and 19.5

3   With the engine off, open the hood and locate the drivebelt at the front of the engine. Using your fingers (and a flashlight, if necessary), move along the belts checking for cracks and separation of the belt plies. Also check for fraying and glazing, which gives the belt a shiny appearance. Both sides of each belt should be inspected, which means you will have to twist the belt to check the underside.
4   Check the ribs on the underside of the belt. They should all be the same depth, with none of the surface uneven **(see illustration)**.
5   The tension of the belt is automatically adjusted by the belt tensioner and does not require any adjustments. Drivebelt wear can be checked visually by inspecting the wear indicator marks located on the side of the tensioner body. Locate the belt tensioner at the front of the engine, then find the tensioner operating marks **(see illustration)**. If the indicator mark is outside the operating range, the belt should be replaced.

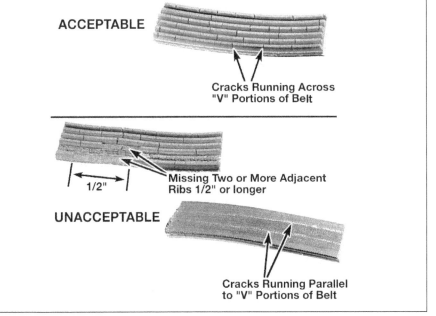

19.4 Here are some of the more common problems associated with drivebelts (check the belts very carefully to prevent an untimely breakdown)

19.5 When the raised stop (A) nears the tensioner body (B), the belt can be considered worn-out

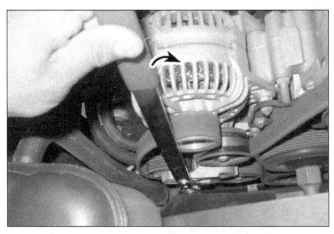

19.6 Rotate the tensioner arm to relieve belt tension

19.8 The routing schematic for the serpentine belt is usually found on the fan shroud

Replacement

Refer to illustrations 19.6 and 19.8

6   To replace the belt, rotate the tensioner to relieve the tension on the belt **(see illustration)**. Some models have a square hole in the tensioner arm that will accept a breaker bar or ratchet. On other models, place a wrench on the tensioner pulley bolt.

7   Remove the belt from the auxiliary components and carefully release the tensioner.

8   Route the new belt over the various pulleys, again rotating the tensioner to allow the belt to be installed, then release the belt tensioner. Make sure the belt fits properly into the pulley grooves - it must be completely engaged. **Note:** *Most models have a drivebelt routing decal on the upper radiator panel to help during drivebelt installation* **(see illustration).**

20   Fuel system check (every 15,000 miles or 12 months)

Warning: *Gasoline and diesel fuels are flammable, so take extra precautions when you work on any part of the fuel system. Don't smoke or allow open flames or bare light bulbs near the work area, and don't work in a garage where a gas-type appliance (such as a water heater or clothes dryer) is present. Since fuel is carcinogenic, wear fuel-resistant gloves when there's a possibility of being exposed to fuel, and, if you spill any fuel on your skin, rinse it off immediately with soap and water. Mop up any spills immediately and do not store fuel-soaked rags where they could ignite. When you perform any kind of work on the fuel system, wear safety glasses and have a Class B type fire extinguisher on hand. The fuel system is under constant pressure, so, before any lines are disconnected, the fuel system pressure must be relieved (see Chapter 4).*

1   If you smell fuel while driving or after the vehicle has been sitting in the sun, inspect the fuel system immediately.

2   Remove the fuel filler cap and inspect it for damage and corrosion. The gasket should have an unbroken sealing imprint. If the gasket is damaged or corroded, install a new cap.

3   Inspect the fuel feed line for cracks. Make sure that the connections between the fuel lines and the fuel injection system and between the fuel lines and the in-line fuel filter are tight. **Warning:** *Your vehicle is fuel injected, so you must relieve the fuel system pressure before servicing fuel system components. The fuel system pressure relief procedure is outlined in Chapter 4.*

4   Since some components of the fuel system - the fuel tank and part of the fuel feed and return lines, for example - are underneath the vehicle, they can be inspected more easily with the vehicle raised on a hoist. If that's not possible, raise the vehicle and support it on jackstands.

5   With the vehicle raised and safely supported, inspect the fuel tank and filler neck for punctures, cracks and other damage. The connection between the filler neck and the tank is particularly critical. Sometimes a rubber filler neck will leak because of loose clamps or deteriorated rubber. Inspect all fuel tank mounting brackets and straps to be sure that the tank is securely attached to the vehicle. **Warning:** *Do not, under any circumstances, try to repair a fuel tank (except rubber components). A welding torch or any open flame can easily cause fuel vapors inside the tank to explode.*

6   Carefully check all rubber hoses and metal lines leading away from the fuel tank. Check for loose connections, deteriorated hoses, crimped lines and other damage. Repair or replace damaged sections as necessary (see Chapter 4).

21   Fuel filter draining and replacement (diesel engine) (every 15,000 miles or 12 months)

Warning: *Diesel fuel is flammable, so take extra precautions when you work on any part*

of the fuel system. Don't smoke or allow open flames or bare light bulbs near the work area, and don't work in a garage where a gas-type appliance (such as a water heater or clothes dryer) is present. Since fuel is carcinogenic, wear fuel-resistant gloves when there's a possibility of being exposed to fuel, and, if you spill any fuel on your skin, rinse it off immediately with soap and water. Mop up any spills immediately and do not store fuel-soaked rags where they could ignite. When you perform any kind of work on the fuel system, wear safety glasses and have a Class B type fire extinguisher on hand.

Fuel filter draining

1   The diesel engine fuel filter incorporates a water separator that removes and traps water in the fuel. This water must be drained from the filter at the specified intervals or when the Water In Fuel (WIF) light is on.

2   Place a small container under the filter drain tube (diesel fuel can damage asphalt paving).

2009 models

3   With the engine off, rotate the drain valve one turn counterclockwise **(see illustration 21.13).** Allow the water and contaminated fuel to drain out. Repeat the procedure until clean fuel flows from the filter drain tube, then place the drain valve in the closed position.

2010 and later models

4   Turn the drain valve knob 1/4-turn counterclockwise. Also unscrew the filter housing cover 1/2 turn. Allow the water and contaminated fuel to drain for two minutes, then close the drain valve and tighten the filter housing cover.

All engines

5   Remove the container and dispose of the fuel/water mixture properly.

6   If more than a couple of ounces had to be drained to remove any accumulated water, the fuel system may have to be primed in order for the engine to start. Refer to Chapter 4B for the priming procedure.

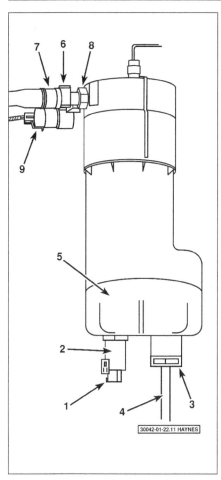

21.13 Fuel filter assembly details (2009 6.7L diesel engine)

1   WIF harness connector
2   WIF sensor
3   Drain valve
4   Drain tube
5   Fuel filter canister
6   Fuel line connector
7   Fuel line
8   Filter screen
9   Fuel heater element connector

Fuel filter replacement

7   At the specified intervals, the fuel filter element (which incorporates a water separator) should be replaced with a new one.
8   Raise the vehicle and support it securely on jackstands.
9   Remove the left front wheel and the fender splash shield (see Chapter 11).
10   Drain the filter into a container and dispose of the fuel safely. Attach a hose extension to avoid spillage. **Note:** *Loosen the filter slightly to speed draining.*

2009 models

Refer to illustration 21.13

11   Disconnect the Water In Fuel (WIF) harness connector.

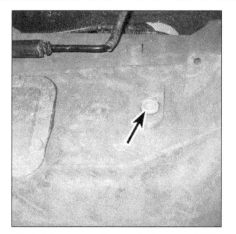

22.1 The manual transmission check/fill plug is located on the side of the case

12   Disconnect the drain hose and extension hose from the drain valve.
13   Use an oil filter wrench to rotate the bottom portion (canister filter) from the fuel filter assembly **(see illustration)**.
14   Spin the fuel filter by hand from the fuel filter/canister assembly and set it aside.
15   Remove the secondary fuel filter screen from the fuel filter assembly. Release the quick-connect fuel line fitting from the fuel filter assembly. Refer to Chapter 4 for detailed information on fuel line disconnection procedures.
16   Unscrew the line fitting from the fuel filter assembly and remove the fuel line along with the O-ring and filter screen.
17   Clean the fuel filter screen and replace the O-ring with a new one.
18   Installation is the reverse of removal. Use a new filter cartridge in the fuel filter assembly and a new O-ring in the fuel line.
19   Turn the key to the On position, allowing the fuel pump to prime the system. Repeat this a few times until the engine starts. Run the engine and check for leaks.

2010 and later models

20   Clean all of the debris from the area of the fuel filter lid.
21   Use a ratchet and socket to remove the fuel filter lid, then discard the O-ring from the canister.
22   Remove the fuel filter element.
23   Wipe out the interior of the canister but don't use any harsh chemicals such as brake cleaner.
24   Install the new filter element. **Caution:** *Don't fill the canister with fuel or try to prime the system.*
25   Lubricate the new O-ring with clean engine oil and install it on the canister.
26   Replace the lid and tighten it to the torque listed in this Chapter's Specifications.
27   Turn the key to the On position, then briefly turn it to Start. This will allow the fuel pump to run for 25 seconds and should prime the system. If the vehicle doesn't start, repeat the procedure.

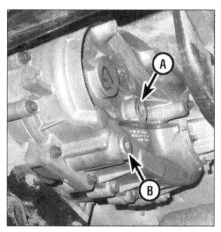

22.5 The transfer case fill (A) and drain (B) plugs are located on the rear of the transfer case

22   Manual transmission, transfer case (4WD models) and differential lubricant level check (every 15,000 miles or 12 months)

Note: *The front and rear differential lubricant should be changed every 15,000 miles or 12 months on vehicles with diesel engines or any vehicle used in rough service conditions such as frequent towing or off-road use.*

Manual transmission

Refer to illustration 22.1

1   The manual transmission has a filler plug which must be removed to check the lubricant level **(see illustration).** If the vehicle is raised to gain access to the plug, support it safely on jackstands - DO NOT crawl under a vehicle that is supported only by a jack! The vehicle must be level, or the check may be inaccurate.
2   Using the appropriate wrench, unscrew the plug from the transmission.
3   Use your little finger to reach inside the housing to feel the lubricant level. The level should be at or near the bottom of the plug hole. If it isn't, add the recommended lubricant through the plug hole with a syringe or squeeze bottle.
4   Install and tighten the plug. Check for leaks after the first few miles of driving.

Transfer case (4WD models)

Refer to illustration 22.5

5   The transfer case lubricant level is checked by removing the fill plug **(see illustration).**
6   After removing the plug, reach inside the hole. The lubricant level should be just at the bottom of the hole. If not, add the appropriate lubricant through the opening.

Differential

Refer to illustrations 22.7a and 22.7b

Note: *4WD vehicles have two differentials; one in the front as well as one in the rear -*

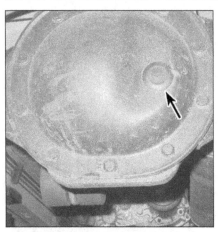

22.7a If the differential has a rubber plug, you can remove the plug by prying it out with a screwdriver

22.7b If the differential has a threaded plug, use a 3/8-inch drive ratchet to unscrew it

24.3a Release the spring clips and lift the air filter housing cover

24.3b Remove the air filter element from the housing

check the lubricant level in both differentials.

7   The differential lubricant level is checked by removing a filler plug from the differential cover **(see illustrations)**. If the vehicle is raised to gain access to the plug, support it safely on jackstands - DO NOT crawl under a vehicle that is supported only by a jack! The vehicle must be level, or the check may be inaccurate.

8   With the differential cold, remove the fill plug. The lubricant should be level with the bottom of the fill plug hole.

9   If the level is low, add the recommended lubricant through the filler plug hole with a pump, syringe or squeeze bottle.

10   Install the plug and check for leaks after the first few miles of driving.

23   Brake fluid change (every 30,000 miles or 24 months)

Warning: *Brake fluid can harm your eyes and damage painted surfaces, so use extreme caution when handling or pouring it. Do not*

use brake fluid that has been standing open or is more than one year old. Brake fluid absorbs moisture from the air. Excess moisture can cause a dangerous loss of braking effectiveness.

1   At the specified intervals, the brake fluid should be drained and replaced. Since the brake fluid may drip or splash when pouring it, place plenty of rags around the master cylinder to protect any surrounding painted surfaces.

2   Before beginning work, purchase the specified brake fluid (see *Recommended lubricants and fluids* in this Chapter's Specifications).

3   Remove the cap from the master cylinder reservoir.

4   Using a hand suction pump or similar device, withdraw the fluid from the master cylinder reservoir.

5   Add new fluid to the master cylinder until it rises to the base of the filler neck.

6   Bleed the brake system at all four brakes until new and uncontaminated fluid is expelled from the bleeder screw (see Chapter 9). Maintain the fluid level in the master cylinder as you perform the bleeding process. If you allow the master cylinder to run dry, air will enter the system.

7   Refill the master cylinder with fluid and check the operation of the brakes. The pedal should feel solid when depressed, with no sponginess. **Warning:** *Do not operate the vehicle if you are in doubt about the effectiveness of the brake system.*

24   Air filter replacement (every 30,000 miles or 24 months)

Refer to illustrations 24.3a and 24.3b

Note: *This service interval is for gasoline engines only. Diesel engines have a sensor that will inform the driver via the EVIC display when the air filter should be replaced.*

1   At the specified intervals, the air filter

element should be replaced with a new one.

2   The air filter is housed in a black plastic box mounted on the inner fenderwell on the right side of the engine compartment.

3   Detach the spring clips and pull the housing cover up, then lift the air filter element out of the housing **(see illustrations)**. Wipe out the inside of the air filter housing with a clean rag.

4   While the cover is off, be careful not to drop anything down into the air filter housing.

5   Place the new filter element in the air filter housing. Make sure it seats properly in the groove of the housing.

6   Installation is the reverse of removal.

25   Spark plug replacement (gasoline engines) (every 30,000 miles or 24 months)

Refer to illustrations 25.2, 25.5a, 25.5b, 25.6a, 25.6b, 25.8, 25.10a and 25.10b

Note: *This service interval is for 3.6L, 3.7L, 5.7L and 6.4L engines. On 4.7L engines, replace the upper (intake) spark plugs every 45,000 miles and the lower (exhaust) spark plugs every 90,000 miles.*

Note: *4.7L engines use different spark plug types on the upper and lower banks.*

1   The spark plugs are threaded into the cylinder heads.

2   In most cases, the tools necessary for spark plug replacement include a spark plug socket which fits onto a ratchet (spark plug sockets are padded inside to prevent damage to the porcelain insulators on the new plugs), various extensions and a gap gauge to check and adjust the gaps on the new plugs **(see illustration)**. A special plug wire removal tool is available for separating the wire boots from the spark plugs, but it isn't absolutely necessary. A torque wrench should be used to tighten the new plugs.

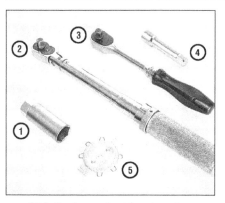

25.2 Tools required for changing spark plugs

1 **Spark plug socket** - *This will have special padding inside to protect the spark plug porcelain insulator*
2 **Torque wrench** - *Although not mandatory, use of this tool is the best way to ensure that the plugs are tightened properly*
3 **Ratchet** - *Standard hand tool to fit the plug socket*
4 **Extension** - *Depending on model and accessories, you may need special extensions and universal joints to reach one or more of the plugs*
5 **Spark plug gap gauge** - *This gauge for checking the gap comes in a variety of styles. Make sure the gap for your engine is included*

3   The best approach when replacing the spark plugs is to purchase the new ones in advance, adjust them to the proper gap and replace them one at a time. When buying the new spark plugs, be sure to obtain the correct plug type for your particular engine. This information can be found on the Emission Control Information label located under the hood, in the factory owner's manual and in this Chapter's Specifications. If differences exist between the plug specified on the emissions label and in the owner's manual, assume that the emissions label is correct.
4   Allow the engine to cool completely before attempting to remove any of the plugs. While you're waiting for the engine to cool, check the new plugs for defects and adjust the gaps.
5   The gap is checked by inserting the proper-thickness gauge between the electrodes at the tip of the plug **(see illustration)**. The gap between the electrodes should be the same as the one specified in this Chapter's Specifications. **Note:** *We recommend using a wire-type thickness gauge when checking the spark plugs.* If the gap is incorrect, use the adjuster on the gauge body to bend the curved side electrode slightly until the proper gap is obtained **(see illustration)**. If the side electrode is not exactly over the center electrode, bend it with the adjuster until it is. Check for cracks in the porcelain insulator (if any are found, the plug should not be used).

25.5a Spark plug manufacturers recommend using a wire-type gauge when checking the gap on platinum or iridium spark plugs

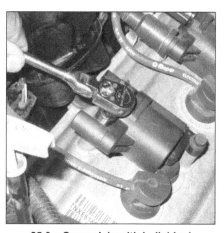

25.6a On models with individual ignition coils, remove the mounting bolts and the individual coil(s) to access the spark plug(s) (one type shown, others similar)

6   Some engines are equipped with individual ignition coils which must be removed first to access the spark plugs **(see illustration)**. On engines equipped with spark plug wires, with the engine cool, remove the spark plug wire from one spark plug. Pull only on the boot at the end of the wire - do not pull on the wire **(see illustration)**. **Note:** *On 3.6L engines, it will be necessary to remove the upper intake manifold (see Chapter 2D) to gain access to the right-side ignition coils.*
7   If compressed air is available, use it to blow any dirt or foreign material away from the spark plug hole. The idea here is to eliminate the possibility of debris falling into the cylinder as the spark plug is removed.
8   Place the spark plug socket over the plug and remove it from the engine by turning counterclockwise **(see illustration)**.
9   Compare the spark plug with the chart on the inside back cover of this manual to get an indication of the general running condition of the engine.

25.5b To change the gap, bend the side electrode only, and be very careful not to crack or chip the porcelain insulator surrounding the center electrode

25.6b Hemi engines have spark plug(s) located under spark plug wire(s) in addition to the individual ignition coils

25.8 Use a socket and extension to unscrew the spark plugs

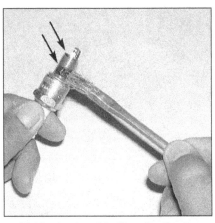

25.10a Apply a thin film of anti-seize compound to the spark plug threads to prevent damage to the cylinder head

25.10b A length of snug-fitting rubber hose will save time and prevent damaged threads when installing the spark plugs

10   Apply a small amount of anti-seize compound to the spark plug threads **(see illustration)**. Thread one of the new plugs into the hole until you can no longer turn it with your fingers, then tighten it with a torque wrench (if available) or the ratchet. It's a good idea to slip a short length of rubber hose over the end of the plug to use as a tool to thread it into place **(see illustration)**. The hose will grip the plug well enough to turn it, but will start to slip if the plug begins to cross-thread in the hole - this will prevent damaged threads and the accompanying repair costs.
11   Repeat the procedure for the remaining spark plugs.

---

26   Spark plug wire replacement (3.7L and 4.7L engines) (see Maintenance schedule for service intervals)

---

1   Remove any interfering air intake components for access to all of the wires.
2   Disconnect and remove only one wire at a time to avoid confusion. Grab the rubber boot, twist slightly and pull it free. Don't pull on the wire itself; only on the rubber boot. A boot pulling tool is useful.
3   Apply a small amount of dielectric ignition grease inside each new wire boot before installing it. Make sure that each wire is routed properly and can't touch any sharp or moving parts.

---

27   EGR valve and EGR cooler cleaning (6.4L Hemi and diesel engines) (every 60,000 miles or 48 months)

Note: *These procedures will be performed at the first 60,000 mile interval by Dodge dealers at no cost in the following states: California, Maine, Massachusetts, New Mexico, New York, Vermont, Connecticut, Oregon and Rhode Island.*

EGR valve cleaning (diesel engines only)

1   Remove the EGR valve (see Chapter 6).
2   Remove the EGR valve motor and its shim from the valve.
3   Press down on the valve keeper retainer using fingers, then remove the keepers. If the retainer is stuck, carefully tap it with a hammer and a 5/8-inch socket to loosen it. Remove the retainer, the keepers and the valve spring.
4   Push on the end of the valve stem to move the valve face away from the valve seat, then use a plastic scrub brush to clean the housing, valve surfaces and all other area of the valve.
5   Obtain Mopar EGR system cleaner or an equivalent and mix it with four parts water. Soak the EGR valve assembly in this solution for an hour. Move the valve occasionally to clean it. Scrub the assembly again to remove all deposits.
6   Rinse the assembly in warm water and dry all parts.
7   Assemble the valve spring, keepers and retainer. Install the EGR valve (see Chapter 6).

EGR cooler cleaning (6.4L Hemi and diesel models)

8   Disconnect the negative cables from both batteries.
9   Remove the engine cover.
10   Drain the cooling system (see Section 32).
11   Remove the air filter housing (see Chapter 4).
12   Detach the breather tube, then remove the turbo outlet tube.
13   Remove the alternator (see Chapter 5).
14   Remove the EGR cooler bypass valve.
15   Remove the clamp from the EGR cooler-to-exhaust transfer manifold, then remove the exhaust transfer manifold.
16   Remove the nut and move the EGR cooler filler tube out of the way.
17   Use large pliers to remove the forward coolant tube from the EGR cooler.
18   Remove the breather tube.

19   Remove the three EGR cooler mounting bolts and the two nuts from the mounting studs.
20   Remove the EGR cooler and its gasket. Discard the gasket.
21   Run hot tap water through the cooler for ten minutes.
22   Obtain a plug and install it into the end outlet of the cooler so that it will hold the cleaning solution.
23   Obtain Mopar EGR system cleaner or an equivalent and mix it with four parts hot water.
24   Stand the cooler upright, then fill it with the cleaning solution and allow it to stand for an hour.
25   Drain the cooler, then rinse it thoroughly with hot tap water and dry it.
26   Install a new cooler gasket and replace the cooler in the reverse of the removal procedure.

Resetting the EVIC emissions system service required light

27   Some models have an "Emissions System Service Required" light. After completing the maintenance, it will be necessary to reset the light.
28   Without starting the engine, turn the key to the On position.
29   Slowly depress the accelerator pedal to the floor two times within 10 seconds.
30   Turn the key to the Off/Lock position.
31   Start the engine and verify on the EVIC that the indicator message is on longer illuminated. Repeat the procedure above if it still appears.

---

28   Manual transmission, transfer case and differential lubricant change (every 60,000 miles or 48 months)

Note: *Rear (and front on 4WD models) differential lubricant should be changed every 15,000 miles or 12 months on vehicles with diesel engines or any vehicle used in rough service conditions such as frequent towing or off-road use.*

Manual transmission

1   This procedure should be performed after the vehicle has been driven so the lubricant will be warm and therefore will flow out of the transmission more easily. Raise the vehicle and support it securely on jackstands.
2   Move a drain pan, rags, newspapers and wrenches under the transmission. Remove the fill plug from the side of the transmission case **(see illustration 22.1)**.
3   Remove the transmission drain plug at the bottom of the case and allow the lubricant to drain into the pan. Some transmissions don't have a drain plug; remove the lower PTO cover bolt to drain the lubricant.
4   After the lubricant has drained completely, reinstall the plug and tighten it securely.
5   Using a hand pump, syringe or funnel, fill

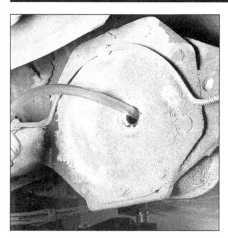

28.18 This is the easiest way to remove the lubricant: Work the end of the hose to the bottom of the differential housing and draw out the old lubricant with a hand pump

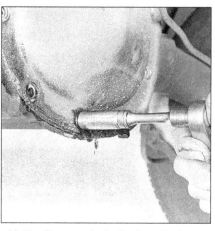

28.19a Remove the bolts from the lower edge of the cover . . .

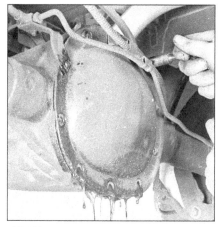

28.19b . . . then loosen the top bolts and allow the lubricant to drain out

the transmission with the specified lubricant until it begins to leak out through the hole. Reinstall the fill plug and tighten it securely.
6   Lower the vehicle.
7   Drive the vehicle for a short distance, then check the drain and fill plugs for leakage.

Transfer case (4WD models)
8   Drive the vehicle for at least 15 minutes to warm the lubricant in the case.
9   Raise the vehicle and support it securely on jackstands.
10   Remove the filler plug from the case.
11   Remove the drain plug from the lower part of the case and allow the old lubricant to drain completely **(see illustration 22.5).**
12   After the lubricant has drained completely, reinstall the plug and tighten it securely.
13   Fill the case with the specified lubricant until it is level with the lower edge of the filler hole.
14   Install the filler plug and tighten it securely.
15   Drive the vehicle for a short distance and recheck the lubricant level. In some instances a small amount of additional lubricant will have to be added.

Differential
Refer to illustrations 28.18, 28.19a, 28.19b, 28.19c and 28.21

16   This procedure should be performed after the vehicle has been driven so the lubricant will be warm and therefore will flow out of the differential more easily.
17   Raise the vehicle and support it securely on jackstands. If the differential has a bolt-on cover at the rear, it is usually easiest to remove the cover to drain the lubricant (which will also allow you to inspect the differential). If there's no bolt-on cover, look for a drain plug at the bottom of the differential housing. If there's not a drain plug and no cover, you'll

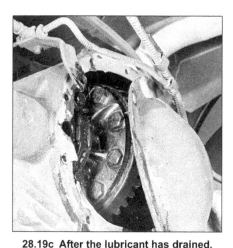

28.19c After the lubricant has drained, remove the remaining cover bolts and the cover

have to remove the lubricant through the filler plug hole with a suction pump. If you'll be draining the lubricant by removing the cover or a drain plug, move a drain pan, rags, newspapers and wrenches under the vehicle.
18   Remove the filler plug from the differential (see Section 22). If a suction pump is being used, insert the flexible hose. Work the hose down to the bottom of the differential housing and pump the lubricant out **(see illustration)**. If you'll be draining the lubricant through a drain plug, remove the plug and allow the lubricant to drain into the pan, then reinstall the drain plug.
19   If the differential is being drained by removing the cover plate, remove the bolts on the lower half of the plate. Loosen the bolts on the upper half and use them to keep the cover loosely attached. Allow the oil to drain into the pan, then completely remove the cover **(see illustrations)**.
20   Using a lint-free rag, clean the inside of the cover and the accessible areas of the dif-

28.21 Carefully scrape the old gasket material off to ensure a leak-free seal

ferential housing. As this is done, check for chipped gears and metal particles in the lubricant, indicating that the differential should be more thoroughly inspected and/or repaired.
21   Thoroughly clean the gasket mating surfaces of the differential housing and the cover plate. Use a gasket scraper or putty knife to remove all traces of the old gasket **(see illustration)**.
22   Apply a thin layer of RTV sealant to the cover flange, then press a new gasket into position on the cover. Make sure the bolt holes align properly.
23   Place the cover on the differential housing and install the bolts. Tighten the bolts securely.
24   Use a hand pump, syringe or funnel to fill the differential housing with the specified lubricant until it's level with the bottom of the plug hole.
25   Install the filler plug and make sure it is secure.

30.2 Remove the PCV screws and pull the valve out of the cover (3.6L engine)

31.6 With the front bolts in place, but loose, pull the rear of the pan down to drain the fluid

29 Closed Crankcase Ventilation (CCV) filter replacement (diesel engine) (every 60,000 miles or 48 months)

Note: *The breather cover is mounted on top of the valve cover.*

1 Remove the engine cover and all interfering heat shields.
2 Detach the breather hoses from the breather cover.
3 Remove the bolts from the edges of the breather cover, then pull the breather off of the valve cover.
4 Remove the filter from the recess in the valve cover.
5 Install a new filter, then replace the breather cover. Don't try to clean the filter.

30 Positive Crankcase Ventilation (PCV) valve replacement (gasoline engines) (every 90,000 miles or 72 months)

3.6L engine

Refer to illustration 30.2

1 The PCV valve is located on the end of the right (passenger's side) valve cover.
2 Detach the hose from the PCV, then remove the v screws and pull the valve from the valve cover **(see illustration)**.
3 Installation is the reverse of removal.

3.7L engine

4 The PCV valve is at the rear of the left cylinder head.
5 Detach the hose from the PCV valve, then unscrew and remove the valve.
6 Replace the valve with a new one and attach the hose.

4.7L engine

7 The PCV valve is at the top rear of the

left valve cover.
8 Detach the hose from the PCV valve.
9 Use a screwdriver to remove the PCV valve from the valve cover.
10 Replace the valve with a new one and attach the hose.

5.7L and 6.4L engines

11 The PCV valve is located in the top of the intake manifold, behind and to the right of the throttle body. **Note:** *It has a small twist handle recessed into the valve cover.*
12 Clean around the PCV valve before removing it to avoid getting debris into the engine.
13 Turn the PCV valve a quarter turn counterclockwise, then pull it out.
14 If you're reusing the old valve, check the condition of the two O-rings on it and replace them if necessary.
15 Clean the sealing surfaces of the valve cover and clean it thoroughly.
16 Apply a little clean engine oil to the O-rings of the PCV valve.
17 Insert the valve, then turn it clockwise to lock it in place.

31 Automatic transmission fluid and filter change (every 90,000 miles or 72 months)

All models except (8HP45/845RE and 8HP70) 8-speed transmissions

Refer to illustrations 31.6, 31.9, 31.11 and 31.12

Note: *This procedure should be done every 60,000 miles if the vehicle is used under severe conditions such as frequent towing or off-road use.*

1 At the specified intervals, the transmission fluid should be drained and replaced. Since the fluid will remain hot long after driving, perform this procedure only after the

engine has cooled down completely.
2 Before beginning work, purchase the specified transmission fluid (see *Recommended lubricants and fluids* in this Chapter's Specifications) and a new filter(s).
3 Other tools necessary for this job include a floor jack, jackstands to support the vehicle in a raised position, a drain pan capable of holding at least four quarts, newspapers and clean rags.
4 Raise the vehicle and support it securely on jackstands.
5 Place the drain pan underneath the transmission pan. Remove the rear and side pan mounting bolts, but only loosen the rear pan bolts approximately four turns.
6 Carefully pry the transmission pan loose with a screwdriver, allowing the fluid to drain **(see illustration)**.
7 Remove the remaining bolts, pan and gasket. Carefully clean the gasket surface of the transmission pan to remove all traces of the old gasket and sealant.
8 Drain the fluid from the transmission pan, clean the pan with solvent and dry it with compressed air, if available. **Note:** *Some models are equipped with magnets in the transmission pan to catch metal debris. Clean the magnet thoroughly. A small amount of metal material is normal at the magnet. If there is considerable debris, consult a dealer or transmission specialist.*
9 Remove the filter and seal from the valve body inside the transmission **(see illustration)**.
10 Use a gasket scraper to remove any traces of old gasket material that remain on the valve body. **Note:** *Be very careful not to gouge the delicate aluminum gasket surface on the valve body.*
11 On 454RFE and 545RFE transmissions, remove the cooler return filter **(see illustration)**. Compare the old filter with the new one to make sure they're the same type. Install the new cooler filter and tighten it to 125 inch-lbs.
12 Install a new gasket and filter. On many replacement filters, the gasket is attached to the filter to simplify installation. **Caution:** *On*

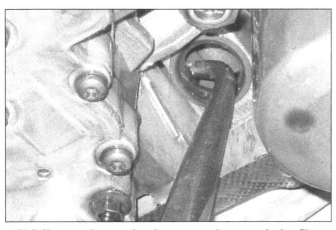

31.9 Use a seal removal tool to remove the transmission filter seal from the valve body

31.11 Use an oil filter wrench to remove the transmission cooler return filter

most models, the seal can be installed on the filter first, then the seal/filter can be pushed in place and secured. On 454RE and 545RFE transmissions, install the filter seal into the valve body first, then install the filter (see illustration).

13   Make sure the gasket surface on the transmission pan is clean, then install a new gasket on the pan. Put the pan in place against the transmission and, working around the pan, tighten each bolt a little at a time to the torque listed in this Chapter's Specifications. **Note:** *On 454RFE or 545RFE transmissions, RTV sealant is used instead of a gasket. Clean the pan and transmission surfaces thoroughly with lacquer thinner and apply a continuous bead of ATF-resistant RTV sealant to the pan, then bolt it in place and tighten to Specifications within five minutes.*
14   Lower the vehicle and add the specified type and amount (minus one quart) of automatic transmission fluid through the filler tube (see Section 4).
15   With the transmission in Park and the parking brake set, run the engine at a fast idle, but don't race it.
16   Move the gear selector through each range and back to Park. Check the fluid level. It will probably be low. Add enough fluid to bring the level between the two holes on the dipstick.
17   Check under the vehicle for leaks during the first few trips. Check the fluid level again when the transmission is hot (see Section 4).

(8HP45/845RE and 8HP70) 8-speed transmissions

Note: *The transmission oil pan and filter are an integrated assembly and cannot be serviced separately.*
18   Raise the vehicle and support it securely on jackstands.
19   Place a drain pan under the drain plug and allow the transmission fluid to drain out, then reinstall the plug.
20   Remove the transmission pan mounting bolts and slowly lower the pan, then remove the gasket. The gasket may be reused if it is not torn or damaged.

31.12 On 454RFE and 545RFE transmissions, install the filter seal into the valve body first

21   If replacing the filter, install a new transmission oil pan.
22   Install the pan and gasket to the transmission and tighten the bolts hand tight.
23   Starting from opposite corners and working in a circular pattern, tighten the pan bolts to the torque listed in this Chapters Specifications.
24   Refer to (Section 4) and follow the procedure on adding fluid and checking the fluid level.
25   The remaining installation is the reverse of removal, making sure to check for leaks.

32   Cooling system servicing (draining, flushing and refilling) (every 60 months)

Warning: *Wait until the engine is completely cool before beginning this procedure.*
Warning: *Do not allow antifreeze to come in contact with your skin or painted surfaces of the vehicle. Rinse off spills immediately with plenty of water. Antifreeze is highly toxic if ingested. Never leave antifreeze lying around in an open container or in puddles on the floor;*

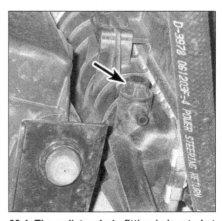

32.4 The radiator drain fitting is located at a lower corner of the radiator

children and pets are attracted by it's sweet smell and may drink it. Check with local authorities about disposing of used antifreeze. Many communities have collection centers which will see that antifreeze is disposed of safely.
1   Periodically, the cooling system should be drained, flushed and refilled to replenish the antifreeze mixture and prevent formation of rust and corrosion, which can impair the performance of the cooling system and cause engine damage. When the cooling system is serviced, all hoses and the pressure cap should be checked and replaced if necessary.

Draining

Refer to illustrations 32.4 and 32.5
2   Apply the parking brake and block the wheels. If the vehicle has just been driven, wait several hours to allow the engine to cool down before beginning this procedure.
3   Once the engine is completely cool, remove the radiator cap.
4   Move a large container under the radiator drain to catch the coolant. Attach a length of hose to the drain fitting to direct the coolant into the container, then open the drain fitting (a pair of pliers may be required to turn it) (see illustration).

32.5 The block drain plugs are generally located about one to two inches above the oil pan - there is one on each side of the engine block

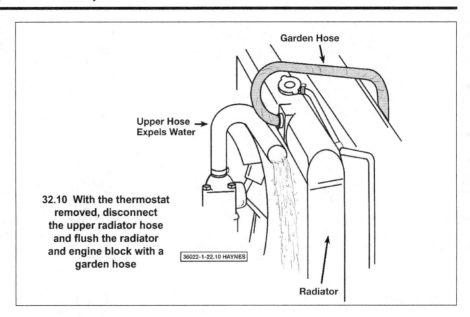

32.10 With the thermostat removed, disconnect the upper radiator hose and flush the radiator and engine block with a garden hose

5   After the coolant stops flowing out of the radiator, close the drain fitting, move the container under the engine block drain plugs and allow the coolant in the block to drain **(see illustration)**.
6   While the coolant is draining, check the condition of the radiator hoses, heater hoses and clamps (see Section 14, if necessary). Replace any damaged clamps or hoses.
7   Reinstall the block drain plugs and tighten them securely.

Flushing

Refer to illustration 32.10
8   Once the system has completely drained, remove the thermostat housing from the engine (see Chapter 3), then reinstall the housing without the thermostat. This will allow the system to be thoroughly flushed.
9   Disconnect the upper hose from the radiator.
10   Place a garden hose in the upper radiator inlet and flush the system until the water runs clear at the upper radiator hose **(see illustration)**.
11   Severe cases of radiator contamination or clogging will require removing the radiator (see Chapter 3) and reverse flushing it. This involves inserting the hose in the bottom radiator outlet to allow the clean water to run against the normal flow, draining out through

the top. A radiator repair shop should be consulted if further cleaning or repair is necessary.
12   When the coolant is regularly drained and the system refilled with the correct coolant mixture there should be no need to employ chemical cleaners or descalers.

Refilling

13   Place the heater temperature control in the maximum heat position.
14   Slowly add new coolant (a 50/50 mixture of water and antifreeze) to the radiator until the level is up to the bottom of the filler neck.
15   Install the radiator cap and run the engine in a well-ventilated area until the thermostat opens (coolant will begin flowing through the radiator and the upper radiator hose will become hot).
16   Turn the engine off and let it cool.
17   Squeeze the upper radiator hose to expel air, then add more coolant mixture if necessary. Reinstall the radiator cap.
18   Start the engine, allow it to reach normal operating temperature and check for leaks. Also, set the heater and blower controls to the maximum setting and check to see that the heater output from the air ducts is warm. This is a good indication that all air has been purged from the cooling system.

33   Power steering fluid replacement (every 90,000 miles or 72 months)

1   Apply the parking brake, raise the front of the vehicle and support it securely with jackstands.
2   Position a drain pan under the power steering pump, then disconnect the return line(s) from the pump (see Chapter 10). Plug the return line port(s) to prevent excessive fluid loss and the entry of contaminants.
3   Position the return line(s) so the fluid can drain into the pan.
4   Start the engine and allow it to run at idle. Turn the wheels from side-to-side, without hitting the stops, while an assistant fills the reservoir with new fluid.
5   Run about a quart of new fluid through the system, then stop the engine and install the line(s).
6   Fill the reservoir (see Section 4), then bleed the system (see Chapter 10).
7   Repeat Steps 2 through 6, making sure all contaminated fluid is removed from the system.
8   Fill the power steering reservoir with the recommended fluid and bleed the system (see Chapter 10).

Chapter 2 Part A
3.7L V6 and 4.7L V8 engines

Contents

Specifications

General

Displacement

3.7L V6	226 cubic inches
4.7L V8	287 cubic inches

Bore and stroke

3.7L V6	3.66 x 3.40 inches
4.7L V8	3.66 x 3.40 inches

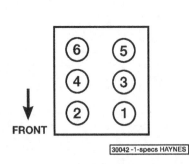

Cylinder identification diagram -
3.7L V6 engine

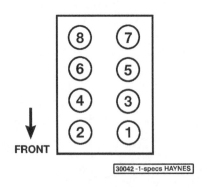

Cylinder identification diagram -
4.7L V8 engine

General (continued)

Cylinder numbers (front-to-rear)
   3.7L V6
      Left (driver's) side 1-3-5
      Right side... 2-4-6
      Firing order 1-6-5-4-3-2
   4.7L V8
      Left (driver's) side 1-3-5-7
      Right side... 2-4-6-8
      Firing order 1-8-4-3-6-5-7-2
Cylinder compression pressure
   Minimum.. 170 psi
   Maximum variation between cylinders .... 40 psi

Camshaft

Endplay .... 0.003 to 0.0079 inch
Camshaft bearing oil clearance
   Standard.. 0.001 to 0.0026 inch
   Service limit.. 0.0026 inch
Camshaft journal diameter .... 1.0227 to 1.0235 inches
Camshaft bore diameter .... 1.0245 to 1.0252 inches

Timing chain

Idler gear endplay.. 0.004 to 0.010 inch

Oil pump

Cover warpage limit (maximum)... 0.001 inch
Inner and outer rotor thickness... 0.473 inch
Outer rotor diameter (minimum) .... 0.400 inch
Outer rotor-to-housing clearance (maximum).. 0.009 inch
Inner rotor-to-outer rotor lobe clearance (maximum)................................. 0.006 inch
Oil pump housing-to-rotor side clearance (maximum).................................. 0.0038 inch

Torque specifications

Ft-lbs (unless otherwise indicated)

Note: *One foot-pound (ft-lb) of torque is equivalent to 12 inch-pounds (in-lbs) of torque. Torque values below approximately 15 ft-lbs are expressed in inch-pounds, since most foot-pound torque wrenches are not accurate at these smaller values.*

Balance shaft retaining bolt (V6 engine)	21
Camshaft sprocket bolts (non-oiled)	90
Camshaft bearing cap bolts	100 in-lbs
Crankshaft pulley/vibration damper bolt	130
Cylinder head bolts	
3.7L V6 **(see illustration 11.19a)**	
Step 1, Bolts 1 through 8	20
Step 2, Bolts 1 through 8	Repeat Step 1 without first loosening
Step 3, Bolts 9 through 12	120 in-lbs
Step 4, Bolts 1 through 8	Tighten an additional 90-degrees
Step 5, Bolts 1 through 8	Tighten an additional 90-degrees
Step 6, Bolts 9 through 12	19
4.7L V8 **(see illustration 11.19b)**	
Step 1, Bolts 1 through 10	20
Step 2, Bolts 1 through 10	Repeat Step 1 without first loosening
Step 3, Bolts 11 through 14	89 in-lbs
Step 4, Bolts 1 through 10	Tighten an additional 90 degrees
Step 5, Bolts 1 through 10	Tighten an additional 90 degrees
Step 6, Bolts 11 through 14	19
Driveplate bolts	
3.7L V6	70
4.7L V8	45
Exhaust manifold bolts	18
Exhaust manifold heat shield nuts	
Step 1	72 in-lbs
Step 2	Loosen 45-degrees
Flywheel bolts	70
Intake manifold bolts **(see illustrations 9.20a or 9.20b)**	105 in-lbs
Oil pan bolts	132 in-lbs
Oil pan drain plug	25
Oil pick-up tube mounting bolt/nut	20
Oil pump/primary chain tensioner mounting bolts	21
Oil pump cover screws	105 in-lbs
Timing chain cover bolts	43
Timing chain guide	
Bolts	21
Access plugs	15
Timing chain idler sprocket bolt	25
Timing chain tensioner arm pivot bolt	
3.7L V6 engine	21
4.7L V8 engine	150 in-lbs
Timing chain tensioner (primary and secondary)	21
Transmission support brace bolts	40
Valve cover bolts	105 in-lbs
Water outlet housing	105 in-lbs

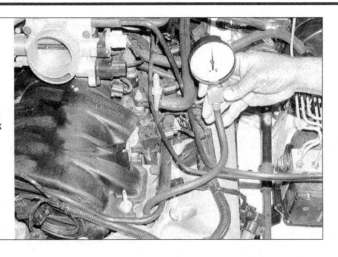

4.5 A compression gauge can be used in the number one spark plug hole to assist in finding TDC

4.8 Align the groove in the damper with the TDC mark on the timing chain cover

1   General information

This Part of Chapter 2 is devoted to in-vehicle repair procedures for the 3.7L V6 and 4.7L V8 single overhead camshaft (SOHC) engines. These engines utilize a cast iron engine block with cylinders arranged in a "V" shape at a 90-degree angle between the two banks. The overhead camshaft aluminum cylinder heads are equipped with replaceable valve guides and seats. Stamped steel rocker arms with an integral roller bearing actuate the valves.

Information concerning engine removal and installation and engine overhaul can be found in Part E of this Chapter.

The following repair procedures are based on the assumption that the engine is installed in the vehicle. If the engine has been removed from the vehicle and mounted on a stand, many of the steps outlined in this Part of Chapter 2 will not apply.

2   Engine identification

Engine identification on the 3.7L V6 and 4.7L V8 engines is accomplished by matching the engine code stamped onto the engine block or by using the VIN number of the vehicle (8th position). Refer to the introductory pages in this manual for additional information.

3   Repair operations possible with the engine in the vehicle

Many major repair operations can be accomplished without removing the engine from the vehicle.

Clean the engine compartment and the exterior of the engine with some type of degreaser before any work is done. It will make the job easier and help keep dirt out of the internal areas of the engine.

Depending on the components involved, it may be helpful to remove the hood to improve access to the engine as repairs are performed (see Chapter 11, if necessary).

Cover the fenders to prevent damage to the paint. Special pads are available, but an old bedspread or blanket will also work.

If vacuum, exhaust, oil or coolant leaks develop, indicating a need for gasket or seal replacement, the repairs can generally be made with the engine in the vehicle. The intake and exhaust manifold gaskets, oil pan gasket, crankshaft oil seals and cylinder head gaskets are all accessible with the engine in place.

Exterior engine components, such as the intake and exhaust manifolds, the oil pan, the oil pump, the water pump (see Chapter 3), the starter motor, the alternator and the fuel system components can be removed for repair with the engine in place.

Since the cylinder heads can be removed without pulling the engine, valve component servicing can also be accomplished with the engine in the vehicle. Replacement of the camshafts, timing chains and sprockets are also possible with the engine in the vehicle.

4   Top Dead Center (TDC) for number one piston - locating

Refer to illustrations 4.5 and 4.8

1   Top Dead Center (TDC) is the highest point in the cylinder that each piston reaches as it travels up the cylinder bore. Each piston reaches TDC on the compression stroke and again on the exhaust stroke, but TDC generally refers to piston position on the compression stroke.

2   Positioning the piston(s) at TDC is an essential part of procedures such as camshaft and timing chain/sprocket removal.

3   Before beginning this procedure, place the transmission in Neutral or Park and apply the parking brake or block the rear wheels. Also remove the spark plugs (see Chapter 1). **Warning:** *If method b) or c) in the next Step will be used to rotate the engine, disable the ignition system by disconnecting the primary electrical connectors at the ignition coils, and disable the fuel system (see Chapter 4, Section 2).*

4   In order to bring any piston to TDC, the crankshaft must be turned using one of the methods outlined below. When looking at the front of the engine, normal crankshaft rotation is clockwise.

 a) *The preferred method is to turn the crankshaft with a socket and ratchet attached to the bolt threaded into the front of the crankshaft. Turn the bolt in a clockwise direction only. Never turn the bolt counterclockwise.*

 b) *A remote starter switch, which may save some time, can also be used. Follow the instructions included with the switch. Once the piston is close to TDC, use a socket and ratchet as described in the previous paragraph.*

 c) *If an assistant is available to turn the ignition switch to the Start position in short bursts, you can get the piston close to TDC without a remote starter switch. Make sure your assistant is out of the vehicle, away from the ignition switch, then use a socket and ratchet as described in Paragraph (a) to complete the procedure.*

5   Install a compression pressure gauge in the number one spark plug hole (see Chapter 2D). It should be a gauge with a screw-in fitting and a hose at least six inches long **(see illustration)**.

6   Rotate the crankshaft using one of the methods described above while observing for pressure on the compression gauge. The moment the gauge shows pressure indicates that the number one cylinder has begun the compression stroke.

7   Once the compression stroke has begun, TDC for the compression stroke is reached by bringing the piston to the top of the cylinder.

8   Continue turning the crankshaft until the notch in the crankshaft damper is aligned with the TDC mark on the timing chain cover **(see illustration)**. At this point, the number one cylinder is at TDC on the compression stroke. If the marks are aligned but there was no compression, the piston was on the exhaust stroke. Continue rotating the crankshaft 360-

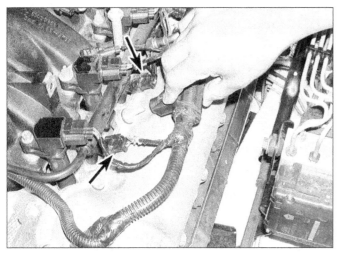

5.10 Detach the wiring harness from the valve cover studs, then disconnect the fuel injector electrical connectors and position the harness aside

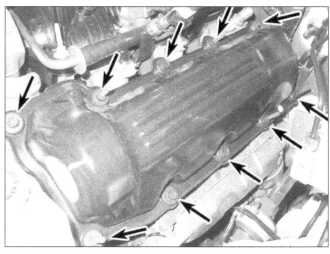

5.15 Remove the valve cover bolts (left side shown)

degrees (1-turn) and realign the marks. **Note:** *If a compression gauge is not available, you can simply place a blunt object over the spark plug hole and listen for compression as the engine is rotated. Once compression at the No.1 spark plug hole is noted, the remainder of the Step is the same.*

9   After the number one piston has been positioned at TDC on the compression stroke, TDC for any of the remaining cylinders can be located by turning the crankshaft in increments of 120-degrees for 3.7L V6 engines or 90-degrees for 4.7L V8 engines and following the firing order (refer to the Specifications). For example on 3.7L V6 engines, rotating the engine 120-degrees past TDC #1 will put the engine at TDC compression for cylinder #6.

10   An even faster way to find TDC for any cylinder other than No. 1 is to make marks on the crankshaft damper 120-degree intervals from the TDC mark on the crankshaft damper (V6 engine) or 90-degree intervals from the TDC mark on the crankshaft damper (V8 engine). Install the compression gauge into the cylinder for which you want to find TDC, rotate the engine until compression begins to register on the gauge, then continue turning the crankshaft until the next mark on the damper aligns with the mark on the timing chain cover.

---

5   Valve covers - removal and installation

Removal

Refer to illustrations 5.10 and 5.15

1   Disconnect the cable from the negative terminal of the battery (see Chapter 5).
2   Remove the air filter housing, the air intake duct and the throttle body resonator (see Chapter 4A).

Right (passenger's) side cover

Warning: *Wait until the engine is completely cool before beginning this procedure.*

3   Drain the cooling system (see Chapter 1).
4   Remove the heater hoses.
5   Remove the drivebelt (see Chapter 1) and the air conditioning compressor (see Chapter 3). Position the A/C compressor to the side without disconnecting the refrigerant lines from the compressor.
6   Detach the heater hoses.
7   Remove the oil filler tube.
8   Remove the rear right breather tube and filter.
9   Remove the PCV hose.
10   On V6 engines, disconnect the injector and coil wiring, then detach the wire harnesses and move them out of the way **(see illustration)**.
11   On V8 engines, remove the spark plug wires.

Left (driver's) side cover

12   On V8 engines, remove the spark plug wires.
13   Move the injector wire harnesses to the front of the valve cover.
14   Remove the left breather tube.

Either cover

15   Remove the valve cover bolts **(see illustration)**. Make a note of the locations of the bolts with studs before removal to ensure correct positioning during installation.
16   Detach the valve cover. **Note:** *If the cover sticks to the cylinder head, use a block of wood and a hammer to dislodge it. If the cover still won't come loose, pry on it carefully, but don't distort the sealing flange.*

Installation

17   The mating surfaces of each cylinder head and valve cover must be perfectly clean when the covers are installed. **Caution:** *Do not use harsh cleaners when cleaning the valve covers or damage to the covers may occur.* **Note:** *The valve cover gasket can be reused if it isn't hardened, cracked or otherwise damaged.*

18   Clean the mounting bolt or stud threads with a wire brush if necessary to remove any corrosion and restore damaged threads. Use a tap to clean the threaded holes in the heads.
19   Place the valve cover and gasket in position, then install the bolts in the correct locations from which they where removed. Tighten the bolts in several steps to the torque listed in this Chapter's Specifications.
20   Complete the installation by reversing the removal procedure. Refill the cooling system, if drained (see Chapter 1). Start the engine and check carefully for oil leaks.

---

6   Rocker arms and hydraulic lash adjusters - removal, inspection and installation

Refer to illustrations 6.5, 6.7 and 6.8

Note: *A special valve spring compressor available from most aftermarket specialty tool manufacturers will be required for this procedure. The only other alternative to accomplishing this task without the use of this special tool is to remove the timing chains and the camshafts, which requires major disassembly of the engine and surrounding components.*

Note: *This engine is a non-freewheeling (interference) engine and the pistons must be down in the cylinder bore before the valve and spring assembly can be compressed to allow rocker arm removal.*

1   Before beginning this procedure, place the transmission in Park and apply the parking brake or block the rear wheels. Also, disable the ignition system by disconnecting the primary electrical connectors at the ignition coils and removing the spark plugs (see Chapter 1).
2   Remove the valve cover(s) (see Section 5).
3   Before the rocker arms and lash adjusters are removed, arrange to label and store them, so they can be kept separate and reinstalled on the same valve they were removed from.

6.5 Using a special type valve spring compressor, depress the valve spring just enough to remove the rocker arm

6.7 Pull the lash adjuster up and out of its bore to remove it from the cylinder head

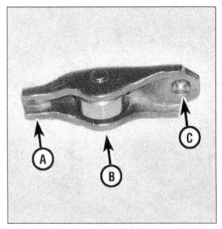

6.8 Inspect the rocker arms at the following locations

A   *Valve stem seat*
B   *Roller*
C   *Lash adjuster pocket*

4   Rotate the engine with a socket and ratchet in a clockwise direction by the crankshaft pulley/vibration damper bolt until the piston(s) are positioned correctly to remove the rocker arms from the corresponding cylinders as follows. Start the sequence from TDC number 1 on the compression stroke (see Section 4).

On V6 engines:
a) *With the No.1 piston at TDC on the compression stroke, remove the rocker arms from cylinders **No. 2 and 6**.*
b) *Rotate the crankshaft another 180-degrees (1/2-turn) from TDC on the compression stroke to bring the No. 1 piston to BDC on the firing stroke. Remove the rocker arms from cylinder **No. 1** with the No. 1 piston at Bottom Dead Center (BDC) on the firing stroke.*
c) *Rotate the crankshaft another 180-degrees (1/2-turn) from BDC on the firing stroke to bring the No.1 piston to TDC on the exhaust stroke. Remove the rocker arms from cylinders **No. 3 and 5** with the No. 1 piston at TDC on the exhaust stroke.*
d) *Rotate the crankshaft another 180-degrees (1/2-turn) from TDC on the exhaust stroke to bring the No.1 piston to BDC on the intake stroke. Remove the rocker arms from cylinder **No. 4** with the No. 1 piston at BDC on the intake stroke.*

On V8 engines:
a) *With the No.1 piston at TDC on the compression stroke, remove the rocker arms from cylinders **No. 2 and 8**.*
b) *Rotate the crankshaft another 360-degrees (1-turn) from TDC on the compression stroke to bring the No.1 piston to TDC on the exhaust stroke. Remove the rocker arms from cylinders **No. 3 and 5** with the No. 1 piston at TDC on the exhaust stroke.*

c) *Position cylinder No. 3 at TDC on the compression stroke. Remove the rocker arms from cylinder **No. 4 and 6**.*
d) *Position cylinder No. 2 at TDC on the compression stroke. Remove the rocker arms from cylinder **No. 1 and 7**.*

5   Hook the valve spring compressor around the base of the camshaft. Depress the valve spring just enough to release tension on the rocker arm to be removed. Once tension on the rocker arm is relieved, the rocker arm can be removed by simply pulling it out **(see illustration)**.
6   If you're removing or replacing only a few of the rocker arms or lash adjusters, locate the cylinder number of the rocker arm or lash adjuster you wish to remove in Step 4, then rotate the crankshaft to the corresponding position. Remember to keep the rocker arm and lash adjuster for each valve together so they can be reinstalled in the same locations. Refer to Section 4 as necessary to help position the designated cylinder at TDC.
7   Once the rocker arms are removed, the lash adjusters can be pulled out of the cylinder head and stored with the corresponding rocker arm **(see illustration)**.
8   Inspect each rocker arm for wear, cracks and other damage. Make sure the rollers turn freely and show no signs of wear, also check the pivot area for wear, cracks and galling **(see illustration)**.
9   Inspect the lash adjuster contact surfaces for wear or damage. Make sure the lash adjusters move up and down freely in their bores on the cylinder head without excessive side-to-side play.
10   Installation is the reverse of removal with the following exceptions: Always install the lash adjuster first and make sure they're at least partially full of oil before installation. This is indicated by little or no lash adjuster plunger travel.

7   Timing chain and sprockets - removal, inspection and installation

Warning: *Wait until the engine is completely cool before beginning this procedure.*
Note: *Special tools are necessary to complete this procedure. Read through the entire procedure and obtain the special tools before beginning work.*
Note: *These engines utilize three timing chains. The primary timing chain runs around the crankshaft sprocket and the idler gear sprocket. This chain synchronizes the crankshaft and pistons with the idler gear, while two secondary timing chains run around the rear of the idler gear and up to the camshaft sprockets to synchronize the valve timing with the crankshaft. Refer to Step 24 in this Section for the in-vehicle timing chain inspection procedure prior to timing chain removal.*

Removal

Refer to illustrations 7.8, 7.11, 7.13, 7.14, 7.15, 7.16a, 7.16b, 7.17 and 7.18
Caution: *The timing system is complex. Severe engine damage will occur if you make any mistakes. Do not attempt this procedure unless you are highly experienced with this type of repair. If you are at all unsure of your abilities, consult an expert. Double-check all your work and be sure everything is correct before you attempt to start the engine.*
1   Disconnect the cable from the negative terminal of the battery (see Chapter 5).
2   Drain the cooling system (see Chapter 1).
3   Remove the engine cooling fan, the accessory drivebelt and the fan shroud from the engine compartment (see Chapter 3).

7.8 Drivebelt tensioner retaining bolt

7.11 When the engine is at TDC on the exhaust stroke, the mark on the crankshaft damper is aligned with the mark on the timing chain cover, and the "V6" or "V8" marks on the camshaft sprockets should be pointing to the 12 o'clock position

7.13 Timing chain cover retaining bolts

7.14 Locking the primary timing chain tensioner in the retracted position

4   Detach the heater hoses and the lower radiator hose from the timing chain cover and position them aside.

5   Unbolt the power steering pump and set it aside without disconnecting the fluid lines (see Chapter 10).

6   Remove the alternator (see Chapter 5).

7   Unbolt the air conditioning compressor (if equipped) and position it aside without disconnecting the refrigerant lines.

8   Remove the accessory drivebelt tensioner from the timing chain cover **(see illustration)**.

9   Remove the valve covers (see Section 5) and the spark plugs (see Chapter 1). Remove all of the rocker arms (see Section 6). **Note:** *These steps are not absolutely necessary, but it will help make alignment of the camshaft sprockets easier upon installation and also eliminate any possibility of the pistons contacting the valves during this procedure, since these are interference engines.*

10   Remove the Camshaft Position (CMP) sensor from the right cylinder head (see Chapter 6).

11   Position the number one piston at TDC on the *exhaust* stroke (one revolution from TDC on the compression stroke - see Section 4). Visually confirm the engine is at TDC on the exhaust stroke, by verifying that the timing mark on the crankshaft damper is aligned with the mark on the timing chain cover and the "V6" or "V8" marks on the camshaft sprockets are pointing straight up in the 12 o'clock position **(see illustration 4.8 and the accompanying illustration)**.

12   Remove the crankshaft damper/pulley (see Section 12).

13   Remove the timing chain cover and the water pump as an assembly **(see illustration)**. Note that various types and sizes of bolts are used. They must be reinstalled in their original locations. Mark each bolt or make a sketch to help remember where they go.

14   Cover the oil pan opening with shop rags to prevent any components from falling into the engine. Collapse the primary timing chain tensioner with a pair of locking pliers and install a locking pin into the holes in the

tensioner body to keep it in the retracted position **(see illustration)**.

15   Remove the secondary timing chain tensioners **(see illustration)**.

7.15 Secondary timing chain tensioner mounting bolts

7.16a Remove the left camshaft sprocket bolt while holding the sprocket with a pin spanner wrench - note the chain guide access plug

7.16b The right camshaft sprocket is identified by the camshaft position sensor ring which is fastened to the rear of the sprocket - be extremely careful not to damage or place a magnetic object of any kind near the camshaft position sensor ring or a no start condition may occur after installation

16   Remove the camshaft sprocket retaining bolts **(see illustrations)**. Pull the camshaft sprockets off the camshaft hubs one at a time. Lower the sprocket(s) into the cylinder head opening until the chain can be displaced from around the sprocket, then remove the camshaft sprockets from the engine and let the secondary chains fall down between the timing chain guides. **Caution:** *If the rocker arms **were not** removed as suggested in Step 9, it will be necessary to hold the camshafts from rotating with a set of locking pliers while the sprocket is being removed. Work on one camshaft and sprocket at a time starting with the left sprocket and proceeding to the right sprocket. After the sprocket is removed from the camshaft(s), let the camshaft slowly rotate to its neutral position. This is typically 15-degrees clockwise on the left camshaft sprocket and 45-degrees counterclockwise on the right camshaft. Force from the valve springs will make the camshafts rotate as the sprockets are removed. Sudden move-*

ment of the camshafts may allow the valves to strike the pistons. **Caution:** *Never install the locking pliers on a camshaft lobe as damage to the camshaft will occur. When using locking pliers, always rotate the camshaft by the shaft.* **Caution:** *Do not rotate the crankshaft or camshafts separately after the secondary timing chains are loosened or removed with the rocker arms installed in the engine as piston or valve damage may occur. The only exception to this rule is when the camshafts must be rotated slightly, to realign the camshaft sprockets with the camshafts during installation.* **Caution:** *The right camshaft sprocket is identified by the camshaft position sensor ring which is fastened to the rear of the sprocket. Be extremely careful not to damage or place a magnetic object of any kind near the camshaft position sensor ring or a no start condition may occur after installation.*

17   Remove the idler sprocket bolt, then detach the idler sprocket, the crankshaft sprocket, the primary timing chain and the

secondary timing chains as an assembly **(see illustration)**.
18   Remove the cylinder head access plugs **(see illustrations 7.16a and 7.16b)**. Also remove the oil fill tube (if not already removed) from the front of the right cylinder head. Detach the timing chain guides and tensioner arms **(see illustration)**.

Inspection

Refer to illustration 7.22

19   Inspect the camshaft and crankshaft sprockets for wear on the teeth and keyways.
20   Inspect the chains for cracks or excessive wear of the rollers.
21   Inspect the facings of the primary chain tensioner, secondary chain guides and tensioner arms for excessive wear. If any of the components show signs of excessive wear or the chain guides are grooved in excess of 0.039 inch deep, they must be replaced.
22   If any of the timing chain guides are excessively grooved or melted, the tensioner

7.17 Remove the primary timing chain and the secondary chains from the engine as an assembly

7.18 Timing chain guide and tensioner arm pivot bolts

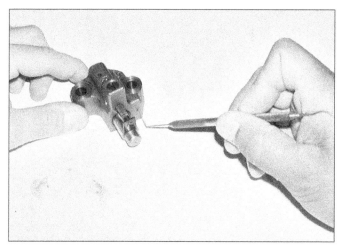

7.22 If excessive wear on the chain guides is evident, check the oil jet on the side of each secondary tensioner for clogging

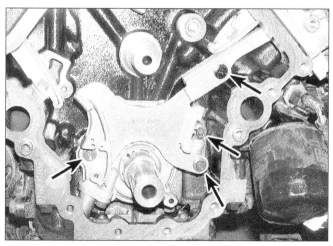

7.26 Primary timing chain tensioner/oil pump mounting bolts

lube jet may be clogged. Remove the jet and clean it with a small metal pick, then blow compressed air through it to remove any debris or foreign material (see illustration).
23   Inspect the idler sprocket bushing, shaft and spline joint for wear.
24   Inspect the tensioner piston and ratchet assembly on both of the secondary chain tensioners. If it appears there has been heavy contact between the piston, and ratchet assembly, replace the tensioner arm and secondary timing chain. Note: Secondary timing chain stretch can be checked by rotating the engine clockwise until the pistons in the secondary tensioners reach their maximum travel or extension. Using a machinist's ruler or a dial caliper, measure the piston protrusion or extension from the stepped ledge on the piston to the tensioner housing on each tensioner. If the maximum extension of either tensioner piston exceeds 0.590 inch, the secondary timing chains are worn beyond their limits and should be replaced.

Installation

Refer to illustrations 7.26, 7.27, 7.31, 7.33, 7.39, 7.48 and 7.50

Caution: Before starting the engine, carefully rotate the crankshaft by hand through at least two full revolutions (use a socket and breaker bar on the crankshaft pulley center bolt). If you feel any resistance, STOP! There is something wrong - most likely, valves are contacting the pistons. You must find the problem before proceeding. Check your work and see if any updated repair information is available.
25   If removed, install the timing chain guides and tensioner pivot arms back onto the engine and tighten the bolts to the torque listed in this Chapter's Specifications. Apply several drops of medium strength thread-locking compound to the tensioner pivot arm bolts before installing them. Note that the silver bolts retain the guides to the cylinder head and the black

colored bolts retain the guides to the engine block.
26   If the primary timing chain tensioner was removed or replaced, install it back onto the engine in the locked position and tighten the lower two bolts to the torque listed in this Chapter's Specifications (see illustration).
27   Working on one secondary timing chain tensioner at a time, compress the tensioner piston in a vise until the stepped edge is flush with the tensioner body (see illustration). Insert a small scribe or other suitable tool into the side of the tensioner body and push the spring loaded ratchet pawl away from the ratchet mechanism, then push the ratchet down into the tensioner body until it's approximately 0.080 inch away from the tensioner body. Insert the end of a paper clip into the hole on the front of the tensioner to lock the tensioner in place.
28   After the two secondary tensioners have been compressed and locked into place, install them on the engine and tighten the bolts to the torque listed in this Chapter's Specifications. Make sure the tensioner with the "R" mark is installed on the right secondary chain (passenger's side) and the tensioner with "L" mark is installed on the left secondary chain (driver's side). The secondary chain tensioners cannot be switched with one another. Also make sure the plate behind the left secondary chain tensioner is installed correctly.
29   If you purchased a new timing chain, verify that you have the correct timing chain for your vehicle by counting the number of links the chain has and comparing the new chain with the old chain. Also compare the position of the colored links in the new chain with the position of the colored links in the old chain.
30   Note that the idler sprocket has three sprockets and a gear incorporated into it. The front or forward facing sprocket (the largest of the three) is for the primary timing chain, the

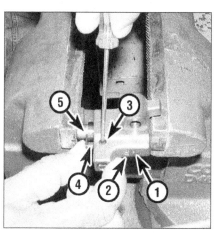

7.27 Locking the secondary tensioner(s) in the retracted position - note the identification mark on the side of the tensioner, as they are not interchangeable

1   Insert locking pin
2   Identification mark
3   Ratchet pawl
4   Ratchet
5   Tensioner piston

second or middle sprocket is for the left secondary chain, the third or rear sprocket is for the right secondary chain and the gear (3.7L V6 engines only) is for the counterbalance shaft (4.7L V8 engines don't have a counterbalance shaft). The next five Steps will involve assembling the timing chains onto the idler sprocket on a workbench.
31   Place the idler sprocket on a workbench with the mark on the front in the 12 o'clock position. Loop the right camshaft chain over the rear gear (farthest away from the primary chain gear) on the idler sprocket and position it so that the two plated links on the chain are visible through the lower (4 o'clock) window in

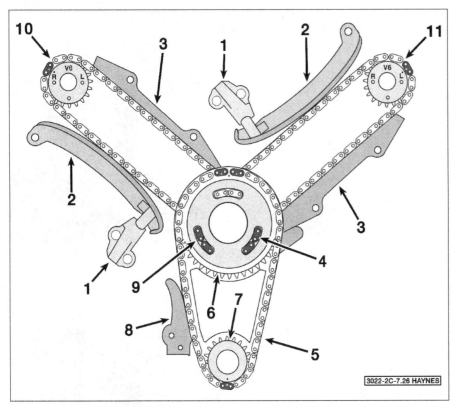

7.31 Timing chain installation details

1 Secondary timing chain tensioner	7 Crankshaft sprocket
2 Secondary tensioner arm	8 Primary chain tensioner
3 Chain guide	9 Two plated links on left camshaft chain
4 Two plated links on right camshaft chain	10 Right camshaft sprocket and secondary chain
5 Primary chain	11 Left camshaft sprocket and secondary chain
6 Idler sprocket	

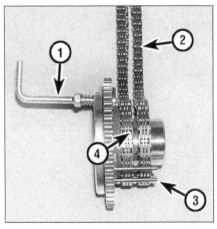

7.33 Install the secondary chain holding tool onto the idler sprocket with the plated links on the right camshaft chain in the 4 o'clock position and the left camshaft chain in the 8 o'clock position (typical)

1 Secondary chain holding tool
2 Right camshaft timing chain
3 Secondary chain holding tool retaining pins
4 Left camshaft timing chain

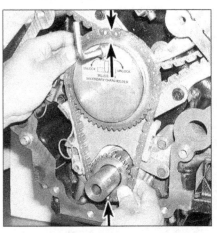

7.39 Align the timing mark on the idler sprocket gear with the timing mark on the counterbalance shaft, then install the idler sprocket, the crankshaft sprocket, the timing chains and the chain holding tool as an assembly onto the engine with the marks aligned as shown - feed the secondary chains up through the timing chain guides and loop them over the camshaft hubs, then push the primary chain assembly the rest of the way until it's seated

the idler sprocket **(see illustration)**.

32   Loop the left camshaft chain over the front of the idler sprocket and position it over the middle gear so that the two plated links on the chain are visible through the lower (8 o'clock) window in the idler sprocket. **Note:** *After the left chain is in position, the two plated links on the right chain will no longer be visible through the 4 o'clock window in the idler sprocket, so be sure that the right camshaft chain is installed correctly before installing the left camshaft chain.*

33   After the secondary (camshaft) chains have been installed properly on the idler sprocket, install the special secondary chain holding tool onto the idler sprocket **(see illustration)**. This tool serves as a third hand, to secure the camshaft chains to the idler sprocket during the installation of the idler sprocket onto the engine.

34   Install the primary timing chain onto the primary chain gear of the idler sprocket and align the double plated links with the mark on the front of the sprocket. The mark on the idler sprocket should still be in the 12 o'clock position.

35   Insert the teeth of the crankshaft sprocket into the primary timing chain with the mark on the crankshaft sprocket pointing down in the 6 o'clock position and aligned with the single plated link on the chain.

36   Lubricate the idler shaft and bushing with clean engine oil.

37   Install the idler sprocket, the crankshaft sprocket and the timing chains with the chain holding tool as an assembly onto the engine. Slide the crankshaft sprocket over the key-way on the crankshaft and position the idler sprocket partially over the idler shaft (just enough to hold the primary chain in place). Then feed the secondary chains up through the chain guides and the cylinder head. **Note:** *It may be easier to bend a hook in the end of a coat hanger to help pull the secondary chains up through the timing chain guides and the cylinder head opening.*

38   Loop the secondary chains over the camshaft hubs and secure them with rubber bands to remove the slack from the chains.

39   Push the idler sprocket, primary timing chain and the crankshaft sprocket assembly back on to the engine until they're fully seated

(see illustration). On 3.7L V6 engines, the timing mark on the idler sprocket gear must align with the timing mark on the counterbalance shaft gear while doing this. **Note:** *On 3.7L V6 engines, make sure the single mark on the counterbalance shaft gear is positioned between the two marks on the idler sprocket gear.*

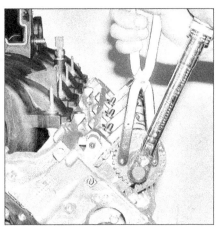

7.48 Tightening the left camshaft
sprocket bolt

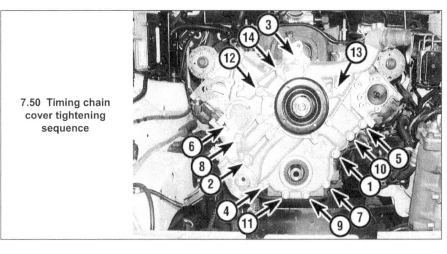

7.50 Timing chain
cover tightening
sequence

40 Thoroughly clean the idler sprocket bolt. Make sure all oil is removed from the bolt threads before installation, as over-tightening of bolts may occur if oil is not removed, then lubricate the idler sprocket washer with small amount of clean engine oil making sure not to get oil on the bolt threads. Remove the secondary timing chain holding tool from the idler sprocket, then install the idler sprocket retaining bolt and tighten it to the torque listed in this Chapter's Specifications.

41 Align the "L" mark on left camshaft sprocket with the plated link on the left camshaft chain and position the camshaft sprocket over the camshaft hub (see illustration 7.31). The camshaft may have to be rotated slightly to align the dowel pin on the camshaft with the slot on the sprocket.

42 Align the "R" mark on right camshaft sprocket with the plated link on the right camshaft chain and position the camshaft sprocket over the camshaft hub. The camshaft may have to be rotated slightly to align the dowel pin on the camshaft with the slot on the sprocket. Note: *If the rocker arms **were not** removed as suggested in Step 9, it will be necessary to rotate and hold the camshafts with a set of locking pliers while the sprocket is being installed. This is typically 15-degrees counterclockwise on the left camshaft sprocket and 45-degrees clockwise on the right camshaft (the exact opposite of removal). Work on one camshaft and sprocket at a time starting with the left sprocket and proceeding to the right sprocket and never rotate the camshaft by a camshaft lobe or damage to the camshaft will occur.*

43 Thoroughly clean the camshaft sprocket bolts. Make sure all oil is removed from the bolt threads before installation, as over-tightening of bolts may occur if oil is not removed, then lubricate the bolt washers with small amount of clean engine oil making sure not to get oil on the threads.

44 Install the camshaft sprocket bolts finger tight.

45 Verify that all the plated timing chain links are aligned with their corresponding marks (see illustration 7.31).

46 Remove the locking pins from the pri-

mary timing chain tensioner and the secondary timing chain tensioners. **Caution:** *Do not manually extend the tensioners; doing so will only over-extend the tensioners and lead to premature timing chain wear.*

47 Rotate the engine two complete revolutions and re-verify the position of the timing marks again. The idler sprocket mark should be located in the 12 o'clock position and the crankshaft sprocket mark should be located in the 6 o'clock position with the "V6" or "V8" marks on the camshaft sprockets located in the 12 o'clock position.

48 Using a spanner wrench to hold the sprockets from turning, tighten the camshaft sprocket bolts to the torque listed in this Chapter's Specifications (see illustration).

49 Remove all traces of old sealant or gasket material from the timing chain cover and the engine block.

50 Place the timing chain cover and gasket in position on the engine and install the bolts in their original locations and tighten the bolts to the torque listed in this Chapter's Specifications. Follow the correct tightening sequence (see illustration).

51 The remainder of installation is the reverse of removal. Use pipe sealant on the cylinder head plugs to prevent oil leaks.

52 Change the engine oil and filter and refill the cooling system (see Chapter 1).

8 Camshafts - removal, inspection and installation

Note: *Special tools are necessary to complete this procedure. Read through the entire procedure and obtain the special tools before beginning work.*

Note: *The camshafts should always be thoroughly inspected before installation and camshaft endplay should always be checked prior to camshaft removal (see Step 17).*

Removal

Refer to illustrations 8.6 and 8.9

1 Disconnect the cable from the negative terminal of the battery (see Chapter 5).

2 Remove the valve covers (see Section 5).

8.6 The timing chain tensioner wedge is pushed down between the chain strands to secure the secondary chain and the tensioner in place while the camshaft is removed - this wedge is fabricated from a block of wood and a piece of wire

3 Rotate the engine with a socket and ratchet (in a clockwise direction only) by the crankshaft pulley/vibration damper bolt until the "V6" or "V8" marks on the camshaft sprockets are located in the 12 o'clock position (see illustration 7.11).

4 Using a permanent marker, apply alignment marks to the secondary timing chain links on either side of the "V6" or "V8" marks on both camshaft sprockets to help aid the installation process (4 marks total).

5 Using a spanner wrench to hold the camshaft sprockets from turning, loosen the camshaft sprocket bolts several turns, then retighten the bolts by hand until they're snug up against the sprocket. If the camshaft sprockets have rotated during the bolt-loosening process, rotate the engine clockwise until the "V6" or "V8" marks on the cam sprockets are realigned in the 12 o'clock position.

6 Install a timing chain tensioner wedge through the opening in the top of the cylinder head and force the wedge down between the narrowest section of the secondary chain (see illustration). If both camshafts are to removed, two timing chain wedges will be

8.9 Verify that the camshaft bearing caps are marked to ensure correct reinstallation - do not mix-up the caps from the left cylinder head with the caps from the right cylinder head

8.14 Inspect the cam bearing surfaces in each cylinder head for pits, score marks and abnormal wear - if wear or damage is noted, the cylinder head must be replaced

8.15 Measure the outside diameter of each camshaft journal and the inside diameter of each bearing to determine the oil clearance measurement

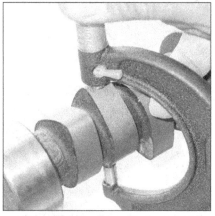

8.18 Measuring cam lobe height with a micrometer, make sure you move the micrometer to get the highest reading (top of cam lobe)

necessary (one for the left camshaft chain and one for the right camshaft chain). The wedge is used to secure the chain and the secondary tensioner in place while the camshaft is removed. **Caution:** *Failure to use a timing chain wedge will allow the secondary tensioner to over-extend and require removal of the timing chain cover to reset the tensioners.* **Caution:** *Never force the wedge past the narrowest section of the secondary timing chain as damage to the tensioner will occur.* If a timing chain wedge is not available, they may be fabricated using a block of wood that is 3/8 to 1/2-inch thick and a piece of wire to pull the wedge out of the cylinder head after installation.

7   Remove the Camshaft Position (CMP) sensor (see Chapter 6).

8   Remove the camshaft sprocket retaining bolt(s) and detach the camshaft sprocket(s) from the camshaft hub(s). Disengage the camshaft chain(s) from the sprocket(s) and remove the camshaft sprocket(s) from the engine.

9   Make note of the markings on the camshaft bearing caps. The caps are marked from 1 to 4 or 5 with arrow marks on the caps indicating the front of the engine **(see illustration)**. If both camshafts are being removed, use a permanent marker to mark each bearing cap on the right cylinder head with an "R" and each bearing cap on the left cylinder head with an "L" to indicate from which cylinder head they came from. Loosen the camshaft bearing caps bolts a little at a time beginning with the bearing caps on the ends, then working inward. **Caution:** *Keep the caps in order. They must go back in the same location they were removed from.*

10   Detach the bearing caps. **Note:** *The rocker arms may slide out of position. Mark the rocker arms so that they will be installed in the same location.*

11   Remove the camshaft(s) from the cylinder head. Mark the camshaft(s) "Left" or "Right" to indicate which cylinder head it came from.

Inspection

Refer to illustrations 8.14, 8.15 and 8.18

12   Inspect the camshaft sprockets for wear on the teeth.

13   Inspect the chains for cracks or excessive wear of the rollers. If any of the components show signs of excessive wear they must be replaced.

14   Visually check the camshaft bearing surfaces on the cylinder head(s) for pitting, score marks, galling and abnormal wear. If the bearing surfaces are damaged, the cylinder head may have to be replaced **(see illustration)**.

15   Measure the outside diameter of each camshaft bearing journal and record your measurements **(see illustration)**. Compare them to the journal outside diameter specified in this Chapter, then measure the inside diameter of each corresponding camshaft bearing and record the measurements. Subtract each cam journal outside diameter from its respec-

tive cam bearing bore inside diameter to determine the oil clearance for each bearing. Compare the results to the specified journal-to-bearing clearance. If any of the measurements fall outside the standard specified wear limits in this Chapter, either the camshaft or the cylinder head, or both, must be replaced.

16   Check camshaft runout by placing the camshaft back into the cylinder head and set up a dial indicator on the center journal. Zero the dial indicator. Turn the camshaft slowly and note the dial indicator readings. Runout should not exceed 0.0010 inch. If the measured runout exceeds the specified runout, replace the camshaft.

17   Check the camshaft endplay by placing a dial indicator with the stem in line with the camshaft and touching the snout. Push the camshaft all the way to the rear and zero the dial indicator. Next, pry the camshaft to the front as far as possible and check the reading on the dial indicator. The distance it moves is the endplay. If it's greater than the Specifications listed in this Chapter, check the bearing caps for wear. If the bearing caps are worn, the cylinder head must be replaced.

18   Compare the camshaft lobe height by measuring each lobe with a micrometer **(see illustration)**. Measure each of the intake lobes and write the measurements and relative positions down on a piece of paper. Then measure of each of the exhaust lobes and record the measurements and relative positions also. This will let you compare all of the intake lobes to one another and all of the exhaust lobes to one another. If the difference between the lobes exceeds 0.005 inch the camshaft should be replaced. Do not compare intake lobe heights to exhaust lobe heights, as lobe lift may be different. Only compare intake lobes to intake lobes and exhaust lobes to exhaust lobes for this comparison.

Installation

19   Apply moly-based engine assembly lubricant to the camshaft lobes and journals

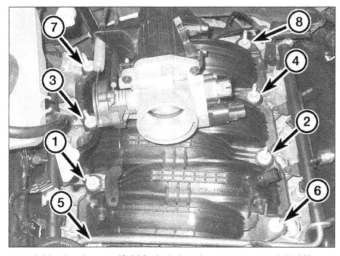

9.20a Intake manifold bolt tightening sequence - 3.7L V6

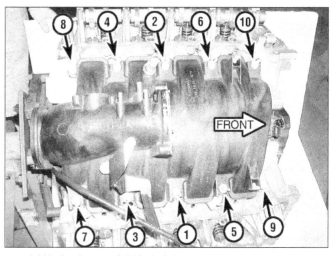

9.20b Intake manifold bolt tightening sequence - 4.7L V8

and install the camshaft(s) into the cylinder head with the dowel pins in the 10 o'clock position. If the old camshafts are being used, make sure they're installed in the exact location from which they came.

20 Install the bearing caps and bolts and tighten them hand tight.

21 Tighten the bearing cap bolts a little at a time, to the torque listed in this Chapter's Specifications, starting with the middle bolts and working outward.

22 Engage the camshaft sprocket teeth with the camshaft drive chain links so that the "V6" or "V8" mark on the sprocket(s) is between the two marks made in Step 5 during removal, then position the sprocket over the dowel on the camshaft hub. At this point the chain marks and the "V6" or "V8" marks on the sprocket should be pointing up in the 12 o'clock position.

23 Thoroughly clean the camshaft sprocket bolts. Make sure all oil is removed from the bolt threads before installation, as over-tightening of bolts may occur if oil is not removed, then lubricate the bolt washers with small amount of clean engine oil making sure not to get oil on the threads.

24 Install the camshaft sprocket bolts and tighten them to the torque listed in this Chapter's Specifications **(see illustration 7.48)**.

25 Remove the timing chain wedge(s).

26 Install the CMP sensor (see Chapter 6)

27 Install the valve covers (see Section 5).

28 Connect the cable to the negative terminal of the battery.

---

9 Intake manifold - removal and installation

Warning: *The engine must be completely cool before beginning this procedure.*

Removal

1 Relieve the fuel pressure (see Chapter 4A).

2 Disconnect the cable from the negative terminal of the battery (see Chapter 5).

3 Remove the air intake duct from the throttle body (see Chapter 4A).

4 Label and disconnect the vacuum hoses leading to the intake manifold for the PCV valve, power brake booster, cruise control servo (if used) and evaporative emission control system.

5 Disconnect all of the electrical connectors leading to the intake manifold for the various sensors, the ignition coils and the Idle Air Control (IAC) motor (if used).

6 Disconnect the electrical connectors for the alternator and the air conditioning compressor.

7 Unbolt the ground straps attached to the intake manifold and the throttle body.

8 Remove the upper dipstick tube mounting bolt.

9 On early models, detach the throttle cable and the cruise control cable from the throttle body and the throttle cable bracket. Remove the throttle body (see Chapter 4A).

10 Remove the ignition coils (see Chapter 5).

11 Remove the fuel rails, then remove the throttle body and mounting bracket (see Chapter 4A).

12 Drain some coolant from the cooling system (see Chapter 1), then remove the heater hoses from the front cover and heater core tubes.

13 On V8 engines, remove the coolant temperature sensor (see Chapter 6). This step is necessary to allow clearance for the intake manifold as it is removed.

14 Remove the intake manifold mounting fasteners in the reverse order of the tightening sequence **(see illustration 9.20a or 9.20b)**.

15 The manifold will probably be stuck to the cylinder heads and force may be required to break the gasket seal. **Caution:** *Don't pry between the manifold and the heads or damage to the gasket sealing surfaces may occur, leading to vacuum or coolant leaks.*

16 Label and detach any remaining hoses or wiring that would interfere with the removal of the intake manifold.

17 Lift the manifold up level with the vehicle and remove it from the engine.

Installation

Refer to illustrations 9.20a and 9.20b

18 Clean and inspect the intake manifold-to-cylinder head sealing surfaces. Inspect the gaskets on the manifold for tears or cracks, replacing them if necessary. The gaskets can be reused if not damaged.

19 Position the manifold on the engine, making sure the gaskets and manifold are aligned correctly over the cylinder heads, then install the intake manifold bolts hand tight.

20 Following the recommended tightening sequence, tighten the bolts to the torque listed in this Chapter's Specifications **(see illustrations)**.

21 The remainder of installation is the reverse of removal.

22 Fill the cooling system, run the engine and check for fuel, vacuum and coolant leaks.

---

10 Exhaust manifold - removal and installation

Warning: *The engine must be completely cool before beginning this procedure.*

Removal

1 Disconnect the cable from the negative terminal of the battery (see Chapter 5).

2 Block the rear wheels, set the parking brake, raise the front of the vehicle and support it securely on jackstands.

3 Unbolt the exhaust pipe from the exhaust manifold. **Note:** *On V8 models, you must soak the nuts for at least five minutes with penetrating oil, then replace the fasteners with new ones.*

Right side exhaust manifold, V8 models

4   Remove the drivebelt (see Chapter 1).
5   Drain the engine coolant below the level of the heater hoses (see Chapter 1).
6   Remove the air inlet duct and air filter housing (see Chapter 4A).
7   Remove the air conditioning compressor (see Chapter 3).
8   Remove the air conditioning accumulator bracket fastener.
9   Disconnect the heater hoses from the engine (see Chapter 3).
10   Remove the starter (see Chapter 5).

Either manifold, all models

11   Remove the exhaust manifold heat shields.
12   Remove the exhaust manifold mounting bolts and remove the manifold from the vehicle. Lower the exhaust manifold and remove it from under the vehicle.

Installation

13   Clean the mating surfaces to remove all traces of old gasket material, then inspect the manifold for distortion and cracks. Warpage can be checked with a precision straightedge held against the mating flange. If a feeler gauge thicker than 0.030-inch can be inserted between the straightedge and flange surface, take the manifold to an automotive machine shop for resurfacing.
14   Place the exhaust manifold in position with a new gasket and install the mounting bolts finger tight.
15   Starting in the middle and working out toward the ends, tighten the mounting bolts in several increments, to the torque listed in this Chapter's Specifications. Tighten the heat shield bolts to the torque listed in this Chapter's Specifications.
16   Install the remaining components in the reverse order of removal. If you're working on a right-side exhaust manifold on a V8 engine, refill the cooling system (see Chapter 1).
17   Start the engine and check for exhaust leaks between the manifold and cylinder head and between the manifold and exhaust pipe.

---

11   Cylinder head - removal and installation

Warning: *The engine must be completely cool before beginning this procedure.*
Note: *The following procedure describes how to remove the cylinder heads with the camshaft(s) and the exhaust manifold(s) still attached to the cylinder head.*

Removal

1   Remove the intake manifold (see Section 9).
2   Remove the timing chains, sprockets and the timing chain guides (see Section 7).
3   Raise the front of the vehicle and sup-

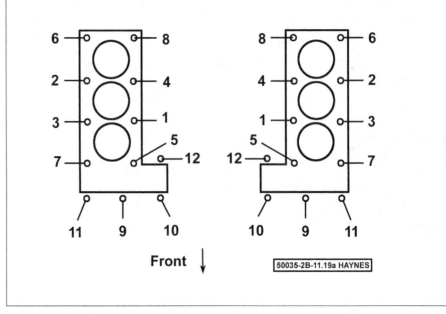

11.19a Cylinder head bolt tightening sequence - 3.7L V6

port it securely on jackstands.
4   Unbolt the exhaust pipes from the exhaust manifolds.
5   Label and remove any remaining items attached to the cylinder head, such as coolant fittings, ground straps, cables, hoses, wires or brackets.
6   Using a breaker bar and the appropriate sized socket, loosen the cylinder head bolts in 1/4-turn increments until they can be removed by hand. Loosen the bolts in the reverse order of the tightening sequence **(see illustration 11.19a or 11.19b)** to avoid warping or cracking the head.
7   Lift the cylinder head off the engine block with the camshaft in place and the exhaust manifold attached. If it's stuck, very carefully pry up at the front end of the cylinder head, beyond the gasket surface, at a casting protrusion.
8   Remove all external components from the head to allow for thorough cleaning and inspection.

Installation

Refer to illustrations 11.19a, 11.19b and 11.19c

9   The mating surfaces of the cylinder head and block must be perfectly clean when the head is installed.
10   Use a gasket scraper to remove all traces of carbon and old gasket material from the cylinder head and engine block being careful not to gouge the aluminum, then clean the mating surfaces with lacquer thinner or acetone. If there's oil on the mating surfaces when the head is installed, the gasket may not seal correctly and leaks could develop. When working on the block, stuff the cylinders with clean shop rags to keep out debris. Use

a vacuum cleaner to remove material that falls into the cylinders.
11   Check the block and head mating surfaces for nicks, deep scratches and other damage. If damage is slight, it can be removed with a file; if it's excessive, machining may be the only alternative.
12   Use a tap of the correct size to chase the threads in the head bolt holes, then clean the holes with compressed air - make sure that nothing remains in the holes. **Warning:** *Wear eye protection when using compressed air!*
13   With a straight-edge, check each cylinder head bolt for stretching. If all threads do not contact the straight-edge, replace the bolt.
14   Check the cylinder head for warpage. Check the head gasket, intake and exhaust manifold surfaces. Consult with an automotive machine shop.
15   Install any components that were removed from the head such as the lash adjusters, the exhaust manifold and the camshaft back onto the cylinder head.
16   Position the new cylinder head gasket over the dowel pins on the block noting which direction on the gasket faces up.
17   Carefully set the head over the dowels on the block without disturbing the gasket.
18   Before installing the M10 head bolts, apply a small amount of clean engine oil to the threads and hardened washers (if equipped). The chamfered side of the washers must face the bolt heads. Before installing the M8 head bolts, apply a small amount of thread sealant to the bolt threads.
19   Install the bolts in their original locations and tighten them finger tight. Then tighten them, following the proper sequence, to the torque and angle of rotation listed in this Chapter's Specifications **(see illustrations)**.

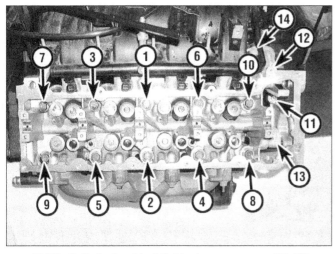

11.19b Cylinder head bolt tightening sequence - 4.7L V8

11.19c Using an angle measurement gauge during the final stages of tightening

20 Install the timing chain guides, the timing chains and the timing chain sprockets (see Section 7). The remainder of installation is the reverse of removal.
21 Refill the cooling system and change the engine oil and filter (see Chapter 1).
22 Start the engine and check for oil and coolant leaks.

12 Crankshaft pulley/vibration damper - removal and installation

Refer to illustrations 12.4 and 12.5

1 Disconnect the cable from the negative terminal of the battery (see Chapter 5).
2 Drain some coolant (see Chapter 1), then remove the upper radiator hose. Remove the cooling fans and shroud together as an assembly (see Chapter 3). **Caution:** *Store the*

fan assembly vertically. If it's horizontal, the fluid can seep into the bearing and damage it.
3 Remove the drivebelt (see Chapter 1).
4 Use a strap wrench around the crankshaft pulley to hold it while using a breaker bar and socket to remove the crankshaft pulley center bolt **(see illustration)**.
5 Pull the damper off the crankshaft with a three-hook puller **(see illustration)**. **Caution:** *The hooks of the puller must only contact the hub of the pulley - not the outer ring. Also, the puller screw must not contact the threads in the nose of the crankshaft; it must either bear on the end of the crankshaft nose or a spacer must be inserted into the nose of the crankshaft to protect the threads.*
6 Check the surface on the pulley hub that the oil seal rides on. If the surface has been grooved from long-time contact with the seal, replace the pulley.
7 Lubricate the pulley hub with clean

engine oil. Align the slot in the pulley with the key on the crankshaft and push the crankshaft pulley on the crankshaft as far as it will go. Use a vibration damper installation tool to press the pulley the rest of the way onto the crankshaft.
8 Install the crankshaft pulley retaining bolt and tighten it to the torque listed in this Chapter's Specifications.
9 The remainder of installation is the reverse of removal.

13 Crankshaft front oil seal - replacement

Refer to illustrations 13.2 and 13.4

1 Remove the crankshaft pulley from the engine (see Section 12).
2 Carefully pry the seal out of the cover with a seal removal tool or a large screwdriver

12.4 Use a strap wrench to hold the crankshaft pulley while removing the center bolt (a chain-type wrench may be used if you wrap a section of old drivebelt or a rag around the crankshaft pulley first)

12.5 The use of a three jaw puller will be necessary to remove the crankshaft pulley - always place the puller jaws around the pulley hub, not the outer ring

13.2 Pry the seal out very carefully with a seal removal tool or screwdriver, being careful not to nick or gouge the seal bore or the crankshaft

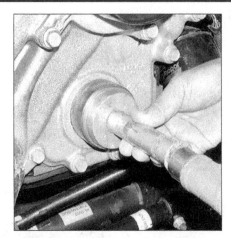

13.4 Use a seal driver or large-diameter socket to drive the new seal into the cover

Installation

11   Clean the oil pan with solvent and remove any gasket material from the block and the pan mating surfaces. Clean the mating surfaces with lacquer thinner or acetone and make sure the bolt holes in the block are clear. Check the oil pan flange for distortion, particularly around the bolt holes.

12   Inspect the oil pan gasket for cuts and tears, replacing it if necessary. If the gasket is in good condition, it can be reused. **Note:** *The oil pan gasket and the windage tray are a one piece design, therefore must be replaced together.*

13   Place the pick-up tube and oil pan gasket/windage tray in the oil pan and position the oil pan on the engine. **Note:** *Always use a new O-ring on the pick-up tube and tighten the pick-up tube-to-oil pump bolt first.* Tighten the oil pump pick-up tube bolt to the torque listed in this Chapter's Specifications.

14   After the oil pan bolts are started, them in several steps in a criss-cross pattern starting from the center and working out to the ends. Tighten them to the torque listed in this Chapter's Specifications.

15   The remainder of installation is the reverse of removal.

16   Refill the engine with oil (see Chapter 1) and replace the oil filter. Run the engine until normal operating temperature is reached and check for leaks.

(see illustration). Caution: *Do not scratch, gouge or distort the area that the seal fits into or an oil leak will develop.*

3   Clean the bore to remove any old seal material and corrosion. Position the new seal in the bore with the seal lip (usually the side with the spring) facing IN (toward the engine). A small amount of oil applied to the outer edge of the new seal will make installation easier.

4   Drive the seal into the bore with a seal driver or a large socket and hammer until it's completely seated **(see illustration)**. Select a socket that's the same outside diameter as the seal and make sure the new seal is pressed into place until it bottoms against the cover flange.

5   Lubricate the seal lips with engine oil and reinstall the crankshaft pulley.

6   The remainder of installation is the reverse of removal. Run the engine and check for oil leaks.

14   Oil pan - removal and installation

Removal

Refer to illustration 14.9

1   Disconnect the cable from the negative

terminal of the battery (see Chapter 5).

2   Apply the parking brake and block the rear wheels. Raise the front of the vehicle and place it securely on jackstands.

3   Drain the engine oil (see Chapter 1).

4   If equipped, remove the skidplate.

5   Remove the transmission oil cooler lines.

6   Remove the front crossmember.

7   Attach an engine hoist or an engine support fixture to the lifting eyes on the engine, raise the engine just enough to take the weight of the engine off the engine mounts, then remove the engine mount through-bolts (see Section 18).

8   Raise the engine until the engine fan is almost in contact with the fan shroud. **Caution:** *Don't allow the fan to contact the fan shroud.*

9   Remove the bolts and nuts, noting the stud locations, then carefully separate the oil pan from the block. Don't pry between the block and the pan or damage to the sealing surfaces and gasket could occur and oil leaks may develop. Instead, tap on the side of the oil pan with a rubber mallet if necessary to break the gasket seal **(see illustration)**.

10   Remove the two nuts and one bolt that secures the oil pump pick-up tube and windage tray. Drop the pick-up tube into the oil pan, then remove the pick-up tube, windage tray and oil pan as a unit.

15   Oil pump - removal, inspection and installation

Removal

Refer to illustration 15.3

1   Remove the timing chains and sprockets (see Section 7).

2   Remove the oil pan, windage tray and pick-up tube (see Section 14).

3   Remove the primary timing chain tensioner/oil pump bolts **(see illustration)**, then remove the tensioner.

4   Remove the remaining oil pump bolts,

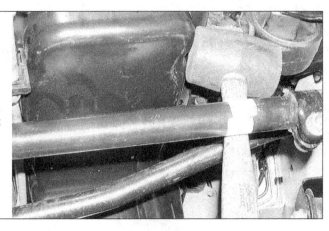

14.9 If the oil pan is stuck to the gasket, gently tap on the side of the oil pan to break the gasket seal

15.3 Oil pump housing/primary timing chain tensioner retaining bolts

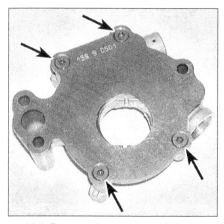

15.5 Remove the screws and lift the cover off

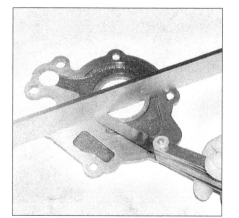

15.7a Place a straightedge across the oil pump cover and check it for warpage with a feeler gauge

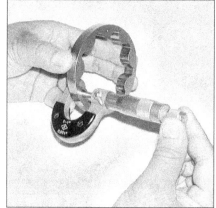

15.7b Use a micrometer or dial caliper to check the thickness and the diameter of the outer rotor

then gently pry the oil pump housing outward enough to clear the flats on the crankshaft and remove it from the engine.

Inspection

Refer to illustrations 15.5, 15.7a, 15.7b, 15.7c, 15.7d, 15.7e and 15.7f

5   Remove the screws holding the front cover on the oil pump housing **(see illustration)**.
6   Clean all components with solvent, then inspect them for wear and damage. **Caution:** *The oil pressure relief valve and spring are an integral part of the oil pump housing. Removal of the relief valve and spring from the oil pump housing will damage the oil pump and require replacement of the entire oil pump assembly.*
7   Check the clearance of the following oil pump components with a feeler gauge and a micrometer or dial caliper **(see illustrations)** and compare the measurement to the clearance specifications listed in this Chapter's Specifications.

a) *Cover flatness*
b) *Outer rotor diameter and thickness*

c) *Inner rotor thickness*
d) *Outer rotor-to-body clearance*
e) *Inner rotor-to-outer rotor tip clearance*
f) *Cover-to-inner rotor side clearance*
g) *Cover-to-outer rotor side clearance*

If any clearance is excessive, replace the entire oil pump assembly.
8   Assemble the oil pump and tighten all fasteners to the torque listed in this Chapter's Specifications. **Note:** *Fill the oil pump rotor cavities with clean engine oil or light white grease to prime it.*

Installation

9   To install the pump, turn the flats in the rotor so they align with the flats on the crankshaft and push the oil pump back into position against the block.
10   Position the primary timing chain tensioner over the oil pump and install the pump-to-block bolts. Tighten the oil pump/primary timing chain tensioner bolts to the torque listed in this Chapter's Specifications.
11   The remainder of installation is the reverse of removal.

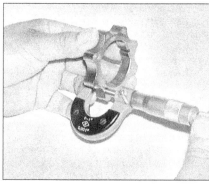

15.7c Use a micrometer or dial caliper to check the thickness of the inner rotor

16   Flywheel/driveplate - removal and installation

1   Raise the vehicle and support it securely on jackstands, then remove the transmission (see Chapter 7).
2   Now would be a good time to check and replace the transmission front pump seal on

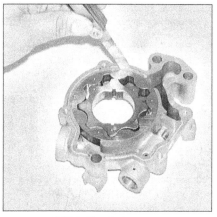

15.7d Check the outer rotor-to-housing clearance

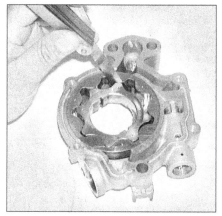

15.7e Check the clearance between the tips of the inner and outer rotors

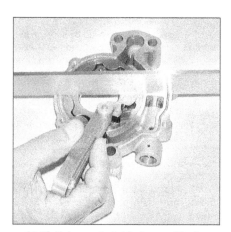

15.7f Using a straightedge and feeler gauge, check the side clearance between the surface of the oil pump and the inner and outer rotors

17.2 Pry the seal out very carefully with a seal removal tool or screwdriver - if the crankshaft is damaged, the new seal will leak!

18.11 Each engine mount is secured by the through-bolt (A) and the mount-to-engine block bolts (B)

18.18 Location of the engine mount through-bolt on a 4WD model

automatic transmissions.

3   Use paint or a center-punch to make alignment marks on the flywheel/driveplate and crankshaft to ensure correct alignment during reinstallation.

4   Remove the bolts that secure the flywheel/driveplate to the crankshaft. If the crankshaft turns, wedge a screwdriver into the ring gear teeth to keep the crankshaft from turning.

5   Pull straight back on the flywheel/driveplate to detach it from the crankshaft.

6   Installation is the reverse of removal. Align the matching paint marks. Use thread locking compound on the bolt threads and tighten them in several steps, in a criss-cross pattern, to the torque listed in this Chapter's Specifications.

---

17   Rear main oil seal - replacement

Refer to illustration 17.2

1   These models use a one-piece rear main seal that is sandwiched between the engine block and the lower main bearing cap assembly (or bed plate, as it's often referred to). Replacing this seal requires removal of the transmission (see Chapter 7) and flywheel/ driveplate (see Section 16).

2   The seal can be removed by prying it out of the engine block with a screwdriver, being careful not to nick the crankshaft surface **(see illustration)**. Wrap the screwdriver tip with tape to avoid damage.

3   Thoroughly clean the seal bore in the block with a shop towel. Remove all traces of oil and dirt.

4   Lubricate the seal lip with clean engine oil and install the seal over the end of the crankshaft. Make sure the lip of the seal points toward the engine.

5   Preferably, a seal installation tool (available at most auto parts stores) should be used to press the new seal back into place. Drive the new seal squarely into the seal bore and flush with the rear of the engine block.

6   The remainder of installation is the reverse of removal.

18   Engine mounts - check and replacement

1   There are three powertrain mounts on the vehicles covered by this manual; left and right engine mounts attached to the engine block and to the frame and a rear mount attached to the transmission and the frame. The rear transmission mount is covered in Chapter 7. Engine mounts seldom require attention, but broken or deteriorated mounts should be replaced immediately or the added strain placed on the driveline components may cause damage or wear.

Check

2   During the check, the engine must be raised slightly to remove the weight from the mounts.

3   Raise the vehicle and support it securely on jackstands, then position a jack under the engine oil pan. Place a large wood block between the jack head and the oil pan, then carefully raise the engine just enough to take the weight off the mounts. **Warning:** *DO NOT place any part of your body under the engine when it's supported only by a jack!*

4   Check for relative movement between the inner and outer portions of the mount (use a large screwdriver or prybar to attempt to move the mounts). If movement is noted, lower the engine and tighten the mount fasteners.

5   Check the mounts to see if the rubber is cracked, hardened or separated from the metal casing which would indicate a need for replacement.

Replacement

Refer to illustrations 18.11 and 18.18

6   Disconnect the cable from the negative terminal of the battery (see Chapter 5).

7   Remove the engine cooling fan and shroud (see Chapter 3). **Caution:** *Raising the engine with the cooling fan in place may damage the viscous clutch.*

8   Raise the front of the vehicle and sup-

port it securely on jackstands.

9   Support the engine with a lifting device from above. **Caution:** *Do not connect the lifting device to the intake manifold.* Raise the engine just enough to take the weight off the engine mounts. If you're removing the driver's side engine mount, removal of the oil filter will be necessary.

2WD models

10   Remove the engine mount-to-frame support bracket through-bolt.

11   Remove the mount-to-engine block bolts, then remove the mount and the heat shield, if equipped **(see illustration)**.

12   Place the heat shield and the new mount in position, install the mount-to-engine block bolts and tighten all the bolts securely.

4WD models

13   Use a floor jack and jackstands to support the front axle. The front axle will be partially lowered to make additional clearance for the engine mount removal.

14   Remove the engine skidplate.

15   Remove the front crossmember (see Chapter 10).

16   Remove the bolts that support the engine mount to the front axle.

17   Remove the bolts that attach the front axle to the engine.

18   Remove the engine mount-to-frame support bracket through-bolt **(see illustration)**.

19   Remove the engine mount-to-engine support bracket bolt and nuts.

20   Install the engine mount onto the engine support bracket and tighten the bolts securely.

All models

21   After the engine mounts have been installed onto the engine, lower the engine while guiding the engine mount and through-bolt into the frame support bracket. Install the through-bolt nut and tighten it securely.

22   The remainder of installation is the reverse of removal. Remove the engine hoist and the jackstands and lower the vehicle.

Chapter 2 Part B
Hemi engine

Contents

Specifications

General

Firing order	1-8-4-3-6-5-7-2
Bore and stroke	
2014 and earlier (non-medium duty models)	3.92 x 3.58 inches
2014 and later (medium duty models)	4.09 x 3.72 inches
Displacement	
2014 and earlier (non-medium duty models)	5.7L (345 cubic inches)
2014 and later (medium duty models)	6.4L (392 cubic inches)
Compression ratio	10.5:1
Cylinder numbers (front-to-rear)	
Left (driver's) side	1-3-5-7
Right side	2-4-6-8
Compression	See Chapter 2E

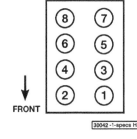

Cylinder locations for the 5.7L and 6.4L Hemi V8 engines

Camshaft

Journal diameters	
No. 1	
5.7L models	2.29 inches
6.4L models	2.67 inches
No. 2	2.28 inches
No. 3	2.26 inches
No. 4	2.24 inches
No. 5	1.72 inches
Journal oil clearance	
No. 1, 3 and 5	0.0015 to 0.0030 inch
No. 2 and 4	0.0019 to 0.0035 inch
Endplay	0.0031 to 0.0114 inch

Oil pump

Minimum pressure at curb idle	4 psi
Operating pressure...	25 to 110 psi at 3,000 rpm

Torque specifications

Ft-lbs (unless otherwise indicated)

Note: *One foot-pound (ft-lb) of torque is equivalent to 12 inch-pounds (in-lbs) of torque. Torque values below approximately 15 ft-lbs are expressed in inch-pounds, since most foot-pound torque wrenches are not accurate at these smaller values.*

Camshaft phaser (sprocket) bolt	63
Camshaft thrust plate bolts..	106 in-lbs
Cylinder head bolts **(see illustration 10.19)**	
5.7L models	
First step	
Large bolts..	25
Small bolts..	15
Second step	
Large bolts..	40
Small bolts..	25
Third step	
Large bolts..	Tighten an additional 90 degrees
Small bolts..	25
6.4L models	
Step 1, bolts 1 though 10..	25
Step 2, bolts 11 though 15..	15
Step 3, bolts 1 though 10..	40
Step 4, bolts 11 though 15..	15
Step 5, bolts 1 though 10..	45
Step 6, bolts 1 through 10..	Tighten an additional 95-degrees
Step 7, bolts 11 though 15..	25
Drivebelt tensioner mounting bolt	30
Drivebelt idler pulley-to-block bolt..	40
Exhaust manifold bolts/nuts..	18
Exhaust manifold heat shield nuts..	132 in-lbs
Exhaust pipe flange nuts	24
Driveplate bolts..	70
Flywheel bolts..	55
Intake manifold bolts...	108 in-lbs
Lifter rail mounting bolts	108 in-lbs
Oil pan bolts/studs...	108 in-lbs
Oil pump pick-up tube bolts...	21
Oil pump mounting bolts..	21
Rear main seal retainer bolts...	132 in-lbs
Rocker arm bolts	16
Rocker arm lifter rail bolts...	106 in-lbs
Timing chain cover bolts..	21
Timing chain cover lifting eye bolt	41
Valve cover nuts/studs..	71 in-lbs
Vibration damper-to-crankshaft bolt	130
Water pump-to-timing chain cover bolts	21

3.4 Use a breaker bar and deep socket to rotate the crankshaft

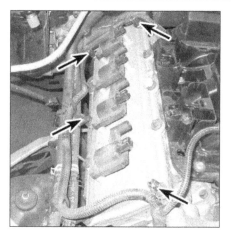

4.2a Location of the harness clips on the right side valve cover

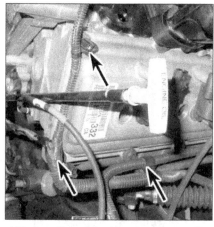

4.2b Location of the harness clips on the left side valve cover - some hidden from view

1 General information

This Part of Chapter 2 is devoted to in-vehicle repair procedures for Hemi engines. All information concerning engine removal and installation and engine overhaul can be found in Part D of this Chapter.

Since the repair procedures included in this Part are based on the assumption that the engine is still installed in the vehicle, if they are being used during a complete engine overhaul (with the engine already out of the vehicle and on a stand) many of the steps included here will not apply.

These Hemi engines use a cast iron engine block with aluminum cylinder heads and intake manifold. The cylinder banks are positioned at a 90-degree angle with the camshaft mounted high in the engine block. The valvetrain includes roller lifters, with pushrods positioned between the camshaft and rocker arms in a nearly horizontal plane. The camshaft can only be removed after the cylinder heads and valvetrain components (rocker arms, pushrods and lifters) have been removed.

2 Repair operations possible with the engine in the vehicle

Many major repair operations can be accomplished without removing the engine from the vehicle.

Clean the engine compartment and the exterior of the engine with some type of pressure washer before any work is done. A clean engine will make the job easier and will help keep dirt out of the internal areas of the engine.

Depending on the components involved, it may be a good idea to remove the hood to improve access to the engine as repairs are

performed (refer to Chapter 11 if necessary).

If oil or coolant leaks develop, indicating a need for gasket or seal replacement, the repairs can generally be made with the engine in the vehicle. The oil pan gasket, the cylinder head gaskets, intake and exhaust manifold gaskets, timing chain cover gaskets and the crankshaft front oil seal are all accessible with the engine in place.

Exterior engine components, such as the water pump, the starter motor, the alternator and the fuel injection components, as well as the intake and exhaust manifolds, can be removed for repair with the engine in place.

Since the cylinder heads can be removed without removing the engine, valve component servicing can also be accomplished with the engine in the vehicle.

Replacement of, repairs to or inspection of the timing chain and sprockets and the oil pump are all possible with the engine in place.

3 Top Dead Center (TDC) for number one piston - locating

Refer to illustration 3.4

1   Top Dead Center (TDC) is the highest point in the cylinder that each piston reaches as it travels up-and-down when the crankshaft turns. Each piston reaches TDC on the compression stroke and again on the exhaust stroke, but TDC generally refers to piston position on the compression stroke.

2   In order to bring any piston to TDC, the crankshaft must be turned using a breaker bar and socket on the vibration damper bolt. When looking at the front of the engine, normal crankshaft rotation is clockwise.

3   Disconnect the cable from the negative terminal of the battery (see Chapter 5).

4   Remove one of the number 1 spark plugs, preferably the closest to the front of the

engine, then thread a compression gauge into the spark plug hole. Turn the crankshaft with a large socket and breaker bar attached to the large bolt that is threaded into the vibration damper **(see illustration)**. When compression registers on the gauge, the number one piston is beginning its compression stroke. Stop turning the crankshaft and remove the gauge.

5   Install a long dowel into the number one spark plug hole until it rests on top of the piston crown. Continue rotating the crankshaft slowly until the dowel levels off (piston reaches top of travel). This will be *approximate* TDC for number 1 piston.

6   These engines are not equipped with external components (vibration damper, flywheel, timing hole, etc.) that are marked to identify the position of number 1 TDC. Therefore the only method to double-check the *exact* location of TDC number 1 on Hemi engines is to remove the timing chain cover to access the timing chain sprockets (see Section 9), or with the use of a degree wheel on the crankshaft vibration damper and a positive stop threaded into the spark plug hole, as you would use in the process of degreeing a camshaft (this procedure is described in detail in the Haynes *Chrysler Engine Overhaul Manual*).

4 Valve covers - removal and installation

Removal

Refer to illustrations 4.2a, 4.2b, 4.3a and 4.3b

1   Disconnect the cable from the negative terminal of the battery (see Chapter 5). Remove the engine cover.

2   Remove the ignition coils (see Chapter 5) and position the wiring harnesses aside **(see illustrations)**.

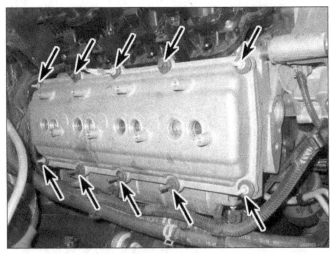

4.3a Location of the valve cover mounting bolts on the right side valve cover

4.3b Location of the valve cover mounting bolts on the left side valve cover

3   Remove the valve cover mounting bolts **(see illustrations)**. Start with the center bolts and move to the outer bolts using a criss-cross pattern.

4   Remove the valve cover. **Note:** *If the cover is stuck to the head, bump the cover with a block of wood and a hammer to release it. If it still will not come loose, try to slip a flexible putty knife between the head and cover to break the seal. Don't pry at the cover-to-head joint, as damage to the sealing surface and cover flange will result and oil leaks will develop.*

Installation

Refer to illustration 4.6

5   The mating surfaces of each cylinder head and valve cover must be perfectly clean when the covers are installed. Wipe the mating surfaces with a cloth saturated with brake cleaner. If there is sealant or oil on the mating

surfaces when the cover is installed, oil leaks may develop.

6   If the valve cover gasket isn't damaged or hardened, it can be re-used. If it is in need of replacement, install a new one into the valve cover perimeter and new rubber seals into the grooves in the valve cover that seal the spark plug tubes **(see illustration)**.

7   Carefully position the cover on the head and install the bolts, making sure the bolts with the studs are in the proper locations.

8   Tighten the bolts in three steps to the torque listed in this Chapter's Specifications. Start with the middle bolts and move to the outer bolts using a criss-cross pattern. **Caution:** *DON'T over-tighten the valve cover bolts.*

9   The remainder of installation is the reverse of removal.

10   Start the engine and check carefully for oil leaks as the engine warms up.

5 Rocker arms and pushrods - removal, inspection and installation

Removal

Refer to illustrations 5.2 and 5.3

1   Remove the valve covers from the cylinder heads (see Section 4).

2   If available, install special pushrod retainer tool #9070 (or equivalent), then loosen the rocker arm shaft bolts a little at a time, starting with the center bolts and working toward the outer bolts. When the bolts have been completely loosened, lift the rocker shaft assembly off the cylinder head **(see illustration)**. Don't remove the retainers (under the bolt heads) from the rocker shafts.

Note: *If special pushrod retainer tool #9070 is not available, the rocker arm shaft bolts must*

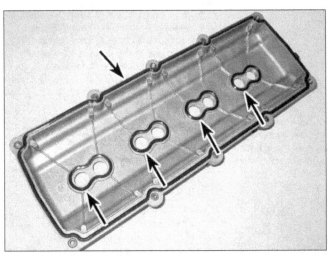

4.6 If they're damaged or hardened, replace the rubber gasket and spark plug tube seals with new ones (if they're OK, they can be re-used)

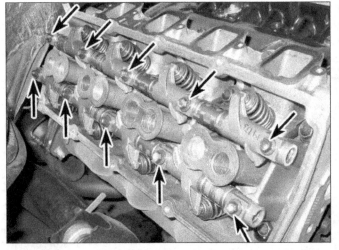

5.2 Remove the rocker arm shaft bolts - mark each rocker arm assembly and note that the upper rocker arms are the intake and the lower rocker arms are the exhaust

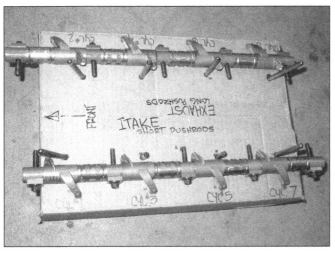

5.3 A perforated sheet of cardboard can be used to store the rocker arms and pushrods to ensure that they're reinstalled in their original locations - note the arrow indicating the front of the engine

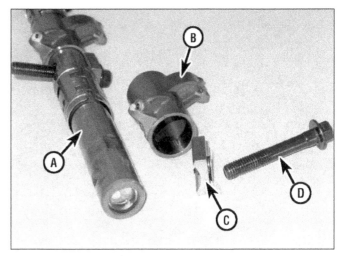

5.4 Rocker arm component details

A	Rocker arm shaft	C	Retainer
B	Rocker arm	D	Rocker shaft bolt

be loosened no more than 1/4-turn at a time.

3   Keep track of the rocker arm positions, since they must be returned to the same locations. **Note:** *The intake rockers are marked with an "I."* Store each set of rocker components in such a way as to ensure that they're reinstalled in their original locations. Remove the pushrods and store them in order as well, to make sure they don't get mixed up during installation **(see illustration)**. **Note:** *The exhaust pushrods are slightly longer than the intake pushrods - they aren't interchangeable.*

Inspection

Refer to illustration 5.4

Caution: *If the cylinder heads have been removed and the surface milled for flatness, install correct length pushrods to compensate for the reduced distances of the rocker arms-to-camshaft dimensions. Consult with a machine shop for the correct length pushrods.*

4   Check each rocker arm for wear, cracks and other damage **(see illustration)**, especially where the pushrods and valve stems contact the rocker arm.

5   Check the rocker arm shafts and bores of the rocker arms for wear. Look for galling, stress cracks and unusual wear patterns. If the rocker arms are worn or damaged, replace them with new ones and install new shafts as well. **Caution:** *Do not remove the rocker arm retainers unless absolutely necessary. Each retainer has tangs at the bottom that can easily break off and get into the engine. If a retainer does break off, retrieve it from the rocker arm shaft prior to reassembly. Replace any retainers with broken tangs.* **Note:** *Keep in mind that there is no valve adjustment on these engines, so excessive wear or damage*

in the valve train can easily result in excessive valve clearance, which in turn will cause valve noise when the engine is running.

6   Make sure the hole at the pushrod end of each rocker arm is open.

7   Inspect the pushrods for cracks and excessive wear at the ends. Roll each pushrod across a piece of plate glass to see if it's bent (if it wobbles, it's bent).

Installation

8   Lubricate the lower end of each pushrod with clean engine oil or engine assembly lube and install them in their original locations. Make sure each pushrod seats completely in the lifter socket.

9   Apply engine assembly lube to the ends of the valve stems and the upper ends of the pushrods to prevent damage to the mating surfaces on initial start-up.

10   Lubricate the rocker shafts with clean engine oil or engine assembly lube, then assemble the rocker shafts, with all of the components in their original positions. Install the rocker shafts onto the cylinder heads.

11   The rocker arm shafts must be tightened starting with the center bolt, then the center right bolt, the center left bolt, the outer right bolt and finally the outer left bolt. Follow this sequence in several steps until the torque listed in this Chapter's Specifications is reached. As the bolts are tightened, make sure the pushrods seat properly in the rocker arms. **Caution:** *Do not continue tightening the rocker arms if the rocker arm bolts become tight before the shaft is seated or the pushrods are binding. Remove the rocker arm shafts and inspect all the components carefully before proceeding.*

12   Install the valve covers (see Section 4). Start the engine, listen for unusual valve train noses and check for oil leaks at the valve cover gaskets.

---

6   Intake manifold - removal and installation

Removal

Refer to illustration 6.11

1   Relieve the fuel system pressure (see Chapter 4A). Disconnect the cable from the negative terminal of the battery (see Chapter 5).

2   Remove the engine cover by pulling up at the front, then sliding it forward.

3   Remove the air filter housing and the intake air duct (see Chapter 4A).

4   Disconnect the brake booster vacuum hose and the PCV hose.

5   Disconnect the electrical connectors at each injector, remove the fuel rail and injectors, and disconnect all connections from the throttle body (see Chapter 4A).

6   If necessary for clearance on some models, remove the air conditioning compressor without disconnecting the refrigerant lines, and set it aside (see Chapter 3).

7   Disconnect the electrical connectors from all sensors that would interfere (see Chapter 6).

8   Remove the wiring harness support from the rear of the manifold. Also remove the air conditioning line support clip and move the line aside.

9   Remove the alternator and set it aside.

10   Remove the throttle body bracket if installed.

11   Loosen the intake manifold mounting

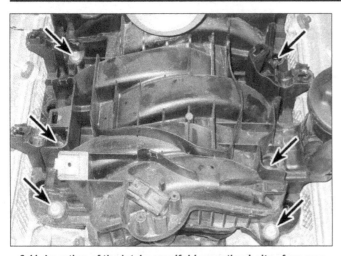

6.11 Location of the intake manifold mounting bolts - four rear mounting bolts hidden from view

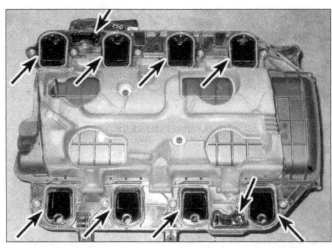

6.14 If necessary, replace the intake manifold O-rings with new ones. Make sure they seat properly in their grooves

bolts in 1/4-turn increments until they can be removed by hand **(see illustration)**. Work in the same pattern as the tightening sequence **(see illustration 6.16)**.

12   Remove the intake manifold by sliding it forward, then disconnect the MAP sensor and any other connectors.

13   As the manifold is lifted from the engine, check for and disconnect anything still attached. The throttle body can be removed now if desired.

Installation

Refer to illustrations 6.14 and 6.16

Note: *The mating surfaces of the cylinder heads and intake manifold must be perfectly clean when the manifold is installed.*

14   Check the O-rings on the intake manifold for damage or hardening. If they're OK, they can be re-used. If necessary, install new O-rings **(see illustration)**.

15   Carefully set the manifold in place. **Caution:** *Do not disturb the rubber seals and DO*

NOT move the manifold fore and aft after it contacts the cylinder heads; the seals could be pushed out of place and the engine may develop vacuum and/or oil leaks.

16   Install the intake manifold bolts and tighten the bolts following the correct sequence **(see illustration)**, to the torque listed in this Chapter's Specifications. Do not overtighten the bolts.

17   The remainder of installation is the reverse of removal. Start the engine and check carefully for vacuum leaks at the intake manifold joints.

7   Exhaust manifolds - removal and installation

Removal

Refer to illustrations 7.6, 7.7a, and 7.7b

Warning: *Allow the engine to cool completely before performing this procedure.*

1   Disconnect the cable from the nega-

tive terminal of the battery (see Chapter 5). Remove the engine cover.

2   If you're removing the right-side manifold, remove the air filter housing (see Chapter 4).

3   Raise the vehicle and support it securely on jackstands. Disconnect the exhaust pipe-to-manifold connections. It's a good idea to apply penetrating oil on the studs/bolts and let it soak in for about 10 minutes before attempting to remove them. **Note:** *It may be easier to disconnect the oxygen sensors and remove the front exhaust pipes completely.*

4   Attach an engine support fixture or an engine hoist to the engine lift points with a chain.

5   Remove the engine mount through-bolts, then raise the engine enough to get access to the exhaust manifold bolts.

6   Lower the vehicle and remove the heat shield nuts and the heat shield **(see illustration)**.

7   Remove the bolts retaining the exhaust manifold to the cylinder head starting with the center ones and working toward the outer

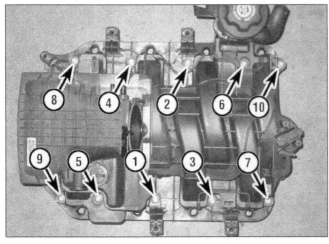

6.16 Tightening sequence for the intake manifold mounting bolts

7.6 Location of the exhaust manifold heat shield nuts (right-side)

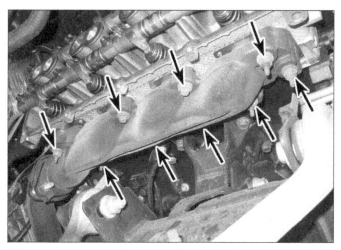

7.7a Location of the exhaust manifold mounting bolts (right-side)

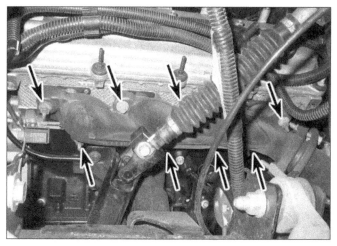

7.7b Location of the exhaust manifold mounting bolts (left-side - not all are visible in this photo)

8.4 Use a chain wrench to lock the vibration damper - position a piece of an old timing belt or drivebelt underneath the chain wrench to prevent damaging the damper

8.5 Use a three-jaw puller to remove the vibration damper from the crankshaft

8.6 Use a seal puller to remove the front seal from the timing chain cover

bolts **(see illustrations). Caution:** *Organize the bolts so they can be installed in their original positions. Some of the bolts are stainless steel alloy and should only be in certain locations.*

8 Remove the exhaust manifold and the gasket.

Installation

9 Installation is the reverse of the removal procedure. Clean the manifold and head gasket surfaces and check for cracks and flatness. Replace the exhaust manifold gaskets.
10 Install the exhaust manifold(s) and fasteners. Tighten the bolts/nuts to the torque listed in this Chapter's Specifications. Start with the center bolts and proceed to the outer ones. Approach the final torque in three steps. Install the heat shields.
11 Apply anti-seize compound to the exhaust manifold-to-exhaust pipe bolts and tighten them securely.

8 Vibration damper and front oil seal - removal and installation

Refer to illustrations 8.4, 8.5, 8.6 and 8.8
Warning: *Wait until the engine is completely cool before beginning this procedure.*
1 Disconnect the cable from the negative terminal of the battery (see Chapter 5).
2 Remove the engine drivebelt (see Chapter 1).
3 Drain the coolant (see Chapter 1), remove the upper radiator hose and remove the fan shroud (see Chapter 3).
4 Remove the large vibration damper-to-crankshaft bolt. To keep the crankshaft from turning, install a chain wrench around the circumference of the pulley. Use a piece of rubber (old drivebelt, old timing belt, etc.) under the chain to protect the vibration damper from nicks or gouges **(see illustration).**
5 Using the proper puller (commonly available from auto parts stores), detach the vibration damper **(see illustration). Caution:** *Do*

not use a puller with jaws that grip the outer edge of the pulley. The puller must be the type that utilizes three arms to apply force to the pulley hub only. Also, the puller screw must not contact the threads in the nose of the crankshaft; it must use an adapter that allows the puller screw to apply force to the end of the crankshaft nose or a spacer must be inserted into the nose of the crankshaft to protect the threads.
6 If the seal is being removed while the cover is still attached to the engine block, carefully pry the seal out of the cover with a seal removal tool or a large screwdriver **(see illustration). Caution:** *Do not scratch, gouge or distort the area that the seal fits into or an oil leak will develop.*
7 Clean the bore to remove any old seal material and corrosion. Position the new seal in the bore with the seal lip (usually the side with the spring) facing IN (toward the engine). A small amount of oil applied to the outer edge of the new seal will make installation easier.
8 Drive the seal into the bore with a seal driver or a large socket and hammer until it's

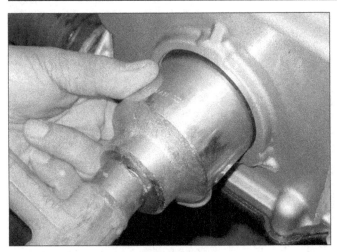

8.8 Use a seal driver or a large socket to drive the new seal into the cover

9.11 Idler pulley mounting bolt (A) and drivebelt tensioner mounting bolt (B) locations on the timing chain cover

completely seated **(see illustration)**. Select a socket that's the same outside diameter as the seal and make sure the new seal is pressed into place until it bottoms against the cover flange.

9   Check the surface of the damper that the oil seal rides on. If the surface has been grooved from long-time contact with the seal, replace the vibration damper.

10   Lubricate the seal lips with engine oil and reinstall the vibration damper, aligning the Woodruff key on the nose of the crankshaft with the keyway in the damper hub. Use a special installation tool (available at most auto parts stores) to press the vibration damper onto the crankshaft.

11   Install the vibration damper bolt and tighten it to the torque listed in this Chapter's Specifications.

12   The remainder of installation is the reverse of removal.

13   Refill the cooling system (see Chapter 1).

---

9   Timing chain cover, chain and sprockets - removal, inspection and installation

---

Warning: *Wait until the engine is completely cool before beginning this procedure.*

Removal

Refer to illustrations 9.11, 9.17, 9.19a, 9.19b, 9.19c, 9.20, 9.21a and 9.21b

Caution: *The timing system is complex. Severe engine damage will occur if you make any mistakes. Do not attempt this procedure unless you are highly experienced with this type of repair. If you are at all unsure of your abilities, consult an expert. Double-check all your work and be sure everything is correct before you attempt to start the engine.*

1   Disconnect the cable from the negative terminal of the battery (see Chapter 5).

2   Remove the engine cover.

3   Drain the cooling system (see Chapter 1).

4   Drain the engine oil (see Chapter 1).

5   Remove the air filter housing and the air intake duct (see Chapter 4).

6   Remove the drivebelt (see Chapter 1).

7   Remove the engine cooling fan, fan shroud and the upper and lower radiator hoses (see Chapter 3).

8   Remove the air conditioning compressor (see Chapter 3). **Warning:** *Do not disconnect the refrigerant lines.*

9   Use rope or wire to tie the compressor away from the front of the engine with the air conditioning lines attached.

10   Remove the coolant reservoir and the washer reservoir (see Chapter 3).

11   Remove the drivebelt tensioner and the idler pulleys **(see illustration)**.

12   Remove the vibration damper (see Section 8).

13   Remove the power steering pump (see Chapter 10) and set it aside without disconnecting the power steering lines.

14   Remove the dipstick tube support bolt.

15   Remove the oil pan and the pickup tube (see Section 12).

16   Disconnect the heater hoses from the front cover.

17   Remove the timing chain cover bolts **(see illustration)** and remove the front cover. Verify that the two upper alignment bushings remain with the cover.

18   Remove the oil pump (see Section 13).

19   Reinstall the bolt into the end of the crankshaft and turn the crankshaft until the crankshaft sprocket timing mark is at the 6 o'clock position (with the crankshaft key at the 2 o'clock) and the camshaft timing mark is at the 12 o'clock position. The crankshaft sprocket mark must be aligned with the single colored link and the camshaft sprocket/phaser mark must straddle the two colored links **(see illustrations)**.

20   Retract the tensioner until the hole in the tensioner aligns with the hole in the camshaft tensioner thrust plate **(see illustration)**. Install a suitable size drill to retain the tensioner in the retracted position. **Note:** *The timing chain tensioner is an integral component of the camshaft thrust plate. If necessary, the tensioner/*

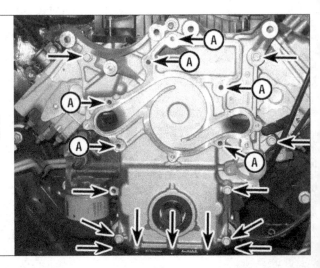

9.17 Location of the timing chain cover mounting bolts - the bolt holes marked with an (A) secure the water pump as well as the timing chain cover

9.19a Use a breaker bar and socket to rotate the crankshaft (clockwise) to TDC number 1 position

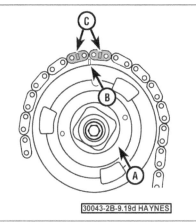

9.19b On 2009 and later models, the two colored timing chain links (C) must straddle the camshaft sprocket mark (B) on the camshaft phaser sprocket (A)

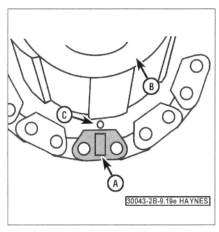

9.19c The single colored link (A) must align with the timing mark (C) on the crankshaft sprocket (B) – the keyway should be pointing to the 2 o'clock position

thrust plate assembly must be replaced as a complete unit.

21   Remove the camshaft sprocket (phaser) bolt and remove the timing chain with the camshaft phaser and the crankshaft sprocket **(see illustrations)**.

Warning: *Do not attempt to disassemble the camshaft phaser. It is a sealed unit and may result in extreme engine damage if it is disassembled.*

22   Remove the camshaft tensioner thrust plate mounting bolts **(see illustration 11.9)** and separate the tensioner from the engine block.

Inspection

23   Inspect the camshaft phase for damage or wear. It should be replaced if there is any suspicion about it.

24   Inspect the crankshaft sprocket for damage or wear. The crankshaft sprocket is a steel sprocket, but these teeth can become grooved or worn enough to cause a poor meshing of the sprocket and the chain.

Installation

Refer to illustration 9.34

Caution: *Before starting the engine, carefully rotate the crankshaft by hand through at least two full revolutions (use a socket and breaker bar on the crankshaft pulley center bolt). If you feel any resistance, STOP! There is something wrong - most likely, valves are contacting the pistons. You must find the problem before proceeding. Check your work and see if any updated repair information is available.*

Note: *Never install a new timing chain onto old sprockets.*

25   Stuff a shop rag into the opening at the front of the oil pan to keep debris out of the engine, then clean off all traces of old gasket material and sealant from the engine block. Wipe the sealing surfaces with a cloth saturated with brake cleaner.

26   If the tensioner is not compressed (retracted position), compress the tensioner until the hole aligns with the bracket hole, then insert a suitable size drill bit through both holes to keep the tensioner locked in

9.20 Use a large pair of pliers to retract the tensioner until the hole in the tensioner aligns with the hole in the bracket (thrust plate), then install a drill bit through the hole to keep it in the retracted position

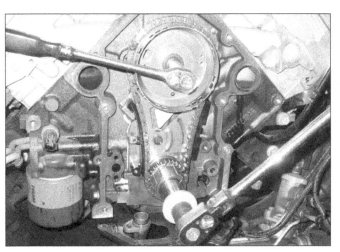

9.21a Use a breaker bar and socket to prevent the crankshaft from rotating while loosening the camshaft sprocket bolt

9.21b Remove the camshaft sprocket, crankshaft sprocket and timing chain as a complete assembly

9.34 Install a new rubber gasket into the timing chain cover groove

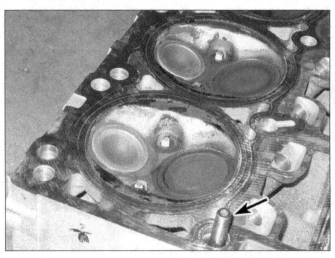

10.12 Make sure the cylinder head surface is perfectly clean, free from old gasket material, carbon deposits and dirt - note that the dowel is part of the cylinder head stand

this position **(see illustration 9.20)**. Install the camshaft thrust plate and tighten the bolts to the torque listed in this Chapter's Specifications.

27 Assemble the timing chain, phaser and sprocket before installing them onto the engine. Loop the new chain over the camshaft phaser with the two colored links straddling the camshaft sprocket/phaser mark **(see illustration 9.19b)**. Mesh the lower half of the chain with the crankshaft sprocket, with the crankshaft sprocket mark aligned with the single colored link **(see illustration 9.19c)**.

28 Align the sprocket with the Woodruff key in the end of the crankshaft and assemble the chain, camshaft phaser and crankshaft sprocket onto the crankshaft and camshaft. Tap it gently into place until it is completely seated. **Caution:** *If resistance is encountered, do not hammer the sprocket onto the crankshaft. It may eventually move onto the shaft, but it may be cracked in the process and fail later, causing extensive engine damage.* When the timing chain components are installed, the timing marks MUST align as shown in **illustrations 9.19b and 9.19c**.

29 Apply a thread locking compound to the camshaft sprocket bolt threads and tighten the bolt to the torque listed in this Chapter's Specifications.

30 Lubricate the chain with clean engine oil.

31 Remove the drill bit from the tensioner and bracket. Make sure the tensioner has released and is pressing against the timing chain. Verify that the timing marks are still aligned properly.

32 Install the oil pump (see Section 13).

33 Install the oil pump pick-up tube to the bottom of the oil pump. Tighten the bolt to the torque listed in this Chapter's Specifications.

34 The timing chain cover rubber gasket can be re-used if it isn't damaged or hard-

ened, but considering the amount of work that you've done to get to this point, it's a good idea to replace it. Double-check the surface of the timing chain cover and engine block. Make sure all old gasket material is removed and the surface is clean. Install a new rubber gasket into the groove in the timing chain cover **(see illustration)**.

35 Check for cracks and deformation of the oil pan gasket. If the gasket has deteriorated or is damaged, replace it (see Section 12).

36 Apply a small amount of RTV sealant to the corner where the timing chain cover, engine block and oil pan meet.

37 Install the timing chain cover on the block **(see illustration 9.17)** and tighten the bolts, a little at a time, until you reach the torque listed in this Chapter's Specifications. **Note:** *Install the water pump onto the timing chain cover and tighten the timing chain cover bolts and the water pump bolts at the same time (see Chapter 3).*

38 Install the oil pan-to-timing chain cover bolts, and tighten them to the torque listed in this Chapter's Specifications.

39 Lubricate the oil seal contact surface of the vibration damper hub with clean engine oil, then install the damper on the end of the crankshaft (see Section 8). Tighten the bolt to the torque listed in this Chapter's Specifications.

40 The remainder of installation is the reverse of removal.

41 Add coolant and engine oil. Run the engine and check for oil and coolant leaks.

---

10 Cylinder heads - removal and installation

Warning: *Wait until the engine is completely cool before beginning this procedure.*

Removal

1 Relieve the fuel system pressure (see Chapter 4A), then disconnect the cable from the negative terminal of the battery (see Chapter 5).

2 Drain the cooling system (see Chapter 1).

3 Remove the drivebelt (see Chapter 1).

4 Remove the PCV system and the EVAP system components (see Chapter 6). Remove the cowl (see Chapter 11).

5 Remove the intake manifold (see Section 6).

6 Remove the exhaust manifolds from the cylinder heads (see Section 7).

7 Working through the holes in the pulley, remove the three steering pump mounting bolts. Move the pump out of the way without disconnecting the hoses.

8 Remove the valve covers (see Section 4).

9 Remove the rocker arms and pushrods (see Section 5). **Caution:** *Keep all the parts in order so they are reinstalled in the same locations.*

10 Loosen the head bolts in 1/4-turn increments in the tightening sequence **(see illustration 10.19)** until they can be removed by hand. **Note:** *There will be different-length head bolts for different locations, so store the bolts in order as they are removed. This will ensure that the bolts are reinstalled in their original holes.*

11 Lift the heads off the engine. If resistance is felt, do not pry between the head and block, as damage to the mating surfaces will result. To dislodge the head, place a block of wood against the end of it and strike the wood block with a hammer, or lift on a casting protrusion. Store the heads on blocks of wood to prevent damage to the gasket sealing surfaces.

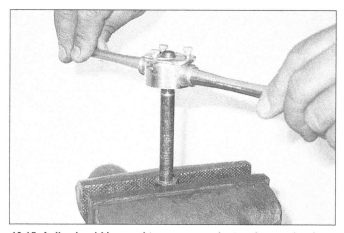

10.15 A die should be used to remove sealant and corrosion from the bolt threads prior to installation

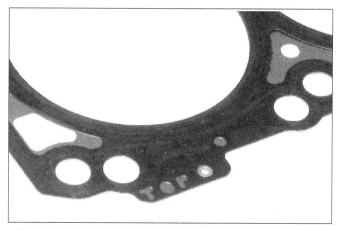

10.16 Be sure the letter designations for the left (L) and right (R) cylinder heads and the top gasket surface (TOP) are correct

Installation

Refer to illustrations 10.12, 10.15, 10.16 and 10.19

12   The mating surfaces of the cylinder heads and block must be perfectly clean when the heads are installed **(see illustration)**. Gasket removal solvents are available at auto parts stores and may prove helpful.

13   Use a gasket scraper to remove all traces of carbon and old gasket material, then wipe the mating surfaces with a cloth saturated with brake cleaner. If there is oil on the mating surfaces when the heads are installed, the gaskets may not seal correctly and leaks may develop. When working on the block, cover the lifter valley with shop rags to keep debris out of the engine. Use a vacuum cleaner to remove any debris that falls into the cylinders.

14   Check the block and head mating surfaces for nicks, deep scratches and other damage. If damage is slight, it can be removed with emery cloth. If it is excessive, machining may be the only alternative.

15   Use a tap of the correct size to chase the threads in the head bolt holes in the block. Mount each bolt in a vise and run a die down the threads to remove corrosion and restore the threads **(see illustration)**. Dirt, corrosion, sealant and damaged threads will affect torque readings.

16   Position the new gaskets over the dowels in the block. Be sure the letter designations for the left (L) and right (R) cylinder heads and the top gasket surface (TOP) are correct **(see illustration)**.

17   Carefully position the heads on the block without disturbing the gaskets.

18   Before installing the head bolts, coat the threads with a small amount of engine oil.

19   Install the bolts in their original locations and tighten them finger tight. Following the recommended sequence **(see illustration)**, tighten the bolts in several steps to the torque listed in this Chapter's Specifications.

20   The remainder of installation is the reverse of removal.

21   Add coolant and change the engine oil and filter (see Chapter 1). Start the engine and check for proper operation and coolant or oil leaks.

11   Camshaft and lifters - removal, inspection and installation

Warning: *Wait until the engine is completely cool before beginning this procedure.*

Removal

Refer to illustrations 11.6, 11.7a, 11.7b and 11.9

1   Relieve the fuel system pressure (see Chapter 4A), then disconnect the cable from the negative terminal of the battery (see Chapter 5).

2   Drain the cooling system (see Chapter 1).

3   Remove the timing chain cover, the timing chain and the phaser (see Section 9). Remove the oil pump.

4   Remove the cylinder heads (see Section 10).

5   Remove the radiator (see Chapter 3).

6   Remove the lifter rail mounting bolts **(see illustration)**.

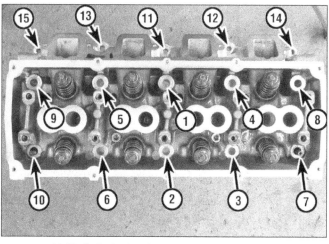

10.19 Cylinder head bolt tightening sequence

11.6 Each lifter rail is secured by one bolt

11.7a Remove the lifters from the lifter bores and install them into the lifter rail - each lifter must be installed into the same lifter bore and in the same direction (roller rotation)

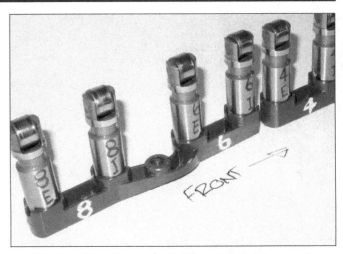

11.7b Install the lifters into the lifter rail and make sure the lifters are marked in their original position (UP) with number designations for the cylinders and letter designations for the valve each one operates (intake or exhaust)

7   Remove the lifters from the lifter bores in the engine block **(see illustration)**. On 5.7L models, carefully place them in the correct location in the lifter rail. Each lifter must be installed into the same lifter bore and in the same direction (roller rotation) **(see illustration)**. Mark each lifter and lifter rail with a felt pen to designate the cylinder number and the type of lifter (intake or exhaust). There are several ways to extract the lifters from the bores. A special tool designed to grip and remove lifters is manufactured by many tool companies and is widely available, but it may not be required in every case. On newer engines without a lot of varnish buildup, the lifters can often be removed with a small magnet or even with your fingers. A machinist's scribe with a bent end can be used to pull the lifters out by positioning the point under the retainer ring inside the top of each lifter. **Caution:** *Do not use pliers to remove the lifters unless you intend to replace them with new ones (along with the camshaft).*

The pliers will damage the precision machined and hardened lifters, rendering them useless. Do not attempt to withdraw the camshaft with the lifters in place.

8   Before removing the camshaft, check the endplay. Mount a dial indicator so that it contacts the nose of the camshaft. Pry the camshaft forward and back using a long screwdriver with the tip taped to prevent damage to the camshaft. Record the movement of the dial indicator and compare it to this Chapter's Specifications. If the endplay is excessive, the camshaft must be replaced.

9   Unbolt and remove the camshaft thrust plate **(see illustration)**.

10   Thread a long bolt into the camshaft sprocket bolt hole to use as a handle when removing the camshaft from the block. Carefully pull the camshaft out while slowly turning it. Support the cam near the block so the lobes do not nick or gouge the bearings as it is withdrawn.

Inspection

Refer to illustrations 11.12, 11.13, 11.14 and 11.18

11   After the camshaft has been removed from the engine, cleaned with solvent and dried, inspect the bearing journals for uneven wear, pitting and evidence of seizure. If the journals are damaged, the bearing inserts in the block are probably damaged as well. Both the camshaft and bearings will have to be replaced. **Note:** *Camshaft bearing replacement requires special tools and expertise that place it beyond the scope of the average home mechanic. The tools for bearing removal and installation are available at stores that carry automotive tools, possibly even found at a tool rental business. It is advisable though, if bearings are bad and the procedure is beyond your ability, remove the engine block and take it to an automotive machine shop to ensure that the job is done correctly.*

12   Measure the bearing journals with a micrometer to determine if they are excessively worn or out-of-round **(see illustration)**.

13   Measure the lobe height of each cam lobe on the intake camshaft and record your measurements **(see illustration)**. Compare the measurements for excessive variations. If the lobe heights vary more than 0.005 inch (0.125 mm), replace the camshaft. Compare the lobe height measurements on the exhaust camshaft and follow the same procedure. Do not compare intake camshaft lobe heights with exhaust camshaft lobe heights, as they are different. Only compare intake lobes with intake lobes and exhaust lobes with other exhaust lobes.

14   Check the camshaft lobes for heat discoloration, score marks, chipped areas, pitting and uneven wear **(see illustration)**. If the lobes are in good condition and if the lobe lift variation measurements recorded earlier are within the limits, the camshaft can be reused.

15   Clean the lifters with solvent and dry

11.9 Location of the camshaft thrust plate mounting bolts

11.12 Check the diameter of each camshaft bearing journal to pinpoint excessive wear and out-of-round conditions

11.13 Measure the camshaft lobe height (greatest dimension) with a micrometer

11.14 Check the cam lobes for pitting, excessive wear and scoring. If scoring is excessive, as shown here, replace the camshaft

11.18 The roller on the roller lifters must turn freely - check for wear and excessive play as well

them thoroughly without mixing them up.

16   Check each lifter wall, pushrod seat and foot for scuffing, score marks and uneven wear. If the lifter walls are damaged or worn (which is not very likely), inspect the lifter bores in the engine block as well. If the pushrod seats are worn, check the pushrod ends.

17   If new lifters are being installed, a new camshaft must also be installed. If a new camshaft is installed, then use new lifters as well. Never install used lifters unless the original camshaft is used and the lifters can be installed in their original locations.

18   Check the rollers carefully for wear and damage and make sure they turn freely without excessive play **(see illustration)**.

Installation

Refer to illustration 11.19

19   Lubricate the camshaft bearing journals and cam lobes with camshaft installation lube **(see illustration)**.

20   Slide the camshaft slowly and gently into the engine. Support the cam near the block and be careful not to scrape or nick the bearings. Only install the camshaft far enough to install the camshaft thrust plate. Pushing it in too far could dislodge the camshaft plug at the rear of the engine, causing an oil leak. Tighten the camshaft thrust plate mounting bolts to the torque listed in this Chapter's Specifications.

21   Install the timing chain and sprockets (see Section 9). Align the timing marks on the crankshaft and camshaft sprockets.

22   Lubricate the lifters with clean engine oil and install them in the block. If the original lifters are being reinstalled, return them to their original locations, and with the numbers facing UP (exactly as they were removed). Install the lifter rail and mounting bolts. Tighten the lifter rail mounting bolts to the torque listed in this Chapter's Specifications. **Note:** *Each lifter rail should be numbered and coincide with the correct cylinder numbers.*

23   The remainder of installation is the reverse of removal.

24   Change the oil and install a new oil filter (see Chapter 1). Fill the cooling system with

the proper type of coolant (see Chapter 1).

25   Start the engine and check for oil pressure and leaks. **Caution:** *Do not run the engine above a fast idle until all the hydraulic lifters have filled with oil and become quiet again.*

26   If a new camshaft and lifters have been installed, the engine should be brought to operating temperature and run at a fast idle for 15 to 20 minutes to break in the new components. Change the oil and filter again after 500 miles of operation.

12   Oil pan - removal and installation

Warning: *Wait until the engine is completely cool before beginning this procedure.*

Removal

Refer to illustrations 12.7 and 12.14

1   Disconnect the cable from the negative terminal of the battery (see Chapter 5).

2   Raise the vehicle and support it securely

11.19 Apply camshaft installation lube to the cam lobes and bearing journals before installing the camshaft

on jackstands (see Chapter 1).

3   Drain the engine oil and replace the oil filter (see Chapter 1).

4   Remove the dust cover/transmission brace.

5   Remove the fan and the fan shroud (see Chapter 3).

6   Detach the automatic transmission cooler lines from the radiator and the clips, then move them out of the way.

7   Remove the crossmember beneath the oil pan **(see illustration)**.

4WD models

8   Unbolt the steering gear and lower it down without disconnecting the fluid lines (see Chapter 10).

9   Mark the front driveshaft position, then detach it from the front axle (see Chapter 8).

10   Use jacks to securely support the front axle at each end.

11   Remove the axle mounting bolts, then slowly and carefully lower the front axle until there's sufficient working room for oil pan removal.

12.7 Remove the front crossmember mounting bolts and take the crossmember from the chassis

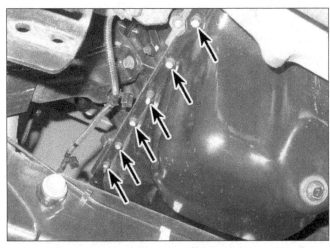

12.14 Remove the oil pan mounting bolts from the perimeter of the oil pan - left side shown

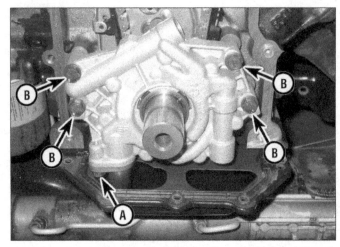

13.4 Location of the oil pump pick-up tube retainer bolt (A) and the oil pump mounting bolts (B)

All models

12   Support the engine from above with an engine hoist or an engine support fixture (see Chapter 2D), take a little weight off the engine with the hoist or fixture, and remove the through-bolts from the engine mounts (see Section 16). Now raise the engine farther.

13   Remove the transmission mount and raise the transmission for additional clearance if needed.

14   Remove the oil pan bolts **(see illustration)**, then lower the pan from the engine. The pan will probably stick to the engine, so strike the pan with a rubber mallet until it breaks the gasket seal. **Caution:** *Before using force on the oil pan, be sure all the bolts have been removed. Don't pry on the gasket - it's part of the windage tray.*

15   When you have enough room, unbolt and remove the oil pump pick-up tube.

16   Remove the windage tray and gasket assembly. Discard it.

Installation

17   Wash out the oil pan with solvent.

18   Thoroughly clean the mounting surfaces of the oil pan and engine block. Wipe the gasket surfaces clean with a rag soaked in brake cleaner.

19   Install the oil pump pick-up tube with new O-rings.

20   Apply a bead of RTV sealant to the four joints where the timing chain cover meets the block and where the rear main oil seal retainer meets the block. Replace the windage tray/gasket assembly with a new one.

21   Make sure the alignment studs are installed in the correct locations in the engine block.

22   Lift the pan into position, slipping it over the alignment studs and being careful not to disturb the windage tray/gasket, install several bolts finger tight. When all the bolts are in place, install the oil pan nuts onto the studs.

23   Tighten the fasteners to the torque listed in this Chapter's Specifications starting with

the center bolts and working outward, from side-to-side.

24   The remainder of installation is the reverse of removal.

25   Add the proper type and quantity of oil and a new oil filter (see Chapter 1), start the engine and check for leaks before placing the vehicle back in service.

13   Oil pump - removal and installation

Warning: *Wait until the engine is completely cool before beginning this procedure.*

Caution: *The oil pressure relief valve must not be removed from the oil pump housing on this engine. If the oil pressure relief valve is removed from the housing, the entire oil pump assembly must be replaced with a new unit.*

Note: *The oil pump on the Hemi engine is available only as a complete unit.*

Removal

Refer to illustration 13.4

1   Disconnect the cable from the negative terminal of the battery (see Chapter 5).

2   Remove the timing chain cover (see Section 9).

3   Remove the oil pan (see Section 12).

4   Remove the pick-up tube and the oil pump mounting bolts and detach the pump from the engine block **(see illustration)**.

Installation

5   Position the pump on the engine. Make sure the pump rotor is aligned with the crankshaft drive. Install the oil pump mounting bolts and tighten them to the torque listed in this Chapter's Specifications.

6   Install a new O-ring onto the oil pump pick-up tube, connect the tube to the pump and tighten the bolt to the torque listed in this Chapter's Specifications.

7   The remainder of installation is the reverse of removal.

8   Fill the crankcase with the proper type and quantity of engine oil, and install a new oil filter (see Chapter 1). Fill the cooling system with the proper type of coolant (see Chapter 1).

9   Run the engine and check for oil pressure and leaks.

14   Flywheel/driveplate - removal and installation

This procedure is essentially the same as for the 3.7L V6 and the 4.7L V8 engines. Refer to Chapter 2A and follow the procedure outlined there, but refer to the Specifications listed in this Chapter for the proper torque.

15   Rear main oil seal - replacement

This procedure is essentially the same as for the 3.7L V6 and the 4.7L V8 engines. Refer to Chapter 2A and follow the procedure outlined there but refer to the Specifications listed in this Chapter for the proper torque.

16   Engine mounts - check and replacement

1   Engine mounts seldom require attention, but broken or deteriorated mounts should be replaced immediately or the added strain placed on the driveline components may cause damage or wear.

Check

2   During the check, the engine must be raised slightly to remove the weight from the mounts.

3   Raise the vehicle and support it securely on jackstands, then position a jack under

the engine oil pan. Place a large wood block between the jack head and the oil pan, then carefully raise the engine just enough to take the weight off the mounts. **Warning:** *DO NOT place any part of your body under the engine when it's supported only by a jack!*

4   Check the mount insulators to see if the rubber is cracked, hardened or separated from the metal in the center of the mount.

5   Check for relative movement between the mount and the engine or frame (use a large screwdriver or prybar to attempt to move the mounts).

6   If movement is noted, lower the engine and tighten the mount fasteners.

Replacement

7   Disconnect the cable from the negative terminal of the battery (see Chapter 5). Raise the vehicle and support it securely on jackstands (if not already done). Support the engine as described in Step 3.

2WD models

Refer to illustrations 16.8a and 16.8b

8   Remove the through-bolts **(see illustrations)**, raise the engine with the jack and detach the mount from the frame bracket and engine.

9   Install the new mount, making sure it is correctly positioned in the bracket. Install the fasteners and tighten them securely.

16.8a Location of the right side engine mount through-bolt

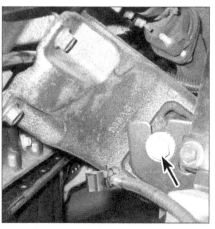

16.8b Location of the left side engine mount through-bolt

4WD models

10   Remove the skidplate from below the engine compartment.

11   Remove the engine crossmember.

12   Remove the engine oil filter (see Chapter 1).

13   Install an engine hoist (see Chapter 2D) and raise the engine slightly to take the weight off the engine mounts.

14   Use a floor jack to support the front axle.

15   Remove the engine mount through-bolts.

16   Remove the bolts that attach the engine mounts to the front axle assembly.

17   Lower the front axle slightly.

18   Remove the bolts that attach the engine mounts to the engine block.

19   Remove the engine mount from the vehicle.

20   Install the new mount, making sure it is correctly positioned in the bracket. Install the fasteners and tighten them securely.

Notes

Chapter 2 Part C
Diesel engine

Contents

Specifications

General

Displacement...	6.7L (409 cu. in.)
Cylinder numbers (front to rear)	1-2-3-4-5-6
Firing order	1-5-3-6-2-4
Bore and stroke	4.21 x 4.88 inches

Cylinder head and block

Cylinder head warpage limit	
End-to-end	0.012 inch
Side-to-side...	0.003 inch
Cylinder block deck warpage limit	
End-to-end	0.003 inch
Side-to-side...	0.002 inch
Cylinder head bolt length (maximum).........................	5.200 inches

Camshaft

Journal diameter	
Journals 1 and 7..	2.127 to 2.128 inches
Journals 2 through 6	2.1245 to 2.1265 inches
Lobe height, minimum	
Intake	1.857 inches
Exhaust	1.797 inches
Endplay..	0.005 to 0.020 inch
Thrust plate thickness, minimum	0.386 inch

Camshaft thrust plate

Thrust plate minimum thickness	0.368 inch
Thrust plate maximum thickness	0.377 inch

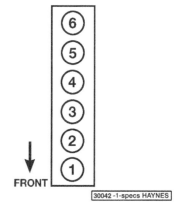

FRONT

30042 -1-specs HAYNES

Cylinder locations

Rocker arms and shafts
Rocker arm bore diameter (maximum)... 0.867 inch
Rocker arm shaft diameter (minimum) .... 0.865 inch

Valve clearance
Intake
    Allowable.. 0.006 to 0.021 inch
    Desired... 0.010 inch
Exhaust
    Allowable.. 0.015 to 0.034 inch
    Desired... 0.026 inch

Oil pump
Oil pressure
    Idle .... 10 psi
    2,500 rpm.. 30 psi
Gerotor-to-planetary tip clearance, maximum .... 0.007 inch
Gerotor planetary-to-body clearance, maximum .... 0.015 inch
Gerotor-to-back plate clearance, maximum .... 0.005 inch
Gear backlash .... 0.006 to 0.010 inch

Torque specifications
Ft-lbs (unless otherwise indicated)

Note: *One foot-pound (ft-lb) of torque is equivalent to 12 inch-pounds (in-lbs) of torque. Torque values below approximately 15 ft-lbs are expressed in inch-pounds, since most foot-pound torque wrenches are not accurate at these smaller values.*

Camshaft thrust plate bolts.. 18
Crankshaft damper bolts
    Step 1 30
    Step 2 .... Tighten an additional 60-degrees
Cylinder head bolts (in sequence **- see illustration 10.45**)
    Step 1 52
    Step 2 .... Loosen 360-degrees
    Step 3 77
    Step 4 .... 77 (re-check)
    Step 5 .... Tighten an additional 90-degrees
Drivebelt tensioner bolt... 32
Exhaust heat shield bolts.. 18
Exhaust manifold bolts 32
Fan support hub assembly bolts.. 24
Flywheel/driveplate bolts 100
Flywheel housing adapter.. 57
Gear housing bolts 18
Gear housing cover bolts... 18
Intake manifold bolts.. 18
Oil pressure relief valve plug 59
Oil pan drain plug 37
Oil pan bolts... 21
Oil pump mounting bolts
    Step 1 .... 72 in-lbs
    Step 2 18
Rear main oil seal retainer bolts.. 80 in-lbs
Rocker arm pedestal bolts... 27
Valve cover bolts 18
Valve adjusting lock nut.. 18

3.2a A special barring tool is available from specialty tool manufacturers

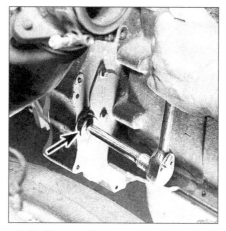

3.2b Remove the cover or rubber plug and insert the special barring tool into the bellhousing - used in conjunction with a ratchet and extension, the engine can be rotated by hand

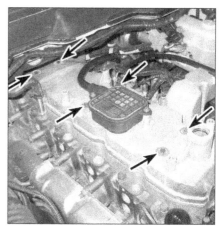

4.9 Valve cover bolt locations

1 General information

This part of Chapter 2 is devoted to in-vehicle repair procedures for the 6.7L Cummins inline six-cylinder diesel engine. Information concerning engine removal and installation and engine overhaul can be found in Part E of this Chapter.

Since the repair procedures included in this Part are based on the assumption that the engine is still installed in the vehicle, if they are being used during a complete engine overhaul (with the engine already out of the vehicle and on a stand) many of the steps included here will not apply.

The 6.7L diesel engine is of an extremely rugged, proven design, incorporating a turbocharger and intercooler for efficient power production and low-end torque for all towing applications, when kept in its proper operating rpm range.

2 Repair operations possible with the engine in the vehicle

Many major repair operations can be accomplished without removing the engine from the vehicle.

Clean the engine compartment and the exterior of the engine with some type of pressure washer before any work is done. A clean engine will make the job easier and will help keep dirt out of the internal areas of the engine.

Depending on the components involved, it may be a good idea to remove the hood to improve access to the engine as repairs are performed (refer to Chapter 11 if necessary).

If oil or coolant leaks develop, indicating a need for gasket or seal replacement, the repairs can generally be made with the engine in the vehicle. The cylinder head gasket, intake and exhaust manifold gaskets, gear

cover gaskets and the crankshaft oil seals are all accessible with the engine in place. The oil pan gasket, however, does require disconnecting the engine mounts and raising the engine.

Exterior engine components, such as the water pump, the starter motor, the alternator, turbocharger and the fuel injection components, as well as the intake manifold cover and exhaust manifold, can be removed for repair with the engine in place.

Since the cylinder head can be removed without pulling the engine, valve component servicing can also be accomplished with the engine in the vehicle.

Replacement of, repairs to or inspection of the gear case components and the oil pump are all possible with the engine in place.

3 Top Dead Center (TDC) for number one piston - locating

Refer to illustrations 3.2a and 3.2b

1   Remove the valve cover (see Section 4).
2   A special barring tool, available from specialty tool manufacturers, fits into a hole in the bellhousing and engages the flywheel ring-gear teeth **(see illustrations)**. Remove the rubber plug from the bellhousing, insert the tool, and the engine can be turned slowly and precisely with the tool, a ratchet and an extension. Rotate the engine until the TDC mark on the crankshaft vibration damper is in the 12 o'clock position.
3   Check the rocker arms for the number one cylinder. If they're both loose, the engine is at TDC compression for cylinder number 1. **Note:** *This is an approximate TDC setting, but it is close enough for checking valve lash.*
4   If both rocker arms for cylinder number 1 are not loose, turn the crankshaft one turn, bringing the mark on the vibration damper to the 12 o'clock position.

4 Valve cover - removal and installation

Refer to illustration 4.9
Note: *The injector wiring harness is integrated into the valve cover gasket. Use care to avoid damaging the wiring.*
1   Disconnect the cables from the negative terminals of the batteries (see Chapter 5).
2   Remove the four bolts and the engine cover.
3   Remove the oil filler cap.
4   Remove the bolts from the breather cover.
5   Remove the Closed Crankcase Ventilation (CCV) cover and the breather filter (see Chapter 1).
6   Disconnect the CCV tube at the Crankcase Depression Regulator (CDR) valve (at the right rear of the valve cover)
7   Disconnect the crankcase pressure sensor from the left rear of the valve cover.
8   Remove the two CCV drain hoses from the left side of the valve cover.
9   Remove the valve cover bolts, then remove the valve cover **(see illustration)**.
10   Remove the nuts that secure the valve cover gasket/wiring harness and disconnect the wiring harness from the injectors. Remove the gasket/wiring harness.
11   Clean the gasket mating surfaces of the cylinder head and the valve cover. Wipe off the gasket and inspect its condition. The gasket and isolators (the rubber grommets for the valve cover bolts) can be reused if they're not hardened, cracked or otherwise damaged. If an isolator or the valve cover gasket is cracked, replace it to avoid leaks.
12   Installation is the reverse of removal. Make sure the gasket is installed properly (some are marked TOP FRONT). Tighten the bolts a little at a time, starting with the center bolts, to the torque listed in this Chapter's Specifications.

6.4 Loosen this clamp attaching the intercooler duct to the air intake housing

5   Rocker arms and pushrods - removal, inspection, installation and adjustment

1   Disconnect the cables from the negative terminals of the batteries (see Chapter 5).
2   Remove the valve cover (see Section 4).

Removal

3   Mark the rocker arm/pedestal assemblies to ensure that they will be reinstalled in their original locations. Remove the rocker arm/pedestal bolts and remove the rocker arm/pedestal assemblies.
4   Mark the crossheads (the pieces that bridge each pair of intake and exhaust valves) to ensure that they will be reinstalled in their original locations. Remove the crossheads. **Caution:** *The sockets may drop from the rocker arms when lifting them from the cylinder head. Secure the rocker arm sockets before lifting them from the cylinder head.*
5   Mark the pushrods to ensure that they will be reinstalled in their original locations. Remove the pushrods. **Note:** *Lift out the intake and exhaust pushrods for the No. 5 and No. 6 cylinders through access holes in the cowl.*

Inspection

6   Disassemble the rocker arms, pedestals and shafts and wash them, along with the crossheads, in clean solvent. Keep each rocker arm pair, the rocker shafts, the pedestal and related parts together. If necessary, use a wire brush to remove deposits. Rinse the parts in hot water and blow them dry with compressed air.
7   Inspect the oil passages in the rocker arms, rocker shafts and pedestals. Blow out all lubrication passages with compressed air to remove any debris.
8   Inspect the bearing surface of each rocker shaft for scoring, cracks and galling, and for any other excessive wear. Inspect the socket and the ball insert for excessive wear. If the retainer is weak, replace it.

9   Measure the inside diameter of each rocker arm bore and the outside diameter of each rocker shaft. Compare your measurements to the maximum allowable diameter for rocker arm bores and to the minimum allowable diameter for rocker shafts. If a rocker arm or rocker shaft is worn excessively, replace it.
10   Inspect each pushrod ball and socket for cracks and any signs of scoring. Roll each pushrod on a clean flat surface with the socket end hanging off the end of the bench. If a pushrod is bent, or excessively worn, replace it.
11   Inspect each crosshead for cracks and excessive wear, especially around the rim of the crosshead/valve tip contact area. If a crosshead is damaged or excessively worn, replace it.

Installation

12   Install the pushrods, with the ball ends facing down, in their original locations. Make sure that the pushrods are fully seated in the tappets.
13   Lubricate the valve tips with clean engine oil and install the crossheads in their original locations.
14   Lubricate the rocker shafts and the rocker shaft bores in the rocker arms with clean engine oil. Reassemble the rocker arm/rocker shaft/pedestal assemblies.
15   Lubricate the crossheads and the pushrod sockets with clean engine oil and install the rocker arm/pedestal assemblies in their original locations. Install the rocker arm/pedestal bolts and tighten them to the torque listed in this Chapter's Specifications.
16   Before installing the valve cover, check and, if necessary, adjust the valve clearance (see below).
17   After the valve clearance has been checked and, if necessary, adjusted, install the valve cover (see Section 4).

Valve clearance check and adjustment

Note: *Valve clearance checking is only necessary after servicing valvetrain components.*
18   Position the engine at Top Dead Center (TDC) for cylinder number 1 (see Section 3).
19   With the engine in this position, measure the valve clearance at the following rocker arms:

> INTAKE - 1, 2 and 4
> EXHAUST - 1, 3 and 5

20   Measure the valve clearance by inserting a feeler gauge between the rocker arm socket and the crosshead. Compare your measurements to the valve clearance ranges listed in this Chapter's Specifications.
21   If the valve clearance is incorrect for any of the above intake or exhaust valves, loosen the locknut and then turn the adjusting screw until the clearance is correct. Set the valve clearances to the dimension listed in this Chapter's Specifications. When the clearance is correct, tighten the locknut securely (see this Chapter's Specifications for the proper torque reading) and then re-check the clearance to make sure that it didn't change as a

result of tightening the nut.
22   Using the barring tool, rotate the engine one complete revolution (360-degrees). With the engine in this position, measure the valve clearance at the following rocker arms:

> INTAKE - 3, 5 and 6
> EXHAUST - 2, 4 and 6

23   Following Steps 20 through 22, measure and adjust the valve clearances for the remaining valves.
24   Install the valve cover (see Section 4).

6   Intake manifold - removal and installation

Refer to illustration 6.4

Warning: *Diesel fuel is flammable, so take extra precautions when you work on any part of the fuel system. Don't smoke or allow open flames or bare light bulbs near the work area, and don't work in a garage where a gas-type appliance (such as a water heater or a clothes dryer) is present. Since diesel fuel is carcinogenic, wear fuel-resistant gloves when there's a possibility of being exposed to fuel, and, if you spill any fuel on your skin, rinse it off immediately with soap and water. Mop up any spills immediately and do not store diesel fuel-soaked rags where they could ignite. When you perform any kind of work on the fuel system, wear safety glasses and have a Class B type fire extinguisher on hand.*
1   Disconnect the cables from the negative terminals of the batteries (see Chapter 5).
2   Disconnect the wiring from the EGR temperature sensor and valve actuator.
3   Unbolt the oil dipstick tube.
4   Remove the intercooler tube **(see illustration)**.
5   Remove the clamps from both ends of the air transfer tube. Remove the bolt from the P-clip for the air conditioning line, then remove the air transfer tube.
6   Remove the six air intake screws, then remove the air intake.
7   Clean debris from both ends of all of the fuel tubes.
8   Loosen the two bolts of the bracket at the rear of the head that holds the injector tubes. Tilt the bracket for access to the rear injector tube.
9   Place rags to absorb spilled fuel, then disconnect the rear four injector tubes and remove them. **Caution:** *Don't bend the lines or allow fuel to run down the sides of the engine.*
10   Disconnect the wiring from the sensor at the rear of the fuel rail.
11   Detach the fuel rail from the dump/overflow valve, then remove it.
12   Detach the remaining interfering wiring connections.
13   Remove the intake manifold bolts and lift off the manifold.
14   Installation is the reverse of removal. Use a new gasket. Tighten the intake manifold cover bolts to the torque listed in this Chapter's Specifications.

7.2 On models so equipped, remove the bolt retaining the heater pipe(s) to the exhaust manifold. Do not lose the spacers

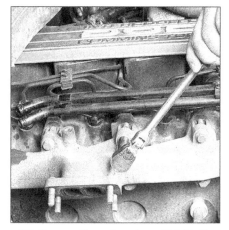

7.6 Remove the exhaust manifold bolts

8.10 Remove the four bolts retaining the vibration damper to the crankshaft

7 Exhaust manifold - removal and installation

Refer to illustrations 7.2 and 7.6

1   Remove the turbocharger (see Chapter 4B).
2   Remove the bolts and brackets retaining the heater line(s) to the exhaust manifold **(see illustration)**.
3   Remove the air filter housing and the air intake hose.
4   On models so equipped, remove the heat shield/noise reduction panel from the exhaust manifold. Remove the EGR cooler (see Chapter 6).
5   Remove the rear bolt lockplates.
6   Remove the exhaust manifold bolts, exhaust manifold, gaskets and spacers **(see illustration)**.
7   Clean the mounting surfaces of the manifold and the cylinder head, and clean the threads of the exhaust manifold mounting bolts.
8   Install the manifold using new gaskets and anti-seize compound on the bolt threads. Tighten the manifold bolts a little at a time, starting with the center bolts and working outward, to the torque listed in this Chapter's Specifications.
9   The remainder of installation is the reverse of removal. **Note:** *The studs go in the holes for the front two cylinders and an additional one goes at the lower rear.*

8 Crankshaft front oil seal and gear housing cover - replacement

Removal

Refer to illustration 8.10

1   Disconnect the cables from the negative terminals of the batteries (see Chapter 5).
2   Remove the drivebelt (see Chapter 1).
3   Drain sufficient coolant into a clean container (see Chapter 1).

4   Remove the upper radiator hose.
5   Detach the coolant reservoir hose from the radiator.
6   Remove the windshield washer bottle.
7   Remove the viscous fan and drive assembly, then remove the fan shroud and fan (see Chapter 3)
8   Remove the fan support from the engine.
9   Raise the vehicle and support it securely on jackstands.
10   Remove the mounting bolts, then remove the crankshaft damper **(see illustration)**. **Note:** *On some models, there is a friction shim behind the damper. Make sure that it is replaced during installation.*
11   Remove the power steering pump (see Chapter 10).
12   Remove the drivebelt tensioner (see Chapter 1).
13   Remove the gear cover mounting bolts. Carefully pry the cover from the front of the engine, taking care to avoid damaging the sealing surface.
14   Place the cover on wooden blocks and use a hammer and punch to drive the seal out from the rear.

Installation

15   Carefully inspect the damper for signs of fluid leakage. Any leakage requires replacement of the damper.
16   Thoroughly clean the new seal and the seal bore in the cover. Remove any old sealer.
17   Straighten the cover if necessary by tapping the sealing surface on a flat area with a hammer. It must be flat before installation. Clean the sealing surfaces of the cover and the block with brake system cleaner when finished.
18   Check the crankcase snout for nicks or burrs. Dress any imperfections with a small file.
19   Place the cover on wooden blocks. Apply stud and bearing mount compound to the outside of the new seal, then tap it into place

using a large socket or a seal installation tool. **Caution:** *The inner sealing lip of the seal and the crankshaft must both be kept dry and free of oil during the entire installation process. Don't try to lubricate the seal or the crankshaft at any time; they must be assembled dry.*
20   Install the alignment pilot tool into the seal. **Note:** *This plastic tool should be provided with the new seal.*
21   Apply a bead of RTV sealant to the sealing surface of the cover.
22   Install the cover and all of the bolts. Hand tighten a couple of the bolts to secure it in position. The seal alignment tool must be in place.
23   Tighten the bolts evenly to the torque listed in this Chapter's Specifications. Make sure that the seal alignment tool is correctly located while this is done, then remove the tool.
24   The remainder of installation is the reverse of removal.

9 Camshaft and tappets - removal, inspection and installation

Warning: *The air conditioning system is under high pressure. DO NOT loosen any fittings or remove any components until after the system has been discharged. Air conditioning refrigerant should be properly discharged into an EPA-approved container at a dealership service department or an automotive air conditioning repair facility. Always wear eye protection when disconnecting air conditioning system fittings.*
Note: *Camshaft and tappet replacement in the diesel engine is a very involved process. The diesel engine tappets have a head that is larger than the tappet body, and they must be inserted into their bores from the bottom of the bore. This is a specialized procedure which requires several special tools. There are two methods for retaining the tappets for camshaft removal; read through the entire Section and obtain the special tools before beginning the procedure.*

9.9a Remove the crankcase vent tube and the two screws, then pull off the plastic cover

9.9b Remove these six bolts and pull off the side cover and its gasket

1   If the vehicle is equipped with air conditioning, and you're planning to remove the camshaft with the engine installed in the vehicle, have the air conditioning system discharged by an automotive air conditioning shop before proceeding (you will have to remove the grille, radiator and condenser to remove the camshaft).

2   Remove the crankshaft damper and gear housing cover (see Section 8). Remove the radiator and the upper radiator support panel, then remove the intercooler and the air conditioning condenser (see Chapters 3 and 4B).

3   If you're planning to retain the tappets using Method Two, remove the fuel transfer pump, the fuel filter/water separator and the fuel injection pump (see Chapter 4B).

4   Position the engine at TDC for cylinder number 1 (see Section 3).

5   Remove the valve cover, rocker arms and pushrods (see Sections 4 and 5). **Note:** There are access holes with rubber plugs in the cowl for the removal of the pushrods for the rear cylinders.

Tappet retention

Note: *The tappets must be retained up in their bores, away from the camshaft lobes, in order to remove the camshaft.*

Method one

6   A set of 12 wooden dowels, four inches longer than the pushrods and the correct diameter to just fit snugly into the top of the tappets, will be necessary for the procedure. If you make your own and size them to fit, make sure that they are well-sanded with fine sandpaper and cleaned of sanding residue before use, so that no wood chips or sawdust contaminates the engine.

7   Insert the dowels down through the pushrod holes until they lodge firmly in the tappets. For each cylinder, pull up the two dowels (intake and exhaust) and secure them to each other with large rubber bands. There must be enough tension on the rubber bands to keep the tappets raised as far as they can go up in their bores.

8   When all of the tappets are secured at the top of their bores, the camshaft can be removed.

Method two

Refer to illustrations 9.9a, 9.9b and 9.10

9   A somewhat easier method of tappet retention can be accomplished if the engine is out of the vehicle for overhaul, or in-vehicle, if the fuel filter/water separator and fuel injection pump are removed allowing removal of the engine side cover **(see illustrations)**.

10   Slip a small hose clamp over each tappet. Pull up the tappet to the top of its travel with a magnetic tool and tighten the hose clamp to keep the tappet at the top of its bore. Repeat this procedure for all of the tappets, tightening the clamps only enough to retain them at the top **(see illustration)**.

Camshaft removal

Refer to illustration 9.11

11   With the engine positioned at TDC, remove the two bolts retaining the camshaft thrust plate **(see illustration)**. **Caution:** *DO NOT let the thrust plate fall into the crankcase.*

12   Pull the camshaft out as straight as possible. Be very careful not to nick any of the camshaft bearings in the block with the lobes or journals. Work slowly - it's a long and heavy camshaft.

Tappet removal

Refer to illustration 9.14

13   A special tool is necessary to extract the tappets from the block. The tool is a long 1/2-section of pipe slightly smaller than the camshaft bore. The top of the pipe is removed, lengthwise, and one end capped. This trough may be available through your local dealer, or you can fabricate your own from copper tubing.

9.10 With the side cover off, use a magnetic tool to lift each tappet up, then tighten a small hose clamp around the tappet to keep it there

9.11 Remove the two bolts retaining the camshaft thrust plate

9.14 Insert the trough-like tool into the camshaft bore with the open side up - then pull the dowels or, as shown, release the hose clamp from one tappet at a time until it falls into the trough and can be withdrawn from the engine

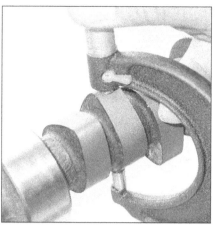

9.17 Measure the camshaft lobe at its greatest dimension to determine the lobe height

9.19 Examine the foot of the tappet for uneven wear or pitting - it should be perfectly flat

14   Insert the trough-like tool the full length of the camshaft bore, with the open side up **(see illustration)**. Remove the rubber bands from two of the dowels, pull the dowel out of one tappet, and re-tie the other dowel with rubber bands to the valve spring or other nearby component. As the dowel is pulled out, that tappet will fall into the trough. Look through the trough with a flashlight to ensure that the tappet is lying on its side in the trough. If not, jiggle the trough to make it fall over and then extract the trough and retrieve the tappet.

15   Repeat this procedure for each tappet, keeping the tappets in order so they can be reinstalled in their original locations, until all of the tappets have been removed for inspection.

Inspection

Refer to illustrations 9.17 and 9.19

16   After the camshaft has been removed from the engine, cleaned with solvent and dried, inspect the bearing journals for uneven wear, pitting and evidence of seizure. If the journals are damaged, the bearing inserts in the block are probably damaged as well. Both the camshaft and bearings will have to be replaced. **Note:** *Camshaft bearing replacement requires special tools and expertise that place it beyond the scope of the average home mechanic. Although the tool for bearing removal/installation is available at stores that carry automotive tools and can possibly even be found at a tool rental business, if the bearings are bad and bearing replacement is beyond your ability, remove the engine and take the block to an automotive machine shop to ensure that the job is done correctly.*

17   Measure the bearing journals with a micrometer to determine if they are excessively worn or out-of-round. Measure the camshaft lobe height **(see illustration)**. Compare your measurements with this Chapter's Specifications to determine if the camshaft is worn.

18   Check the camshaft lobes for heat discoloration, score marks, chipped areas, pitting and uneven wear. If the lobes are in good condition and if the journal diameters and lobe height measurements are as specified, the camshaft can be reused.

19   Check each tappet wall, pushrod seat and foot for scuffing, score marks and uneven wear. Each tappet foot (the surface that rides on the cam lobe) should be perfectly flat, although it may be slightly concave in normal wear **(see illustration)**. If there are signs of uneven wear or scoring, the tappets and camshaft must be replaced. If the tappet walls are damaged or worn (which is not very likely), inspect the tappet bores in the engine block as well, using a small inspection mirror with a long handle. If the pushrod seats are worn, check the pushrod ends.

20   If new tappets are being installed, a new camshaft must also be installed. If a new camshaft is installed, then use new tappets as well. Never install used tappets unless the original camshaft is used and the tappets can be installed in their original locations.

21   Measure the thickness of the camshaft thrust plate with a micrometer. Refer to the Specifications listed in this Chapter. If the camshaft thrust plate is too thin or thick, replace it with a new one.

Camshaft and tappet installation

Refer to illustrations 9.22, 9.26 and 9.27

22   A special tappet installation tool is needed to fish the new tappets up into their bores before the camshaft is installed. The installation tool is basically a long wire attached to a short plug that fits inside the tappet. This special tool is available through your dealer, or you can make your own. Insert the tappet trough tool fully into the camshaft bore and drop the tappet retrieval tool down through the pushrod hole from above until it hits the trough. Using the hooked retrieval tool (included with the tappet installation tool

kit), pull the tappet installation tool out of the trough, then push the plug into the tappet, lubricate the sides and foot of the tappet with engine assembly lube and pull the wire back out the tappet hole until the tappet is seated in its bore. If you don't have the factory installation tools, make your own **(see illustration)**. You can get by without the hooked retrieval tool; instead, carefully pull the trough out of the engine until the tappet installation tool is exposed.

23   Push the trough back into the engine fully, if removed, then from above, pull up on the wire until the tappet is at the top of its bore.

24   Now rotate the trough around until its closed side is UP, which will keep the tappet from falling out of its bore. The fishing tool can now be pulled out of the tappet and the dowel inserted and secured, retaining that tappet in

9.22 You can make your own tappet installation tools: Our trough was made from a length of copper tubing, and the retrieval tool is a short section of hose that just fits into the tappet and is attached to a long piece of wire (you can get by without the hooked retrieval tool - simply pull the trough out to retrieve each tappet)

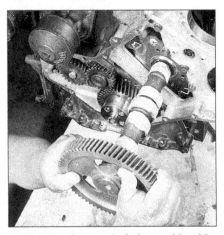

9.26 Lube the camshaft thoroughly with camshaft installation lube

9.27 Align the timing marks on the crankshaft and camshaft gears

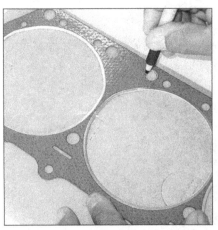

10.32 Use a new head gasket to trace the bolt pattern on a piece of cardboard - punch holes for the head bolts and use the cardboard to keep track of the locations of the bolts

place. If the side cover is off, secure the tappet with a hose clamp, as in Step 10.

25   Repeat Steps 22 through 24 for the remaining tappets until all are held high in their proper bores with the dowels and rubber bands, or hose clamps.

26   Lubricate the camshaft lobes and journals thoroughly with camshaft installation lubricant **(see illustration)** and insert the camshaft into the block, again being careful to insert it straight without nicking the bearings with the lobes or journals.

27   As the camshaft is close to being fully inserted, align the timing marks on the camshaft gear with the timing mark on the crankshaft gear **(see illustration)**. When the camshaft is fully inserted and the gears properly meshed, install the camshaft thrust plate bolts and tighten them to the torque listed in this Chapter's Specifications. **Caution:** *Do not push the camshaft any further into the block than is necessary, or the camshaft plug at the back of the block could be loosened, creating an oil leak.*

28   All of the tappet retaining dowels or clamps may now be removed, and the pushrods, rocker arms and other components installed. Adjust the valve clearance (see Section 5).

29   The remainder of installation is the reverse of removal. If the side cover had been removed, install it with a new gasket. Change the engine oil and oil filter (see Chapter 1).

30   Have the air conditioning system evacuated, recharged and leak tested by the shop that discharged it.

10   Cylinder head - removal, inspection and installation

Warning: *Make sure the engine is completely cool before beginning this procedure.*
Warning: *Diesel fuel is flammable, so take extra precautions when you work on any part of the fuel system. Don't smoke or allow open flames or bare light bulbs near the work area, and don't work in a garage where a gas-*

type appliance (such as a water heater or a clothes dryer) is present. Since diesel fuel is carcinogenic, wear fuel-resistant gloves when there's a possibility of being exposed to fuel, and, if you spill any fuel on your skin, rinse it off immediately with soap and water. Mop up any spills immediately and do not store diesel fuel-soaked rags where they could ignite. When you perform any kind of work on the fuel system, wear safety glasses and have a Class B type fire extinguisher on hand.

Removal

Refer to illustration 10.32

1   Disconnect the cables from the negative terminals of the batteries (see Chapter 5).
2   Drain the coolant (see Chapter 1).
3   Disconnect the exhaust pipe from the turbocharger.
4   Remove the viscous fan and drive assembly (see Chapter 3).
5   Install an engine support fixture or attach a hoist to the engine (see Chapter 2E). Raise the engine enough to take weight from the engine mounts, then remove the through-bolt from the right engine mount. Remove the engine mount.
6   Detach the turbo oil drain fitting from the bottom of the turbocharger. Seal the openings to prevent contamination.
7   Disconnect the wiring from the air intake temperature sensor, then remove the air filter housing and snorkel. Tape the open turbo port to prevent contamination.
8   Disconnect the heater hoses from the head and the pipe.
9   Disconnect the oil supply line and the coolant lines from the turbocharger. Seal the openings to prevent contamination.
10   Remove the EGR cooler (see Chapter 1).
11   Unbolt the exhaust manifold and remove it with the turbo attached.
12   Remove the fan and shroud assembly (see Chapter 3).
13   Remove the drivebelt (see Chapter 1).
14   Remove the fan support from the engine block.

15   Remove the top mounting bolt from the alternator. Loosen the lower bolt, then lower the alternator away.
16   Disconnect the upper radiator hose from the engine.
17   Label and disconnect all wiring that attaches to the cylinder head.
18   Remove the engine oil dipstick tube bolt.
19   Disconnect and remove the air intake housing.
20   Remove the pump-to-fuel rail line.
21   Remove the fuel rail-to-head lines and the fuel line shield.
22   On 2010 and later models, remove the fuel rail.
23   Remove the engine lift bracket from the back of the head.
24   On 2009 models, disconnect and remove the fuel filter assembly.
25   Remove the P-clip from the wiring harness behind the fuel filter housing.
26   Remove the valve cover (see Section 4).
27   Remove the injector harness nuts from the fuel injectors.
28   Remove the rockers, crossheads and pushrods (see Section 5).
29   Remove the banjo bolt, sealing washers and the fuel line from the rear of the head.
30   Remove the fuel injectors (see Chapter 4B).
31   Remove the rocker housing and the gasket.
32   Using a new head gasket, make a template of the bolt hole locations on a piece of cardboard **(see illustration)**. Loosen the head bolts, working in the order opposite that of the tightening sequence **(see illustration 10.45)**. As each head bolt is removed, insert it into its location on the cardboard template.
33   Make a thorough inspection of the head for any wiring harnesses and/or ground cables that might still be attached to the head. Make sure that all components have been removed or detached from the head. When you're sure that nothing more is attached to the head, carefully pry on a casting protrusion between

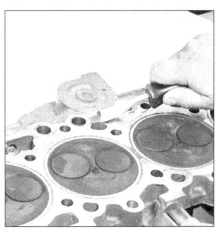

10.37 Use a gasket scraper and gasket removing solvent to clean the cylinder head and block sealing surfaces

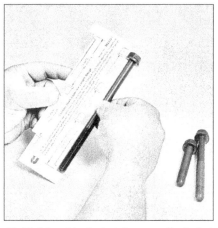

10.40 A head bolt stretch gauge (included with the gasket set) is used to determine if the cylinder head bolts can be reused

the head and the block to break the head gasket seal. The cylinder head is very heavy, so have an assistant help you lift it off the engine. If an assistant is not available, reattach the lift bracket to the rear of the head, attach an engine lifting hoist to the lifting brackets at the front and rear of the cylinder head and lift out the head. **Warning:** *Do NOT attempt to remove the head by yourself!*

Inspection

Refer to illustrations 10.37 and 10.40

34   The cylinder head is a vital part of the engine's efficiency. If you have gone to the trouble to remove it - to replace the head gasket, for example - and the engine has accumulated many, many miles, consider having the valves and seats refaced at an automotive machine shop to restore full sealing of the valves.
35   Remove the fuel injectors and rocker housing from the cylinder head. **Note:** *Fuel injector removal requires special tools. The fuel injectors can be removed after the cylinder head is removed and set on the work bench. Have the injectors removed by a*

dealer service department or other diesel specialist equipped with the proper tools.
36   The mating surfaces of the cylinder head and block must be perfectly clean when the heads are installed. Gasket removal solvents are available at auto parts stores and may prove helpful.
37   Use a gasket scraper to remove all traces of carbon and old gasket material **(see illustration)**, then wipe the mating surfaces with a cloth saturated with brake system cleaner. If there is oil on the mating surfaces when the head is installed, the gasket may not seal correctly and leaks may develop. When working on the block, fill the cylinders with shop rags to keep debris out of the engine. Use a vacuum cleaner to remove any debris that falls into the cylinders. **Note:** *Clean any carbon from the injector nozzle seats with a brass or nylon brush.*
38   Check the block and head mating surfaces for nicks, deep scratches and other damage. If imperfections are slight, they can be removed with emery cloth. If excessive, machining may be the only alternative. Use a straightedge and feeler gauge to check the cylinder head for warpage. If it is not within Specifications, the head must be machined to restore flatness. Consult

with an automotive machine shop. **Note:** *There is a raised pad at the rear of the cylinder head. There may be stampings there to indicate whether the head has been machined before. More than 0.010-inch removal may require valve regrinding as well. Consult with an automotive machine shop.*
39   The block deck surface should also be checked for warpage by using a precision straightedge and feeler gauges. Compare your findings with the limits listed in this Chapter's Specifications.
40   Use a tap of the correct size to chase the threads in the head bolt holes in the block. Mount each bolt in a vise and run a die down the threads to remove corrosion and restore the threads. Dirt, corrosion, sealant and damaged threads will affect torque readings. Measure the length of each cylinder head bolt, from under the bolt head to the end of the threads, comparing your measurements to the maximum bolt length. Most engine gasket sets will include a bolt stretch gauge for the head bolts **(see illustration)**. If the head bolts are longer than the allowable length indicated on the gauge, they are stretched and must be replaced with new bolts.
41   If removed, have the rocker housing and fuel injectors reinstalled by the shop that removed them.

Installation

Refer to illustrations 10.42 and 10.45

42   Position the new gasket over the dowels in the block **(see illustration)**.
43   Carefully position the cylinder head on the block without disturbing the gasket.
44   Apply engine oil to the threads and underneath the bolt heads.
45   Install the bolts in their original locations and tighten them finger tight. Following the recommended sequence **(see illustration)**, tighten the bolts in several steps to the torque listed in this Chapter's Specifications.
46   The remainder of installation is the reverse of removal. Change the oil and filter and refill the cooling system (see Chapter 1).
47   Prime the fuel filter before attempting to start the engine (see Chapter 1).

10.42 Position the new head gasket over the dowels in the block - note any markings on the gasket that indicate Top or Front

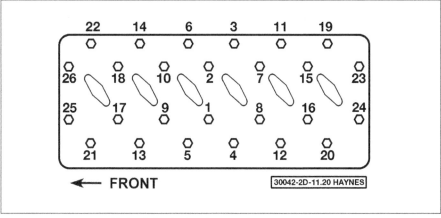

10.45 Cylinder head bolt tightening sequence

11.3 With the dial indicator tip positioned against the idler gear, rotate the idler gear and compare its backlash to the Specifications

11.4 Check the backlash at the drive gear while holding the idler gear

11.5 Remove the four mounting bolts to remove the oil pump from the gear case

11.6 Remove the two screws on the back plate, then mark the top of the gerotor planetary with a felt marker

11 Oil pump - removal, inspection and installation

Refer to illustrations 11.3, 11.4, 11.5, 11.6, 11.7a, 11.7b and 11.7c

1   Disconnect the cables from the negative terminals of the batteries (see Chapter 5).
2   Remove the gear housing cover (see Section 8).
3   Measure the backlash between the gears using a dial indicator. Install the dial indicator tip against one of the teeth on the idler gear and twist the idler gear back and forth **(see illustration)**. This will give the backlash measurement between the crankshaft gear and the idler gear. Compare it to this Chapter's Specifications.
4   Place the dial indicator tip on the oil pump drive gear and check the backlash between it and the idler gear while securely holding the idler gear from moving **(see illustration)**. If

the backlash is greater than that listed in this Chapter's Specifications for either gear, the gears must be replaced.
5   Remove the four mounting bolts and remove the oil pump assembly from the gear case **(see illustration)**.
6   Remove the back plate of the oil pump assembly **(see illustration)**. Clean the top of the gerotor planetary of oil and use a felt pen or paint to mark it with the word TOP.
7   Remove the components, clean them with solvent and dry them thoroughly. Assemble the components back in the oil pump housing. **Note:** *The chamfer on the outside diameter of the gerotor planetary must face down, into the housing.* Using a feeler gauge, measure the planetary-to-body clearance, the tip clearance and the gerotor-to-back plate clearance **(see illustrations)**. If any clearance is beyond that listed in this Chapter's Specifications, replace the oil pump assembly.

11.7a Measure the gerotor-to-planetary tip clearance with a feeler gauge

11.7b Measure the gerotor planetary-to-body clearance

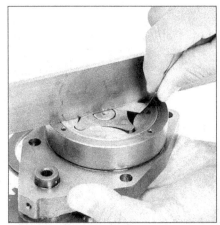

11.7c Using the feeler gauge and a straightedge, measure the gerotor-to-back plate clearance

12.12 Support the engine with an engine hoist, two large wood blocks, and a floor jack placed under the transmission; an engine support fixture can also be used in place of the hoist

12.14 With the engine safely supported, remove the oil pan bolts

8   If the original pump is being reused, fill the pump cavity with engine oil and reinstall the back cover, then install the pump assembly in the gear case and tighten the mounting bolts to the torque listed in this Chapter's Specifications.
9   The remainder of installation is the reverse of removal. Refill the crankcase with oil, start the engine, and check for leaks and proper oil pressure.

12   Oil pan - removal and installation

Removal

Refer to illustrations 12.12, 12.14 and 12.15
1   Disconnect the cables from the negative terminals of the batteries (see Chapter 5).
2   Drain the engine oil (see Chapter 1).
3   Raise the vehicle and support it securely on jackstands.
4   Remove the oil dipstick.

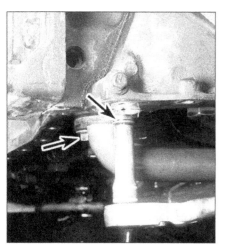

12.15 Lower the oil pan and remove the suction tube bolts at the front of the engine

2WD models
5   Remove the fan and fan shroud (see Chapter 3).
6   Remove the starter (see Chapter 5).
7   Remove the transmission (see Chapter 7).
8   Remove the flywheel housing and the driveplate.
9   Remove the steering gear mounting bolts and lower the steering gear.

4WD models
10   Unbolt the radiator shroud but don't remove it.
11   Attach an engine support fixture or a hoist to the engine lift brackets with a chain.
12   Raise the engine slightly, then remove the front engine mount through-bolts. Continue raising the engine until sufficient clearance is obtained under it **(see illustration)**.
13   Remove the two bolts from the front of the engine block support.

All models
14   Remove the oil pan bolts and use a putty knife between the block and the pan to break the gasket seal **(see illustration)**. **Caution:** *Do not gouge the sealing surfaces with a screwdriver or pry tool.*
15   Lower the pan to access and remove the two bolts retaining the oil suction tube near the front of the block **(see illustration)**. Lower the suction tube into the oil pan and remove the oil pan and suction tube together.

Installation
16   Clean the interior of the oil pan with rags and solvent and clean the oil suction tube, blowing it out with compressed air if available. Also clean the pan mounting surface of the block of any sealant, gasket material or oil.
17   On the engine, apply a small bead of RTV sealant to the block and front cover joints, and the block and the rear seal retainer joints.

18   Attach a new gasket to the oil suction tube mounting flange and place the suction tube into the pan, with the mounting flange at the front.
19   Raise the oil pan up next to the block and attach the oil suction tube to the block. Install the oil pan. Tighten the bolts, starting in the center and working towards the ends, to the torque listed in this Chapter's Specifications.
20   Install the oil pan drain plug with a new sealing washer and tighten it to the torque listed in this Chapter's Specifications.
21   Refill the engine with new oil (see Chapter 1).
22   The remainder of installation is the reverse of removal. Start the engine and check for oil leaks.

13   Oil pressure relief valve - replacement

1   Disconnect the cables from the negative terminals of the batteries (see Chapter 5).
2   Remove the threaded plug, spring and valve from the oil filter assembly.
3   Clean the oil pressure relief valve bore thoroughly with solvent and compressed air.
4   Install the plunger, spring and plug and torque the oil pressure relief valve plug to the Specifications listed in this Chapter.

14   Flywheel/driveplate - removal and installation

Flywheel/driveplate removal and installation for the diesel engine is essentially the same as for the gasoline engines. Refer to Chapter 2A for procedures, but use the torque Specifications in this Chapter. Always use new bolts when the flywheel/driveplate is reinstalled, and apply Loctite 242 on the threads.

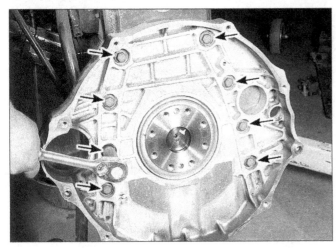

15.2 This adapter plate can be removed for easier access to the seal retainer plate

15.3 Rear crankshaft seal retainer plate

15   Rear main oil seal - replacement

Refer to illustrations 15.2 and 15.3

1   Rear seal removal and installation for the diesel engine is similar to that of the gasoline engines (see Chapter 2A).

2   Remove the transmission (see Chapter 7) and the clutch (see Chapter 8) if working with a manual transmission. Refer to Chapter 2A and mark the position of the flywheel or driveplate, then remove it and any adapter plate **(see illustration)**.

3   Carefully drill a few holes through the seal, avoiding contact with the crankshaft **(see illustration)**.

4   Install sheet metal or wood screws into the holes and use a slide hammer, pry bar or a claw hammer to pull the seal out.

5   Thoroughly clean the seal bore and the crankshaft and wipe off all traces of oil. **Caution:** *DO NOT use oil on the seal lip or crankshaft during installation - the crankshaft sealing surface and the oil seal must be clean and dry of oil or leaks may result. Soapy water can be used as an assembly lubricant if necessary.*

6   If the crankshaft is worn at the seal contact area, install a crankshaft sleeve over it.

7   Install the new seal along with the installation pilot tool (included with the seal) onto the crankshaft, then remove the pilot.

8   Use a seal installation tool, a large socket or a steel plate and a hammer to drive the seal into place. Drive it in evenly and don't allow it to tilt.

9   The remainder of installation is the reverse of removal.

16   Engine mounts - inspection and replacement

Refer to Chapter 2A for this procedure. Tighten all the engine mount bolts and through-bolts securely.

Chapter 2 Part D
3.6L V6 engine

Contents

Specifications

General

Displacement	220 cubic inches
Bore	3.779 inches
Stroke	3.268 inches
Compression ratio	10.2:1
Cylinder numbers	
Right bank	1-3-5
Left bank	2-4-6
Firing order	1-2-3-4-5-6
Oil pressure	
At idle speed	5 psi (minimum)
At 3,000 rpm	30 (warm) – 139 (cold) psi

Cylinder locations

Camshaft

Bore diameter	
Cam tower 1	1.2606 to 1.2615 inches
Cam tower 2, 3, and 4	0.9457 to 0.9465 inch
Bearing journal diameter	
No. 1	1.2589 to 1.2596 inches
No. 2, 3, and 4	0.9440 to 0.9447 inch
Bearing clearance	
No. 1	0.0001 to 0.0026 inch
No. 2, 3, and 4	0.0009 to 0.0025 inch
Endplay	0.003 to 0.010 inch

Cylinder head

Gasket thickness (compressed)	0.019 to 0.024 inch
Warpage limit	0.0035 inch

Torque specifications

Ft-lbs (unless otherwise indicated)

Note: *One foot-pound (ft-lb) of torque is equivalent to 12 inch-pounds (in-lbs) of torque. Torque values below approximately 15 ft-lbs are expressed in inch-pounds, since most foot-pound torque wrenches are not accurate at these smaller values.*

Camshaft bearing cap bolts	84 in-lbs
Crankshaft pulley bolt (M16 bolt)	
Step 1	30
Step 2	Tighten an additional 105-degrees
Cylinder head bolts* (in sequence - see illustrations **10.29a and 10.29b**)	
2013 and 2014 models	
Step 1	22
Step 2	33
Step 3	Tighten an additional 75 degrees
Step 4	Tighten an additional 50 degrees
Step 5	Loosen all in reverse of tightening sequence
Step 6	22
Step 7	33
Step 8	Tighten an additional 70 degrees
Step 9	Tighten an additional 70 degrees

* *Use new bolts*

Torque specifications (continued)   Ft-lbs (unless otherwise indicated)

Note: *One foot-pound (ft-lb) of torque is equivalent to 12 inch-pounds (in-lbs) of torque. Torque values below approximately 15 ft-lbs are expressed in inch-pounds, since most foot-pound torque wrenches are not accurate at these smaller values.*

Cylinder head bolts* (in sequence - see illustrations **10.29a and 10.29b**)(continued)

2015 and later models**

Step 1	22
Step 2	33
Step 3	33
Step 4	Tighten an additional 125 degrees
Step 5	Loosen all in reverse of tightening sequence
Step 6	22
Step 7	33
Step 8	33
Step 9	Tighten an additional 130 degrees
Drivebelt idler sprocket bolt	18
Driveplate-to-crankshaft bolts	70
Catalytic converter to cylinder head fasteners	17
Engine mounts	
Bolts	47
Through-bolt and nut	96
Heat shield nut	41
Exhaust crossover bolts	21
Intake manifold bolts	
2013 models (upper and lower)	71 in-lbs
2014 and later models	
Upper	89 in-lbs
Lower	106 in-lbs
Oil cooler	
Bolts	106 in-lbs
Screws	35 in-lbs
Oil pan drain plug	20
Oil pan	
Lower pan-to-upper pan nut/bolts	93 in-lbs
Upper pan-to-rear main seal housing (M6 bolts)	89 in-lbs
Upper pan-to-cylinder block (M8 bolts)	18
Oil pump pick-up tube mounting bolts	106 in-lbs
Oil pump cover (plate) screws	106 in-lbs
Oil pump-to-engine block fasteners	106 in-lbs
Rear main oil seal retainer bolts	106 in-lbs
Timing chain cover bolts	106 in-lbs
Camshaft chain tensioner bolts (primary)	106 in-lbs
Camshaft chain guide bolts	106 in-lbs
Camshaft chain idler sprocket bolt	18
Camshaft chain tensioner bolts (secondary)	106 in-lbs
Camshaft oil control valve (sprocket bolt)	111
Variable valve timing solenoid bolt	35 in-lbs
Oil pump timing chain sprocket (T45)	18
Valve cover-to-cylinder head bolts	106 in-lbs
Water pump bolts	See Chapter 3

* Use new bolts

** If using a new engine block, follow all nine Steps. If using the existing block, follow Steps 6 to 9.

1   General information

1   This Part of Chapter 2 is devoted to in-vehicle repair procedures for the 3.6L V6 engine.

2   The 3.6 liter engine utilizes Variable Valve Timing (VVT), Dual Overhead Camshafts (DOHC), four timing chains, an aluminum cylinder block, steel cylinder sleeves or liners with six cylinders arranged in a "V"-shape, with 60-degrees between the two banks. The 3.6 liter engine has a chain-driven oil pump with a multi-stage pressure regulator to increase fuel economy. The exhaust manifolds are integral with the cylinder heads to make the engine lighter.

Caution: *This engine is of an interference design and severe engine damage will occur if the timing chain breaks.*

3   The cylinders are numbered from front to rear. The right bank is numbered 1, 3, 5 and the left bank is numbered 2, 4, 6. The firing order is 1–2–3–4–5–6.

4   Information concerning engine removal and installation can be found in Chapter 2E. The following repair procedures are based on the assumption that the engine is installed in the vehicle. If the engine has been removed from the vehicle and mounted on a stand, many of the steps outlined in this Part of Chapter 2 do not apply.

2   Repair operations possible with the engine in the vehicle

1   Many major repair operations can be done without removing the engine from the vehicle.

2   Clean the engine compartment and the exterior of the engine with degreaser before any work is done. It'll make the job easier and help keep dirt out of internal parts of the engine.

3   It may be helpful to remove the hood to improve engine access when repairs are performed (see Chapter 11). Cover the fenders to prevent damage to the paint. Special pads are available, but an old bedspread or blanket will also work.

4   If vacuum, exhaust, oil, or coolant leaks develop, indicating a need for gasket or seal replacement, the repairs can generally be done with the engine in the vehicle. The intake and exhaust manifold gaskets, timing chain cover gasket, oil pan gasket, crankshaft oil seals, and cylinder head gaskets are all accessible with the engine in the vehicle.

5   Exterior engine components, such as the intake and exhaust manifolds, the oil pan, the oil pump, the timing chain cover, the water pump, the starter motor, the alternator, and fuel system components can be removed for repair with the engine in the vehicle.

6   Cylinder heads can be removed without pulling the engine. Valve component servicing can also be done with the engine in the vehicle. Replacement of the timing chain and sprockets is also possible with the engine in the vehicle, as is camshaft and valve train removal and installation.

7   Repair or replacement of piston rings, pistons, connecting rods, and rod bearings is possible with the engine in the vehicle, however, this practice is not recommended because of the cleaning and preparation work that must be done to the components.

3   Top Dead Center (TDC) for number one piston - locating

1   Top Dead Center (TDC) is the highest point in the cylinder that each piston reaches as it travels up the cylinder bore. Each piston reaches TDC on the compression stroke and again on the exhaust stroke, but TDC generally refers to piston position on the compression stroke.

2   Positioning the piston(s) at TDC is an essential part of certain procedures such as camshaft and timing chain/sprocket removal.

3   Before beginning this procedure, be sure to place the transaxle in Neutral and apply the parking brake or block the rear wheels. Disconnect the negative battery cable from the remote ground terminal (see Chapter 5). Remove the ignition coils (see Chapter 5) and the spark plugs (see Chapter 1).

4   Install a compression pressure gauge in the number one spark plug hole (see Chapter 2E). It should be a gauge with a screw-in fitting and a hose at least six inches long.

5   Rotate the crankshaft using a socket and breaker bar on the crankshaft pulley bolt while observing for pressure on the compression gauge. The moment the gauge shows pressure, indicates that the number one cylinder has begun the compression stroke.

6   Once the compression stroke has begun, TDC for the compression stroke is reached by bringing the piston to the top of the cylinder.
Note: *If a compression gauge is not available, you can simply place a blunt object over the spark plug hole and listen for compression as the engine is rotated. Once compression at the No.1 spark plug hole is noted, the remainder of the Step is the same.*

7   This engine is not equipped with exter-

nal components (crankshaft pulley, flywheel, timing hole, etc.) that are marked to identify the position of number 1 TDC. Therefore, the only method to double-check the location of TDC number 1 is to remove the valve cover to access the camshaft sprockets and alignment marks (see Section 8).

8   After the number one piston has been positioned at TDC on the compression stroke, TDC for any of the remaining cylinders can be located by turning the crankshaft 120-degrees clockwise and following the firing order (refer to this Chapter's Specifications). For example, rotating the engine 120-degrees past TDC number 1 will put the engine at TDC compression for cylinder number 2.

4   Valve covers - removal and installation

Removal

Refer to illustration 4.11

1   Disconnect the negative battery cable from the remote ground terminal (see Chapter 5).

2   Remove the engine cover.

3   Remove the upper intake manifold (see Section 5).
Note: *Cover open ports on the intake to prevent debris from entering the engine.*
Caution: *Once the valve covers are removed, the magnetic timing wheels are exposed* (see illustration 9.9). *The magnetic timing wheels on the camshafts must not come in contact with any type of magnet or magnetic field. If contact is made, the timing wheels will have to be replaced.*

4   Remove insulator from the left valve cover by lifting the insulator up and off of the retaining posts, then remove it from the front valve cover.

5   Before removing the variable valve timing solenoid connectors from the front of each valve cover, mark them appropriately so they can be reinstalled in their original locations.

6   Disconnect the wiring harness retainers from the valve cover and move the harnesses

out of the way.

7   Remove the ignition coils on both sides of the engine (see Chapter 5).

8   Mark the Camshaft Position (CMP) sensors to each valve cover so they can be reinstalled in their original locations, then remove the sensor(s) (see Chapter 6).

9   Remove the PCV valve from the rear cover (see Chapter 1).

10   Remove the bolts for the transmission dipstick tube and engine oil dipstick tube.

11   Remove the valve cover fasteners **(see illustration)** and remove the cover(s).
Caution: *If the cover is stuck to the cylinder head, tap one end with a block of wood and a hammer to jar it loose. If that doesn't work, slip a flexible putty knife between the cylinder head and cover to break the gasket seal. Don't pry at the cover-to-cylinder head joint or damage to the sealing surfaces may occur (leading to future oil leaks).*

12   Remove and discard the valve cover gasket, then remove the spark plug tube seals.
Note: *The cover gaskets can be reused if they are not damaged.*

Installation

13   The mating surfaces of each cylinder head and valve cover must be perfectly clean when the covers are installed. Use a gasket scraper to remove all traces of sealant and old gasket material, then clean the mating surfaces with brake system cleaner. If there's sealant or oil on the mating surfaces when the cover is installed, oil leaks may develop.

14   Inspect the spark plug tube seals; if damaged, carefully remove the seals using an appropriate pry tool. Position the new seal with the part number facing the valve cover, then use a socket that contacts the outer edge to drive the seal in place.

15   Apply a dab of RTV sealant at the joints where the timing chain cover meets the cylinder head.

16   Install the valve cover and bolts, then tighten the bolts to the torque listed in this Chapter's Specifications.

17   The remainder of installation is the reverse of removal.

4.11 Left side valve cover mounting bolts

5.5 Loosen the resonator hose clamp, remove the attaching pin and remove the resonator

5   Intake manifolds - removal and installation

Warning: *Wait until the engine is completely cool before beginning this procedure.*

Removal

1   If you will be removing the lower intake manifold, relieve the fuel system pressure (see Chapter 4A).

2   Disconnect the negative battery cable from the remote ground terminal (see Chapter 5).

3   If equipped, remove the engine cover.

Upper intake manifold

Refer to illustrations 5.5, 5.6, 5.8, 5.9, 5.11 and 5.12

4   Disconnect the IAT sensor electrical connector then remove the top half of the air filter housing (see Chapter 1).

5   Remove the intake resonator **(see illustration)**.

6   Disconnect the wiring harness from the MAP sensor and the Electronic Throttle Control (ETC) **(see illustration)**.

7   Disconnect the PCV valve hose (see Chapter 1), vapor purge hose and brake booster hoses.

8   Disconnect the wiring harness retainers from the upper intake support bracket and the retainer from the stud bolt **(see illustration)**.

9   Remove the nuts and stud bolt, then remove the upper intake manifold bracket **(see illustration)**.

10   Remove the brake booster hose-to-support bracket retainer.

11   Remove the support bracket-to-upper manifold nuts **(see illustration)**.

12   Loosen the upper intake manifold bolts

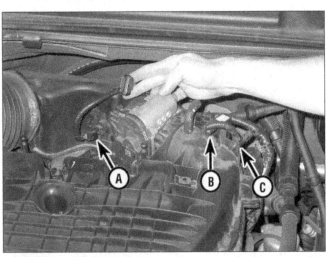

5.6 Disconnect the ETC connectors (A), the MAP sensor (B) and the electrical harness (C) retainer

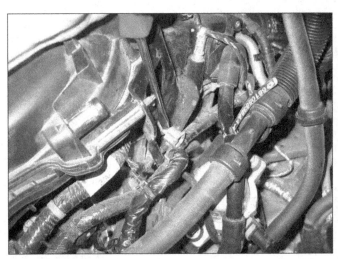

5.8 Pry the wiring harness retainer off of the bracket stud

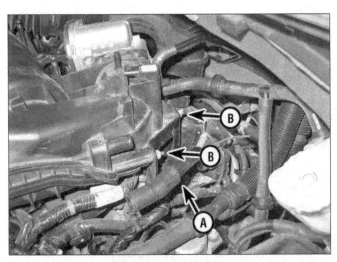

5.9 Remove the stud bolt (A) and bracket nuts (B), and remove the bracket

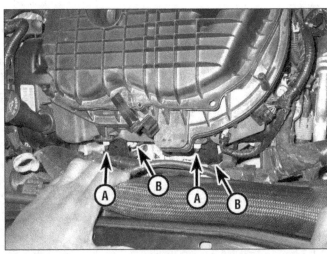

5.11 Remove the support bracket upper nuts (A), loosen the lower nuts (B) and remove the brackets from the upper intake manifold

and remove the upper intake manifold **(see illustration)**.
Note: *The bolts are captive – don't attempt to remove them from the manifold.*
13 Discard the six upper-to-lower intake manifold seals, and cover the open intake ports to prevent debris from entering the engine.
14 If required, remove the insulator from the left valve cover **(see Section 4)**.

Lower intake manifold

Refer to illustration 5.19

15 Remove the upper intake manifold (see Steps 4 through 14).
16 Disconnect the fuel line to the fuel rail (see Chapter 4A).
17 Remove the fuel injectors and fuel rail (see Chapter 4A).
Note: *The lower intake manifold can be removed with the injectors and fuel rail in place, if desired. Be careful not to damage the fuel injectors once the manifold is removed.*
18 Pry the wiring harness retainer from the end of the manifold and move the harness out of the way.

19 Remove the lower intake manifold bolts **(see illustration)**, and remove the manifold from the cylinder heads.
20 Discard the six manifold-to-cylinder head seals.

Installation

Lower intake manifold

Refer to illustrations 5.19, 5.24 and 5.25

Note: *The mating surfaces of the cylinder heads, cylinder block, and the intake manifold must be perfectly clean when the lower intake manifold is installed. Gasket removal solvents are available at most auto parts stores and may be helpful when removing old gasket material that's stuck to the cylinder heads, cylinder block and lower intake manifold (the lower intake manifold is made of aluminum - aggressive scrapping can cause damage). Be sure to follow the instructions printed on the solvent container.*

21 Use a gasket scraper to remove all traces of sealant and old gasket material, then clean the mating surfaces with brake system cleaner. If there's old sealant or oil on the mating surfaces when the lower intake manifold is installed, oil or vacuum leaks may develop. Use a vacuum cleaner to remove gasket material that falls into the intake ports or the lifter valley.
22 If removed, install the fuel injectors and the fuel rail (see Chapter 4A).
23 Install new intake manifold seals to the manifold.
Note: *Remove any rags or towels used in the manifold ports.*
24 Carefully lower the lower intake manifold into place **(see illustration)** and install the mounting bolts finger-tight.
25 Tighten the mounting bolts in steps, following the tightening sequence **(see illustration)**, to the torque listed in this Chapter's Specifications.
26 Install the upper intake manifold.

Upper intake manifold

Refer to illustration 5.30

27 Check the condition of the rubber seals that are installed into each intake runner on

5.12 Location of the upper intake manifold bolts

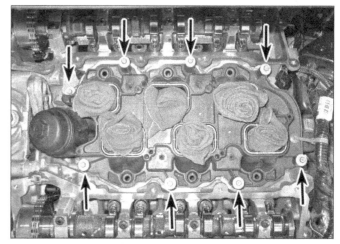

5.19 Lower intake manifold bolt locations

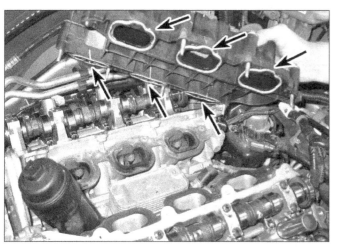

5.24 Install the manifold, making sure the new intake seals do not fall out of the manifold

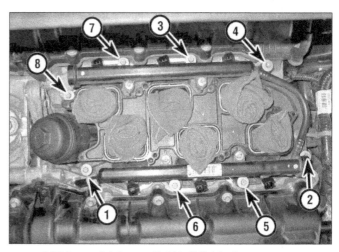

5.25 Lower intake manifold bolt tightening sequence

the upper intake manifold. If they are damaged, replace them.

28   Place the insulator on the mounting pins, if removed.

29   Install the upper intake manifold onto the lower intake manifold while pulling the bolts up.

Note: *The bolts are specially made for the composite material and should be turned slowly to prevent damage to the upper intake manifold.*

30   Tighten the mounting bolts in sequence **(see illustration)** to the torque listed in this Chapter's Specifications.

31   The remainder of installation is the reverse of removal.

---

6   Crankshaft pulley - removal and installation

Removal

Refer to illustration 6.3

1   Disconnect the negative battery cable from the remote ground terminal (see Chapter 5).

2   Remove the drivebelt (see Chapter 1).

3   The crankshaft pulley bolt is incredibly tight; using a breaker bar, socket and special tool #10198 or equivalent **(see illustration)**, hold the pulley from turning while loosening the bolt.

4   Pull the crankshaft pulley off the crankshaft.

Installation

5   Apply clean engine oil or multi-purpose grease to the seal contact surface of the pulley hub (if it isn't lubricated, the seal lip could be damaged and oil leakage would result).

6   Install the crankshaft pulley, aligning the keyway on the crankshaft with the slot in the pulley hub. Install the bolt and tighten it by hand.

7   Prevent the engine from rotating (see Step 3) then tighten the bolt to the torque listed in this Chapter's Specifications.

8   The remainder of installation is the reverse of removal.

7   Crankshaft front oil seal - replacement

Refer to illustrations 7.2 , 7.3 and 7.5

1   Remove the crankshaft pulley (see Section 6).

2   Use a screwdriver or hook tool to carefully pry out the seal **(see illustration)**.

Note: *Be careful not to damage the oil pump cover bore where the seal is seated, or the nose and sealing surface of the crankshaft.*

3   Another method for removing the seal is to drill a small hole on each side of the seal and place a self-tapping screw in each hole **(see illustration)**. Use these screws as a means of pulling the seal out without having to pry on it.

4   If the seal is being replaced when the timing chain cover is removed, support the cover on top of two blocks of wood and drive the seal out from the backside with a hammer and punch.

Caution: *Be careful not to scratch, gouge or distort the area that the seal fits into or a leak will develop.*

5.30   Upper intake manifold bolt tightening sequence

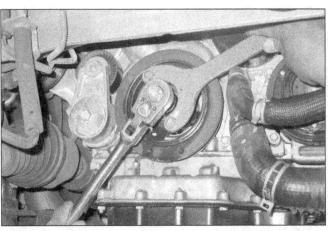

6.3   Using a special holding tool to prevent the crankshaft from turning, loosen, then remove the bolt

7.2   Use a hook tool to pry the seal from the timing chain cover

7.3   Another way of removing an old oil seal is to screw a self-tapping screw partially into the seal, then use pliers as a lever to pull it from the engine

5   Apply clean engine oil or multi-purpose grease to the outer edge of the new seal, then install it in the cover with the lip (spring side) facing IN. Drive the seal into place with a seal driver or a large socket and a hammer **(see illustration)**. Make sure the seal enters the bore squarely, then stop when the front face is at the proper depth.

6   Check the surface on the pulley hub that the oil seal rides on. If the surface has been grooved from long-time contact with the seal, the pulley will have to be replaced.

7   Lubricate the pulley hub with clean engine oil, then install it (see Section 6).

8   Timing chain cover, chain and sprockets - removal, inspection and installation

Warning: Wait until the engine is completely cool before beginning this procedure.

Caution: *The timing system is complex, and severe engine damage will occur if you make any mistakes. Do not attempt this procedure unless you are highly experienced with this type of repair. If you are at all unsure of your abilities, be sure to consult an expert. Double-check all your work and be sure everything is correct before you attempt to start the engine.*

Caution: *Do not rotate the crankshaft or cam-shafts separately during this procedure (with the timing chains removed), as damage to the valves will occur.*

Note: *Several special tools are required to complete these procedures, so read through the entire Section and obtain the special tools before beginning work.*

Removal

Timing chain cover

Refer to illustration 8.13

1   Disconnect the negative battery cable from the remote ground terminal (see Chapter 5).

2   Drain the engine coolant and engine oil (see Chapter 1).

3   Remove the air intake resonator (see Chapter 4A).

4   Remove the brake vacuum pump (see Chapter 9).

5   Remove the drivebelt, drivebelt tensioner and both idler pulleys (see Chapter 1).

6   Remove the thermostat housing and radiator hoses, and disconnect the heater hose from the water pump (see Chapter 3).

7   Remove the heater core supply pipe fasteners from the rear cylinder head and move the pipe out of the way.

8   Remove the alternator (see Chapter 5).

9   Remove the crankshaft pulley (see Section 6).

10   Remove the upper intake manifold (see Section 5).

11   Remove the valve covers (see Section 4).

Caution: *Once the valve covers are removed, the magnetic timing wheels are exposed* **(see illustration 9.9)***. The magnetic timing wheels on the camshafts must not come in contact with any type of magnet or magnetic field. If contact is made, the timing wheels will have to be replaced.*

12   Remove the upper and lower oil pans (see Section 11).

13   Remove the timing chain cover (M6-size) mounting bolts **(see illustration)**. There are seven indented prying points, one on top and three on each side; carefully pry the cover free of the engine block and cylinder heads. If it still sticks, slip a putty knife between the engine block and cover to break the bond (but be careful not to scratch the surfaces).

14   If necessary, remove the coolant outlet housing and water pump from the timing chain cover (see Chapter 3).

Timing chain

Refer to illustrations 8.15 and 8.17

Warning: *When the timing chains are removed, do not rotate the camshafts or crankshaft; the valves and pistons can be damaged if contact is made.*

15   Temporarily install the crankshaft pulley bolt. Turn the crankshaft with the bolt to TDC number 1, on the exhaust stroke to align the timing marks on the crankshaft and camshaft sprockets. Rotate the engine clockwise only, until the mark on the crankshaft aligns with the line made where the engine block and bearing cap meets **(see illustration)**.

7.5 Drive the seal squarely into the cover using a socket and hammer

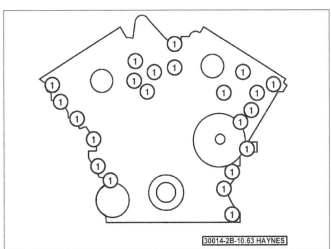

8.13 Timing chain cover bolt locations (1) - *M6 size bolts*

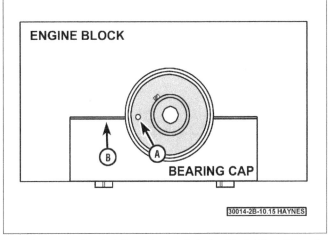

8.15 Align the dimple (A) on the crankshaft with the line (B) made where the engine block and bearing cap meets

16   On the left-side camshaft phaser, the machined scribe lines should be facing away from each other, and the arrows should be pointing towards each other in a parallel line with the gasket surface of the cylinder head. On the right-side camshaft phaser, the arrows should be facing away from each other and the machined scribe lines should be pointing towards each other in a parallel line with the gasket surface of the cylinder head **(see illustrations 9.10a and 9.10b)**. If, when you align the crankshaft mark with the bearing cap parting line, the camshaft marks are not in alignment as shown in **illustration 8.59**, rotate the engine one full revolution, realign the crankshaft mark, and verify that the camshaft marks are in proper alignment.

17   Verify the phaser marks are aligned with the plated links; if the plated links cannot be distinguished, make sure there are 12 pins between the two marks **(see illustration)**.

Note: *Use paint or a permanent marker to mark the direction of rotation on all chains before removing them so they can be installed in the same direction.*

18   Starting with the right side chain tensioner, press the tensioner plunger in until special tool #8514 or a 3 mm Allen wrench can be inserted through both small holes in the top and bottom of the tensioner body, holding the plunger in the compressed position.

19   Working on the left side chain tensioner, locate the access hole on the side of the tensioner. Working through the hole, lift and hold the pawl off of the rack of the plunger in the tensioner. Press the plunger in until special tool #8514 or a 3 mm Allen wrench can be inserted through both small holes in the top and bottom of the tensioner body, holding the plunger in the compressed position.

20   Remove the timing gear splash shield fasteners, then remove the shield from the oil pump housing.

21   Remove the oil pump tensioner and sprocket (see Section 12), then remove the oil pump chain from the crankshaft gear.

Note: *The oil pump chain and sprocket do not have to be timed, but the chain should be marked to make sure it is installed in the same direction of rotation.*

22   Starting with the right-side chain, slide camshaft phaser lock tool #10202-1 from the

front, between the two camshaft phasers, towards the chain (with the tool number facing up).

Note: *It may be necessary to rotate the intake camshaft a few degrees using a wrench on the camshaft flat when installing the phaser lock tool.*

23   Using a large wrench on the camshaft flats and a socket and ratchet on the oil control valves, loosen, but do not remove, the oil control valves.

24   Remove the right side camshaft phaser lock tool, then unscrew the intake camshaft oil control valve from the center of the phaser.

25   Slide the intake camshaft phaser off of the end of the camshaft, then remove the right-side timing chain.

Note: *If necessary, remove the exhaust camshaft oil control valve from the center of the phaser and remove the phaser.*

26   Working on the left-side chain, slide camshaft phaser lock tool #10202-2 from the front, between the two camshaft phasers, towards the chain (with the tool number facing up).

Note: *It may be necessary to rotate the intake camshaft a few degrees using a wrench on the camshaft flat when installing the phaser lock tool.*

27   Using a large wrench on the camshaft flats and a socket and ratchet on the oil control valves, loosen, but do not remove, the oil control valves

28   Remove the left-side camshaft phaser lock tool, then unscrew the exhaust camshaft oil control valve from the center of the phaser.

29   Slide the exhaust camshaft phaser off of the end of the camshaft, then remove the left side timing chain.

Note: *If necessary, remove the intake camshaft oil control valve from the center of the phaser and remove the phaser.*

30   Locate the primary chain tensioner to the side of the crankshaft chain and press the tensioner plunger in until special tool #8514 or a 3 mm Allen wrench can be inserted through the small hole in the side of the tensioner body, holding the plunger in the compressed position.

31   With the tensioner in the compressed position, remove the TORX (T30) mounting fasteners and the tensioner.

32   Remove the primary chain guide TORX (T30) mounting fasteners and the guide.

33   Remove the idler sprocket TORX (T45) mounting fastener and washer, then remove the idler sprocket, primary chain and crankshaft sprocket.

Note: *The chain should be marked to make sure it is installed in the same direction of rotation.*

34   If necessary, remove the chain tensioner (T30) fasteners and remove the tensioner(s), keeping the tensioners in the compressed position.

35   If necessary, remove the chain guide fasteners and guides for both chains.

Inspection

36   Inspect the timing chain dampener (guide) for cracks and wear and replace it, if necessary.

37   Clean the timing chain and sprockets with solvent and dry them with compressed air (if available).

Warning: *Wear eye protection when using compressed air.*

38   Inspect the components for wear and damage. Look for teeth that are deformed, chipped, pitted, and cracked.

39   The timing chain and sprockets should be replaced with new ones if the engine has high mileage, the chain has visible damage, or total freeplay midway between the sprockets exceeds one inch. Failure to replace a worn timing chain and sprockets may result in erratic engine performance, loss of power, and decreased fuel mileage. Loose chains can jump timing. In the worst case, chain jumping or breakage will result in severe engine damage.

Installation

Refer to illustrations 8.46 and 8.59

Caution: *Before starting the engine, carefully rotate the crankshaft by hand through at least two full revolutions (use a socket and breaker bar on the crankshaft pulley center bolt). If you feel any resistance, STOP! There is something wrong - most likely, valves are contacting the pistons. You must find the problem before proceeding.*

40   Use a plastic gasket scraper to remove all traces of old gasket material and sealant from the cover, engine block and cylinder heads. The cover is made of aluminum, so be careful not to nick or gouge it. Only clean the gasket sealing surfaces with rubbing alcohol (isopropyl) - do not use any oil based fluids.

41   If removed, install the chain guides and tensioners (still in the compressed position).

42   Make sure the keyway is installed on the crankshaft and the dimple on the crankshaft is aligned with the line made where the engine block and bearing cap meets **(see illustration 8.15)**.

43   Verify the camshafts are at TDC, with the alignment holes pointing up **(see illustration 9.29)**.

44   Place the primary chain on the crankshaft sprocket, with the plated link of the primary chain aligned with the arrow on the bottom of the sprocket. Insert the idler sprocket into the chain, aligning the other plated link

8.17 Verify that there are 12 pins between the mark on each phaser

with the machined mark on the idler sprocket.

45   Using clean engine oil, coat the sprockets and chain. Install the assembly while keeping the marks aligned, then install the idler sprocket mounting fastener finger tight.

46   Check the alignment of the marks; the plated link on the idler sprocket should be on top (12 o'clock) and the machined mark on the crankshaft should be aligned with the line made where the engine block and bearing cap meet **(see illustration)**. If the marks are all aligned, tighten the idler sprocket fastener to the torque listed in this Chapter's Specifications.

47   Install the primary chain guide and tensioner, then tighten the fasteners to the torque listed in this Chapter's Specifications. Remove the special tool from the tensioner plunger.

48   Starting with the left-side chain, install the intake camshaft phaser and oil control valve, then tighten the valve finger tight.

49   Place the left side chain over the intake phaser and around the inside cogs of the idler sprocket so that the plate link of the chain is aligned with the machined arrow on the sprocket.

50   With the chain aligned at the idler sprocket, install the exhaust camshaft phaser so that the arrows are pointing towards each other and in a parallel line with the cylinder head gasket surface **(see illustration 9.10a)**, then install the oil control valve finger tight.

51   Slide camshaft phaser lock tool #10202-2 from the front, between the two camshaft phasers towards the chain with the tool number facing up.

52   Using a large wrench on the camshaft flats and a socket and ratchet on the oil control valves, tighten both valves to the torque listed in this Chapter's Specifications.

53   Working on the right-side chain, install the exhaust camshaft phaser and oil control valve, tightening the valve finger tight.

54   Place the right side chain over the intake phaser and around the outside cogs of the idler sprocket so that the plate link of the chain is aligned with the machined circle on the sprocket.

55   With the chain aligned at the idler sprocket, install the intake camshaft phaser so that the machined lines are pointing towards each other and in a parallel line with the gasket surface of the cylinder head **(see illustration 9.10b)**, then install the oil control valve finger tight.

56   Slide camshaft phaser lock tool #10202-1 from the front, between the two camshaft phasers, towards the chain (with the tool number facing up).

57   Using a large wrench on the camshaft flats and a socket and ratchet on the oil control valves, tighten both valves to the torque listed in this Chapter's Specifications.

58   Install the oil pump timing chain, tensioner, sprocket and splash shield, if equipped (see Section 12).

Note: *There are no timing marks on the oil pump chain or sprocket.*

59   Verify all the marks are aligned **(see illustration)**, then remove the special tool

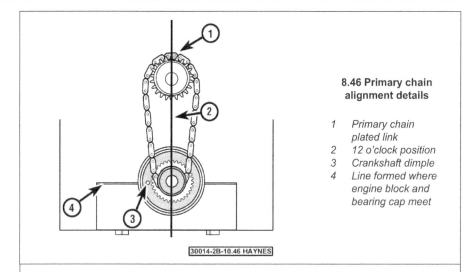

8.46 Primary chain alignment details

1   *Primary chain plated link*
2   *12 o'clock position*
3   *Crankshaft dimple*
4   *Line formed where engine block and bearing cap meet*

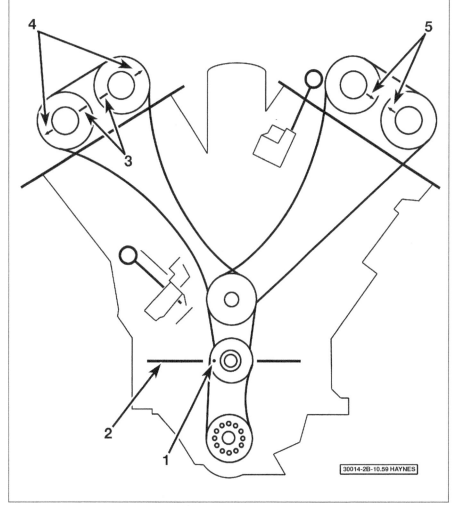

8.59 Timing mark alignment details

1   *Dimple on crankshaft*
2   *Junction of main bearing cap and cylinder block*
3   *Lines on right bank cam phasers - must be pointing toward each other and parallel with cylinder head*
4   *Arrows on right bank cam phasers - must be pointing away from each other*
5   *Arrows on left bank cam phasers - must be pointing toward each other*

or Allen wrenches from the primary and secondary tensioners. Also remove the camshaft phaser lock tools.

60   Rotate the engine two complete turns using the machined mark on the crankshaft with the line made where the engine block and bearing caps meet as the reference. Verify all the marks are aligned and there are 12 pins between the phaser marks **(see illustration 8.17)**; if the marks are off, rotate the engine two more complete turns and check again.

61   Once the timing marks are correct, install the new coolant housing and water pump housing gaskets into the grooves on the back side of the timing chain cover.

62   Apply a 1/8-inch wide by 1/16-inch high, bead of RTV sealant to the sealing surface of the cover, then install the cover on the alignment dowels.

63   Install the cover bolts **(see illustration 8.13)** and tighten them in a criss-cross pattern, in three steps, to the torque listed in this Chapter's Specifications.

64   The remainder of installation is the reverse of removal.

65   Add oil and coolant (see Chapter 1), start the engine and check for leaks.

9   Camshaft(s) – removal, inspection and installation

Caution: *The timing system is complex, and severe engine damage will occur if you make any mistakes. Do not attempt this procedure unless you are highly experienced with this type of repair. If you are at all unsure of your abilities, be sure to consult an expert. Double-check all your work and be sure everything is correct before you attempt to start the engine.*
Caution: *Once the valve covers are removed, the magnetic timing wheels are exposed. The magnetic timing wheels on the camshafts must not come in contact with any type of magnet or magnetic field. If contact is made, the timing wheels will need to be replaced.*
Note: *The timing chain for each camshaft can be removed from the camshafts individually, without removing all the timing chains, using the tools outlined in this Section. If the tools are not available, the timing chain cover and all chains will need to be removed before the camshafts can be removed (see Section 8).*

Removal

Refer to illustrations 9.9, 9.10a and 9.10b

1   Disconnect the cable from the negative terminal of the battery (see Chapter 5).

2   Drain the engine oil and coolant, then remove the drivebelt (see Chapter 1).

3   Remove the air filter housing (see Chapter 4) and resonator **(see illustration 5.5)**.

4   Remove the intake manifolds (see Section 5).

5   Disconnect all wires and vacuum hoses from the cylinder heads. Label them to simplify reinstallation.

6   Label the ignition coils to simplify reinstallation then disconnect and remove the ignition coils (see Chapter 1).

7   Remove the spark plugs (see Chapter 1).

8   Remove the valve covers (see Section 4).

9   Once the valve covers are removed, the magnetic timing wheels are exposed **(see illustration)**. The magnetic timing wheels on the camshafts must not come in contact with any type of magnet or magnetic field. If contact is made, the timing wheels will need to be replaced.

10   Rotate the crankshaft clockwise and place the #1 piston at TDC on the exhaust stroke. On the left side camshaft phaser, the machined scribe lines should be facing away from each other, and the arrows should be pointing towards each other in a parallel line with the gasket surface of the cylinder head. On the right side camshaft phaser, the arrows should be facing away from each other, and the machined scribe lines should be pointing towards each other in a parallel line with the gaskets surface of the cylinder head **(see illustrations)**.

11   Using a permanent marker or paint, mark the camshaft phasers to the timing chains for reinstallation.

12   Working from the top of the timing chain cover, insert special tool #10200-3 down the

9.9 Location of the magnetic timing wheels

9.10a With the engine at TDC #1, the left camshaft phaser scribe marks (A) should be pointing away from each other, the arrow marks (B) should be pointing towards each other in a straight line and that line should be parallel with the cylinder head surface

9.10b The right phaser scribe marks should be pointing towards each other in a straight line (and that line should be parallel with the cylinder head surface)

side of the tensioner to the access hole on the side of the tensioner. Working through the small hole in the side of the tensioner, lift and hold the pawl off of the rack of the plunger in the tensioner. Slide chain holding tool #10200-1 between the cylinder head and the back side of the chain against the chain guide, forcing the rack and plunger back into the tensioner body.

Caution: *The chain holding tool must remain in place while the phasers are removed or the timing chain will fall off into the timing chain cover.*

13   Slide camshaft phaser lock tool #10202-1 (right side) or 10202-2 (left side), from the front, between the two camshaft phasers, towards the chain.

Note: *It may be necessary to rotate the intake camshaft a few degrees using a wrench on the camshaft flat when installing the phaser lock tool.*

14   Using a large wrench on the camshaft flats and a socket and ratchet on the oil control valves, loosen, then remove each of the oil control valves from the phaser end of the camshaft.

15   At the same time, carefully slide both the intake and exhaust phaser (with the phaser lock securely between them) forward until they are off the end of the camshafts.

Caution: *Do not remove the phaser lock or try to disassemble the phasers.*

16   Using the alignment holes in the camshaft as a reference point, slowly rotate both camshafts counterclockwise approximately 30-degrees Before Top Dead Center (BTDC). In this position, the camshafts are in a neutral or no load position.

Note: *The camshaft bearing caps are marked with a number and letter code; " 1I " is for the number one Intake camshaft bearing cap. The notch on the caps should always be installed towards the front.*

17   Loosen the camshaft bearing cap bolts in the reverse of the tightening sequence **(see illustration 9.28)**.

18   Remove the camshaft bearing caps and

carefully lift the camshafts from the cylinder head.

19   Mark the rocker arms so they can be installed in their original locations, then remove them.

20   Mark the hydraulic lash adjusters so they can be installed in their original locations, then remove them from the cylinder head.

Inspection

Refer to illustration 9.22

21   Check the camshaft bearing surfaces for pitting, score marks, galling, and abnormal wear. If the bearing surfaces are damaged, the cylinder head will have to be replaced.

22   Compare the camshaft lobe height by measuring each lobe with a micrometer **(see illustration)**. Measure each of the intake lobes and record the measurements and relative positions. Then measure each of the exhaust lobes and record the measurements and relative positions also. This will let you compare all of the intake lobes to one another and all of the exhaust lobes to one another. If the difference between the lobes exceeds 0.005 inch, the camshaft should be replaced. Do not compare intake lobe heights to exhaust lobe heights as lobe lift may be different. Only compare intake lobes to intake lobes and exhaust lobes to exhaust lobes for this comparison.

23   Check the rocker arms and shafts for abnormal wear, pits, galling, score marks, and rough spots. Don't attempt to restore rocker arms by grinding the pad surfaces. Replace defective parts.

Installation

Refer to illustrations 9.28 and 9.29

Caution: *Before starting the engine, carefully rotate the crankshaft by hand through at least two full revolutions (use a socket and breaker bar on the crankshaft pulley center bolt). If you feel any resistance, STOP! There is something wrong - most likely, valves are contacting the pistons. You must find the problem before proceeding.*

24   Dip the hydraulic lash adjusters in clean engine oil and install them into their original locations.

25   Apply moly-base grease or engine assembly lube to the rocker arm contact points and rollers and install them into their original locations.

26   Lubricate the camshaft bearing journals and lobes with moly-base grease or engine assembly lube, then install them carefully in the cylinder head about 30-degrees before (counterclockwise of) TDC. Don't scratch the bearing surfaces with the cam lobes!

Caution: *Do not rotate the camshafts more than a few degrees to prevent the valves from contacting the pistons.*

27   Install the camshaft bearing caps, then install the mounting bolts and finger tighten them.

28   Tighten the bearing caps in sequence **(see illustration)** to the torque listed in this Chapter's Specifications.

29   Rotate the camshafts clockwise 30-degrees, and verify the alignment holes in the camshafts are in the 12 o'clock (pointing straight up) or neutral position **(see illustration)**.

9.22 Use a micrometer to measure cam lobe height

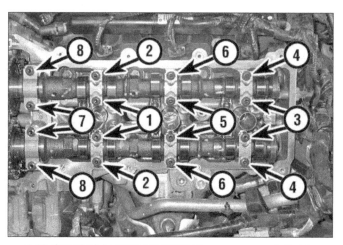

9.28 Camshaft bearing cap tightening sequence – left side shown, right side is identical

9.29 Locate the alignment holes on the camshafts and make sure they are in the neutral position (pointing straight up) – right side shown, left side is identical

30   Carefully slide both the intake and exhaust phaser (with the phaser lock tool securely between them) onto the camshafts and verify the marks are aligned.

31   Install the oil control valves onto the camshaft phasers and install the bolts, then tighten the bolts to the torque listed in this Chapter's Specifications. Remove the chain holding tool and release the tensioner plunger.

Caution: *Be sure to prevent the camshafts from turning by holding the camshaft with a large wrench on the camshaft flats.*

32   Slowly rotate the engine two complete turns (720-degrees) and verify the alignment marks are correct **(see illustrations 9.10a and 9.10b).**

33   The remainder of installation is the reverse of removal.

10   Cylinder heads - removal and installation

Warning: *Wait until the engine is completely cool before beginning this procedure.*

Removal

1   Relieve the fuel system pressure, then disconnect the cable from the negative terminal of the battery (see Chapter 5).

2   Drain the engine oil and coolant (see Chapter 1).

3   Remove the drivebelt (see Chapter 1) and the idler pulleys.

4   Remove the air filter housing and resonator (see Chapter 4A).

5   Remove the intake manifolds (see Section 5).

6   Disconnect all wires and vacuum hoses from the cylinder heads. Label them to simplify reinstallation.

7   Disconnect the ignition coils and remove the spark plugs (see Chapter 1).

8   Remove the catalytic converter(s) (see Chapter 6).

9   Remove the valve covers (see Section 4).

10   Remove the crankshaft pulley (see Section 6).

11   Remove the oil pans (see Section 11).

12   If you're working on the left cylinder head:

   • *Remove the alternator (see Chapter 5).*
   • *Remove the transmission and oil dipstick tube fasteners and remove the oil dipstick tube from the oil pan.*
   • *Unbolt the air conditioning compressor and support it out of the way (see Chapter 3)*

Warning: *Do not disconnect the refrigerant lines.*

   • *Disconnect the main engine harness connectors at the rear of the cylinder and move the harness and retainers out of the way.*

13   If you're working on the right cylinder head:

   • *Remove the heater core tube fasteners and move the tube away from the cylinder head.*

14   Remove the timing chain cover (see Section 8).

15   Rotate the crankshaft clockwise and place the #1 piston at TDC on the exhaust stroke. When the crankshaft is at TDC, the dimple on the crankshaft will be in line with the line made where the bearing cap meets the engine block. The left cylinder bank cam phaser arrows should be pointing toward each other and be parallel with where the cylinder head and valve cover meets. The right side cam phaser arrows should point away from each other and the lines on the phasers should be pointing towards each other **(see illustration 8.59).**

16   Remove the timing chain for the cylinder head or, if both cylinder heads are being removed, remove both chains (see Section 8).

17   Remove the oil control valves from the cam phasers (sprockets) (see Section 9).

18   Remove the timing chain tensioner and chain guides (see Section 8).

19   Remove the camshafts, rocker arms and lash adjusters (see Section 9).

Caution: *Once the valve covers are removed the magnetic timing wheels are exposed. The magnetic timing wheels on the camshafts must not come in contact with any type of magnet or magnetic field. If contact is made the timing wheels will have to be replaced (see Section 9).*

Note: *Keep the rocker arms and lash adjusters in order so that they can be installed in their original locations.*

20   Loosen the cylinder head bolts in the reverse of the tightening sequence **(see illustrations 10.29a and 10.29b).**

21   Lift the cylinder head off the block. If resistance is felt, dislodge the cylinder head by striking it with a wood block and hammer. If prying is required, pry only on a casting protrusion - be very careful not to damage the cylinder head or block!

Caution: *Do not set the cylinder head on its gasket side; the sealing surface can be easily damaged.*

22   Have the cylinder head inspected and serviced by a qualified automotive machine shop.

Installation

Refer to illustrations 10.29a and 10.29b

23   The mating surfaces of each cylinder head and the engine block must be perfectly clean when the cylinder head is installed.

24   Carefully use a gasket scraper to remove all traces of carbon and old gasket material, then clean the mating surfaces with brake system cleaner. If there's oil on the mating surfaces when the cylinder head is installed, the gasket may not seal correctly and leaks may develop.

25   When working on the engine block, it's a good idea to cover the lifter valley with shop rags to keep debris out of the engine. Use a shop rag or vacuum cleaner to remove any debris that falls into the cylinders.

26   Check the engine block and cylinder head mating surfaces for nicks, deep scratches, and other damage. If damage is slight, it can be removed with a file; if it's excessive, machining may be the only alternative.

27   Position the new gasket over the dowel pins in the engine block. Some gaskets are marked TOP or FRONT to ensure correct installation.

28   Carefully position the cylinder head on the engine block without disturbing the gasket.

29   Install NEW cylinder head bolts and tighten them in the recommended sequence **(see illustrations)** to the torque steps listed in this Chapter's Specifications.

Caution: *Do not use a torque wrench for steps requiring additional rotation or turns; apply a paint mark to the bolt head or use a torque-angle gauge (available at most auto parts stores) and a socket and breaker bar.*

30   The remainder of installation is the reverse of removal.

31   Change the engine oil and filter (see Chapter 1).

32   Refill the cooling system (see Chapter 1). Start the engine and check for leaks and proper operation.

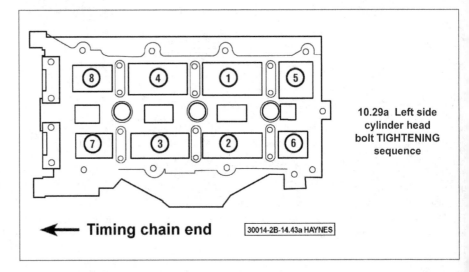

10.29a Left side cylinder head bolt TIGHTENING sequence

← **Timing chain end**   30014-2B-14.43a HAYNES

11 Oil pans - removal and installation

Removal

1   Disconnect the cable from the negative terminal of the battery (see Chapter 5).
2   Raise the front of the vehicle and support it securely on jackstands. Apply the parking brake and block the rear wheels to keep it from rolling off the stands.
3   Drain the engine oil (see Chapter 1).
4   Remove the lower splash shield fasteners and skid plate fasteners, if equipped.

Lower oil pan

5   Remove the bolts and nuts around the perimeter of the pan, then carefully separate the lower oil pan from the upper oil pan. Don't pry between the upper pan and the lower pan or damage to the sealing surfaces could occur and oil leaks may develop. Tap the pan with a soft-face hammer to break the gasket seal. If it still sticks, slip a putty knife between the upper pan and lower pan to break the bond (but be careful not to scratch the surfaces).

Upper oil pan

6   Remove the power steering gear (see Chapter 10).
7   Remove the dipstick tube bracket mounting bolt. Using a twisting motion, pull the dipstick tube out of the upper oil pan.
8   Remove the lower oil pan (see Step 5).
9   Remove the coolant tube-to-upper pan fastener and move the tube back.
10   If you're working on the 4WD models:
  • *Separate the front differential from the engine mounts (see Chapter 8).*
Note: *The front differential assembly and CV joints do not have to be completely removed from the vehicle.*
  • *Connect an engine hoist to the engine (see Chapter 2E), then remove the engine mounts (see Section 16) and brackets.*

  • *Raise the engine enough to access the oil pan bolts.*
11   Remove the four lower oil pan-to-transmission mounting bolts.
12   Remove the torque converter access plate, and the rubber plugs just below the plate.
13   Remove the two rubber plugs then the two upper pan-to-rear main seal housing bolts (M6 size).
Caution: *The oil pan-to-rear main seal bolts are hard to see and can easily be missed. If they are not removed, the rear main seal housing will be severely damaged when the pan is lowered.*
14   Remove the nineteen upper oil pan bolts (M8 size) around the perimeter of the pan, then carefully separate the oil pan from the engine block. Use the two indented prying points on each side of the oil pan to carefully pry the pan free of the engine block. If it still sticks, slip a putty knife between the engine block and oil pan to break the bond (but be careful not to scratch the surfaces).
Note: *On 4WD models, then engine may have to be raised or lowered slightly to allow the oil pan to be removed.*

Installation

15   Clean the pan(s) with solvent and remove all old sealant and gasket material from the engine block and pan mating surfaces. Clean the mating surfaces with lacquer thinner or acetone and make sure the bolt holes in the engine block are clear. Check the oil pan flange(s) for distortion, particularly around the bolt holes. If necessary, place the pan(s) on a wood block and use a hammer to flatten and restore the gasket surface.

Upper oil pan

16   Apply a 1/8-inch wide by 1/16-inch high bead of RTV sealant to the sealing surface of the pan. Install the upper pan and the bolts, then tighten the bolts finger-tight.

17   Tighten the pan-to-transmission bolts to the torque listed in this Chapter's Specifications.
18   Tighten the remaining bolts in a circular pattern, starting from the middle and working your way outwards, to the torque listed in this Chapter's Specifications.
19   The remainder of installation is the reverse of removal.
20   Refill the engine with oil (see Chapter 1). Start and run the engine until normal operating temperature is reached, then check for leaks.

Lower oil pan

21   Apply a 1/8-inch wide by 1/16-inch high bead of RTV sealant to the sealing surface of the pan. Install the lower pan and the bolts.
22   Tighten the bolts in a circular pattern, starting from the middle and working your way outwards, to the torque listed in this Chapter's Specifications.
23   The remainder of installation is the reverse of removal.
24   Refill the engine with oil (see Chapter 1). Start and run the engine until normal operating temperature is reached, then check for leaks.

12 Oil pump - removal, inspection and installation

Removal

1   Disconnect the cable from the negative terminal of the battery (see Chapter 5).
2   Raise the front of the vehicle and support it securely on jackstands. Apply the parking brake and block the rear wheels to keep it from rolling off the stands.
3   Drain the engine oil (see Chapter 1).
4   Remove the skid plate, if equipped, and the lower splash shield fasteners then remove the splash shield.
5   Remove the lower and upper oil pans (see Section 11).
6   Remove the oil pump pick-up tube fastener, and remove the tube from the pump. Discard the pick-up tube O-ring.
7   Disconnect the oil pump solenoid electrical connector from the side of the engine.
8   Working from the side of the block, depress the oil pump solenoid electrical connector locking tab and push the connector into the block.
Note: *The connector will have to be maneuvered around the tensioner mounting bolt.*
9   Remove the oil pump timing gear splash shield bolts and remove the splash shield.
10   Press the oil pump chain tensioner away from the chain until a 3 mm Allen wrench can be inserted into the housing to hold the tensioner back.
11   Using a permanent marker or paint, make reference marks on the chain and oil pump gear.
12   Hold the oil pump gear from moving, then remove the T45 Torx mounting bolt and

10.29b Right side cylinder head bolt TIGHTENING sequence

30014-2B-14.43b HAYNES   **Timing chain end →**

the oil pump gear.

13   Hold the tensioner and remove the Allen wrench, allowing the tensioner to release. Remove the spring from the dowel pin and slide the tensioner from the oil pump.

14   Remove the oil pump mounting bolts and remove the pump.

Inspection

Refer to illustrations 12.16, 12.17, 12.19, 12.20 and 12.21

Note: *The oil pump is not serviceable; if there is a problem, the pump assembly must be replaced.*

15   Clean all parts thoroughly in solvent and carefully inspect the rotors, pump cover, and timing chain cover for nicks, scratches, or burrs. Replace the assembly if it is damaged.

16   Use a straightedge and a feeler gauge to measure the oil pump cover for warpage **(see illustration)**. If it's warped more than the

limit listed in this Chapter's Specifications, the pump should be replaced.

17   Measure the thickness of the outer rotor **(see illustration)**. If the thickness is less than the value listed in this Chapter's Specifications, the pump should be replaced.

18   Measure the thickness of the inner rotor. If the thickness is less than the value listed in this Chapter's Specifications, the pump should be replaced.

19   Insert the outer rotor into the oil pump housing and measure the clearance between the rotor and housing **(see illustration)**. If the measurement is more than the maximum allowable clearance listed in this Chapter's Specifications, the pump should be replaced.

20   Install the inner rotor in the oil pump assembly and measure the clearance between the lobes on the inner and outer rotors **(see illustration)**. If the clearance is more than the value listed in this Chapter's

Specifications, the pump should be replaced.

21   Place a straightedge across the face of the oil pump assembly **(see illustration)**. If the clearance between the pump surface and the rotors is greater than the limit listed in this Chapter's Specifications, the pump should be replaced.

Installation

22   Place the oil pump onto the engine block using the aligning dowels. Install the mounting bolts and tighten them to the torque listed in this Chapter's Specifications.

23   Slide the oil pump chain tensioner onto the pivot, then push the tensioner back against the spring. Insert a 3 mm Allen wrench into the tensioner to hold it in place.

24   Place the oil pump timing chain gear into the chain, center it onto the oil pump shaft and install the T45 mounting bolt. Tighten the bolt to the torque listed in this Chapter's Specifications.

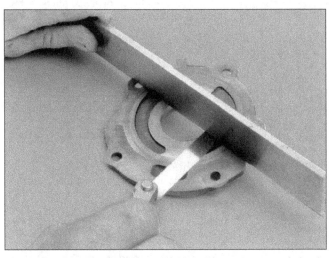

12.16 Place a straightedge across the oil pump cover and check it for warpage with a feeler gauge

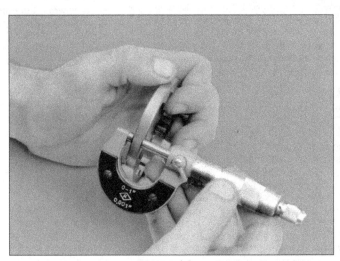

12.17 Use a micrometer to measure the thickness of the outer rotor

12.19 Check the outer rotor-to-housing clearance with a feeler gauge

12.20 Check the clearance between the lobes of the inner and outer rotors

Note: *Make sure the gear is facing the same way as when it was removed (see Step 11). There are no timing marks on the pump gear or chain, and no timing is necessary.*

25 Maneuver the oil pump solenoid into position and insert it through the block opening until it snaps in place.

26 Install the timing gear splash shield and bolts, then tighten the bolts to the torque listed in this Chapter's Specifications.

27 The remainder of installation is the reverse of removal.

28 Refill the engine with oil and change the oil filter (see Chapter 1).

---

13 Oil cooler - removal and installation

Warning: *Wait until the engine is completely cool before beginning this procedure.*

Removal

Refer to illustration 13.4

1 Relieve the fuel system pressure, then disconnect the cable from the negative terminal of the battery (see Chapter 5).

2 Drain the coolant (see Chapter 1).

3 Remove the lower intake manifold (see Section 5).

4 Remove the oil cooler mounting fasteners **(see illustration)**.

5 Remove the oil cooler and discard the seals.

Installation

6 Install new O-ring seals to the oil cooler.
Caution: *Do not lubricate the O-ring seals – install them dry.*

7 Place the oil cooler onto the block and install the two mounting screws.

8 Install the mounting bolts and tighten the screws and bolts to the torque listed in this

Chapter's Specifications.

9 The remainder of installation is the reverse of removal.

10 Change the engine oil and filter and refill the cooling system (see Chapter 1). Run the engine until normal operating temperature is reached, and check for leaks.

---

14 Driveplate - removal and installation

Removal

1 Raise the vehicle and support it securely on jackstands.

2 Remove the transmission (see Chapter 7B).

3 To ensure correct alignment during reinstallation, match-mark the driveplate and backing plate to the crankshaft so they can be reassembled in the same position.

4 Remove the bolts that hold the driveplate to the crankshaft. A special tool is available at most auto parts stores to hold the driveplate while loosening the bolts. If the tool is not available, wedge a screwdriver in the starter ring gear teeth to jam the driveplate.

5 Remove the driveplate from the crankshaft. The driveplate is fairly heavy; be sure to support it while removing the last bolt.

6 Clean the driveplate to remove grease and oil. Inspect the driveplate for damage or other defects.

7 Clean and inspect the mating surfaces of the driveplate and the crankshaft.

8 If the crankshaft rear main seal is leaking, replace it before reinstalling the driveplate (see Section 15).

Installation

9 Position the driveplate and backing plate against the crankshaft. Align the previously

applied matchmarks. Before installing the bolts, apply thread-locking compound to the threads.

10 Hold the driveplate with the holding tool, or wedge a screwdriver in the starter ring gear teeth to keep the driveplate from turning. Tighten the bolts to the torque listed in this Chapter's Specifications in a criss-cross pattern.

11 The remainder of installation is the reverse of removal.

---

15 Rear main oil seal - replacement

1 Remove the oil pans (see Section 11).

2 Remove the driveplate (see Section 14).

3 Unbolt the seal retainer from the engine block and slide the retainer and seal off the end of the crankshaft.

Note: *The rear main oil seal has been incorporated into the seal retainer and must be replaced as a unit.*

4 Clean the engine block, oil pan and crankshaft.

5 The new seal and retainer assembly comes with a plastic installation sleeve; make sure it's in place.

6 Apply a 1/4-inch bead of RTV sealant where the lower corners of the seal retainer meet the oil pan.

7 Place the assembly over the crankshaft and push it squarely into place, making sure the dowels in the seal retainer engage the locating holes in the engine block.

8 Install, but don't fully tighten, the seal retainer bolts.

9 Remove the plastic installation sleeve, then tighten the bolts, using an alternating pattern, to the torque listed in this Chapter's Specifications.

10 The remainder of installation is the reverse of removal.

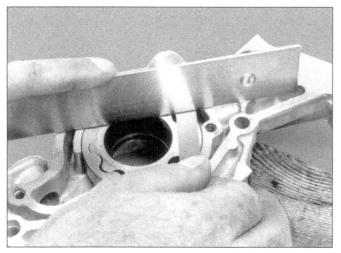

12.21 Using a straightedge and feeler gauge, check the clearance between the surface of the oil pump cover and the rotors

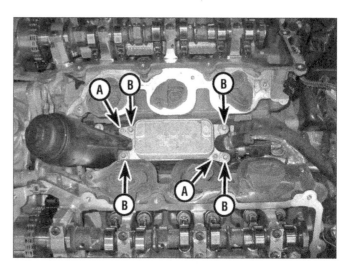

13.4 Remove the oil cooler mounting screws (A) and mounting bolts (B)

16 Engine mounts - check and replacement

1   The engine mounting system on these models consists of three molded mounts. The front (right and left) mounts support the engine assembly while the rear mount support the transmission and restricts torqueing action of the powertrain.

2   Engine mounts seldom require attention, but broken or deteriorated mounts should be replaced immediately, or the added strain placed on driveline components may cause damage or accelerated wear.

Check

3   During the check, the engine must be raised slightly to remove the weight from the mounts.

4   Raise the vehicle and support it securely on jackstands, then position a jack under the engine oil pan. Place a large wood block between the jack head and the oil pan to prevent oil pan damage, then carefully raise the engine just enough to take the weight off the mounts.

Warning: *DO NOT place any part of your body under the engine when it's supported only by a jack!*

5   Check the mounts to see if the rubber is cracked, hardened or separated from the metal backing. Sometimes the rubber will split right down the center.

6   Check for relative movement between the mount plates and the engine or frame (use a large screwdriver or pry bar to attempt to move the mounts). If movement is noted, lower the engine and tighten the mount fasteners.

7   Rubber preservative may be applied to the mounts to slow deterioration.

Replacement

Front mounts

This procedure is essentially the same as for the 3.7L V6 and the 4.&L V8 engines. Refer to Chapter 2A and follow the procedure outlined there but refer to the Specifications listed in this Chapter for the proper torque.

Rear mounts

Refer to Chapter 7B and follow the procedures outlined there, including torque specifications.

Chapter 2 Part E
General engine overhaul procedures

Contents

Specifications

General

Displacement

3.6L V6 engine	220 cubic inches
3.7L V6 engine	226 cubic inches
4.7L V8 engine	287 cubic inches
5.7L Hemi engine	345 cubic inches
6.4L Hemi engine	392 cubic inches
6.7L diesel engine	409 cubic inches

Bore and stroke

3.6L V6 engine	3.779 x 3.268 inches
3.7L V6 engine	3.66 x 3.40 inches
4.7L V8 engine	3.66 x 3.40 inches
5.7L Hemi engine	3.91 x 3.58 inches
6.4L Hemi engine	4.09 x 3.72 inches
6.7L diesel engine	4.21 x 4.88 inches

Cylinder compression pressure

3.6L V6 engine

Minimum	100 psi
Maximum variation between cylinders	25 psi

3.7L V6 and 4.7L V8 engines

Minimum	170 psi
Maximum variation between cylinders	40 psi

5.7L Hemi engine	Pressure should not vary more than 40 psi between cylinders

6.4L Hemi engine

Minimum	100 psi
Maximum variation between cylinders	25 psi

6.7L diesel engine

Minimum	350 psi
Maximum variation between cylinders	70 psi

Note: *Checking compression on the diesel engine requires several special tools in addition to a special compression gauge, therefore the procedure is not included in this manual.*

Oil pump pressure

3.6L V6 engine

Minimum operating pressure	5 psi
Operating pressure	30 (warm) – 139 (cold) between 1,201 to 3,500 rpm

3.7L V6, 4.7L V8, 5.7L and 6.4L Hemi engines

Minimum pressure at curb idle	5 psi
Operating pressure	25 to 110 psi at 3,000 rpm

6.7L diesel engine

Minimum pressure at curb idle	10 psi
Operating pressure	30 psi at 2,500 rpm

Torque specifications

Ft-lbs (unless otherwise indicated)

Note: *One foot-pound (ft-lb) of torque is equivalent to 12 inch-pounds (in-lbs) of torque. Torque values below approximately 15 ft-lbs are expressed in inch-pounds, since most foot-pound torque wrenches are not accurate at these smaller values.*

Connecting rod cap bolts
   3.6L V6 engine
      Step 1 15
      Step 2 .... Tighten an additional 90-degrees
   3.7L V6 and 4.7L V8 engines
      Step 1 20
      Step 2 .... Tighten an additional 90-degrees
   5.7L Hemi engine
      Step 1 15
      Step 2 .... Tighten an additional 90-degrees
   6.4L Hemi engine
      Step 1 30
      Step 2 .... Tighten an additional 90-degrees
   6.7L diesel engine
      Step 1 22
      Step 2 44
      Step 3 .... Tighten an additional 60-degrees
Main bearing cap bolts/bedplate assembly
   3.6L V6 engine **(see illustrations 10.19g 10.19h and 10.19i)**
      Inner bolts (M11 bolts)
         Step 1 15
         Step 2, bolts 1 through 8 .... Tighten an additional 90-degrees
      Outer bolts and windage tray (M8 bolts)
         Step 1 16
         Step 2, bolts 1 through 8 .... Tighten an additional 90-degrees
      Side bolts (crossbolts) 20
   3.7L V6 engine bedplate fasteners* **(see illustration 10.19a)**
      Step 1, bolts 4, 7 and 6.. Hand-tighten until bedplate contacts block mating surface
      Step 2, bolts 11 through 22.. 40
      Step 3, bolts 11 through 18.. 60 in-lbs
      Step 4, bolts 11 through 18.. Tighten an additional 90-degrees
      Step 5, bolts 23 through 28.. 20
   4.7L V8 engine bedplate fasteners** **(see illustration 10.19b)**
      Step 1, bolts 1 through 22... 26
      Step 2, bolts 11 through 22.. 40
      Step 3, bolts 1 through 10... Tighten an additional 90-degrees
      Step 4, bolts 23 through 28.. 20
   5.7L and 6.4L Hemi engine*** **(see illustration 10.19f)**
      Step 1, bolts 1 through 10... 120 in-lbs
      Step 2, bolts 1 through 10... 20
      Step 3, bolts 1 through 10... Tighten an additional 90-degrees
      Step 4, crossbolts 11 through 20 21
      Step 5, crossbolts 11 through 20 .... 21 (re-check)
   6.7L diesel engine
      Used bolts
         Step 1 44
         Step 2 59
         Step 3 .... Tighten an additional 90-degrees
      New bolts
         Step 1 89
         Step 2 .... Loosen completely
         Step 3 44
         Step 4 63
         Step 5 .... Tighten an additional 120-degrees

Install studs at positions 1 and 2
**Install studs at positions 3, 5 and 6*
***Install a new washer/seal onto each crossbolt before tightening. First, install all the main bearing bolts and main bearing crossbolts finger tight. Next torque the main bearing bolts completely (Steps 1 and 2) before torquing the crossbolts*

1.1 An engine block being bored. An engine rebuilder will use special machinery to recondition the cylinder bores

1.2 If the cylinders are bored, the machine shop will normally hone the engine on a machine like this

1 General information - engine overhaul

Refer to illustrations 1.1, 1.2, 1.3, 1.4, 1.5 and 1.6

Included in this portion of Chapter 2 are general information and diagnostic testing procedures for determining the overall mechanical condition of your engine.

The information ranges from advice concerning preparation for an overhaul and the purchase of replacement parts and/or components to detailed, step-by-step procedures covering removal and installation.

The following Sections have been written to help you determine whether your engine needs to be overhauled and how to remove and install it once you've determined it needs to be rebuilt. For information concerning in-vehicle engine repair, see Chapter 2A, 2B, 2C or 2D.

The Specifications included in this Part are general in nature and include only those necessary for testing the oil pressure and checking the engine compression. Refer to Chapter 2A, 2B, 2C or 2D for additional engine Specifications.

It's not always easy to determine when, or if, an engine should be completely overhauled, because a number of factors must be considered.

High mileage is not necessarily an indication that an overhaul is needed, while low mileage doesn't preclude the need for an overhaul. Frequency of servicing is probably the most important consideration. An engine that's had regular and frequent oil and filter changes, as well as other required maintenance, will most likely give many thousands of miles of reliable service. Conversely, a neglected engine may require an overhaul very early in its service life.

Excessive oil consumption is an indication that piston rings, valve seals and/or valve guides are in need of attention. Make sure that oil leaks aren't responsible before deciding that the rings and/or guides are bad. Perform a cylinder compression check to determine

the extent of the work required (see Section 3). Also, on gasoline engines, check the vacuum readings under various conditions (see Section 4).

Check the oil pressure with a gauge installed in place of the oil pressure sending unit and compare it to this Chapter's Specifications (see Section 2). If it's extremely low, the bearings and/or oil pump are probably worn out.

Loss of power, rough running, knocking or metallic engine noises, excessive valve train noise and high fuel consumption rates may also point to the need for an overhaul, especially if they're all present at the same time. If a complete tune-up doesn't remedy the situation, major mechanical work is the only solution.

An engine overhaul involves restoring the internal parts to the specifications of a new engine. During an overhaul, the piston rings are replaced and the cylinder walls are reconditioned (rebored and/or honed) **(see illustrations 1.1 and 1.2)**. If a rebore is done by an automotive machine shop, new oversize pistons will also be installed. The main bearings, connecting rod bearings and camshaft

bearings are generally replaced with new ones and, if necessary, the crankshaft may be reground to restore the journals **(see illustration 1.3)**. Generally, the valves are serviced as well, since they're usually in less-than-perfect condition at this point. While the engine is being overhauled, other components, such as the distributor, starter and alternator, can be rebuilt as well. The end result should be similar to a new engine that will give many trouble free miles. **Note:** *Critical cooling system components such as the hoses, drivebelts, thermostat and water pump should be replaced with new parts when an engine is overhauled. The radiator should be checked carefully to ensure that it isn't clogged or leaking (see Chapter 3). If you purchase a rebuilt engine or short block, some rebuilders will not warranty their engines unless the radiator has been professionally flushed. Also, we don't recommend overhauling the oil pump - always install a new one when an engine is rebuilt.*

Overhauling the internal components on today's engines is a difficult and time-consuming task which requires a significant amount of specialty tools and is best left to a professional engine rebuilder **(see illustrations 1.4, 1.5 and**

1.3 A crankshaft having a main bearing journal ground

1.4 A machinist checks for a bent connecting rod, using specialized equipment

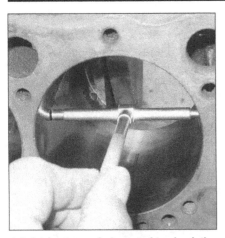

1.5 A bore gauge being used to check the main bearing bore

1.6 Uneven piston wear like this indicates a bent connecting rod

2.2a Location of the oil pressure sending unit on 3.7L V6 and 4.7L V8 engines

1.6). A competent engine rebuilder will handle the inspection of your old parts and offer advice concerning the reconditioning or replacement of the original engine, never purchase parts or have machine work done on other components until the block has been thoroughly inspected by a professional machine shop. As a general rule, time is the primary cost of an overhaul, especially since the vehicle may be tied up for a minimum of two weeks or more. Be aware that some engine builders only have the capability to rebuild the engine you bring them while other rebuilders have a large inventory of rebuilt exchange engines in stock. Also be aware that many machine shops could take as much as two weeks time to completely rebuild your engine depending on shop workload. Sometimes it makes more sense to simply exchange your engine for another engine that's already rebuilt to save time.

2   Oil pressure check

Refer to illustrations 2.2a, 2.2b, 2.2c and 2.3

1   Low engine oil pressure can be a sign of an engine in need of rebuilding. A low oil pressure indicator (often called an "idiot light") is not a test of the oiling system. Such indicators only come on when the oil pressure is dangerously low. Even a factory oil pressure gauge in the instrument panel is only a relative indication, although much better for driver information than a warning light. A better test is with a mechanical (not electrical) oil pressure gauge.

2   Locate the oil pressure indicator sending unit on the engine block. The oil pressure sending unit is located in various locations depending upon engine type:

a) *On 3.6L engines, the oil pressure sending unit (sensor) is located on the oil filter housing under the intake manifold; it's the sensor protruding straight out from the housing. To access the sending unit, the intake manifold must be removed (see Chapter 2D).*

b) *On 3.7L V6 and 4.7L V8 engines, the oil pressure sending unit is located near the timing chain cover next to the oil filter* **(see illustration).**

c) *On Hemi engines, the oil pressure sending unit is located on the oil filter housing above the oil filter* **(see illustration).**

d) *On diesel engines, the oil pressure sending unit is located on the side of the engine block, under the fuel transfer pump* **(see illustration).**

3   Unscrew and remove the oil pressure sending unit, then screw in the hose for your oil pressure gauge **(see illustration).** If necessary, install an adapter fitting. Use Teflon tape or thread sealant on the threads of the adapter and/or the fitting on the end of your gauge's hose.

4   If you're working on a 3.6L V6 engine, reinstall the intake manifolds, guiding the pressure gauge hose out next to the sensor's wiring harness.

5   Connect an accurate tachometer to the engine, according to the tachometer manufacturer's instructions.

6   Check the oil pressure with the engine running (normal operating temperature) at the specified engine speed, and compare it to this Chapter's Specifications. If it's extremely low, the bearings and/or oil pump are probably worn out.

2.2b Location of the oil pressure sending unit on Hemi engines

2.2c Location of the oil pressure sending unit on diesel engines

2.3 Install an oil pressure gauge into the block after removing the oil pressure sending unit - Hemi engine shown

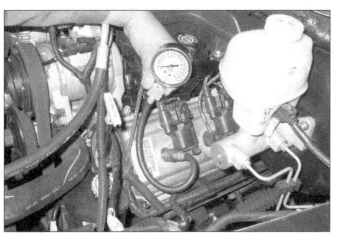

3.6 Use a compression gauge with a threaded fitting for the spark plug hole, not the type that requires hand pressure to maintain the seal

4.4 A simple vacuum gauge can be handy in diagnosing engine condition and performance

3   Cylinder compression check

Gasoline engines

Refer to illustration 3.6

1   A compression check will tell you what mechanical condition the upper end of your engine (pistons, rings, valves, head gaskets) is in. Specifically, it can tell you if the compression is down due to leakage caused by worn piston rings, defective valves and seats or a blown head gasket. **Note:** *The engine must be at normal operating temperature and the battery must be fully charged for this check.*

2   Begin by cleaning the area around the spark plugs before you remove them (compressed air should be used, if available). The idea is to prevent dirt from getting into the cylinders as the compression check is being done.

3   Disable the ignition and fuel systems by unplugging the primary (low voltage) electrical connector from the ignition coil assemblies (see Chapter 5) and by removing the fuel pump fuse (see Chapter 4A, Section 2).

4   On engines with two spark plugs per cylinder, remove one spark plug from each cylinder. On engines with one spark plug per cylinder, remove all of the spark plugs (see Chapter 1).

5   Block the throttle wide open.

6   Install a compression gauge in the spark plug hole **(see illustration)**.

7   Crank the engine over at least seven compression strokes and watch the gauge. The compression should build up quickly in a healthy engine. Low compression on the first stroke, followed by gradually increasing pressure on successive strokes, indicates worn piston rings. A low compression reading on the first stroke, which doesn't build up during successive strokes, indicates leaking valves or a blown head gasket (a cracked head could also be the cause). Deposits on the undersides of the valve heads can also cause low compression. Record the highest gauge reading obtained.

8   Repeat the procedure for the remaining cylinders and compare the results to this Chapter's Specifications.

9   Add some engine oil (about three squirts from a plunger-type oil can) to each cylinder, through the spark plug hole, and repeat the test.

10   If the compression increases after the oil is added, the piston rings are definitely worn. If the compression doesn't increase significantly, the leakage is occurring at the valves or head gasket. Leakage past the valves may be caused by burned valve seats and/or faces or warped, cracked or bent valves.

11   If two adjacent cylinders have equally low compression, there's a strong possibility that the head gasket between them is blown. The appearance of coolant in the combustion chambers or the crankcase would verify this condition.

12   If one cylinder is slightly lower than the others, and the engine has a slightly rough idle, a worn lobe on the camshaft could be the cause.

13   If the compression is unusually high, the combustion chambers are probably coated with carbon deposits. If that's the case, the cylinder head(s) should be removed and decarbonized.

14   If compression is way down or varies greatly between cylinders, it would be a good idea to have a leak-down test performed by an automotive repair shop. This test will pinpoint exactly where the leakage is occurring and how severe it is.

Diesel engine

15   To test the compression on the diesel engine, the fuel injectors must be removed using special tools. A special compression gauge must be installed to register the higher compression readings. Have the compression checked by a dealer service department or other qualified automotive repair facility.

4   Vacuum gauge diagnostic checks

Refer to illustrations 4.4 and 4.6

Note: *This procedure applies to gasoline engines only (diesel engines don't create manifold vacuum).*

1   A vacuum gauge provides inexpensive but valuable information about what is going on in the engine. You can check for worn rings or cylinder walls, leaking head or intake manifold gaskets, incorrect carburetor adjustments, restricted exhaust, stuck or burned valves, weak valve springs, improper ignition or valve timing and ignition problems.

2   Unfortunately, vacuum gauge readings are easy to misinterpret, so they should be used in conjunction with other tests to confirm the diagnosis.

3   Both the absolute readings and the rate of needle movement are important for accurate interpretation. Most gauges measure vacuum in inches of mercury (in-Hg). The following references to vacuum assume the diagnosis is being performed at sea level. As elevation increases (or atmospheric pressure decreases), the reading will decrease. For every 1,000 foot increase in elevation above approximately 2,000 feet, the gauge readings will decrease about one inch of mercury.

4   Connect the vacuum gauge directly to the intake manifold vacuum, not to ported (throttle body) vacuum **(see illustration)**. Some models are equipped with a vacuum fitting built into the brake booster vacuum hose grommet at the brake booster. Other models are equipped with a vacuum hose fitting on the intake manifold. Use a T-fitting to access the vacuum signal. Be sure no hoses are left disconnected during the test or false readings will result.

5   Before you begin the test, allow the engine to warm up completely. Block the wheels and set the parking brake. With the

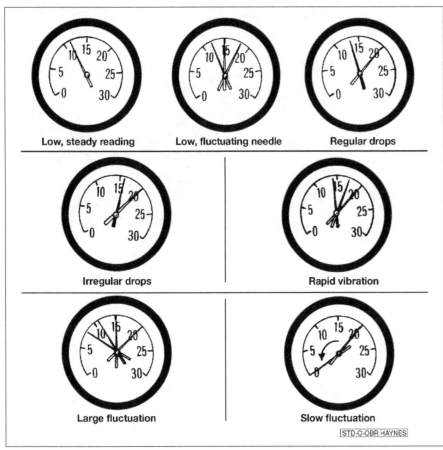

Low, steady reading   Low, fluctuating needle   Regular drops

Irregular drops   Rapid vibration

Large fluctuation   Slow fluctuation

STD-O-OBR HAYNES

4.6 Typical vacuum gauge readings

transmission in Park, start the engine and allow it to run at normal idle speed. **Warning:** *Keep your hands and the vacuum gauge clear of the fans.*

6   Read the vacuum gauge; an average, healthy engine should normally produce about 17 to 22 in-Hg with a fairly steady needle **(see illustration)**. Refer to the following vacuum gauge readings and what they indicate about the engine's condition:

7   A low, steady reading usually indicates a leaking gasket between the intake manifold and cylinder head(s) or throttle body, a leaky vacuum hose, late ignition timing or incorrect camshaft timing. Check ignition timing with a timing light and eliminate all other possible causes, utilizing the tests provided in this Chapter before you remove the timing chain cover to check the timing marks.

8   If the reading is three to eight inches below normal and it fluctuates at that low reading, suspect an intake manifold gasket leak at an intake port or a faulty fuel injector.

9   If the needle has regular drops of about two-to-four inches at a steady rate, the valves are probably leaking. Perform a compression check or leak-down test to confirm this.

10   An irregular drop or down-flick of the needle can be caused by a sticking valve or an ignition misfire. Perform a compression check or leak-down test and read the spark plugs.

11   A rapid vibration of about four in-Hg

vibration at idle combined with exhaust smoke indicates worn valve guides. Perform a leak-down test to confirm this. If the rapid vibration occurs with an increase in engine speed, check for a leaking intake manifold gasket or head gasket, weak valve springs, burned valves or ignition misfire.

12   A slight fluctuation, say one inch up and down, may mean ignition problems. Check all the usual tune-up items and, if necessary, run the engine on an ignition analyzer.

13   If there is a large fluctuation, perform a compression or leak-down test to look for a weak or dead cylinder or a blown head gasket.

14   If the needle moves slowly through a wide range, check for a clogged PCV system, incorrect idle fuel mixture, throttle body or intake manifold gasket leaks.

15   Check for a slow return after revving the engine by quickly snapping the throttle open until the engine reaches about 2,500 rpm and let it shut. Normally the reading should drop to near zero, rise above normal idle reading (about 5 in-Hg over) and then return to the previous idle reading. If the vacuum returns slowly and doesn't peak when the throttle is snapped shut, the rings may be worn. If there is a long delay, look for a restricted exhaust system (often the muffler or catalytic converter). An easy way to check this is to temporarily disconnect the exhaust ahead of the suspected part and redo the test.

5   Engine rebuilding alternatives

The do-it-yourselfer is faced with a number of options when purchasing a rebuilt engine. The major considerations are cost, warranty, parts availability and the time required for the rebuilder to complete the project. The decision to replace the engine block, piston/connecting rod assemblies and crankshaft depends on the final inspection results of your engine. Only then can you make a cost effective decision whether to have your engine overhauled or simply purchase an exchange engine for your vehicle.

Some of the rebuilding alternatives include:

Individual parts - If the inspection procedures reveal that the engine block and most engine components are in reusable condition, purchasing individual parts and having a rebuilder rebuild your engine may be the most economical alternative. The block, crankshaft and piston/connecting rod assemblies should all be inspected carefully by a machine shop first.

Short block - A short block consists of an engine block with a crankshaft and piston/connecting rod assemblies already installed. All new bearings are incorporated and all clearances will be correct. The existing camshafts, valve train components, cylinder head and external parts can be bolted to the short block with little or no machine shop work necessary.

Long block - A long block consists of a short block plus an oil pump, oil pan, cylinder head, valve cover, camshaft and valve train components, timing sprockets and chain or gears and timing cover. All components are installed with new bearings, seals and gaskets incorporated throughout. The installation of manifolds and external parts is all that's necessary.

Low mileage used engines - Some companies now offer low mileage used engines which is a very cost effective way to get your vehicle up and running again. These engines often come from vehicles which have been in totaled in accidents or come from other countries which have a higher vehicle turn over rate. A low mileage used engine also usually has a similar warranty like the newly remanufactured engines.

Give careful thought to which alternative is best for you and discuss the situation with local automotive machine shops, auto parts dealers and experienced rebuilders before ordering or purchasing replacement parts.

6   Engine removal - methods and precautions

Refer to illustrations 6.1, 6.2, and 6.3

If you've decided that an engine must be removed for overhaul or major repair work, several preliminary steps should be taken. Read all removal and installation procedures carefully prior to committing to this job.

6.1 After tightly wrapping water-vulnerable components, use a spray cleaner on everything, with particular concentration on the greasiest areas, usually around the valve cover and lower edges of the block. If one section dries out, apply more cleaner

6.2 Depending on how dirty the engine is, let the cleaner soak in according to the directions and then hose off the grime and cleaner. Get the rinse water down into every area you can get at; then dry important components with a hair dryer or paper towels

Locating a suitable place to work is extremely important. Adequate work space, along with storage space for the vehicle, will be needed. If a shop or garage isn't available, at the very least a flat, level, clean work surface made of concrete or asphalt is required.

Cleaning the engine compartment and engine before beginning the removal procedure will help keep tools clean and organized **(see illustrations 6.1 and 6.2)**.

An engine hoist will also be necessary. Make sure the hoist is rated in excess of the weight of the engine. Safety is of primary importance, considering the potential hazards involved in removing the engine from the vehicle. If you're removing a diesel engine, you'll need a heavy duty engine hoist and a heavy duty engine stand.

If you're a novice at engine removal, get at least one helper. One person cannot eas-

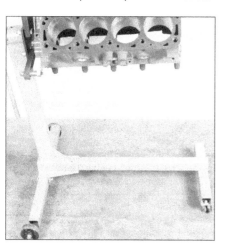

6.3 Get an engine stand sturdy enough to firmly support the engine while you're working on it. Stay away from three-wheeled models; they have a tendency to tip over more easily, so get a four-wheeled unit

ily do all the things you need to do to remove a big heavy engine from the engine compartment. It's also helpful to seek advice and assistance from someone who's experienced in engine removal.

Plan the operation ahead of time. Arrange for or obtain all of the tools and equipment you'll need prior to beginning the job **(see illustration 6.3)**. Some of the equipment necessary to perform engine removal and installation safely and with relative ease are (in addition to an engine hoist) a heavy duty floor jack (preferably fitted with a transmission jack head adapter), complete sets of wrenches and sockets as described in the front of this manual, wooden blocks, plenty of rags and cleaning solvent for mopping up spilled oil, coolant and gasoline.

Plan for the vehicle to be out of use for quite a while. A machine shop can do the work that is beyond the scope of the home mechanic. Machine shops often have a busy schedule, so before removing the engine, consult the shop for an estimate of how long it will take to rebuild or repair the components that may need work.

---

7   Engine - removal and installation

Warning: *Gasoline and diesel fuel are extremely flammable, so take extra precautions when you work on any part of the fuel system. Don't smoke or allow open flames or bare light bulbs near the work area, and don't work in a garage where a gas-type appliance (such as a water heater or clothes dryer) is present. Since fuel is carcinogenic, wear fuel-resistant gloves when there's a possibility of being exposed to fuel, and, if you spill any fuel on your skin, rinse it off immediately with soap and water. Mop up any spills immediately and do not store fuel-soaked rags where they could ignite. The fuel system is under constant pressure, so, if any*

fuel lines are to be disconnected, the fuel pressure in the system must be relieved first (see Chapter 4A or 4B for more information). When you perform any kind of work on the fuel system, wear safety glasses and have a Class B type fire extinguisher on hand.
Warning: *The air conditioning system is under high pressure. Do not loosen any hose fittings or remove any components until after the system has been discharged. Air conditioning refrigerant must be properly discharged into an EPA-approved recovery/recycling unit at a dealer service department or an automotive air conditioning repair facility. Always wear eye protection when disconnecting air conditioning system fittings.*
Warning: *The engine must be completely cool before beginning this procedure.*

Removal

Refer to illustrations 7.14, 7.41 and 7.51

1   Have the air conditioning system discharged by an automotive air conditioning technician.
2   On gasoline models, relieve the fuel system pressure (see Chapter 4A).
3   Disconnect the cable(s) from the negative battery terminal(s) (see Chapter 5).
4   Remove the hood (see Chapter 11) or disconnect the hood support struts and secure it all the way back with a strap. Use padding if there's danger of scratching painted areas.
5   Remove the air filter housing and the air intake duct (see Chapter 4A or 4B).
6   Drain the cooling system and remove the drivebelt (see Chapter 1).
7   Remove the radiator, shroud(s) and engine cooling fan (see Chapter 3). Also remove the upper crossmember and the radiator support.
8   Detach the radiator and heater hoses from the engine.
9   If equipped, remove the air conditioning condenser, automatic transmission oil cooler

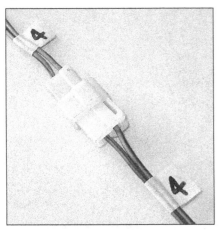

7.14 Label both ends of each wire or vacuum connection before disconnecting them

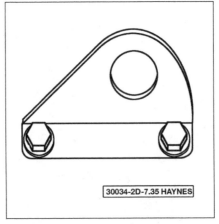

30034-2D-7.35 HAYNES

7.41 A bracket like this will be required for diesel engine removal

(see Chapter 3) and turbo intercooler (see Chapter 4B).

10   On diesel engines, remove the intercooler and piping (see Chapter 4B).

11   Detach all connections from the throttle body.

12   Remove the power steering pump without disconnecting the hoses and tie it out of the way (see Chapter 10).

13   Remove the alternator (see Chapter 5). **Note:** *On models equipped with dual alternators, remove the secondary drivebelt and alternator.*

14   Label and disconnect all wires from the engine **(see illustration).** Masking tape and/or a touch-up paint applicator work well for marking items. **Note:** *Take instant photos or sketch the locations of components and brackets to help with reassembly.*

15   Disconnect the fuel line from the engine (see Chapter 4A or 4B) and plug the openings.

16   Label and remove all vacuum lines between the engine and the firewall (or other components in the engine compartment).

17   Raise the vehicle and support it securely on jackstands. **Note:** *This step may not be necessary, because some models already have sufficient ground clearance to allow disconnection of the exhaust system, the engine mounts, etc. from underneath the vehicle. Raising these vehicles any higher might even make engine removal more difficult because it might position the vehicle too high to lift the engine out of the engine compartment with a hoist.*

18   On automatic transmission-equipped vehicles, detach the transmission cooler lines from the engine.

19   Drain the engine oil (see Chapter 1).

3.6L models

20   Remove the upper intake manifold and support bracket (see Chapter 2D).

21   Remove the vacuum pump and bracket (see Chapter 9).

22   Remove the coolant heater tubes and oil dipstick tube mounting bolts.

23   Disconnect and remove the stop/start solenoid (see Chapter 5).

24   Install the special side engine lifting brackets #10242-1 on each side of the engine using the bolts provided with the lifting brackets. Tighten the bolts to 15 ft-lbs.

3.7L V6 and 4.7L V8 models

25   On 3.7L V6 4WD models, disconnect the front driveshaft from the differential, then disconnect the exhaust cross pipe.

26   On 4.7L V8 4WD models, remove the axle bracket from the axle, engine and transmission.

27   Remove the oil pan support plate.

28   Remove both crankcase breather tubes from the valve covers.

29   Remove the intake manifold (see Chapter 2A).

30   Attach an engine lifting plate or brackets to the intake manifold mounting location, using the intake manifold bolts.

5.7L and 6.4L Hemi models

31   Remove the intake manifold (see Chapter 2B).

32   Remove the oil control valve and its connector.

33   On 6.4L models, remove the EGR cooler (see Chapter 6).

34   Attach engine lifting brackets to the front of the block and the rear of the intake manifold valley.

6.7L diesel models

35   Remove the valve cover (see Chapter 2C).

36   Remove the EGR cooler (see Chapter 1).

37   Remove the rocker arms, pedestals and pushrods for the rear three cylinders.

38   Remove the rear high pressure fuel line shield.

39   Remove the fuel lines from cylinders five and six.

40   Remove the fuel injectors from cylinders five and six.

41   Obtain or fabricate a lifting bracket that can be attached by using the head bolts at the rear of the engine **(see illustration).** Attach one lifting chain to this bracket and another to the existing lift bracket.

All models

42   Disconnect the exhaust pipes from the exhaust manifolds or, if you're working on a diesel, from the turbocharger outlet.

43   Remove the starter.

44   Support the engine from above with a hoist. Attach the hoist chain to the engine lifting brackets. If no brackets are present, you'll have to fasten the chains to some substantial part of the engine - one that is strong enough to take the weight, but in a location that will provide good balance. If you're attaching the chain to a stud on the engine, or are using a bolt passing through the chain and into a threaded hole, place a washer between the nut or bolt head and the chain, and tighten the nut or bolt securely. **Caution:** *Do not lift the engine by the intake manifold.*

45   On automatic transmission models, remove the torque converter-to-driveplate bolts (see Chapter 7B).

46   On manual transmission diesel models, remove the flywheel bolts.

47   Use the hoist to take the weight off the engine mounts, then remove the engine mount through-bolts (see Chapter 2A, 2B or 2C).

48   Support the transmission with a floor jack. Place a block of wood on the jack head to protect the transmission.

49   Unbolt the engine from the bellhousing.

50   Check to make sure everything is disconnected. All wires, hoses and other components must be labeled and secured so they won't interfere with lifting the engine.

51   Slowly lift the engine out of the vehicle while checking for anything that could stop it **(see illustration).** The engine will probably need to be tilted and/or maneuvered as it's lifted out, so have an assistant handy. **Warning:** *Do not place any part of your body under the engine when it is supported only by a hoist or other lifting device.*

52   Remove the flywheel/driveplate and mount the engine on an engine stand or set the engine on the floor and support it so it doesn't tip over. Disconnect the engine hoist.

Installation

53   Check the engine mounts. If they're worn or damaged, replace them.

54   On manual transmission models, inspect the clutch components (see Chapter 8). On automatic transmission-equipped models, inspect the front seal and bushing.

55   Apply a dab of grease to the pilot bearing on manual transmission models.

56   Attach the hoist to the engine, remove the engine from the engine stand and install the flywheel or driveplate (see Chapter 2A, 2B, 2C or 2D).

57   Carefully guide the engine into place, lowering it slowly and moving it back into the engine compartment until the engine mounts can be secured.

58   Tighten the transmission-to-engine bolts to the torque listed in the Chapter 7A or 7B Specifications.

59   On automatic transmission models, install the torque converter-to-driveplate bolts (see Chapter 7B). On manual transmission

7.51 Lift the engine carefully up and forward with the engine hoist, making sure everything is disconnected

9.1 Before you try to remove the pistons, use a ridge reamer to remove the raised material (ridge) from the top of the cylinders

diesel models, bolt the clutch to the flywheel.

60   Tighten all the bolts on the engine mounts and remove the hoist and jack.

61   Reinstall the remaining components in the reverse order of removal.

62   Add coolant, oil, power steering and transmission fluid as needed (see Chapter 1).

63   Run the engine and check for proper operation and leaks. Shut off the engine and recheck the fluid levels.

8   Engine overhaul - disassembly sequence

1   It's much easier to remove the external components if it's mounted on a portable engine stand. A stand can often be rented quite cheaply from an equipment rental yard. Before the engine is mounted on a stand, the flywheel/driveplate should be removed from the engine.

2   If a stand isn't available, it's possible to remove the external engine components with it blocked up on the floor. Be extra careful not to tip or drop the engine when working without a stand.

3   If you're going to obtain a rebuilt engine, all external components must come off first, to be transferred to the replacement engine. These components include:

   Flywheel/driveplate
   Ignition system components
   Emissions-related components
   Engine mounts and mount brackets
   Engine rear cover (spacer plate between flywheel/driveplate and engine block)
   Intake/exhaust manifolds
   Fuel injection components
   Oil filter
   Spark plug wires (if used) and spark plugs
   Thermostat and housing assembly
   Water pump

Note: *When removing the external components from the engine, pay close attention to details that may be helpful or important during installation. Note the installed position of gaskets, seals, spacers, pins, brackets, washers, bolts and other small items.*

4   If you're going to obtain a short block

9.3 Checking the connecting rod endplay (side clearance)

(assembled engine block, crankshaft, pistons and connecting rods), then remove the timing chain, cylinder head, oil pan, oil pump pick-up tube, oil pump and water pump from your engine so that you can turn in your old short block to the rebuilder as a core. See *Engine rebuilding alternatives* for additional information regarding the different possibilities to be considered.

9   Pistons and connecting rods - removal and installation

Removal

Refer to illustrations 9.1, 9.3 and 9.4

Note: *Prior to removing the piston/connecting rod assemblies, remove the cylinder head and oil pan (see Chapter 2A, 2B, 2C or 2D).*

1   Use your fingernail to feel if a ridge has formed at the upper limit of ring travel (about 1/4-inch down from the top of each cylinder). If carbon deposits or cylinder wear have produced ridges, they must be completely removed with a special tool **(see illustration)**. Follow the manufacturer's instructions provided with the tool. Failure to remove the ridges before attempting to remove the piston/connecting rod assemblies may result in piston breakage.

2   After the cylinder ridges have been

9.4 If the connecting rods or caps are not marked, use permanent ink or paint to mark the caps to the rods by cylinder number (for example, this would be number 4 cylinder connecting rod)

removed, turn the engine so the crankshaft is facing up.

3   Before the main bearing cap assembly and connecting rods are removed, check the connecting rod endplay with feeler gauges. Slide them between the first connecting rod and the crankshaft throw until the play is removed **(see illustration)**. Repeat this procedure for each connecting rod. The endplay is equal to the thickness of the feeler gauge(s). Check with an automotive machine shop for the endplay service limit (a typical end play limit should measure between 0.005 to 0.015 inch [0.127 to 0.381 mm]). If the play exceeds the service limit, new connecting rods will be required. If new rods (or a new crankshaft) are installed, the endplay may fall under the minimum allowable. If it does, the rods will have to be machined to restore it. If necessary, consult an automotive machine shop for advice.

4   Check the connecting rods and caps for identification marks. If they aren't plainly marked, use paint or marker **(see illustration)** to clearly identify each rod and cap (1, 2, 3, etc., depending on the cylinder they're associated with). Do not interchange the rod caps. Install the exact same rod cap onto the

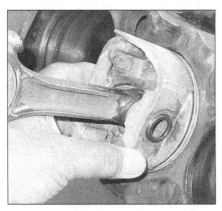

9.13 Install the piston ring into the cylinder then push it down into position using a piston so the ring will be square in the cylinder

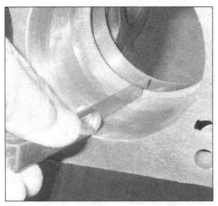

9.14 With the ring square in the cylinder, measure the ring end gap with a feeler gauge

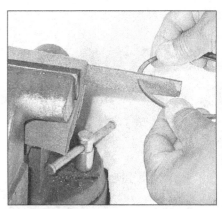

9.15 If the ring end gap is too small, clamp a file in a vise as shown and file the piston ring ends - remove all raised material

same connecting rod. **Caution:** *Do not use a punch and hammer to mark the connecting rods or they may be damaged.*

5   Loosen each of the connecting rod cap bolts or nuts 1/2-turn at a time until they can be removed by hand. **Note:** *On 3.7L, 4.7L and Hemi engines, new connecting rod cap bolts must be used when reassembling the engine, but save the old bolts for use when checking the connecting rod bearing oil clearance.*

6   Remove the number one connecting rod cap and bearing insert. Don't drop the bearing insert out of the cap.

7   Remove the bearing insert and push the connecting rod/piston assembly out through the top of the engine. Use a wooden or plastic hammer handle to push on the upper bearing surface in the connecting rod. If resistance is felt, double-check to make sure that all of the ridge was removed from the cylinder.

8   Repeat the procedure for the remaining cylinders.

9   After removal, reassemble the connecting rod caps and bearing inserts in their respective connecting rods and install the cap bolts finger tight. Leaving the old bearing inserts in place until reassembly will help prevent the connecting rod bearing surfaces from being accidentally nicked or gouged.

10   The pistons and connecting rods are now ready for inspection and overhaul at an automotive machine shop.

Piston ring installation

Refer to illustrations 9.13, 9.14, 9.15, 9.19a, 9.19b and 9.22

11   Before installing the new piston rings, the ring end gaps must be checked. It's assumed that the piston ring side clearance has been checked and verified correct.

12   Lay out the piston/connecting rod assemblies and the new ring sets so the ring sets will be matched with the same piston and cylinder during the end gap measurement and engine assembly.

13   Insert the top (number one) ring into the first cylinder and square it up with the cylinder walls by pushing it in with the top of the piston

(see illustration). The ring should be near the bottom of the cylinder, at the lower limit of ring travel.

14   To measure the end gap, slip feeler gauges between the ends of the ring until a gauge equal to the gap width is found **(see illustration)**. The feeler gauge should slide between the ring ends with a slight amount of drag. A typical ring gap should fall between 0.010 and 0.020 inch [0.25 to 0.50 mm] for compression rings (0.033 to 0.045 inch [0.85 to 1.15 mm] for the second compression ring on the diesel engine) and up to 0.030 inch [0.76 mm] for the oil ring steel rails. If the gap is larger or smaller than specified, double-check to make sure you have the correct rings before proceeding.

15   If the gap is too small, it must be enlarged or the ring ends may come in contact with each other during engine operation, which can cause serious damage to the engine. If necessary, increase the end gaps by filing the ring ends very carefully with a fine file. Mount the file in a vise equipped with soft jaws, slip the ring over the file with the ends contacting the file face and slowly move the ring to remove material from the ends. When performing this operation, file only by pushing the

ring from the outside end of the file towards the vise **(see illustration)**.

16   Excess end gap isn't critical unless it's greater than 0.040 inch (1.01 mm) (the exception to this is the second compression ring on the diesel). Again, double-check to make sure you have the correct ring type.

17   Repeat the procedure for each ring that will be installed in the first cylinder and for each ring in the remaining cylinders. Remember to keep rings, pistons and cylinders matched up.

18   Once the ring end gaps have been checked/corrected, the rings can be installed on the pistons.

19   The oil control ring (lowest one on the piston) is usually installed first. It's composed of three separate components. Slip the spacer/expander into the groove **(see illustration)**. If an anti-rotation tang is used, make sure it's inserted into the drilled hole in the ring groove. Next, install the upper side rail in the same manner **(see illustration)**. Don't use a piston ring installation tool on the oil ring side rails, as they may be damaged. Instead, place one end of the side rail into the groove between the spacer/expander and the ring land, hold it firmly in place and slide a finger around the piston while pushing the rail into the groove.

9.19a Installing the spacer/expander in the oil ring groove

9.19b DO NOT use a piston ring installation tool when installing the oil control side rails

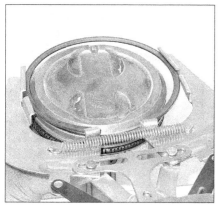

9.22 Use a piston ring installation tool to install the compression rings - on some engines the number two compression ring has a directional mark that must face toward the top of the piston

Finally, install the lower side rail.

20   After the three oil ring components have been installed, check to make sure that both the upper and lower side rails can be rotated smoothly inside the ring grooves.

21   The number two (middle) ring is installed next. It's usually stamped with a mark which must face up, toward the top of the piston. Do not mix up the top and middle rings, as they have different cross-sections. **Note:** *Always follow the instructions printed on the ring package or box - different manufacturers may require different approaches.*

22   Use a piston ring installation tool and make sure the identification mark is facing the top of the piston, then slip the ring into the middle groove on the piston **(see illustration)**. Don't expand the ring any more than necessary to slide it over the piston.

23   Install the number one (top) ring in the same manner. Be careful not to confuse the number one and number two rings.

24   Repeat the procedure for the remaining pistons and rings.

Installation

25   Before installing the piston/connecting rod assemblies, the cylinder walls must be perfectly clean, the top edge of each cylinder bore must be chamfered, and the crankshaft must be in place.

26   Remove the cap from the end of the number one connecting rod (refer to the marks made during removal). Remove the original bearing inserts and wipe the bearing surfaces of the connecting rod and cap with a clean, lint-free cloth. They must be kept spotlessly clean.

Connecting rod bearing oil clearance check

Refer to illustrations 9.30a, 9.30b, 9.35, 9.37 and 9.41

27   Clean the back side of the new upper bearing insert, then lay it in place in the connecting rod.

28   Make sure the tab on the bearing fits into the recess in the rod. Don't hammer the bearing insert into place and be very careful not to nick or gouge the bearing face. Don't lubricate the bearing at this time.

29   Clean the back side of the other bearing insert and install it in the rod cap. Again, make sure the tab on the bearing fits into the recess in the cap, and don't apply any lubricant. It's critically important that the mating surfaces of the bearing and connecting rod are perfectly clean and oil free when they're assembled.

30   Position the piston ring gaps at the intervals around the piston as shown **(see illustrations)**.

31   Lubricate the piston and rings with clean engine oil and attach a piston ring compressor to the piston. Leave the skirt protruding about 1/4-inch to guide the piston into the cylinder. The rings must be compressed until they're flush with the piston.

32   Rotate the crankshaft until the number one connecting rod journal is at BDC (bottom dead center) and apply a liberal coat of engine oil to the cylinder walls. Refer to the TDC locating procedure in Chapter 2A, 2B, 2C or 2D for additional information.

33   With the "front" mark on the piston facing the front (timing chain end) of the engine, gently insert the piston/connecting rod assembly into the number one cylinder bore and rest the bottom edge of the ring compressor on the engine block. **Note:** *Some engines have a letter "F" marking on the side of the piston near the wrist pin, others have an arrow, an "F" or a dimple or groove on the top of the piston. All of these are marks that indicate the front of the piston.*

34   Tap the top edge of the ring compressor to make sure it's contacting the block around its entire circumference.

35   Gently tap on the top of the piston with the end of a wooden or plastic hammer handle **(see illustration)** while guiding the end of the connecting rod into place on the crankshaft

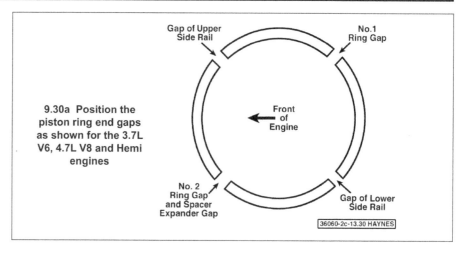

9.30a Position the piston ring end gaps as shown for the 3.7L V6, 4.7L V8 and Hemi engines

36060-2c-13.30 HAYNES

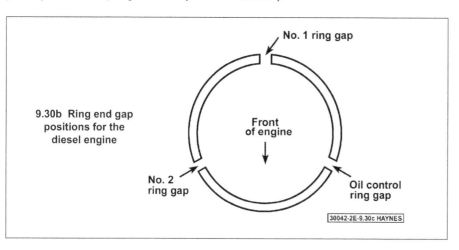

9.30b Ring end gap positions for the diesel engine

30042-2E-9.30c HAYNES

9.35 Use a plastic or wooden hammer handle to push the piston into the cylinder

ENGINE BEARING ANALYSIS

Debris

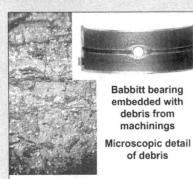

Babbitt bearing embedded with debris from machinings

Microscopic detail of debris

Microscopic detail of gouges

Overplated copper alloy bearing gouged by cast iron debris

Aluminum bearing embedded with glass beads

Microscopic detail of glass beads

Damaged lining caused by dirt left on the bearing back

Misassembly

Result of a lower half assembled as an upper - blocking the oil flow

Excessive oil clearance is indicated by a short contact arc

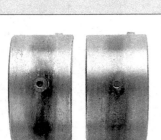

Polished and oil-stained backs are a result of a poor fit in the housing bore

Result of a wrong, reversed, or shifted cap

Overloading

Damage from excessive idling which resulted in an oil film unable to support the load imposed

Damaged upper connecting rod bearings caused by engine lugging; the lower main bearings (not shown) were similarly affected

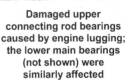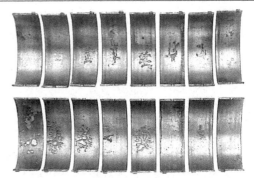

The damage shown in these upper and lower connecting rod bearings was caused by engine operation at a higher-than-rated speed under load

Misalignment

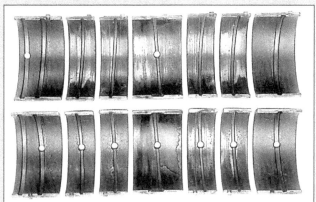

A warped crankshaft caused this pattern of severe wear in the center, diminishing toward the ends

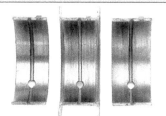

A poorly finished crankshaft caused the equally spaced scoring shown

A tapered housing bore caused the damage along one edge of this pair

A bent connecting rod led to the damage in the "V" pattern

Lubrication

Result of dry start: The bearings on the left, farthest from the oil pump, show more damage

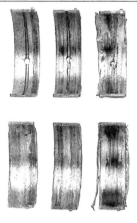

Result of a low oil supply or oil starvation

Severe wear as a result of inadequate oil clearance

Corrosion

Microscopic detail of corrosion

Corrosion is an acid attack on the bearing lining generally caused by inadequate maintenance, extremely hot or cold operation, or inferior oils or fuels

Microscopic detail of cavitation

Example of cavitation - a surface erosion caused by pressure changes in the oil film

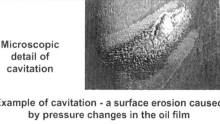

Damage from excessive thrust or insufficient axial clearance

Bearing affected by oil dilution caused by excessive blow-by or a rich mixture

9.37 Place Plastigage on each connecting rod bearing journal parallel to the crankshaft centerline

9.41 Use the scale on the Plastigage package to determine the bearing oil clearance - measure the widest part of the Plastigage and use the correct scale; it comes with both standard and metric scales

journal. The piston rings may try to pop out of the ring compressor just before entering the cylinder bore, so keep some downward pressure on the ring compressor. Work slowly, and if any resistance is felt as the piston enters the cylinder, stop immediately. Find out what's hanging up and fix it before proceeding. Do not, for any reason, force the piston into the cylinder - you might break a ring and/or the piston.

36   Once the piston/connecting rod assembly is installed, the connecting rod bearing oil clearance must be checked before the rod cap is permanently installed.

37   Cut a piece of the appropriate size Plastigage slightly shorter than the width of the connecting rod bearing and lay it in place on the number one connecting rod journal, parallel with the journal axis **(see illustration)**.

38   Clean the connecting rod cap bearing face and install the rod cap. Make sure the mating mark on the cap is on the same side as the mark on the connecting rod **(see illustration 9.4)**.

39   Install the old rod bolts, at this time, and tighten them to the torque listed in this Chapter's Specifications. **Note:** *Use a thin-wall socket to avoid erroneous torque readings that can result if the socket is wedged between the rod cap and the bolt or nut. If the socket tends to wedge itself between the fastener and the cap, lift up on it slightly until it no longer contacts the cap. DO NOT rotate the crankshaft at any time during this operation.*

40   Remove the fasteners and detach the rod cap, being very careful not to disturb the Plastigage. Discard the cap bolts at this time as they cannot be reused. **Note:** *You MUST use new connecting rod bolts.*

41   Compare the width of the crushed Plastigage to the scale printed on the Plastigage envelope to obtain the oil clearance **(see illustration)**. The connecting rod bearing oil clearance is usually about 0.001 to 0.002 inch. Consult an automotive machine shop for the clearance specified for the rod bearings on your engine.

42   If the clearance is not as specified, the bearing inserts may be the wrong size (which

means different ones will be required). Before deciding that different inserts are needed, make sure that no dirt or oil was between the bearing inserts and the connecting rod or cap when the clearance was measured. Also, recheck the journal diameter. If the Plastigage was wider at one end than the other, the journal may be tapered. If the clearance still exceeds the limit specified, the bearing will have to be replaced with an undersize bearing. **Caution:** *When installing a new crankshaft always use a standard size bearing.*

Final installation

43   Carefully scrape all traces of the Plastigage material off the rod journal and/or bearing face. Be very careful not to scratch the bearing - use your fingernail or the edge of a plastic card.

44   Make sure the bearing faces are perfectly clean, then apply a uniform layer of clean moly-base grease or engine assembly lube to both of them. You'll have to push the piston into the cylinder to expose the face of the bearing insert in the connecting rod.

45   Slide the connecting rod back into place on the journal, install the rod cap, install the nuts or bolts and tighten them to the torque

listed in this Chapter's Specifications. Again, work up to the torque in three steps.

46   Repeat the entire procedure for the remaining pistons/connecting rods.

47   The important points to remember are:

a) *Keep the back sides of the bearing inserts and the insides of the connecting rods and caps perfectly clean when assembling them.*

b) *Make sure you have the correct piston/rod assembly for each cylinder.*

c) *The mark on the piston must face the front (timing chain end) of the engine.*

d) *Lubricate the cylinder walls liberally with clean oil.*

e) *Lubricate the bearing faces when installing the rod caps after the oil clearance has been checked.*

48   After all the piston/connecting rod assemblies have been correctly installed, rotate the crankshaft a number of times by hand to check for any obvious binding.

49   As a final step, check the connecting rod endplay again. If it was correct before disassembly and the original crankshaft and rods were reinstalled, it should still be correct. If new rods or a new crankshaft were installed, the endplay may be inadequate. If so, the rods will have to be removed and taken to an automotive machine shop for resizing.

10   Crankshaft - removal and installation

Removal

Refer to illustrations 10.1 and 10.3

Note: *The crankshaft can be removed only after the engine has been removed from the vehicle. It's assumed that the flywheel or driveplate, crankshaft pulley, timing chain, oil pan, oil pump, oil filter and piston/connecting rod assemblies have already been removed. The rear main oil seal retainer must be unbolted and separated from the block before proceeding with crankshaft removal.*

1   Before the crankshaft is removed, measure the endplay. Mount a dial indicator with the indicator in line with the crankshaft and touching the end of the crankshaft **(see illustration)**.

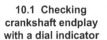

10.1 Checking crankshaft endplay with a dial indicator

10.3 Checking crankshaft endplay with feeler gauges at the thrust bearing journal

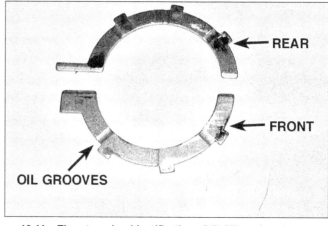

10.11a Thrust washer identification - 3.7L V6 engine shown, 4.7L V8 similar

2   Pry the crankshaft all the way to the rear and zero the dial indicator. Next, pry the crankshaft to the front as far as possible and check the reading on the dial indicator. The distance traveled is the endplay. A typical crankshaft endplay will fall between 0.003 to 0.010 inch (0.076 to 0.254 mm). If it is greater than that, check the crankshaft thrust washer/bearing assembly surfaces for wear after it's removed. If no wear is evident, new main bearings should correct the endplay. Refer to Step 11 for the location of the thrust washer/bearing assembly on each engine.
3   If a dial indicator isn't available, feeler gauges can be used. Gently pry the crankshaft all the way to the front of the engine. Slip feeler gauges between the crankshaft and the front face of the thrust bearing or washer to determine the clearance **(see illustration)**.
4   Loosen the main bearing cap/bedplate bolts 1/4-turn at a time each, until they can be removed by hand. **Note:** *The main bearing caps on the Hemi engine are each secured by four bolts, two of which are accessed from the sides of the engine block.*
5   Gently tap the main bearing caps/bedplate assembly with a soft-face hammer around the perimeter of the assembly. Pull the main bearing cap/bedplate assembly straight up and off the cylinder block. Try not to drop the bearing inserts if they come out with the assembly. **Note:** *The bedplate on 3.7L V6 and 4.7L V8 engines has built in pry points; don't pry anywhere else or damage to the bedplate will occur.*
6   Carefully lift the crankshaft out of the engine. It may be a good idea to have an assistant available, since the crankshaft is quite heavy and awkward to handle. With the bearing inserts in place inside the engine block and main bearing caps, reinstall the main bearing cap assembly onto the engine block and tighten the bolts finger tight. Make sure you install the main bearing cap assembly with the arrow facing the front end (timing chain) of the engine.

Installation

7   Crankshaft installation is the first step in engine reassembly. It's assumed at this point that the engine block and crankshaft have been cleaned, inspected and repaired or reconditioned.
8   Position the engine block with the bottom facing up.
9   Remove the mounting bolts and lift off the main bearing cap assembly.
10   If they're still in place, remove the original bearing inserts from the block and from the main bearing cap assembly. Wipe the bearing surfaces of the block and main bearing cap assembly with a clean, lint-free cloth. They must be kept spotlessly clean. This is critical for determining the correct bearing oil clearance.

Main bearing oil clearance check

Refer to illustrations 10.11a, 10.11b, 10.17, 10.18a, 10.18b, 10.19a through 10.19i and 10.21
11   Without mixing them up, clean the back sides of the new upper main bearing inserts (with grooves and oil holes) and lay one in each main bearing saddle in the block. Each upper bearing has an oil groove and oil hole in it. **Caution:** *The oil holes in the block must line up with the oil holes in the upper bearing inserts. If the main bearing cap is not centered, the crankshaft counterweights can contact the cap and cause severe engine damage.* Locate the thrust washers.

a) On 3.6L and 3.7L V6 engines, the thrust washers are located on the number 2 main journal **(see illustrations)**.
b) On V8 engines, the thrust washers are located on the number 3 main journal.
c) On diesel engines, the thrust washer/main bearing is located on the number 6 main journal.

The thrust washers must be installed in the correct journal. Clean the back sides of the

10.11b Insert the thrust washer into the machined surface between the crankshaft and the upper bearing saddle, then rotate it down into the block until it's flush with the parting line on the main bearing saddle - make sure the oil grooves on the thrust washer face the crankshaft

lower main bearing inserts and lay them in the corresponding location in the main bearing cap. Make sure the tab on the bearing insert fits into the recess in the block or main bearing cap. The upper bearings with the oil holes are installed into the engine block while the lower bearings without the oil holes are installed in the caps or bedplate. **Caution:** *Do not hammer the bearing insert into place and don't nick or gouge the bearing faces. DO NOT apply any lubrication at this time.*
12   Clean the faces of the bearing inserts in the block and the crankshaft main bearing journals with a clean, lint-free cloth.
13   Check or clean the oil holes in the crankshaft, as any dirt here can go only one way - straight through the new bearings.
14   Once you're certain the crankshaft is clean, carefully lay it in position in the cylinder block. **Note:** *On 3.7L and 4.7L engines, install the thrust washers with the groove in the thrust washer facing the crankshaft with the smooth sides facing the main bearing saddle.*

10.17   Place the Plastigage onto the crankshaft bearing journal as shown

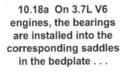

10.18a   On 3.7L V6 engines, the bearings are installed into the corresponding saddles in the bedplate . . .

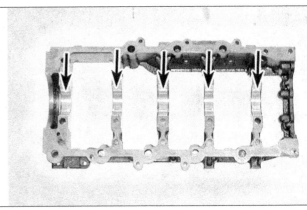

15   Before the crankshaft can be permanently installed, the main bearing oil clearance must be checked.
16   Cut several strips of the appropriate size of Plastigage. They must be slightly shorter than the width of the main bearing journal.

17   Place one piece on each crankshaft main bearing journal, parallel with the journal axis as shown **(see illustration)**.
18   Clean the faces of the bearing inserts in the main bearing caps or bedplate assembly **(see illustrations)**. Hold the bearing inserts in place and install the assembly onto the crankshaft and cylinder block. DO NOT disturb the Plastigage. Make sure you install the main bearing cap assembly with the arrow facing

the front (timing chain end) of the engine.
19   Apply clean engine oil to all bolt threads prior to installation, then install all bolts finger-tight. Tighten main bearing caps/bedplate assembly bolts in the sequence shown **(see illustrations)** progressing in steps, to the torque listed in this Chapter's Specifications. DO NOT rotate the crankshaft at any time during this operation.
20   Remove the bolts in the *reverse* order of

10.18b   . . . then the bedplate is set over the crankshaft onto the dowels on the engine block

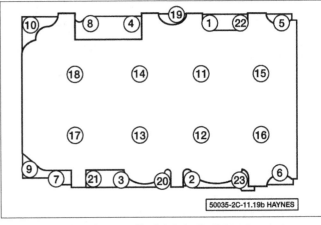

10.19a   Main bearing caps/bedplate bolts tightening sequence - 3.7L V6 engine

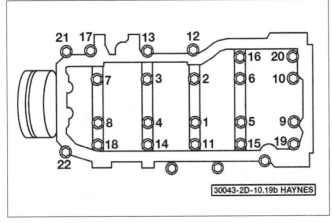

10.19b   4.7L V8 engine, tightening step 1 - after lubricating and installing all bedplate bolts, tighten them in this order (install the studs in locations 3, 5 and 6)

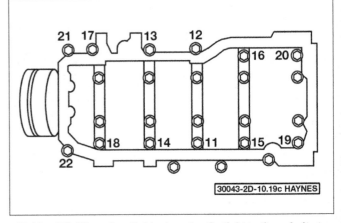

10.19c   4.7L V8 engine, tightening step 2 - tighten these bolts to the listed torque

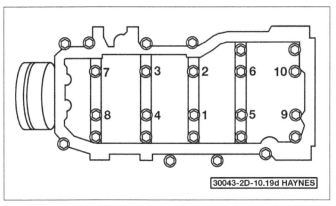

10.19d 4.7L V8 engine, tightening step 3 - mark these bolts with a dab of paint or use a torque-angle gauge, then rotate them 90 degrees in this order

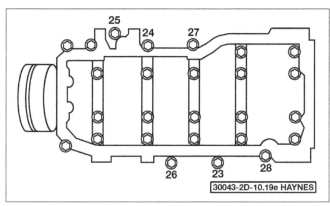

10.19e 4.7L V8 engine, tightening step 4 - tighten these bolts to the torque listed in this Chapter's Specifications

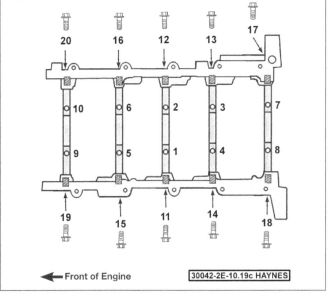

10.19f Main bearing cap and crossbolt tightening sequence - Hemi engine

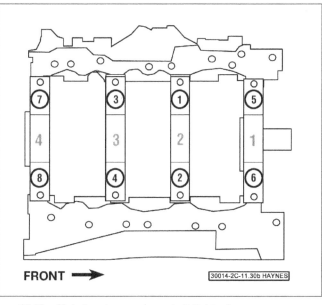

10.19g Main bearing cap inner bolt tightening sequence – 3.6L engine

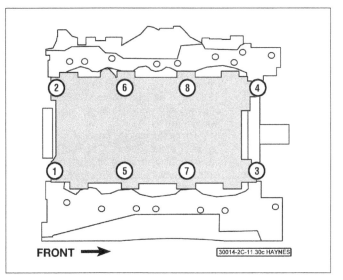

10.19h Main bearing cap outer bolt and windage tray tightening sequence – 3.6L engine

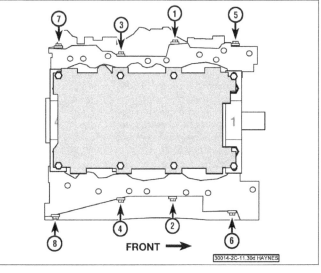

10.19i Main bearing cap side bolt (crossbolt) tightening sequence – 3.6L engine

10.21 Use the scale on the Plastigage package to determine the bearing oil clearance - measure the widest part of the Plastigage and use the correct scale; it comes with both standard and metric scales

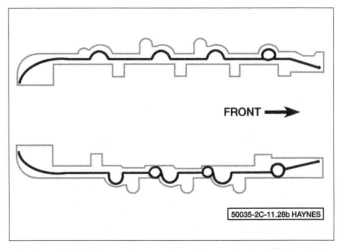

10.28a On 3.7L V6 engines, apply a 2.5 mm bead of RTV sealant to the engine block-to-bedplate sealing surface as shown

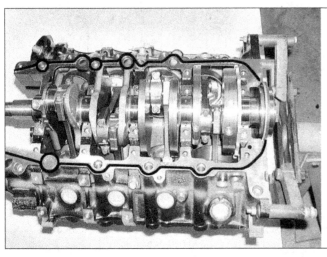

10.28b On 4.7L V8 engines, apply a 2.5 mm bead of RTV sealant to the engine block-to-bedplate sealing surface as shown

the tightening sequence and carefully lift the main bearing cap assembly straight up and off the block. Do not disturb the Plastigage or rotate the crankshaft. If the main bearing cap assembly is difficult to remove, tap it gently from side-to-side with a soft-face hammer to loosen it.

21 Compare the width of the crushed Plastigage on each journal to the scale printed on the Plastigage envelope to determine the main bearing oil clearance **(see illustration)**. Check with an automotive machine shop for the crankshaft endplay service limits.

22 If the clearance is not as specified, the bearing inserts may be the wrong size (which means different ones will be required). Before deciding if different inserts are needed, make sure that no dirt or oil was between the bearing inserts and the cap assembly or block when the clearance was measured. If the Plastigage was wider at one end than the other, the crankshaft journal may be tapered. If the clearance still exceeds the limit speci-

fied, the bearing insert(s) will have to be replaced with an undersize bearing insert(s). **Caution:** *When installing a new crankshaft always install a standard bearing insert set.*

23 Carefully scrape all traces of the Plastigage material off the main bearing journals and/or the bearing insert faces. Remove all residue from the oil holes. Use your fingernail or the edge of a plastic card - don't nick or scratch the bearing faces.

Final installation

Refer to illustrations 10.28a and 10.28b

24 Carefully lift the crankshaft out of the cylinder block.

25 Clean the bearing insert faces in the cylinder block, then apply a thin, uniform layer of moly-base grease or engine assembly lube to each of the bearing surfaces. Coat the thrust faces as well as the journal face of the thrust washers. **Note:** *Install the thrust washers on*

3.6L, 3.7L, 4.7L and Hemi engines after the crankshaft has been installed.

26 Make sure the crankshaft journals are clean, then lay the crankshaft back in place in the cylinder block.

27 Clean the bearing insert faces and apply the same lubricant to them. Clean the engine block thoroughly. The surfaces must be free of oil residue. On 3.6L, 3.7L, 4.7L and Hemi engines, install the thrust washers.

28 On 3.7L and 4.7L models, apply a 2.5 mm bead of Mopar Engine RTV sealant or equivalent to the bedplate sealing area on the block **(see illustrations)**. Install each main bearing cap (or the bedplate) onto the crankshaft and cylinder block. On engines with individual main bearing caps, make sure the arrow faces the front (timing chain) of the engine.

29 Prior to installation, apply clean engine oil to all bolt threads wiping off any excess, then install all bolts finger-tight.

30 Tighten the main bearing cap bolts or bedplate assembly bolts to the torque listed in this Chapter's Specifications (and in the proper sequence on 3.6L, 3.7L, 4.7L and Hemi engines) **(see illustration[s] 10.19a, 10.19b through 10.19e, or 10.19i)**. On diesel engines, start with the center cap bolts and work toward the ends.

31 Recheck crankshaft endplay with a feeler gauge or a dial indicator. The endplay should be correct if the crankshaft thrust faces aren't worn or damaged and if new bearings have been installed.

32 Rotate the crankshaft a number of times by hand to check for any obvious binding. It should rotate with a running torque of 50 in-lbs or less. If the running torque is too high, identify and correct the problem at this time.

33 Install the new rear main oil seal (see Chapter 2A, 2B, 2C or 2D).

11 Engine overhaul - reassembly sequence

1 Before beginning engine reassembly, make sure you have all the necessary new parts, gaskets and seals as well as the following items on hand:

Common hand tools
A 1/2-inch drive torque wrench
New engine oil
Gasket sealant
Thread locking compound

2 If you obtained a short block it will be necessary to install the cylinder head, the oil pump and pick-up tube, the oil pan, the water pump, the timing chain and timing cover, and the valve cover (see Chapter 2A, 2B or 2C). In order to save time and avoid problems, the external components must be installed in the following general order:

Thermostat and housing cover
Water pump
Intake and exhaust manifolds
Fuel injection components
Emission control components
Spark plugs
Ignition coils
Oil filter
Engine mounts and mount brackets
Flywheel/driveplate

12 Initial start-up and break-in after overhaul

Warning: *Have a fire extinguisher handy when starting the engine for the first time.*

1 Once the engine has been installed in the vehicle, double-check the engine oil and coolant levels.

2 With the spark plugs out of the engine and the ignition system and fuel pump disabled, crank the engine until oil pressure registers on the gauge or the light goes out.

3 Install the spark plugs, hook up the plug wires and restore the ignition system and fuel pump functions. On diesel engines, prime the fuel system (see Chapter 4B).

4 Start the engine. It may take a few moments for the fuel system to build up pressure, but the engine should start without a great deal of effort.

5 After the engine starts, it should be allowed to warm up to normal operating temperature. While the engine is warming up, make a thorough check for fuel, oil and coolant leaks.

6 Shut the engine off and recheck the engine oil and coolant levels.

7 Drive the vehicle to an area with minimum traffic, accelerate from 30 to 50 mph, then allow the vehicle to slow to 30 mph with the throttle closed. Repeat the procedure 10 or 12 times. This will load the piston rings and cause them to seat properly against the cylinder walls. Check again for oil and coolant leaks.

8 Drive the vehicle gently for the first 500 miles (no sustained high speeds) and keep a constant check on the oil level. It is not unusual for an engine to use oil during the break-in period.

9 At approximately 500 to 600 miles, change the oil and filter.

10 For the next few hundred miles, drive the vehicle normally. Do not pamper it or abuse it.

11 After 2000 miles, change the oil and filter again and consider the engine broken in.

COMMON ENGINE OVERHAUL TERMS

B

Backlash - The amount of play between two parts. Usually refers to how much one gear can be moved back and forth without moving the gear with which it's meshed.

Bearing Caps - The caps held in place by nuts or bolts which, in turn, hold the bearing surface. This space is for lubricating oil to enter.

Bearing clearance - The amount of space left between shaft and bearing surface. This space is for lubricating oil to enter.

Bearing crush - The additional height which is purposely manufactured into each bearing half to ensure complete contact of the bearing back with the housing bore when the engine is assembled.

Bearing knock - The noise created by movement of a part in a loose or worn bearing.

Blueprinting - Dismantling an engine and reassembling it to EXACT specifications.

Bore - An engine cylinder, or any cylindrical hole; also used to describe the process of enlarging or accurately refinishing a hole with a cutting tool, as to bore an engine cylinder. The bore size is the diameter of the hole.

Boring - Renewing the cylinders by cutting them out to a specified size. A boring bar is used to make the cut.

Bottom end - A term which refers collectively to the engine block, crankshaft, main bearings and the big ends of the connecting rods.

Break-in - The period of operation between installation of new or rebuilt parts and time in which parts are worn to the correct fit. Driving at reduced and varying speed for a specified mileage to permit parts to wear to the correct fit.

Bushing - A one-piece sleeve placed in a bore to serve as a bearing surface for shaft, piston pin, etc. Usually replaceable.

C

Camshaft - The shaft in the engine, on which a series of lobes are located for operating the valve mechanisms. The camshaft is driven by gears or sprockets and a timing chain. Usually referred to simply as the cam.

Carbon - Hard, or soft, black deposits found in combustion chamber, on plugs, under rings, on and under valve heads.

Cast iron - An alloy of iron and more than two percent carbon, used for engine blocks and heads because it's relatively inexpensive and easy to mold into complex shapes.

Chamfer - To bevel across (or a bevel on) the sharp edge of an object.

Chase - To repair damaged threads with a tap or die.

Combustion chamber - The space between the piston and the cylinder head, with the piston at top dead center, in which air-fuel mixture is burned.

Compression ratio - The relationship between cylinder volume (clearance volume) when the piston is at top dead center and cylinder volume when the piston is at bottom dead center.

Connecting rod - The rod that connects the crank on the crankshaft with the piston. Sometimes called a con rod.

Connecting rod cap - The part of the connecting rod assembly that attaches the rod to the crankpin.

Core plug - Soft metal plug used to plug the casting holes for the coolant passages in the block.

Crankcase - The lower part of the engine in which the crankshaft rotates; includes the lower section of the cylinder block and the oil pan.

Crank kit - A reground or reconditioned crankshaft and new main and connecting rod bearings.

Crankpin - The part of a crankshaft to which a connecting rod is attached.

Crankshaft - The main rotating member, or shaft, running the length of the crankcase, with offset throws to which the connecting rods are attached; changes the reciprocating motion of the pistons into rotating motion.

Cylinder sleeve - A replaceable sleeve, or liner, pressed into the cylinder block to form the cylinder bore.

D

Deburring - Removing the burrs (rough edges or areas) from a bearing.

Deglazer - A tool, rotated by an electric motor, used to remove glaze from cylinder walls so a new set of rings will seat.

E

Endplay - The amount of lengthwise movement between two parts. As applied to a crankshaft, the distance that the crankshaft can move forward and back in the cylinder block.

F

Face - A machinist's term that refers to removing metal from the end of a shaft or the face of a larger part, such as a flywheel.

Fatigue - A breakdown of material through a large number of loading and unloading cycles. The first signs are cracks followed shortly by breaks.

Feeler gauge - A thin strip of hardened steel, ground to an exact thickness, used to check clearances between parts.

Free height - The unloaded length or height of a spring.

Freeplay - The looseness in a linkage, or an assembly of parts, between the initial application of force and actual movement. Usually perceived as slop or slight delay.

Freeze plug - See Core plug.

G

Gallery - A large passage in the block that forms a reservoir for engine oil pressure.

Glaze - The very smooth, glassy finish that develops on cylinder walls while an engine is in service.

H

Heli-Coil - A rethreading device used when threads are worn or damaged. The device is installed in a retapped hole to reduce the thread size to the original size.

I

Installed height - The spring's measured length or height, as installed on the cylinder head. Installed height is measured from the spring seat to the underside of the spring retainer.

J

Journal - The surface of a rotating shaft which turns in a bearing.

K

Keeper - The split lock that holds the valve spring retainer in position on the valve stem.

Key - A small piece of metal inserted into matching grooves machined into two parts fitted together - such as a gear pressed onto a shaft - which prevents slippage between the two parts.

Knock - The heavy metallic engine sound, produced in the combustion chamber as a result of abnormal combustion - usually detonation. Knock is usually caused by a loose or worn bearing. Also referred to as detonation, pinging and spark knock. Connecting rod or main bearing knocks are created by too much oil clearance or insufficient lubrication.

L

Lands - The portions of metal between the piston ring grooves.

Lapping the valves - Grinding a valve face and its seat together with lapping compound.

Lash - The amount of free motion in a gear train, between gears, or in a mechanical assembly, that occurs before movement can begin. Usually refers to the lash in a valve train.

Lifter - The part that rides against the cam to transfer motion to the rest of the valve train.

M

Machining - The process of using a machine to remove metal from a metal part.

Main bearings - The plain, or babbit, bearings that support the crankshaft.

Main bearing caps - The cast iron caps, bolted to the bottom of the block, that support the main bearings.

O

O.D. - Outside diameter.

Oil gallery - A pipe or drilled passageway in the engine used to carry engine oil from one area to another.

Oil ring - The lower ring, or rings, of a piston; designed to prevent excessive amounts of oil from working up the cylinder walls and into the combustion chamber. Also called an oil-control ring.

Oil seal - A seal which keeps oil from leaking out of a compartment. Usually refers to a dynamic seal around a rotating shaft or other moving part.

O-ring - A type of sealing ring made of a special rubberlike material; in use, the O-ring is compressed into a groove to provide the sealing action.

Overhaul - To completely disassemble a unit, clean and inspect all parts, reassemble it with the original or new parts and make all adjustments necessary for proper operation.

P

Pilot bearing - A small bearing installed in the center of the flywheel (or the rear end of the crankshaft) to support the front end of the input shaft of the transmission.

Pip mark - A little dot or indentation which indicates the top side of a compression ring.

Piston - The cylindrical part, attached to the connecting rod, that moves up and down in the cylinder as the crankshaft rotates. When the fuel charge is fired, the piston transfers the force of the explosion to the connecting rod, then to the crankshaft.

Piston pin (or wrist pin) - The cylindrical and usually hollow steel pin that passes through the piston. The piston pin fastens the piston to the upper end of the connecting rod.

Piston ring - The split ring fitted to the groove in a piston. The ring contacts the sides of the ring groove and also rubs against the cylinder wall, thus sealing space between piston and wall. There are two types of rings: Compression rings seal the compression pressure in the combustion chamber; oil rings scrape excessive oil off the cylinder wall.

Piston ring groove - The slots or grooves cut in piston heads to hold piston rings in position.

Piston skirt - The portion of the piston below the rings and the piston pin hole.

Plastigage - A thin strip of plastic thread, available in different sizes, used for measuring clearances. For example, a strip of plastigage is laid across a bearing journal and mashed as parts are assembled. Then parts are disassembled and the width of the strip is measured to determine clearance between journal and bearing. Commonly used to measure crankshaft main-bearing and connecting rod bearing clearances.

Press-fit - A tight fit between two parts that requires pressure to force the parts together. Also referred to as drive, or force, fit.

Prussian blue - A blue pigment; in solution, useful in determining the area of contact between two surfaces. Prussian blue is commonly used to determine the width and location of the contact area between the valve face and the valve seat.

R

Race (bearing) - The inner or outer ring that provides a contact surface for balls or rollers in bearing.

Ream - To size, enlarge or smooth a hole by using a round cutting tool with fluted edges.

Ring job - The process of reconditioning the cylinders and installing new rings.

Runout - Wobble. The amount a shaft rotates out-of-true.

S

Saddle - The upper main bearing seat.

Scored - Scratched or grooved, as a cylinder wall may be scored by abrasive particles moved up and down by the piston rings.

Scuffing - A type of wear in which there's a transfer of material between parts moving against each other; shows up as pits or grooves in the mating surfaces.

Seat - The surface upon which another part rests or seats. For example, the valve seat is the matched surface upon which the valve face rests. Also used to refer to wearing into a good fit; for example, piston rings seat after a few miles of driving.

Short block - An engine block complete with crankshaft and piston and, usually, camshaft assemblies.

Static balance - The balance of an object while it's stationary.

Step - The wear on the lower portion of a ring land caused by excessive side and back-clearance. The height of the step indicates the ring's extra side clearance and the length of the step projecting from the back wall of the groove represents the ring's back clearance.

Stroke - The distance the piston moves when traveling from top dead center to bottom dead center, or from bottom dead center to top dead center.

Stud - A metal rod with threads on both ends.

T

Tang - A lip on the end of a plain bearing used to align the bearing during assembly.

Tap - To cut threads in a hole. Also refers to the fluted tool used to cut threads.

Taper - A gradual reduction in the width of a shaft or hole; in an engine cylinder, taper usually takes the form of uneven wear, more pronounced at the top than at the bottom.

Throws - The offset portions of the crankshaft to which the connecting rods are affixed.

Thrust bearing - The main bearing that has thrust faces to prevent excessive endplay, or forward and backward movement of the crankshaft.

Thrust washer - A bronze or hardened steel washer placed between two moving parts. The washer prevents longitudinal movement and provides a bearing surface for thrust surfaces of parts.

Tolerance - The amount of variation permitted from an exact size of measurement. Actual amount from smallest acceptable dimension to largest acceptable dimension.

U

Umbrella - An oil deflector placed near the valve tip to throw oil from the valve stem area.

Undercut - A machined groove below the normal surface.

Undersize bearings - Smaller diameter bearings used with re-ground crankshaft journals.

V

Valve grinding - Refacing a valve in a valve-refacing machine.

Valve train - The valve-operating mechanism of an engine; includes all components from the camshaft to the valve.

Vibration damper - A cylindrical weight attached to the front of the crankshaft to minimize torsional vibration (the twist-untwist actions of the crankshaft caused by the cylinder firing impulses). Also called a harmonic balancer.

W

Water jacket - The spaces around the cylinders, between the inner and outer shells of the cylinder block or head, through which coolant circulates.

Web - A supporting structure across a cavity.

Woodruff key - A key with a radiused backside (viewed from the side).

Notes

Chapter 3
Cooling, heating and air conditioning systems

Contents

Specifications

General

Coolant capacity..	See Chapter 1
Drivebelt tension..	See Chapter 1
Radiator cap pressure rating	14 to 18 psi (94 to 124 kPa)
Thermostat opening temperature	
Diesel engines...	186-degrees F (85-degrees C)
All other engines	195-degrees F (90-degrees C)

Torque specifications

Ft-lbs (unless otherwise indicated)

Note: *One foot-pound (ft-lb) of torque is equivalent to 12 inch-pounds (in-lbs) of torque. Torque values below approximately 15 ft-lbs are expressed in inch-pounds, since most foot-pound torque wrenches are not accurate at these smaller values.*

Thermostat housing bolts	
3.6L V6 engine	106 in-lbs
3.7L V6 and 4.7L V8 engines	122 in-lbs
5.7L Hemi engine	21
6.4L Hemi engine	22
Diesel engine	89 in-lbs
Water pump bolts	
3.6L V6 engine – in sequence **(illustration 8.28a)**	
M6 bolts..	108 in-lbs
M8 bolts..	18
M10 bolt...	41
3.7L V6 and 4.7L V8 engines	43
5.7L Hemi engine	18
6.4L Hemi engine	21
Diesel engines...	18
Fan clutch-to-engine cooling fan mounting bolts	18
Engine cooling fan clutch nut	
Diesel engines...	85
All other engines	37

2.3 An inexpensive hydrometer can be used to test the condition of your coolant

1   General information

Engine cooling system

All vehicles covered by this manual employ a pressurized engine cooling system with thermostatically controlled coolant circulation. An impeller-type water pump mounted on the front of the engine pumps coolant through the engine. The coolant flows around each cylinder and toward the rear of the engine. Cast-in coolant passages direct coolant around the intake and exhaust ports, near the spark plug areas and in close proximity to the exhaust valve guides. Because of the design of the accessories, pulleys and drivebelts, some engines are equipped with reverse rotating water pumps, cooling fans and viscous fan clutches. Always check with a parts department and install only components marked REVERSE.

A wax-pellet type thermostat is located in a housing near the front of the engine. During warm up, the closed thermostat prevents coolant from circulating through the radiator. As the engine nears normal operating temperature, the thermostat opens and allows hot coolant to travel through the radiator, where it's cooled before returning to the engine.

The radiator is sealed by a pressure cap. This raises the boiling point of the coolant and this higher boiling point increases the efficiency of the radiator. If the system pressure exceeds the cap's pressure relief value, then the excess pressure in the system forces the spring-loaded valve in the cap off its seat and allows the coolant to escape through the overflow tube and into the coolant reservoir. When the system cools, the coolant is drawn from the reservoir back into the radiator. This type of system is known as a closed design because coolant that escapes past the pressure cap is saved and reused.

The radiator cooling fan is mounted on the front of the water pump on gasoline engines or the fan pulley on diesel engines. Gasoline models are equipped with a viscous fan clutch

while diesel models are equipped with an electronic viscous clutch that is controlled by the Powertrain Control Module (PCM).

Heating system

The heating system consists of a blower fan and heater core located in the heater/air conditioning unit under the dash, the hoses connecting the heater core to the engine cooling system and the heater/air conditioning control head on the instrument panel. Hot engine coolant is circulated through the heater core. When the heater mode is activated, a flap door opens to expose the heater box to the passenger compartment. A fan switch on the control head activates the blower motor, which forces air through the core, heating the air.

Air conditioning system

The air conditioning system consists of a condenser mounted either in front of (diesel models) or adjacent to the radiator (all other models), an evaporator mounted adjacent to the heater core under the dash, a compressor mounted on the engine, an accumulator mounted near the firewall and the plumbing that connects all of these components.

A blower fan forces the warmer air of the passenger compartment through the evaporator core (sort of a radiator-in-reverse), transferring the heat from the air to the refrigerant. The liquid refrigerant boils off into low-pressure vapor, taking the heat with it when it leaves the evaporator.

2   Antifreeze/coolant - general information

Refer to illustration 2.3

Warning: *Do not allow antifreeze to come in contact with your skin or painted surfaces of the vehicle. Rinse off spills immediately with plenty of water. Antifreeze is highly toxic if ingested. Never leave antifreeze lying around in an open container or in puddles on the floor; children and pets are attracted by its sweet smell and may drink it. Check with local authorities about disposing of used antifreeze. Many communities have collection centers which will see that antifreeze is disposed of safely. Never dump used antifreeze on the ground or pour it into drains. Keep antifreeze containers covered and repair leaks in your cooling system as soon as they are noticed.* **Caution:** *These models use only Mopar 5 year/100,000 mile coolant or equivalent. Never mix coolant types or damage to the cooling system could occur. Refer to the Specifications listed in Chapter 1 for the correct coolant type.*

These models are equipped with a hybrid organic acid technology (HOAT) type antifreeze. Do not mix different types or colors of coolants or damage to the cooling system can occur. Refer to the Specifications listed in

Chapter 1 for the correct type of coolant.

Before adding antifreeze, check all hose connections, because antifreeze tends to leak through very minute openings. Engines don't normally consume coolant, so if the level goes down, find the cause and correct it.

The exact mixture of antifreeze-to-water which you should use depends on the manufacturer's recommendation. Consult the mixture ratio chart on the antifreeze container before adding coolant. Hydrometers are available at most auto parts stores to test the coolant **(see illustration)**.

3   Thermostat - check and replacement

Warning: *Do not remove the radiator cap or expansion tank cap, drain the coolant or replace the thermostat until the engine has cooled completely.*

Check

1   Before assuming the thermostat is to blame for a cooling system problem, check the coolant level, drivebelt tension (see Chapter 1) and temperature gauge operation.
2   If the engine seems to be taking a long time to warm up (based on heater output or temperature gauge reading), the thermostat is probably stuck open. Replace the thermostat with a new one.
3   If the engine runs hot, use your hand to check the temperature of the radiator hose that leads from the thermostat to the radiator. The thermostat is in different locations for each engine. If the hose isn't hot, but the engine is, the thermostat is probably stuck closed, preventing the coolant inside the engine from escaping to the radiator. Replace the thermostat. **Caution:** *Don't drive the vehicle without a thermostat. On some models, the computer may stay in open-loop, potentially causing oil-sludge, excessive exhaust emissions/fuel consumption. On other models, the engine will operate in radiator bypass mode, which will cause it to overheat.*
4   If the inlet radiator hose is hot, it means that the coolant is flowing and the thermostat is open. Consult the *Troubleshooting* section at the front of this manual for cooling system diagnosis.

Replacement

5   Disconnect the cable(s) from the negative battery terminal(s) (see Chapter 5).
6   Drain the cooling system to a level below the thermostat (see Chapter 1). If the coolant is relatively new or in good condition (see Section 2), save it and reuse it.

3.6L engine

7   Remove the air filter housing and resonator (see Chapter 4A).
8   Follow the upper radiator hose to the engine to locate the thermostat housing.

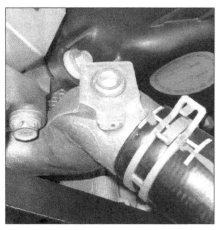

3.10 Thermostat housing location (4.7L V8 engine shown, 3.7L V6 similar)

3.11 Thermostat housing location (Hemi engine) - the air conditioning compressor is removed for clarity

3.19 Location of the jiggle valve on the thermostat (Hemi engine)

3.7L V6 and 4.7L V8 engines

Refer to illustration 3.10

9   Raise the vehicle and support it securely on jackstands, then remove the splash shield.
10   Follow the lower radiator hose to the engine to locate the thermostat housing **(see illustration)**.

5.7L and 6.4L Hemi engine

Refer to illustration 3.11

11   Follow the upper radiator hose to the engine to locate the thermostat housing **(see illustration)**.

Diesel engine

12   Follow the upper radiator hose to the engine to locate the thermostat housing.
13   Remove the cooling system vent plug; it's located next to the EGR cooler.
14   Disconnect the wiring from the exhaust gas pressure sensor, then remove the exhaust pressure tube from the thermostat housing.

15   Remove the interfering EGR crossover tube and the heat shield.

All models

Refer to illustrations 3.19, 3.21a, 3.21b and 3.21c

16   Squeeze the tabs on the hose clamp to loosen it from the hose(s), then reposition the clamp several inches back up the hose. Detach the hose(s) from the thermostat housing. If the hose is stuck, grasp it near the end with a pair of adjustable pliers and twist it to break the seal, then pull it off. If the hose is old or deteriorated, cut it off and install a new one.
17   If the outer surface of the thermostat housing that mates with the hose is deteriorated (corroded, pitted, etc.) it may be damaged further by hose removal. If it is, the thermostat housing will have to be replaced.
18   Remove the thermostat housing from the engine. If the housing is stuck, tap it with a soft-face hammer to jar it loose. Be prepared for some coolant to spill as the gasket seal is broken.

19   Note how the thermostat is installed (which end is facing up, or out, and the position of the air bleed jiggle valve, if equipped) and remove it from the engine **(see illustration)**.
Note: *On 3.6L engines, the thermostat is an integral part of the housing and must be replaced as an assembly.*
20   Remove the rubber gasket from around the thermostat.
21   Install a new rubber gasket around the thermostat. Make sure to align the rubber tab on the inside of the O-ring groove with the notch on the thermostat **(see illustration)**. Then align the rubber tab on the outside of the gasket with the notch on the thermostat housing and insert the thermostat and gasket into the thermostat housing **(see illustration)**.
Note: *Some models are not equipped with alignment notches on the thermostat. Simply install the rubber gasket around the thermostat* **(see illustration)**.
22   Install the thermostat housing and bolts onto the engine. Tighten the bolts to the

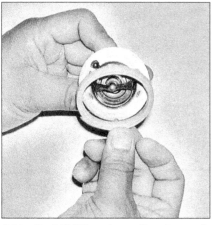

3.21a On 3.7L V6 engines, align the notch on the thermostat with the rubber tab on the gasket's inner groove . . .

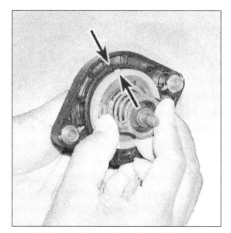

3.21b . . . then align the tab (lower arrow) on the outer edge of the gasket with the notch in the thermostat housing (upper arrow) and insert the thermostat into the housing

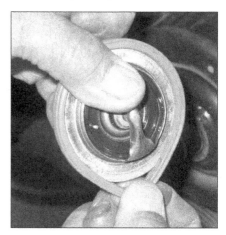

3.21c On models without alignment notches, install a new rubber gasket around the perimeter of the thermostat

4.6 Location of the electric fan electrical connector (4.7L V8 engine)

torque listed in this Chapter's Specifications.
23   Reattach the hose(s) to the fitting(s) and tighten the hose clamp(s) securely.
24   Refill the cooling system (see Chapter 1).
25   Start the engine and allow it to reach normal operating temperature, then check for leaks and proper thermostat operation.

---

4   Engine-mounted and radiator-mounted cooling fans - check and component replacement

---

Check

Warning: *Keep hands, tools and clothing away from the fan when the engine is running. To avoid injury or damage DO NOT operate the engine with a damaged fan. Do not attempt to repair fan blades - replace a damaged fan with a new one.*

Engine-mounted cooling fan

Warning: *While checking the fan, make sure that the engine is NOT started. If it is, you could be severely injured.*

Warning: *Before the fan clutch operation can be checked in Step 5, the engine must be warmed up to its normal operating temperature and then turned off. Even though the engine won't be running during this check, it's HOT! Make sure that you don't touch the engine itself during this check, or you could be burned.*
Note: *Diesel engines are equipped with an electronically controlled viscous fan that uses the PCM to control the operation of the fan during warm-up and driving conditions. If the engine is overheating because the fan isn't coming on, have the electronically controlled viscous fan checked by a dealer service department or other qualified repair shop.*
1   Symptoms of fan clutch failure are continuous noisy operation, looseness, vibration and/or silicone fluid leaking from the clutch.

Cold engine checks
2   Rock the fan back and forth by hand to check for excessive bearing play.
3   Turn the blades by hand. The fan should turn freely.
4   Visually inspect for substantial fluid leakage from the fan clutch assembly, a deformed bi-metal spring or grease leakage from the cooling fan bearing. If any of these conditions exist, replace the viscous fan clutch.

Hot engine check
5   Start the engine and allow it to warm up to its normal operating temperature. When the engine is fully warmed up, turn off the engine. Turn the fan by hand. Some resistance should be felt. If the fan turns easily, replace the fan clutch.

Radiator-mounted electric cooling fan
Refer to illustration 4.6
6   If the cooling fan is not coming on when the A/C is selected or the engine temperature rises to an excessive level, unplug the fan motor electrical connector, then connect

one of the two terminals of the motor directly to the battery with a fused jumper wire **(see illustration)**. Connect the other terminal to ground using another fused jumper wire. If the fan motor doesn't come on, replace the motor. **Caution:** *Do not apply battery power to the harness side of the connector. Test ONLY the cooling fan motor.*
7   If the fan motor is okay, but it isn't coming on when the A/C is selected or the engine gets hot, the fan relay might be defective. A relay is used to control a circuit by turning it on and off in response to a control decision by the Powertrain Control Module (PCM). These control circuits are fairly complex, and checking them should be left to a dealer service department or other qualified repair shop. Sometimes, the control system can be fixed by simply identifying and replacing a bad relay.
8   If the relay is okay, check all wiring and connections to the fan motor. Refer to the wiring diagrams at the end of Chapter 12.

Component replacement
Gasoline engines
Refer to illustrations 4.12, 4.13, 4.14, 4.18, 4.20a and 4.20b
Note: *The viscous (engine-mounted) and electric (radiator-mounted) cooling fans are removed together.*
Note: *Some models do not have an engine-mounted viscous fan.*
9   Disconnect the cable from the negative battery terminal (see Chapter 5).
10   Remove the rivets and push-pin clips, then remove the radiator top cover.
11   Hold the engine fan pulley. You can use a pin spanner or a strap wrench. If using a strap wrench, you'll have to first remove the drivebelt (see Chapter 1).
12   Unscrew the viscous fan mounting nut with a 36mm wrench or adjustable wrench **(see illustration)**.
13   Remove the fan shroud bolts **(see illustration)**.

4.12 Use a pin spanner wrench or a strap wrench to prevent the water pump pulley from turning, then loosen the fan clutch nut

4.13 Typical fan shroud mounting bolt

4.14 Disconnect this plug when removing the electric fan assembly - a shroud mounting bolt is directly above it

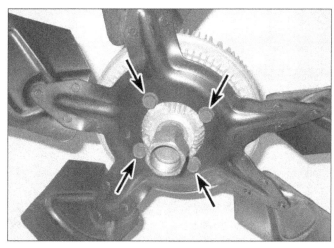

4.18 Remove the four bolts retaining the cooling fan to the fan clutch (Hemi engine shown, other models similar)

14   Disconnect the wiring from the fans **(see illustration)**.
15   Detach the clips at the lower section of the fan shroud.
16   Release the clips from the fan.
17   Lift the complete viscous fan, electric fan and shroud assembly up and out of the vehicle. **Caution**: *Set the viscous fan in a vertical position. If it's allowed to lie flat, it may leak fluid into the bearing and damage it.*
18   Working on the bench, remove the cooling fan mounting bolts from the viscous fan clutch **(see illustration)**.
19   Installation is the reverse of removal. Tighten the fan clutch-to-cooling fan bolts to the torque Specifications listed in this Chapter. Tighten the fan clutch securely.
20   Remove the clip to detach the motor from the fan, then unbolt the fan from its frame **(see illustrations)**. **Warning:** *To avoid possible injury or damage, DO NOT operate the engine with a damaged fan. Do not attempt to repair fan blades - replace a damaged fan with a new one.* **Caution:** *Some models are equipped with REVERSE engine cooling fans. ONLY install cooling fans and fan clutches marked REVERSE, or the engine may be overheated and damaged.*

Diesel engines

Engine-mounted viscous fan

21   Disconnect the cables from the negative battery terminals (see Chapter 5).
22   On 2009 models, detach the fan shroud from the engine by removing the nuts.
23   Disconnect the viscous fan wiring at the lower shroud. Detach the wiring harness from the shroud.
24   On 2010 and later models, remove the fan shroud bolts and push-pins that attach it to the radiator shroud.
25   Install a large screwdriver between the fan pulley bolts while a large (36 mm) open-end wrench is used to loosen the fan drive nut on the cooling fan. Turn the large nut counter-clockwise. **Caution:** *It's a right-hand thread.*

26   On 2010 and later models, release the lower fan shroud clips.
27   Collapse the fan shroud toward the radiator, then remove the shroud and fan assembly together. DO NOT damage the radiator fins while doing so. **Caution:** *Do not remove any of the cooling fan pulley bolts. The pulley is under spring tension and could release.*
28   On 2010 and later models, remove the wiring bracket from the fan shroud unit by detaching the clip and sliding it off.
29   Remove the fan blade mounting bolts.
30   Installation is the reverse of removal.

Radiator-mounted electric fan (2010 and later models)

31   Disconnect the cables from the negative battery terminals (see Chapter 5).
32   Remove the radiator top cover plate.
33   Detach the viscous fan from the pulley (see Steps 23 through 25).
34   Disconnect the wiring from the electric fan.
35   Unbolt the upper and lower fan shrouds

from the radiator, unclip it at the bottom and slide it to the rear.
36   Release the electric fan clips from the radiator.
37   Pull the electric fan, viscous fan and shroud up and out of the vehicle as a complete assembly.
38   Installation is the reverse of removal.

5   Coolant reservoir - removal and installation

Warning: *Wait until the engine is completely cool before beginning this procedure.*
Note: *All models are equipped with a coolant reservoir in the engine compartment (see Chapter 1 for location). These systems incorporate the radiator cap on top of the radiator to create pressure, and the coolant reservoir stores excess coolant that is transferred through the overflow hose.*
1   On gasoline engines, remove the grille (see Chapter 11).

4.20a Slide the clip off the stud in the direction of the arrow and separate the cooling fan from the electric motor

4.20b Remove the fan motor mounting bolts

6.5 Use large pliers to remove the radiator hose clamps

6.6 2012 models have a lower fan shroud that must be removed

2   Disconnect the hose from the coolant reservoir.
3   Remove the coolant reservoir. **Note:** *On gasoline engines, pull the reservoir upward to disengage it.*
4   Installation is the reverse of removal.
5   Add coolant to the reservoir as necessary (see Chapter 1).

6   Radiator - removal and installation

Warning: *Wait until the engine is completely cool before beginning this procedure.*

Removal

Gasoline engines

Refer to illustrations 6.5, 6.6, 6.9, 6.10, 6.22 and 6.30

1   Disconnect the cable from the negative battery terminal (see Chapter 5).

2   Raise the vehicle and support it securely on jackstands.
3   Drain the cooling system (see Chapter 1).
4   Remove the cover from beneath the radiator, if so equipped.
5   Remove the lower radiator hose **(see illustration)**.
6   On 2012 and later models, remove the lower fan shroud **(see illustration)**.
7   Remove the lower electric fan wiring harness clip.
8   Lower the truck.
9   Remove the upper radiator hose **(see illustration)**.
10   Remove the upper radiator cover **(see illustration)**.
11   On 2009 through 2011 models, remove the engine-mounted viscous fan (see Section 4).
12   Remove the grille (see Chapter 11).
13   On automatic transmission-equipped vehicles, unbolt the transmission cooler lines from the radiator.

2009 through 2011 models

14   Remove the coolant reservoir (see Section 5) and the windshield washer reservoir.
15   Remove the bolts from the front of the transmission cooler and air conditioning condenser.
16   Remove the top center clip that secures the electric fan to the radiator.
17   On automatic transmission-equipped vehicles, disconnect the transmission cooler lines from the cooler. Seal the openings to prevent contamination.
18   Detach the air conditioning compressor and transmission cooler from the radiator at the left side. Tie the cooler securely so it can't fall.
19   Remove the fan shroud mounting bolts, detach it from the retainers at the bottom, then slide it to the rear as far as possible.
20   Release the fan clips from the radiator. Pull the viscous engine-mounted fan, the shroud and the electric fan out together.

2012 and later models

21   Unbolt the jumper line tapping block from the center right side of the radiator.

6.9 Typical 4.7L V8 upper radiator hose connections

6.10 Remove the radiator upper cover for access to the radiator and air conditioning condenser

6.22 Typical condenser brackets to be removed from 2012 models

6.30 Inspect the radiator mounts for damage

22   Detach the three brackets from the air conditioning condenser and transmission cooler **(see illustration)**.
23   Without disconnecting the air conditioning lines, carefully release the cooler from the clip at the left end of the radiator.
24   Disconnect the wiring from the electric fan, then release the upper fan clips from the radiator.
25   Remove the fan shroud mounting bolts, detach it from the retainers at the bottom, then slide it to the rear as far as possible.
26   Remove the fan shroud and the electric fan together.
27   Disconnect the overflow hose from the filler neck of the radiator.

All models
28   Remove the upper radiator mounting bolts.
29   Lift the radiator from the engine compartment. Don't spill coolant on the vehicle or scratch the paint. Also be careful not to damage the cooling fins of the transmission cooler or power steering cooler.
30   Whenever the radiator is removed from the vehicle, make note of the location of all rubber mounting cushions **(see illustration)**. If they're cracked, hardened or otherwise deteriorated, replace them.

Diesel engine
31   Disconnect the cables from the negative battery terminals (see Chapter 5).
32   Drain the cooling system (see Chapter 1).
33   Disconnect the sensor wiring harnesses, then remove the air filter housing and intake tube (see Chapter 4).
34   Detach the coolant reservoir hose from the top of the radiator. On 2009 models, also detach the windshield washer reservoir hose and the battery cable, then remove the power steering cooler mounting bolts and move it aside.
35   On 2009 models, remove the fan shroud mounting nuts from the mounting brackets,

then push it forward and clear of the brackets. Turn and push the shroud to the rear as far as possible for clearance.
36   On 2010 and later models, remove the engine-mounted viscous fan and the fan shroud (see Section 4).
37   Disconnect the transmission cooler lines from the cooler. Plug the openings to prevent contamination.
38   Unbolt the power steering cooler and move it aside.
39   Remove the radiator top mounting bolts.
40   Angle the radiator to the front, then lift it out.

Installation
41   With the radiator removed, it can be inspected for leaks and damage. If it needs repair, have a radiator shop or dealer service department perform the work as special techniques are required.
42   Bugs and dirt can be removed from the front of the radiator with a garden hose, followed by compressed air and a soft brush. Don't bend the cooling fins as this is done. When blowing out the core, direct the hose or air line only from the engine side out.
43   Inspect the radiator mounts for deterioration and make sure there's nothing in them when the radiator is installed.
44   Installation is the reverse of removal.
45   Fill the cooling system with the proper mixture of antifreeze and water (see Chapter 1).
46   Start the engine and check for leaks. Allow the engine to reach normal operating temperature, indicated by the inlet radiator hose becoming hot. Recheck the coolant level and add more if required.

---

7   Water pump - check

Refer to illustration 7.3
1   A failure in the water pump can cause serious engine damage due to overheating.
2   There are several ways to check the

7.3 The water pump weep hole will drip coolant when the seal on the pump shaft fails (typical)

operation of the water pump while it's installed on the engine. If the pump is defective, it should be replaced with a new or rebuilt unit.
3   Water pumps are equipped with weep or vent holes. If a failure occurs in the pump seal, coolant will leak from the hole. In most cases you'll need a flashlight to find the hole on the water pump from underneath to check for leaks **(see illustration)**. **Note:** *The weep hole on some models is difficult to see because of the design of the pulleys and drivebelts.*
4   If the water pump shaft bearings fail there may be a howling sound at the front of the engine while it's running. Shaft wear can be felt if the water pump pulley is rocked up and down. Don't mistake drivebelt slippage, which causes a squealing sound, for water pump bearing failure.
5   It is possible for a water pump to be bad, even if it doesn't howl or leak water. Sometimes the fins on the back of the impeller can corrode away until the pump is no longer effective. The only way to check for this is to remove the pump for examination (see Section 8).

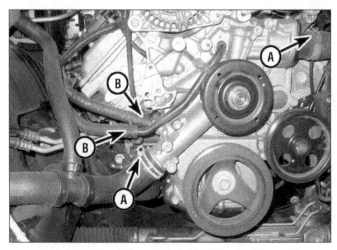

8.14 Location of the radiator (A) and heater hoses (B) (Hemi engine)

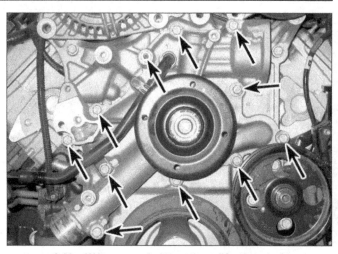

8.20a Water pump bolt locations (Hemi engine)

8.20b Water pump removal (diesel engine)

8.23 First remove the rubber O-ring from the water pump flange, then carefully scrape all gasket material and sealer from the face of the pump

---

8 Water pump - removal and installation

Warning: *Wait until the engine is completely cool before beginning this procedure.*

Removal

Refer to illustrations 8.14, 8.20a and 8.20b

1   Disconnect the cable(s) from the negative battery terminal(s) (see Chapter 5).
2   Drain the cooling system (see Chapter 1). If the coolant is relatively new or in good condition, save it and reuse it.
3   Remove the drivebelt (see Chapter 1).

3.6L engine

4   Disconnect the IAT sensor electrical connector and remove the air filter housing (see Chapter 4A).
5   Disconnect the heater and radiator hoses at the water pump.
6   Remove the idler pulley mounting bolt and idler pulley.
7   Remove the water pump mounting bolts, making sure to identify the four bolts that bolt

directly to the timing chain cover.
8   Remove the pump and discard the gasket.

3.7L V6 and 4.7L V8 engines

9   Detach the engine-mounted viscous fan assembly from the water pump (see Section 4). Don't try to remove it from the truck at this point.
10   Remove the radiator-mounted electric fan (see Section 4).
11   Disconnect the lower radiator hose from the water pump. If a hose sticks, grasp it near the end with a pair of adjustable pliers and twist it to break the seal, then pull it off. If the hose is deteriorated, cut it off and install a new one.

5.7L and 6.4L Hemi engines

12   Remove the air filter housing, the air inlet ducts and the resonator brackets (see Chapter 4A).
13   Remove the fan assembly (see Section 4).
14   Disconnect the radiator hoses from the

engine **(see illustration)**, then move them out of the way. If a hose sticks, grasp it near the end with a pair of adjustable pliers and twist it to break the seal, then pull it off. If the hose is deteriorated, cut it off and install a new one.
15   Remove the drivebelt idler pulley and the tensioner.
16   Remove the top metal heater tube from the cylinder head.
17   On 6.4L models, disconnect EGR cooler hose then disconnect the temperature sensor electrical connector and move the harness out of the way.

6.7L diesel engine

18   Remove the air filter housing and the air intake pipe (see Chapter 4B).
19   Remove the alternator (see Chapter 5).

All engines

20   Remove the bolts and detach the water pump from the engine. Note the locations of the various lengths and different types of bolts as they're removed to ensure correct installation **(see illustrations)**.

Installation

Refer to illustrations 8.23, 8.25a, 8.25b, 8.28a and 8.28b

21   Clean the bolt threads and the threaded holes in the engine to remove corrosion and sealant.
22   Compare the new pump to the old one to make sure they're identical. If the old pump is being reused, check the impeller blades on the rear for corrosion. If any vanes are missing or badly corroded, replace the pump with a new one.
23   Remove all traces of old gasket material from the engine (and water pump, if the same one is to be installed) **(see illustration)**.
24   Clean the engine and water pump mating surfaces with brake system cleaner.
25   On water pumps equipped with an O-ring seal, position it in the groove on the back of

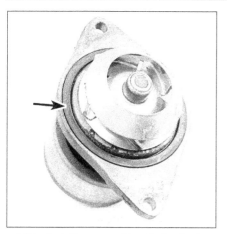

8.25a Install a new rubber O-ring in the groove on the back of the pump (diesel engine shown)

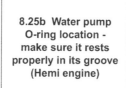

8.25b Water pump O-ring location - make sure it rests properly in its groove (Hemi engine)

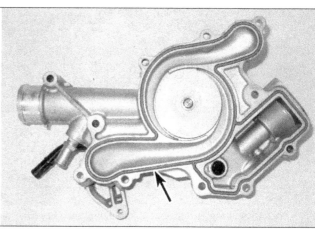

the water pump (see illustrations). Caution: *Make sure the O-ring is correctly seated in the water pump groove to avoid a coolant leak.*

26 On water pumps equipped with a gasket, apply a thin film of RTV sealant to the gasket mating surface of the new pump, and position the gasket on the pump. Apply a thin film of RTV sealant to the engine-side of the gasket and slip a couple of bolts through the pump mounting holes to hold the gasket in place.

27 Carefully attach the pump and O-ring/gasket to the engine and thread the bolts into the holes finger tight. On diesel engines, make sure the weep hole is pointing down.

28 Install the remaining bolts and tighten them in a criss-cross pattern to the torque listed in this Chapter's Specifications in 1/4-turn increments (see illustrations). Turn the water pump by hand to make sure it rotates freely.

29 Reinstall all parts removed for access to the pump.

30 Refill the cooling system (see Chapter 1). Run the engine and check for leaks.

9 Coolant temperature sending unit - check and replacement

Warning: *Wait until the engine is completely cool before beginning this procedure.*

Check

1 The coolant temperature indicator system consists of a warning light or a temperature gauge on the dash and a coolant temperature sending unit mounted on the engine. On the models covered by this manual, the Engine Coolant Temperature (ECT) sensor, which is an information sensor for the Powertrain Control Module (PCM), also functions as the coolant temperature sending unit (see Chapter 6).

2 If an overheating indication occurs, check the coolant level in the system and make sure all connectors in the wiring harness between the sensor and the indicator light or gauge are tight.

3 When the ignition switch is turned to START and the starter motor is turning, the indicator light (if equipped) should come on. This doesn't mean the engine is overheated; it just means that the bulb is good.

4 If the light doesn't come on when the ignition key is turned to START, the bulb might be burned out, the ignition switch might be faulty or the circuit might be open.

5 As soon as the engine starts, the indicator light should go out and remain off, unless the engine overheats. If the light doesn't go out, check for any stored trouble codes in the PCM (see Chapter 6).

6 If the engine tends to overheat easily, check the coolant to make sure it's correctly mixed (see Chapter 1).

Replacement

7 See Chapter 6 for the ECT sensor replacement procedure.

10 Blower motor resistor and blower motor - replacement

Warning: *The models covered by this manual are equipped with a Supplemental Restraint System (SRS), more commonly known as airbags. Always disarm the airbag system before working in the vicinity of any airbag system component to avoid the possibility of*

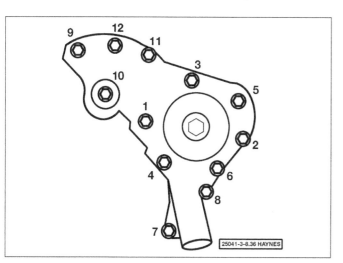

8.28a 3.6L engine water pump bolt tightening sequence

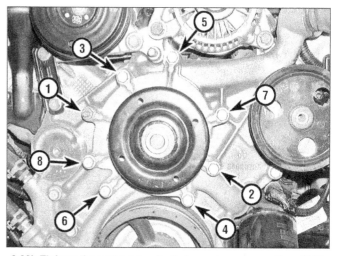

8.28b Tighten the water pump bolts in a criss-cross pattern (3.7L V6/4.7L V8 engine shown)

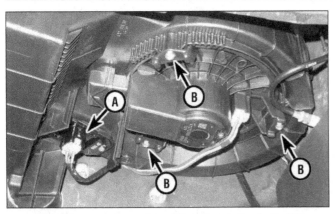

10.2 Blower motor resistor (A) and mounting bolts (B)

11.4 Pull the center instrument panel out, then depress the tabs and detach the electrical connectors from the back of the heater/ air conditioning control assembly

accidental deployment of the airbag, which could cause personal injury (see Chapter 12). Do not use a memory saving device to preserve the PCM's memory when working on or near airbag system components.

Blower motor resistor

Refer to illustration 10.2

1   Disconnect the cable(s) from the negative battery terminal(s) (see Chapter 5).
2   Disconnect the electrical connector from the blower motor resistor **(see illustration)**.
3   Remove the blower motor resistor mounting screws and remove it from the blower housing.
4   Installation is the reverse of removal.

Blower motor

5   Disconnect the cable(s) from the negative battery terminal(s) (see Chapter 5).
6   Disconnect the electrical connector from the blower motor, then remove the mounting screws and lower the blower motor out of the housing **(see illustration 10.2)**.
7   Installation is the reverse of removal.

11   Heater/air conditioning control assembly - removal and installation

Refer to illustration 11.4

Warning: *The models covered by this manual are equipped with a Supplemental Restraint System (SRS), more commonly known as airbags. Always disarm the airbag system before working in the vicinity of any airbag system component to avoid the possibility of accidental deployment of the airbag, which could cause personal injury (see Chapter 12). Do not use a memory saving device to preserve the PCM's memory when working on or near airbag system components.*

1   Disconnect the cable(s) from the negative battery terminal(s) (see Chapter 5).
2   Remove the floor console trim panel (see Chapter 11), if equipped.
3   Remove the center trim bezel (see Chapter 11).
4   Pull the bezel forward and disconnect the electrical connectors **(see illustration)**.
5   Remove the four mounting screws and detach the air conditioning and heater control assembly.
6   Installation is the reverse of removal.

12   Heater core - removal and installation

Warning: *The models covered by this manual are equipped with a Supplemental Restraint System (SRS), more commonly known as airbags. Always disarm the airbag system before working in the vicinity of any airbag system component to avoid the possibility of accidental deployment of the airbag, which could cause personal injury (see Chapter 12). Do not use a memory saving device to preserve the PCM's memory when working on or near airbag system components.*
Warning: *The air conditioning system is under high pressure. Do not loosen any hose fittings or remove any components until after the system has been discharged. Air conditioning refrigerant must be properly discharged into an EPA-approved recovery/recycling unit at a dealer service department or an automotive air conditioning repair facility. Always wear eye protection when disconnecting air conditioning system fittings.*
Note: *Heater core removal is a difficult task for the home mechanic. It can be done with slow, careful attention to detail, but many fasteners and wiring connectors are difficult to get at behind the instrument panel. The entire instrument panel must be removed to allow the heater/air conditioning unit to be removed from the vehicle.*

Removal

Refer to illustrations 12.4, 12.5 and 12.6

1   Have the air conditioning system discharged (see **Warning** above).
2   Disconnect the cable(s) from the negative battery terminal(s) (see Chapter 5).
3   Drain the cooling system (see Chapter 1).
4   Disconnect the heater hoses at the heater core on the right side of the engine compartment at the firewall **(see illustration)**. Tape or plug all openings.
5   Disconnect both air conditioning lines from the firewall **(see illustration)**.

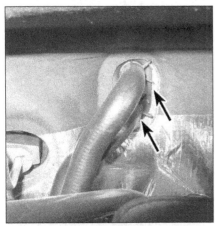

12.4 Slide the clamps away and detach the hoses from the heater core at the engine compartment firewall

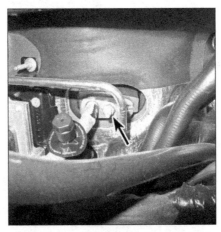

12.5 Air conditioning line connection at the firewall

6   Remove the two nuts on the firewall in the engine compartment that retain the heater housing **(see illustration)**. On some models it will be necessary to first remove the Powertrain Control Module (PCM) for access (see Chapter 6).

7   Remove the instrument panel (see Chapter 11). On models with a center console, also remove the air duct that goes to the console.

8   Remove the rear floor air ducts, if equipped.

9   Remove the lower bolt and the two nuts that retain the heater housing, then remove the complete heater housing assembly from the vehicle.

10   Remove the foam insulator from the water pipe flange, then remove the two screws that attach the flange.

11   Release the tab and remove the flange.

12   Remove the heater core from the housing.

Installation

13   Before installing the heater core, make sure all foam seals are in place on the heater core and evaporator coil tubes.

14   Slide the heater core into the housing, engage the tab and install the screws.

15   Reinstall the remaining components in the reverse order of removal. **Note:** When installing the heater/air conditioning unit, avoid pinching any wiring harnesses between the body and the housing.

16   Refill the cooling system (see Chapter 1).

17   Start the engine and check for proper operation.

18   Have the air conditioning system evacuated, recharged and leak-tested by the shop that discharged it.

---

13   Air conditioning and heating system - check and maintenance

Air conditioning system

Refer to illustration 13.1

Warning: *The air conditioning system is under high pressure. Do not loosen any hose fittings or remove any components until after the system has been discharged. Air conditioning refrigerant must be properly discharged into an EPA-approved recovery/recycling unit at a dealer service department or an automotive air conditioning repair facility. Always wear eye protection when disconnecting air conditioning system fittings.*

Caution: *All models covered by this manual use environmentally friendly R-134a. This refrigerant (and its appropriate refrigerant oils) are not compatible with R-12 refrigerant system components and must never be mixed or the components will be damaged.*

Caution: *When replacing entire components, additional refrigerant oil should be added equal to the amount that is removed with the component being replaced. Read the can*

12.6 Remove the two heater/air conditioning unit nuts from the studs in the engine compartment (components removed for clarity)

before adding any oil to the system, to make sure it is compatible with the R-134a system.

1   The following maintenance checks should be performed on a regular basis to ensure that the air conditioning continues to operate at peak efficiency.

a) *Inspect the condition of the drivebelt. If it is worn or deteriorated, replace it (see Chapter 1).*

b) *Inspect the system hoses. Look for cracks, bubbles, hardening and deterioration. Inspect the hoses and all fittings for oil bubbles or seepage. If there is any evidence of wear, damage or leakage, replace the hose(s).*

c) *Inspect the condenser fins for leaves, bugs and any other foreign material that may have embedded itself in the fins. Use a fin comb or compressed air to remove debris from the condenser.*

d) *Make sure the system has the correct refrigerant charge.*

e) *If you hear water sloshing around in the dash area or have water dripping on the carpet, check the evaporator housing drain tube and insert a piece of wire into the opening to check for blockage.* **Note:** *The drain tube is at the bottom of the firewall on the passenger side* **(see illustration).**

2   It's a good idea to operate the system for about 10 minutes at least once a month, particularly during the winter. Long-term non-use can cause hardening, and subsequent failure, of the seals.

3   Leaks in the air conditioning system are best spotted when the system is brought up to temperature and pressure, by running the engine with the air conditioning ON for five minutes. Shut the engine off and inspect the air conditioning hoses and connections. Traces of oil usually indicate refrigerant leaks.

4   Because of the complexity of the air conditioning system and the special equipment necessary to service it, in-depth troubleshooting and repairs are not included in this man-

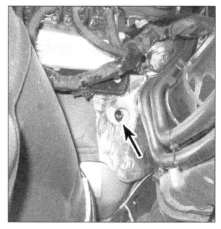

13.1 Check the evaporator housing drain tube for blockage - the view here is from under the engine compartment

ual. However, simple checks and component replacement procedures are provided in this Chapter.

5   If the air conditioning system doesn't operate at all, check the fuse panel and the air conditioning relay, located in the relay box in the engine compartment (see Chapter 12).

6   The most common cause of poor cooling is simply a low system refrigerant charge. If a noticeable drop in cool air output occurs, the following quick check will help you determine if the refrigerant level is low. For more complete information on the air conditioning system, refer to the *Haynes Automotive Heating and Air Conditioning Manual*.

Checking the refrigerant charge

7   Warm the engine up to normal operating temperature.

8   Place the air conditioning temperature selector at the coldest setting and put the blower at the highest setting. Open the doors (to make sure the air conditioning system doesn't cycle off as soon as it cools the passenger compartment).

9   With the compressor engaged, the clutch will make an audible click and the center of the clutch will rotate. Feel the evaporator inlet pipe at the firewall with one hand while placing your other hand on the outlet pipe.

10   One should be colder than the other. If it isn't, the system charge is probably low. The earliest warning that a system is low on refrigerant is the air temperature coming out of the ducts inside the vehicle. If the air isn't as cold as it used to be, the system probably needs a charge.

11   Further inspection or testing of the system requires special tools and techniques and is beyond the scope of this manual.

Adding refrigerant

Refer to illustrations 13.12, 13.15 and 13.18

12   Buy an automotive charging kit at an auto parts store. A charging kit includes a can of refrigerant, a tap valve and a short section

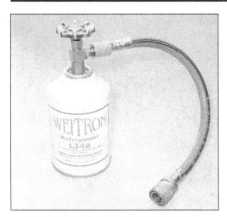

13.12 A basic charging kit for 134a systems is available at most auto parts stores - it must say 134a (not R-12) and so must the can of refrigerant

13.15 Add refrigerant to the system at the low-pressure port (near the accumulator) - Hemi model shown

13.18 If you have an accurate thermometer, you can place it in the center air conditioning duct inside the vehicle

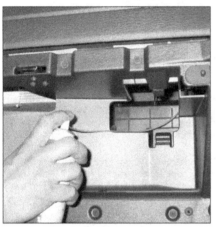

13.24 To disinfect the evaporator housing, insert the nozzle of the disinfectant can through the opening in the recirculation housing and point it toward the evaporator core (which is toward the left of the opening) - glove box removed for access

of hose that can be attached between the tap valve and the system low side service valve **(see illustration). Caution:** *Although the system will hold more than one can of refrigerant, don't add more than one can (you could overfill the system).*

13   Hook up the charging kit by following the manufacturer's instructions. **Warning:** *DO NOT try to connect the charging kit hose to the system high side!* The fittings on the charging kit are designed to fit **only** on the low side of the system.

14   Back off the valve handle on the charging kit and screw the kit onto the refrigerant can, making sure first that the O-ring or rubber seal inside the threaded portion of the kit is in place. **Warning:** *Wear protective eyewear when dealing with pressurized refrigerant cans.*

15   Remove the dust cap from the low-side charging port and attach the quick-connect fitting on the kit hose **(see illustration).**

16   Warm up the engine and turn on the air conditioner. Keep the charging kit hose away from the fan and other moving parts. The charging process requires the compressor to be running.

17   Turn the valve handle on the kit until the stem pierces the can, then back the handle out to release the refrigerant. You should be able to hear the rush of gas. Add refrigerant to the low side of the system until both the accumulator surface and the evaporator inlet pipe feel about the same temperature. Allow stabilization time between each addition.

18   If you have an accurate thermometer, you can place it in the center air conditioning duct inside the vehicle **(see illustration)** and keep track of the outlet air temperature. A charged system that is working properly should cool to approximately 40-degrees F. If the ambient (outside) air temperature is very high, say 110-degrees F, or if the relative humidity is high, the duct air temperature may be as high as 60- to 70-degrees F, but generally the air conditioning is 30 to 50-degrees F cooler than the ambient air.

19   When the can is empty, turn the valve handle to the closed position and release the connection from the low-side port. Replace the dust cap.

20   Remove the charging kit from the can and store the kit for future use with the piercing valve in the UP position to prevent inadvertently piercing the can on the next use.

Eliminating air conditioning odors

Refer to Illustration 13.24

21   Unpleasant odors that often develop in air conditioning systems are caused by the growth of a fungus, usually on the surface of the evaporator core. The warm, humid environment there is a perfect breeding ground for mildew to develop.

22   The evaporator core on most vehicles Is difficult to access, and factory dealerships have a lengthy, expensive process for eliminating the fungus by opening up the evaporator case and using a powerful disinfectant and rinse on the core until the fungus is gone. You

can service your own system at home, but it takes something much stronger than basic household germ-killers or deodorizers.

23   Aerosol disinfectants for automotive air conditioning systems are available in most auto parts stores, but remember when shopping for them that the most effective treatments are also the most expensive. The basic procedure for using these sprays is to start by running the system in the Recirculating mode for ten minutes with the blower on its highest speed. Use the highest heat mode to dry out the system - keep the compressor from engaging by disconnecting the wiring connector at the compressor (see Section 14).

24   Make sure that the disinfectant can comes with a long spray hose. Work the nozzle through the opening in the heater/air conditioning recirculation housing so that it protrudes inside and points toward the evaporator housing (towards the center of the instrument panel) **(see illustration)** and then spray according to the manufacturer's recommendations. Try to cover the whole surface of the evaporator core, by aiming the spray up, down and sideways. Follow the manufacturer's recommendations for the length of spray and waiting time between applications.

25   Once the evaporator has been cleaned, the best way to prevent the mildew from coming back again is to make sure your evaporator housing drain tube is clear.

Heating system

26   If the carpet under the heater core is damp, or if antifreeze vapor or steam is coming through the vents, the heater core is leaking. Remove it (see Section 12) and install a new unit (most radiator shops will not repair a leaking heater core).

27   If the air coming out of the heater vents isn't hot, the problem could stem from any of the following causes:

   a) *The thermostat is stuck open, preventing the engine coolant from warming up enough to carry heat to the heater core. Replace the thermostat (see Section 3).*

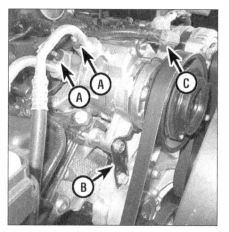

14.4 Air conditioning compressor details (4.7L V8 engine)

A Suction and discharge line attaching nuts
B Mounting bracket
C Wiring connector

b) There is a blockage in the system, preventing the flow of coolant through the heater core. Feel both heater hoses at the firewall. They should be hot. If one of them is cold, there is an obstruction in one of the hoses or in the heater core. Detach the hoses and back flush the heater core with a water hose. If the heater core is clear but circulation is still restricted, remove the two hoses and flush them out with a water hose.

c) If flushing fails to remove the blockage from the heater core, the core must be replaced (see Section 12).

14 Air conditioning compressor - removal and installation

Refer to illustration 14.4

Warning: *The air conditioning system is under high pressure. Do not loosen any hose fittings or remove any components until after the system has been discharged. Air conditioning refrigerant must be properly discharged into an EPA-approved recovery/recycling unit at a dealer service department or an automotive air conditioning repair facility. Always wear eye protection when disconnecting air conditioning system fittings.*

Note: *The receiver-drier (which is integral with the condenser) should be replaced whenever the compressor is replaced (see Section 15).*

1 Have the air conditioning system discharged (see **Warning** above).
2 Disconnect the cable(s) from the negative battery terminal(s) (see Chapter 5).
3 On 2010 and later diesel models, remove the air filter housing and the turbo outlet pipe (see Chapter 4B).
4 Disconnect the compressor clutch electrical connector **(see illustration)**.

15.6 Remove these nuts to detach the air conditioning lines from the short condenser jumper lines

5 Remove the drivebelt (see Chapter 1).
6 Disconnect the refrigerant lines from the compressor **(see illustration 14.4)**. Plug the open fittings to prevent entry of dirt and moisture.
7 Unbolt the compressor from the mounting brackets and lift it out of the vehicle. **Note:** *On Hemi engines, remove the alternator bracket mounting bolts and the compressor bracket.*
8 If a new compressor is being installed, follow the directions with the compressor regarding the draining of excess oil prior to installation. **Note:** *Any replacement compressor used must be designated as compatible with R-134a refrigerant.*
9 The clutch may have to be transferred from the original to the new compressor.
10 Installation is the reverse of removal. Replace all O-rings with new ones specifically made for air conditioning system use and compatible with R-134a refrigerant. Lubricate them with refrigerant oil. Any refrigerant oil added must also be compatible with R-134a refrigerant.
11 Have the system evacuated, recharged and leak-tested by the shop that discharged it.

15 Air conditioning condenser - removal and installation

Warning: *The air conditioning system is under high pressure. Do not loosen any hose fittings or remove any components until after the system has been discharged. Air conditioning refrigerant must be properly discharged into an EPA-approved recovery/recycling unit at a dealer service department or an automotive air conditioning repair facility. Always wear eye protection when disconnecting air conditioning system fittings.*

Note: *The receiver-drier is an integral part of the condenser.*

1 Have the air conditioning system dis-

15.8a The lower condenser jumper line is secured by a bolt . . .

15.8b . . . the upper line is attached with a clip that's under a plastic cover

charged (see **Warning** above).
2 Disconnect the cable(s) from the negative battery terminal(s) (see Chapter 5).

Gasoline models

Refer to illustrations 15.6, 15.8a, 15.8b and 15.9

3 Drain the cooling system (see Chapter 1).
4 On automatic transmission models, disconnect the transmission cooler lines from the jumper lines near the right of the radiator. Seal the openings to prevent contamination.
5 On 2012 and later models, remove the radiator grille (see Chapter 11).
6 Disconnect the air conditioning lines from the condenser jumper line tapping block **(see illustration)**. Seal the openings to prevent contamination.
7 On 2011 and earlier models, remove the air conditioning condenser and the radiator as a single unit (see Section 6). On 2012 and later models, remove the condenser-to-radiator line bracket bolts and condenser mount-

15.9 The left side of the condenser can be separated from the radiator by sliding it out of this bracket

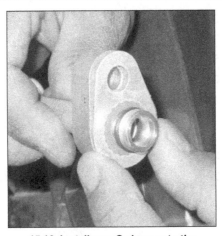

15.16 Install new O-rings onto the refrigerant lines

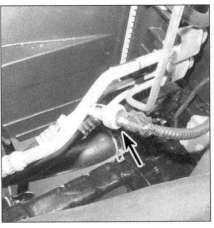

16.2 Air conditioning system pressure switch location (4.7L V8 engine)

ing bolts and separate the condenser from the radiator.

8   On 2011 and earlier models remove the jumper line(s) from the condenser **(see illustrations)**. Seal the openings to prevent contamination.

9   On 2011 and earlier models, remove the condenser mounting bolts from the right side of the radiator. Disengage the brackets from the left end **(see illustration)** and separate the condenser from the radiator.

Diesel models

10   On 2009 models, remove the cover from the right end of the condenser for access to the fittings.

11   On 2010 and later models, remove the grille (see Chapter 11), then disconnect the automatic transmission cooler lines from the cooler (if installed). Seal the openings to prevent contamination.

12   Disconnect the refrigerant lines from the condenser. Seal the openings to prevent contamination.

13   Remove the mounting bolts, then remove the condenser.

All models

Refer to illustration 15.16

14   If the original condenser will be reinstalled, store it with the line fittings on top to prevent oil from draining out.

15   If a new condenser is being installed, pour 1 oz (30 ml) of refrigerant oil into it prior to installation (an oil designated as compatible with R-134a refrigerant). **Note:** *New R-134a compatible O-rings should be used in each fitting during reassembly.*

16   Install new O-rings onto the refrigerant and

transmission cooler lines **(see illustration)**.

17   Reinstall the components in the reverse order of removal.

18   Have the system evacuated, recharged and leak-tested by the shop that discharged it.

16   Air conditioning pressure switch - replacement

Refer to illustration 16.2

Note: *The air conditioning pressure switch is mounted on the high pressure line. It's not necessary to evacuate the system in order to replace the switch.*

1   Disconnect the cable(s) from the negative battery terminal(s) (see Chapter 5).

2   Unplug the electrical connector from the air conditioning pressure switch **(see illustration)**.

3   Unscrew the pressure switch.

4   Lubricate the new switch O-ring with clean refrigerant oil of the correct type. **Caution:** *Use an O-ring that's compatible with R-134a refrigerant.*

5   Screw the new part onto the refrigerant line until hand tight, then tighten it securely.

6   Reconnect the electrical connector and the negative battery cable(s).

17   Air conditioning expansion valve - removal and installation

Warning: *The air conditioning system is under high pressure. Do not loosen any hose fittings or remove any components until after the system has been discharged. Air conditioning*

refrigerant must be properly discharged into an EPA-approved recovery/recycling unit at a dealer service department or an automotive air conditioning repair facility. Always wear eye protection when disconnecting air conditioning system fittings.

Note: *After operating a fully-charged air conditioner for five minutes, the liquid line should be hot near the condenser and it should be cold near the evaporator (be careful - it can get very hot!). If there isn't a significant temperature difference, the expansion valve may be plugged. If the system is checked with the appropriate gauges, and the high-pressure reads extremely high and the low-pressure reads almost a vacuum, the expansion valve is plugged. In either case, the expansion valve must be replaced.*

1   The expansion valve is a metering device that is located where the refrigerant lines connect to the evaporator at the firewall.

2   Have the air conditioning system discharged (see **Warning** above).

3   Disconnect the cable(s) from the negative battery terminal(s) (see Chapter 5).

4   Disconnect the air conditioning line fitting block from the firewall **(see illustration 12.5)**. Remove the O-rings and seal the openings to prevent contamination.

5   Remove the two counter-sunk bolts from the expansion valve, then remove the valve. Remove the O-rings and seal the openings to prevent contamination.

6   Lubricate the new expansion valve O-ring with clean refrigerant oil of the correct type. **Caution:** *Use an O-ring that's compatible with R-134a refrigerant.*

7   Installation is the reverse of removal.

8   Have the system evacuated, recharged and leak-tested by the shop that discharged it.

Chapter 4 Part A
Fuel and exhaust systems - gasoline engines

Contents

Specifications

General

Fuel pressure (engine running at idle speed).........	56 to 60 psi
Fuel injector resistance.........	Not specified, but the injectors can still be checked with an ohmmeter to verify that the solenoid coils are not open (very high resistance) or shorted (near zero resistance)

Torque specifications

Note: *One foot-pound (ft-lb) of torque is equivalent to 12 inch-pounds (in-lbs) of torque. Torque values below approximately 15 ft-lbs are expressed in inch-pounds, since most foot-pound torque wrenches are not accurate at these smaller values.*

Throttle body mounting bolts	
3.6L V6 engine	62 in-lbs
3.7L V6 and 4.7L V8.........	65 in-lbs
5.7L and 6.4L Hemi.........	50 in-lbs
Fuel rail mounting nuts/bolts	
3.6L V6 engine	62 in-lbs
All other models	100 in-lbs

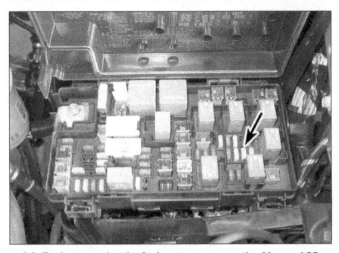

2.2 To depressurize the fuel system, remove the 20 amp ASD fuse (there may be more than one ASD fuse, but only one of them is a 20 amp). This is the location on our project 2012 Ram 1500 with a 4.7L V8, but you'll have to use the guide on the underside of the fuse/relay box cover to locate the proper fuse on your model

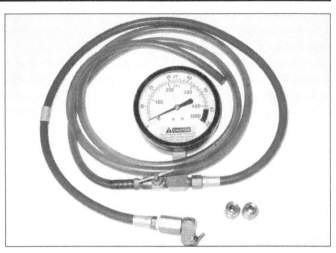

3.3 To check the fuel pressure, you'll need to obtain a fuel pressure gauge capable of reading the fuel pressure within the specified operating system pressure, a hose to connect the gauge to the fuel pressure test port and an adapter suitable for connecting the hose to the test port

1   General information

Warning: *Gasoline is extremely flammable, so take extra precautions when you work on any part of the fuel system. Don't smoke or allow open flames or bare light bulbs near the work area, and don't work in a garage where a gas-type appliance (such as a water heater or a clothes dryer) is present. Since gasoline is carcinogenic, wear fuel-resistant gloves when there's a possibility of being exposed to fuel, and, if you spill any fuel on your skin, rinse it off immediately with soap and water. Mop up any spills immediately and do not store fuel-soaked rags where they could ignite. The fuel system is under constant pressure, so, if any fuel lines are to be disconnected, the fuel pressure in the system must be relieved first. When you perform any kind of work on the fuel system, wear safety glasses and have a Class B type fire extinguisher on hand.*

All models covered by this manual are equipped with a sequential multi-port fuel injection system. This type of fuel injection system uses timed impulses to sequentially inject the fuel directly into the intake ports of each cylinder in the same sequence as the firing order. The Powertrain Control Module (PCM) controls the injectors. The PCM monitors various engine parameters and delivers the exact amount of fuel, in firing order sequence, into the intake ports. For more information about the fuel injection system, see Section 9.

The fuel pump is located inside the fuel tank, and can be accessed by removing a locking ring on top of the tank. You must lower the fuel tank before you can remove this ring and pull the fuel pump out of the fuel tank. The fuel level sending unit is an integral component of the fuel pump and it must be accessed in the same manner.

There are two fuel filters. One is the nylon mesh fuel sock or strainer at the lower (inlet) end of the fuel pump/fuel level sending unit module, which is mounted in the roof of the fuel tank and protrudes down into the fuel inside. The main fuel filter is an integral part of the fuel pressure regulator, which is located on top of the fuel pump/fuel level sending unit module. No external fuel filters are used on any of the gasoline-powered vehicles covered by this manual. The fuel filters are extended-life parts and do not need to be replaced at scheduled maintenance intervals. They should only be replaced if diagnostic testing indicates the need to do so.

The exhaust system consists of the two exhaust manifolds, catalytic converters, exhaust pipes and mufflers. For further information regarding the catalytic converter, refer to Chapter 6.

2   Fuel pressure relief procedure

Refer to illustration 2.2

Warning: *Gasoline is extremely flammable, so take extra precautions when you work on any part of the fuel system. See the* **Warning** *in Section 1.*

1   Remove the fuel filler cap (this will relieve any pressure that has built-up in the tank).

2   Remove the cover from the fuse and relay center in the engine compartment, then remove the 20 amp ASD fuse **(see illustration)**.

3   Turn the ignition key to START and crank over the engine for several seconds. It will either start momentarily and immediately stall, or it won't start at all.

4   Turn the ignition key to the OFF position.

5   Disconnect the cable from the negative battery terminal before beginning work on the fuel system (see Chapter 5).

6   After all work on the fuel system has been completed, install the fuse.

7   Reconnect the battery cable.

3   Fuel pump/fuel pressure - check

Warning: *Gasoline is extremely flammable, so take extra precautions when you work on any part of the fuel system. See the* **Warning** *in Section 1.*

Preliminary check

1   The fuel pump is located inside the fuel tank, which muffles its sound when the engine is running. Turn the ignition key to ON (*not* START) and listen for the sound made by the fuel pump as it's briefly turned on by the Powertrain Control Module (PCM) to pressurize the fuel system. You will only hear it for a second or two, but that sound tells you that the pump is working. If you can't hear the pump from inside the vehicle, remove the fuel filler cap and have an assistant turn the ignition switch to ON while you listen for the sound of the pump. If the pump does not come on when the ignition key is turned to ON, check the fuses in the underhood fuse/relay box. **Note:** *There is no serviceable fuel pump relay; it is an integral part of the underhood fuse/relay box, which the manufacturer calls the Totally Integrated Power Module (TIPM).* If the fuses are okay, check the wiring back to the fuel pump (see Section 5 if you need help locating the fuel pump electrical connector). If the fuse, relay and wiring are okay, the fuel pump is probably defective. If the pump runs *continuously* with the ignition key in its ON position, the PCM or TIPM is probably defec-

3.4a Fuel pressure test port location (4.7L V8 engines shown, 3.7L V6 similar)

3.4b Fuel pressure test port location (Hemi engine)

tive. Have the PCM and TIPM checked by a dealer service department or other qualified repair shop.

Pressure check

Refer to illustration 3.3, 3.4a, 3.4b and 3.5

Note: *In order to perform the fuel pressure test, you will need a fuel pressure gauge capable of measuring high fuel pressure. You'll also need the correct fittings or adapters to attach it to the fuel rail.*

2   Relieve the fuel system pressure (see Section 2).

3   For this check, you'll need to obtain a fuel pressure gauge with a hose and an adapter suitable for connecting it to the Schrader valve type test port on the injector fuel rail **(see illustration)**. **Note:** *If your engine is not equipped with a test port, you'll have to fabricate a gauge adapter or have the system checked by a shop equipped with the proper tools.*

4   The test port is located on the fuel rail **(see illustrations)**. If you can't access the fuel pressure test port, remove the air intake duct and, if necessary, the air resonator box.

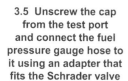

3.5 Unscrew the cap from the test port and connect the fuel pressure gauge hose to it using an adapter that fits the Schrader valve

5   Unscrew the threaded cap from the test port and connect the fuel pressure gauge hose to the test port **(see illustration)**.

6   Start the engine and check the pressure on the gauge, comparing your reading with the pressure listed in this Chapter's Specifications.

7   If the fuel pressure is not within specifications, check the following:

a) *Check for a restriction in the fuel system (kinked fuel line or plugged fuel pump inlet strainer). If no restrictions are found, replace the fuel pump module (see Section 7).*

b) *If the fuel pressure is higher than specified, replace the fuel pump module (see Section 7).*

4   Fuel lines and fittings - repair and replacement

Warning: *Gasoline is extremely flammable, so take extra precautions when you work on any part of the fuel system. See the* **Warning** *in Section 1.*

1   Always relieve the fuel pressure before servicing fuel lines or fittings on fuel-injected vehicles (see Section 2).

2   The fuel supply and return lines connect the fuel tank to the fuel rail on the engine. Inspect the fuel and the evaporative emission control (EVAP) system lines for leaks, kinks and dents whenever you're servicing something underneath the vehicle. All of these lines are secured to the vehicle underbody by plastic clips. To disengage fuel and/or vapor line(s) from the simpler style clip, simply pry the retainers apart and pull out the line(s). To disengage the lines from a clamshell style clip, pry open the clamshell door (the outer half of the clip) with a small screwdriver and swing the door open.

3   Whenever you're working under the vehicle, Inspect all fuel and EVAP lines for leaks, kinks, dents and other damage. Always replace a damaged fuel line or EVAP line immediately. Leaking fuel and EVAP lines will result in loss of fuel and excessive air pollution (leaking raw fuel emits unburned hydrocarbon vapors into the atmosphere).

4   If you find signs of dirt in the lines during disassembly, disconnect all lines and blow them out with compressed air. Inspect the fuel strainer on the fuel pump (see Section 7) for damage and deterioration.

Steel tubing

5   Because fuel lines used on fuel-injected vehicles are under fairly high pressure, it is critical that they be replaced with lines of equivalent specification. If you have to replace a steel line, make sure that you use steel tubing that meets the manufacturer's specifications. Don't use copper or aluminum tubing to replace steel tubing. These materials cannot withstand normal vehicle vibration.

6   Some steel fuel lines have threaded fittings. When loosening these fittings to service or replace components:

a) *Use a backup wrench on the stationary portion of the fitting while loosening and tightening the fitting nuts.*

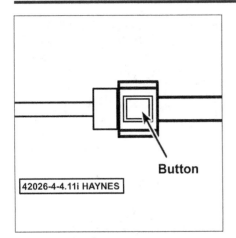

4.11 Some quick-connect fittings have two buttons (one on each side of the fitting) that must be depressed to release the fitting. No tools are necessary

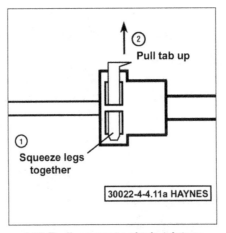

4.15 To disconnect a single-tab type fitting, squeeze the legs of the tab together, pull up the tab and pull the fitting and fuel line or fuel component apart. Discard the old tab and substitute a new one for reassembly

it's connected.

12   To reconnect a two-button fitting, push the fitting onto the fuel line or fuel system component until the raised stop on the fuel line or component is fully seated against the back of the fitting, then keep pushing until you hear/feel a click. Verify that the fitting is correctly reconnected by trying to pull the fuel line or component and the fitting in opposite directions.

Pinch-type fitting

13   Instead of push-buttons, this type of fitting has two finger tabs. Again, no special tools are needed to disconnect it. Simply pinch both tabs together and pull the fitting off the fuel line or fuel component to which it's connected.

14   To reconnect a pinch-type fitting, push it onto the fuel line or fuel component until the raised stop on the fuel line or component is fully seated against the back of the fitting. Then keep pushing until you feel a click. Verify that the fitting is correctly reconnected by trying to pull the fuel line or component and the fitting in opposite directions.

Single-tab type fitting

Refer to illustration 4.15

15   This type of fitting is locked into place by a single removable pull tab **(see illustration)**. The only tool you might need to disconnect it is a small screwdriver. To disconnect the fitting, squeeze the legs of the tab together, pull up the tab (pry it up with a small screwdriver if necessary) and pull the fitting off the fuel line or fuel component to which it's connected. Remove and discard the old pull tab, then install a new tab before reconnecting the fitting.

16   To reconnect a single-tab type fitting, install a new pull tab (but don't push it down into its locked position), push the fitting onto the fuel line or fuel component until the raised stop on the fuel line or component is seated against the back of the fitting, then keep pushing until you hear/feel a click. When the fitting is fully seated, push the new tab down until it locks into place. Verify that the fitting is locked by firmly pulling the fuel line or component and the fitting in opposite directions.

Two-tab type fitting

Refer to illustrations 4.17a, 4.17b, and 4.18

17   This type of fitting is one of the more common ones used on Dodge pick-ups. It uses a plastic retainer with two release tabs protruding from opposite sides of the fitting. To disconnect it, depress both release tabs with your *fingers*, then pull the fitting off the fuel line or fuel component **(see illustration)**. **Caution:** *Do NOT use a tool, such a pair of pliers, to squeeze the tabs together. Using anything besides your fingers might damage the plastic retainer.* The retainer will remain on the fuel line or fuel component and the O-ring **(see illustration)** will remain inside the fitting. Inspect the condition of the O-ring. If it's cracked, torn, deteriorated or otherwise damaged in any way, replace it.

b) *If you're going to replace one of these fittings, use original equipment parts or parts that meet original equipment standards.*

Plastic tubing

7   If you ever have to replace a plastic line, use only the original equipment plastic tubing. **Caution:** *When removing or installing plastic fuel line tubing, be careful not to bend or twist it too much, which can damage it. And damaged fuel lines MUST be replaced! Also, be aware that the plastic fuel tubing is NOT heat resistant, so keep it away from excessive heat. Nor is it acid-proof, so don't wipe it off with a shop rag that has been used to wipe off battery electrolyte. If you accidentally spill or wipe electrolyte on plastic fuel or emissions tubing, replace the tubing.*

Flexible hoses

Warning: *Use only original equipment replacement hoses or their equivalent. Unapproved hoses might fail when subjected to the high operating pressures of the fuel system.*

8   Don't route fuel hoses within four inches of exhaust system components or within ten inches of a catalytic converter. Make sure that no flexible hoses are installed directly against the vehicle, particularly in places where there is any vibration. If allowed to touch some vibrating part of the vehicle, a hose can easily become chafed and it might start leaking. A good rule of thumb is to maintain a minimum of 1/4-inch clearance around a hose (or metal line) to prevent contact with the vehicle underbody.

Fuel line and EVAP line quick-connect fittings

Warning: *ALWAYS relieve the fuel system pressure (see Section 2) before disconnecting a fuel line fitting.*

9   The vehicles covered in this manual use several kinds of fuel line quick-connect fittings for connections at the fuel pump, the EVAP

canister and the fuel rail.

10   The procedure for releasing each type of fuel line fitting is different. But a few rules of thumb apply to all fittings:

a) *ALWAYS relieve the fuel system pressure (see Section 2) before disconnecting a fuel line fitting.*

b) *Inspect the fitting for dirt. If the fitting is dirty, clean it off before disassembling it. The seals in the fitting will stick to the fuel line as they age. Twist the fitting on the line, then push and pull the fitting until it moves freely.*

c) *Always disconnect all fuel line fittings from a fuel system component before removing the component.*

d) *When disconnecting a quick-connect fitting, inspect the condition of the retainer before reconnecting the fitting. The best strategy with respect to retainers is to simply replace the retainer every time that you disconnect the fitting.*

e) *When you disconnect a fitting with an O-ring inside, inspect the O-ring before reconnecting the fitting. Fuel line fittings are under the same pressure as the rest of the fuel system, so to avoid leaks (and fires!) make VERY SURE that the O-ring is good condition. Even better, simply replace it.*

f) *In most cases, the fitting itself is a non-removable part of the fuel line, so you might have to replace an entire fuel line if a fitting is damaged or defective.*

Two-button type fitting

Refer to illustration 4.11

11   This type of fitting has a pair of push-buttons located on opposite sides of the fitting **(see illustration)**. No special tools are needed to disconnect a two-button fitting. Simply press on both buttons and pull the fitting off the fuel line or fuel component to which

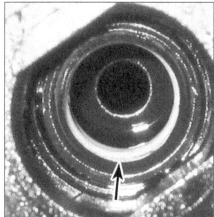

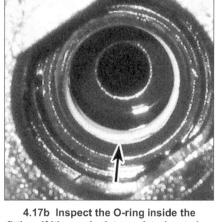

4.17a To disconnect a two-tab type fitting, depress both tabs with your fingers then pull the fuel line and the fitting apart (the plastic retainer comes off with the fuel line)

4.17b Inspect the O-ring inside the fitting. If it's cracked, torn, deteriorated or otherwise damaged in any way, replace it

4.18 Before reconnecting a two-tab type fitting, position the plastic retainer correctly:

A Plastic retainer release tabs
B Plastic retainer locking tabs (must be aligned with window in fitting)
C Raised stop (must be visible through opening in plastic retainer)
D Fitting window (must be aligned with locking tabs on plastic retainer)

18   To reconnect a two-tab type fitting, make sure that the retainer is correctly positioned on the fuel line or component, with the locking tabs of the retainer aligned with the windows in the fitting **(see illustration)**. When everything's correctly aligned, push the fitting onto the fuel line or fuel component until the raised stop on the fuel line or fuel component seats against the back of the fitting, then keep pushing until you feel a click. Verify that the fitting is locked into place by firmly pulling the fuel line or component and the fitting in opposite directions.

Plastic retainer ring type fitting

Refer to illustration 4.19
19   This type of fitting has a round plastic retainer ring, usually black in color, located inside the fitting. To disconnect a plastic retainer ring type fitting, push the fitting body toward the fuel line or fuel system component to which it's connected while firmly pushing the plastic retainer ring into the fitting, then while holding the retainer in its depressed position, pull the fuel line or fuel component out of the fitting **(see illustration)**. **Caution:**

When disconnecting a plastic retainer type fitting, make sure that the plastic retainer ring is pressed squarely into the fitting body. If the retainer becomes cocked in the fitting body it will be difficult or impossible to disconnect the fuel line or fuel system component from the fitting. One way to ensure that the retainer is depressed squarely into the fitting body is to depress the shoulder of the retainer with the flat face of an open-end wrench. After disconnecting the fuel line or fuel system component from the fitting, the plastic retainer ring will remain with quick-connect fitting connector body. Inspect the condition of the fitting body, the plastic retainer ring and the end of the fuel line or fuel system component for damage. Replace any damaged parts.
20   To reconnect a plastic retainer type fitting, push the fitting onto the fuel line or fuel system component until the raised stop on the fuel line or fuel component rests against the back of the fitting, then continue pushing until you feel a click. Verify that the two halves of the fitting are locked together by firmly pulling the fuel line and the fitting in opposite directions.

Metal collar type fittings with latch clips

Refer to illustrations 4.21, 4.22a, 4.22b and 4.23
Note: *You'll find these fittings at the connections between the fuel supply and return lines and the fuel rail.*
Note: *You'll need a special tool, available at most auto parts stores, to disconnect these fittings.*
21   To remove a latch clip, pull or pry the end of the latch off the fuel line, then disengage the other end of the latch from the female end of the metal collar fitting **(see illustration)**.

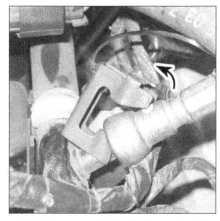

4.21 To remove latch clip from a metal collar fitting, pull or pry the end of the clip off the fuel line, then disengage the other end from the female side of the fitting

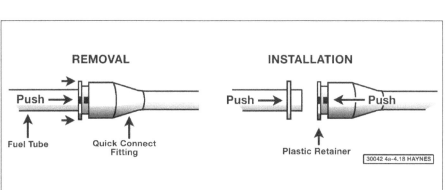

4.19 To disconnect a plastic retainer type fitting, push the plastic retainer ring and fitting body together and pull the line from the fitting. To reconnect, push the line into the fitting until you hear/feel a click

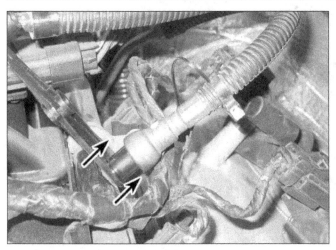

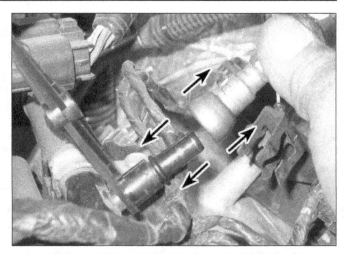

4.22a To release the locking fingers inside a metal collar type fitting, insert a special fuel line disconnection tool into the female side of the fitting until it releases the locking fingers . . .

4.22b . . . then pull the two halves of the fitting apart

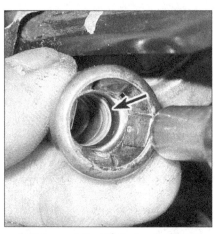

4.23 Inspect the old O-ring inside the female side of the metal collar fitting. If it's cracked, torn or otherwise deteriorated, replace it

5.4 To disconnect the fuel filler neck hose and the fuel EVAP vent hose from their metal pipes, loosen the clamps on these hoses

5.7 To detach the fuel tank from the bottom of the vehicle, remove these two fuel tank strap nuts, then allow the fuel tank straps to swing down (they're hinged) and disengage the left end of each strap from its hinge by lifting up the end of the strap

22   To disconnect a metal collar fitting (with either type of latch clip), release the locking fingers inside the fitting by inserting a special fuel line disconnection tool (available at most auto parts stores) into the metal collar **(see illustration)**. With the special tool still inserted, disconnect the metal collar from the fuel line or from the fuel system component **(see illustration)**.

23   Inspect the O-ring inside the metal collar fitting **(see illustration)**. If it's dried out, cracked, torn or otherwise deteriorated, replace it.

24   Apply a few drops of clean engine oil to the end of the fuel line or fuel system component, push the fitting onto the fuel line or component until the locking fingers snap into place. Verify that the fitting is correctly connected by pulling the fuel line or component and the fitting in opposite directions.

25   Install the latch clip until it snaps into position.

5   Fuel tank - removal and installation

Refer to illustrations 5.4, 5.7, 5.8 and 5.9

Warning: *Gasoline is extremely flammable, so take extra precautions when you work on any part of the fuel system. See the* **Warning** *in Section 1.*

1   Relieve the fuel system pressure (see Section 2).

2   Disconnect the cable from the negative battery terminal (see Chapter 5).

3   Raise the vehicle and place it securely on jackstands.

4   Loosen the hose clamps for the fuel filler neck hose and for the EVAP fuel vent hose **(see illustration)** and disconnect both hoses from their metal pipes. **Note:** *If the fuel tank still has a lot of fuel in it, now is the time to siphon the remaining fuel from the tank*

through the rubber fuel filler neck hose. Using a siphoning kit (available at most auto part stores), siphon the fuel from the tank, through the filler neck hose, into an approved gasoline container. **Warning:** *Never start the siphoning action by mouth!*

5   Disconnect the wiring and the quick-connect fuel lines from the fuel tank.

6   Support the fuel tank with a transmission jack or with a floor jack. If you're going to use a floor jack, put a piece of wood between the jack head and the fuel tank to protect the tank.

7   Remove the fuel tank retaining strap nuts **(see illustration)** and remove both retaining straps. The hinged straps are secured with fasteners on the right side of the tank. To disengage the left end of each strap from its hinge, lift up the left end of the strap, then move it to the left.

8   Carefully lower the tank until you have access to the top of it, then disconnect the fuel pump module wiring and the quick-connect fitting **(see illustration)**.

9   Disconnect any remaining components **(see illustration)**, then lower the tank to the floor.

10   Installation is basically the reverse of removal. Note the following guidelines:

a) *If the fuel tank is being replaced, remove the necessary components from the old fuel tank and install them on the new tank. If you need help with any of the EVAP hoses, refer to Chapter 6.*

b) *Tighten the fuel tank strap nuts securely.*

6   Fuel tank cleaning and repair - general information

1   The fuel tank installed in the vehicles covered by this manual is not repairable. If the fuel tank becomes damaged, it must be replaced.

2   Cleaning the fuel tank (due to fuel contamination) should be performed by a professional with the proper training to carry out this critical and potentially dangerous work. Even after cleaning and flushing, explosive fumes may remain inside the fuel tank.

3   If the fuel tank is removed from the vehicle, it should not be placed in an area where sparks or open flames could ignite the fumes coming out of the tank. Be especially careful inside a garage where a gas-type appliance is located.

7   Fuel pump and fuel level sending unit module - removal and installation

Refer to illustrations 7.8a, 7.8b and 7.10
Warning: *Gasoline is extremely flammable,*

5.8 Lower the fuel tank just enough to access the fuel pump/fuel level sending unit module, then disconnect the electrical connector from the module

so take extra precautions when you work on any part of the fuel system. See the **Warning** *in Section 1.*

1   The fuel pump/fuel level sending unit module includes the fuel filter/fuel pressure regulator, the electric fuel pump assembly (including the fuel pump inlet filter) and the fuel level sending unit. This section covers the removal and installation of the complete module, which is removed as a complete assembly. None of the components can be replaced separately; if any part fails, the entire assembly must be replaced.

2   Relieve the fuel system pressure (see Section 2).

3   Disconnect the cable from the negative battery terminal (see Chapter 5).

4   Raise the vehicle and place it securely on jackstands.

5   Remove the fuel tank (see Section 5).

6   To prevent dirt from entering the fuel tank, clean the area surrounding the fuel

5.9 Disconnect the EVAP line fitting (the EVAP line fitting has already been disconnected in this photo. If you don't know how to disconnect this type of fitting, see Section 4)

pump/fuel level sending unit.

7   Mark the position of the arrow on the fuel pump module so the new one can be installed in the same orientation.

8   Using a pair of large pliers to turn the lock-ring, or using a hammer and a brass drift **(see illustration)** to tap on the lock-ring, turn it counterclockwise. When the lock-ring is loose, unscrew it. Carefully remove the fuel pump/fuel level sending unit module from the tank **(see illustration)**.

9   Inspect the large rubber gasket **(see illustration 7.8b)**. If it's cracked, torn, deteriorated or otherwise damaged, replace it.

10   When installing the fuel pump/fuel level sending unit module, rotate the module until the embossed alignment arrow on the module is aligned properly **(see illustration)**.

11   Installation is otherwise the reverse of removal.

7.8a Use a large pair of pliers or a hammer and drift to loosen the lock-ring that secures the fuel pump/fuel level sending unit module

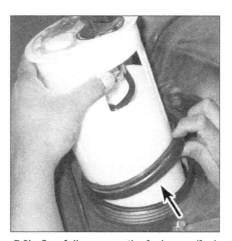

7.8b Carefully remove the fuel pump/fuel level sending unit module from the fuel tank. Inspect the large rubber gasket that seals the mounting hole for the pump/sending unit

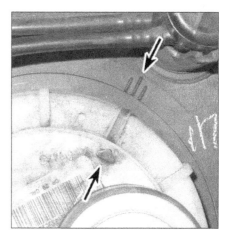

7.10 When installing the fuel pump/fuel level sending unit module, make sure the alignment arrow on top of the module is aligned with the center mark of the three hash marks on top of the fuel tank

8.1 To remove the air intake duct, loosen the hose clamps at each end and disconnect the duct from the air filter housing and the air resonator box

8.5 To detach the air filter housing from the vehicle, grasp it firmly and lift it straight up to disengage the grommets (A) from their corresponding locator pins (B) (typical air filter housing shown)

8   Air filter housing and air intake duct - removal and installation

Air intake duct

Refer to illustration 8.1

1   Loosen the hose clamps and remove the air intake duct **(see illustration)**.
2   Inspect the condition of the air intake duct. Look for cracks, tears, deterioration and other damage. If the air intake duct is damaged in any way, replace it.
3   Installation is the reverse of removal.

Air filter housing

Refer to illustration 8.5

4   Remove the air intake duct (see Step 1).
5   Grasp the air filter housing firmly and lift it straight up **(see illustration)**. The air filter housing is secured to the vehicle by grommets that fit onto locator pins.
6   Inspect the condition of the four mounting grommets. If they're cracked, torn, deteriorated or otherwise damaged, replace them.
7   Installation is the reverse of removal.

9   Fuel injection system - general information

The fuel injection systems used on all vehicles covered by this manual are the sequential multi-port type. This means that there is a fuel injector in each intake port, and that they inject fuel into the intake ports in the cylinder firing order. The injectors are turned on and off by the Powertrain Control Module (PCM). When the engine is running, the PCM constantly monitors engine operating conditions with an array of information sensors, calculates the correct amount of fuel, then varies the interval of time during which the injectors

are open. Sequential multi-port systems provide much better control of the air/fuel mixture ratio than earlier fuel injection systems, and are therefore able to produce more power, better mileage and lower emissions.

The fuel injection system uses the PCM and an array of information sensors to determine and deliver the correct air/fuel ratio under all operating conditions. The fuel injection system consists of three sub-systems: air induction, electronic control and fuel delivery. The fuel injection system is also closely inter-related with PCM-controlled emission control systems. For additional information about the PCM, the information sensors and the emission control systems, refer to Chapter 6.

Air induction system

The air induction system consists of the air filter housing assembly, the air intake duct, the throttle body and the intake manifold. The throttle body contains a throttle plate that regulates the amount of air entering the intake manifold. The throttle plate is opened and closed by the PCM in response to input from the Accelerator Pedal Position Sensor (APPS). The lower part of the throttle body on some engines is heated by engine coolant to prevent icing in cold weather. The throttle body is also the location of the Throttle Position (TP) sensor; this monitors the opening angle of the throttle plate and sends a signal to the Powertrain Control Module (PCM). The Manifold Absolute Pressure (MAP) sensor is also located on the throttle body. The MAP sensor measures intake manifold pressure/vacuum and generates a signal that's proportional to the pressure or vacuum. The PCM uses this data to calculate the load on the engine. Another information sensor, the Intake Air Temperature (IAT) sensor, is located on the intake manifold. The IAT sensor relays a signal to the PCM that varies in accordance with the temperature of the incoming air. The

PCM uses this data to calculate how rich the air/fuel mixture should be. All of the air induction components (air filter housing, air intake duct and throttle body) are covered in this Chapter, except for the intake manifold, which is covered in Chapter 2, and the information sensors, which are covered in Chapter 6.

Electronic control system

For more information about the electronic control system, its information sensors and output actuators, refer to Chapter 6.

Fuel delivery system

The fuel delivery system consists of the fuel pump, the fuel filter/fuel pressure regulator, the fuel rail and fuel injectors, and the hoses, lines and pipes that carry fuel between all of these components. For more information about the fuel lines and the various types of fittings used on different models, refer to Section 4.

The fuel pump is an in-tank design. Fuel is drawn through a strainer at the pump inlet, then pumped out the other end of the pump and through a fuel filter, which is an integral part of the fuel pressure regulator. The fuel filter/fuel pressure regulator, which is mounted on top of the fuel pump/fuel level sending unit module, maintains the fuel pressure within the specified operating range. When the operating pressure exceeds the specified operating range, the pressure regulator opens and sends the excess fuel back into the fuel tank (there is no return fuel line from the fuel rail back to the tank). After the fuel has been filtered, it's pumped through a fuel supply line to the fuel rail on the engine.

The fuel rail, which is bolted to the intake manifold, functions as a reservoir for pressurized fuel so that there's always enough fuel available. The upper end of each injector is inserted into the fuel rail and the lower end of each injector is inserted into the intake mani-

10.7 Use a stethoscope to determine if the injectors are working properly - they should make a steady clicking sound that rises and falls with engine speed changes

10.8 If you discover an injector that doesn't appear to be operating, measure the resistance across the two terminals of the injector to see if the injector itself is defective

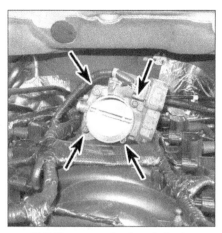

11.5 To remove the throttle body, disconnect the electrical connector, label all vacuum lines connected to the throttle body and disconnect them, then remove the throttle body mounting bolts (4.7L V8 engine shown)

fold. Each end of each injector is sealed by an O-ring.

Each fuel injector is a solenoid-actuated, pintle-type design consisting of a solenoid, plunger, valve, and housing. When the engine is running, there is always voltage on the hot side of each injector terminal. Injector drivers inside the PCM turn the injectors on and off by switching their ground paths on and off. When the ground path for an injector is closed by the PCM, current flows through the solenoid coil, the ball or needle valve rises and pressurized fuel inside the injector housing squirts out the nozzle into the intake port directly above the intake valve(s). The quantity of fuel injected each time an injector opens is determined by its pulse width, which is the length of time during which the valve is open.

10 Fuel injection system - check

Refer to illustrations 10.7 and 10.8

Note: *The following procedure is based on the assumption that the fuel pressure is adequate (see Section 3).*

1   Inspect all electrical connectors that are related to the system. Check the ground wire connections on the intake manifold for tightness. Loose connectors and poor grounds can cause many problems that resemble more serious malfunctions.
2   Verify that the battery is fully charged, as the control unit and sensors depend on an accurate supply of voltage in order to properly meter the fuel.
3   Inspect the air filter element (see Chapter 1). A dirty or partially blocked filter will severely impede performance.
4   Check the related fuses. If a blown fuse is found, replace it and see if it blows again. If it does, search for a grounded wire in the harness.
5   Inspect the condition of all vacuum hoses

connected to the intake manifold.
6   Remove the air intake duct and air resonator box and inspect the mouth of the throttle body for dirt, carbon or other residue build-up. If it's dirty, clean with a shop towel and a solvent specifically made for throttle bodies.
7   With the engine running, place an automotive stethoscope against each injector, one at a time, and listen for a clicking sound, indicating operation **(see illustration)**. If you don't have a stethoscope, place the tip of a screwdriver against the injector and listen through the handle.
8   If an injector doesn't make a clicking sound, disconnect the electrical connector and measure the resistance of the injector **(see illustration)**. Compare the measurement with the resistance values of the other injectors. If it varies significantly from the others, it is probably defective. If the resistance is similar, the PCM or the injector wiring harness could be the cause of the injector not operating.
9   Any further testing of the fuel injection system should be performed at a dealer service department or other qualified repair shop.
10   For more information about the engine control system, refer to Chapter 6.

11 Throttle body - removal and installation

Removal

Refer to illustration 11.5

1   Disconnect the cable from the negative battery terminal (see Chapter 5).
2   Remove the air inlet tube.
3   Disconnect the wiring harness from the throttle body.
4   Disconnect any vacuum lines from the throttle body.
5   Remove the four mounting bolts and remove the throttle body **(see illustration)**.

Installation

6   Thoroughly clean all sealing surfaces.
7   Inspect the gasket/O-ring and replace it if necessary.
8   Install the throttle body and tighten the bolts finger tight, then tighten them with a torque wrench in a criss-cross pattern to the torque listed in this Chapter's Specifications.
9   The remainder of installation is the reverse of removal.
10   On 5.7L Hemi engines, take the vehicle to a dealer or other qualified service center that has the appropriate scan tool and have them perform the ETC Relearn function.

12 Fuel rail and injectors - removal and installation

Warning: *Gasoline is extremely flammable, so take extra precautions when you work on any part of the fuel system. See the* **Warning** *in Section 1.*
Warning: *The engine must be completely cool before beginning this procedure.*
1   Relieve the fuel system pressure (see Section 2).
2   Disconnect the cable from the negative battery terminal (see Chapter 5).

3.6L V6 engines

3   Remove the upper intake manifold (see Chapter 2D).
4   Disconnect the fuel delivery line quick-connect fitting from the fuel rail (if you're unfamiliar with quick-connect fittings, see Section 4).
5   Disconnect the fuel injector electrical connectors, then detach the injector wiring harness mounting clips from the fuel rail and set the harness aside.
6   Remove the fuel rail mounting bolts, pull up on the fuel rail to disengage the injectors from their bores in the intake manifold,

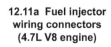

12.11a Fuel injector wiring connectors (4.7L V8 engine)

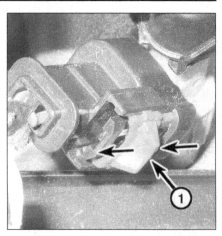

12.11b To disconnect the electrical connector from the injector on a 3.7L V6 or 4.7L V8, move the slider (1) up (away from the injector) . . .

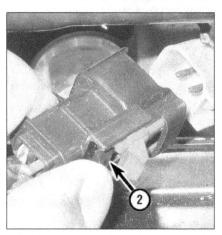

12.11c . . . then depress the tab (2) and pull the connector off of the injector

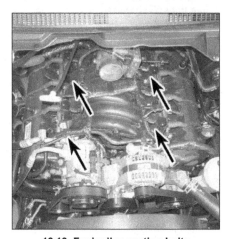

12.13 Fuel rail mounting bolts (4.7L V8 engine)

10   Disconnect the electrical connectors for any wiring harnesses that will interfere with fuel rail removal and set the wiring harnesses aside.

11   Disconnect the fuel injector electrical connectors **(see illustrations)** and set the injector wiring harness aside. **Note:** *Each connector should be numbered with the corresponding cylinder number. If the number tag is obscured or missing, renumber the connectors.*

12   Remove all of the ignition coils (see Chapter 5).

13   Clean any debris from around the injectors. Remove the fuel rail mounting nuts/bolts **(see illustration)**. Gently rock the fuel rail and injectors to loosen the injectors, then remove the fuel rail and fuel injectors as an assembly. **Caution:** *Do not attempt to separate the left and right fuel rails. Both sides are serviced together as an assembly.*

3.6L V6, 3.7L V6 and 4.8L engines

14   Remove the injectors from the fuel rail, then remove and discard the O-rings **(see illustrations)**. **Note:** *Whether you're replacing an injector or a leaking O-ring, it's a good*

then remove the fuel rail and injectors as an assembly. The injectors might initially stick in their bores, but they'll pull free when sufficient force is applied. Proceed to Step 14.

3.7L V6 and 4.7L V8 engines

Refer to illustrations 12.11a, 12.11b, 12.11c, 12.13, 12.14a, 12.14b and 12.14c

7   Remove the air intake duct (see Section 8).

8   Remove the bracket at the front of the throttle body. Disconnect the fuel supply line from the fuel rail (see Section 4).

9   Clearly label any vacuum hoses that will interfere with fuel rail removal, then disconnect them from the throttle body and the intake manifold.

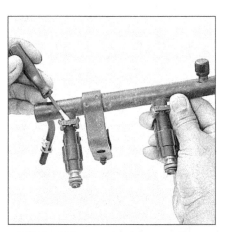

12.14a Using a screwdriver or pliers, remove the injector retaining clip . . .

12.14b . . . and withdraw the injector from the fuel rail

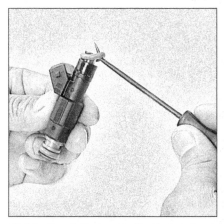

12.14c Carefully remove the O-rings from the injectors

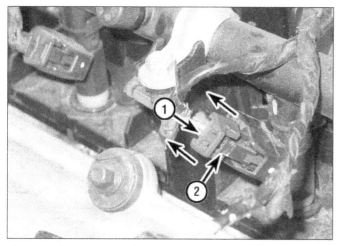

12.23 To disconnect an electrical connector from an injector on a Hemi engine, move the red slider (1) away from the injector, then depress the release tab (2) and pull off the connector

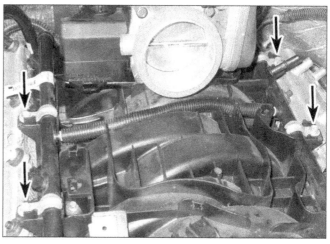

12.25 To detach the fuel rail from the intake manifold, remove these four bolts and hold-down clamps

idea to remove all the injectors from the fuel rail and replace all the O-rings.

15 Coat the new O-rings with clean engine oil and install them on the injector(s), then insert each injector into its corresponding bore in the fuel rail. Install the injector retaining clip.

16 Clean the injector bores on the intake manifold.

17 Guide the injectors/fuel rail assembly into the injector bores on the intake manifold. Make sure the injectors are fully seated, then tighten the fuel rail mounting nuts/bolts to the torque listed in this Chapter's Specifications.

18 The remainder of installation is the reverse of removal.

19 After the injector/fuel rail assembly installation is complete, turn the ignition switch to ON, but don't operate the starter. This activates the fuel pump for about two seconds, which builds up fuel pressure in the fuel lines and the fuel rail. Repeat this step two or three times, then check the fuel lines, fuel rail and injectors for fuel leakage.

5.7L and 6.4L Hemi engine

Refer to illustrations 12.23, 12.25 and 12.26

20 Remove the engine cover by pulling up at the front, then sliding it off.

21 Remove the air inlet tube from the air filter housing and the throttle body.

22 Disconnect the fuel supply line from the fuel rail (see Section 4).

23 Look at the wiring harness for the ignition coils and for the fuel injectors. Note how each branch of the main harness branches into two leads - one for the coil and one for the injector - for each cylinder. To make sure that you don't accidentally plug the wrong connector into the wrong injector, each injector lead is identified by a numeric designation - 1, 2, 3, 4, etc. - imprinted on the electrical tape protecting the lead. If the harness does not have these numbers on the eight injector leads, label them now, before disconnecting the electrical connectors from the fuel injec-

tors. Then disconnect the electrical connectors from all eight fuel injectors (see illustration).

24 Detach the PCV hose and the air make-up hose and move them out of the way.

25 Remove the four fuel rail mounting bolts (see illustration).

26 Starting with the left side of the fuel rail, carefully pull on the injectors while wiggling them from side to side at the same time until all four injectors start to clear their mounting holes. Repeat this step on the right side. Go back and forth between the two sides of the engine, gradually working the injectors out of their mounting holes, until all eight injectors are free. When all of the injectors are free, remove the fuel rail and injectors as a single assembly (see illustration).

27 To remove an injector from the fuel rail, remove the injector retainer clip (see illustration 12.14a) and pull the injector out of the fuel rail. Note: Whether you're replacing an injector or a leaking O-ring, it's a good idea to remove all the injectors from the fuel rail and

replace all the O-rings.

28 Remove the O-rings from the injector (see illustration 12.14c), discard them and install new O-rings. Lubricate the new O-rings with clean engine oil before installing the injector into the fuel rail.

29 Installation is the reverse of removal. Tighten the fuel rail mounting bolts to the torque listed in this Chapter's Specifications.

13 Exhaust system servicing - general information

Refer to illustrations 13.1 and 13.4

Warning: The vehicle's exhaust system generates very high temperatures and must be allowed to cool down completely before touching any of the components. Be especially careful around the catalytic converter, which stays hot longer than other exhaust components.

1 The exhaust system consists of the

12.26 Carefully lift the fuel rail and injectors off the engine as a single assembly

13.1 Inspect the exhaust system rubber hangers for damage

13.4 Exhaust manifold-to-exhaust pipe fasteners (shown) and exhaust pipe-to-catalytic converter fasteners can be extremely difficult to loosen because they're subjected to intense heat, water, mud and road grime. If you're unable to loosen a fastener, apply a liberal amount of penetrant, wait for it to soak in, then try again

exhaust manifolds, the exhaust pipes, the catalytic converter(s), an extension pipe (on some models), the muffler, the tailpipe, various exhaust heat shields and all connecting flanges and clamps. The exhaust system is isolated from the vehicle body and from chassis components by a series of rubber hangers **(see illustration)**. Inspect these hangers periodically for cracks or other signs of deterioration, and replace them as necessary. Some exhaust components are also supported by brackets bolted to the underside of the vehicle. Make sure that these brackets are tightly fastened to the exhaust system and to the vehicle and that they're neither cracked nor corroded.

2   Conduct regular inspections of the exhaust system to keep it safe and quiet. Look for any damaged or bent parts, open seams, holes, loose connections, excessive corrosion or other defects which could allow exhaust fumes to enter the vehicle. Do not repair deteriorated exhaust system components; replace them with new parts.

3   If the exhaust system components are extremely corroded, or rusted together, you'll need welding equipment and a cutting torch to remove them. The convenient strategy at this point is to have a muffler repair shop remove the corroded sections with a cutting torch. If you want to save money by doing it yourself, but you don't have a welding outfit and cutting torch, simply cut off the old components with a hacksaw. If you have compressed air, there are special pneumatic cutting chisels (available from specialty tool manufacturers) that can also be used. If you decide to tackle the job at home, wear safety goggles to protect your eyes from metal chips and wear work gloves to protect your hands.

4   Replacement of exhaust system components is basically a matter of removing the heat shields, disconnecting the component and installing a new one. The heat shields and exhaust system hangers must be reinstalled in the original locations or damage could result. Due to the high temperatures and exposed locations of the exhaust system components, rust and corrosion can seize parts together. Penetrating oils are available to help loosen frozen fasteners. However, in some cases it may be necessary to cut the pieces apart with a hacksaw or cutting torch. **Caution:** *Only persons experienced in this work should employ this latter method.* Here are some simple guidelines to follow when repairing the exhaust system:

a) *Work from the back to the front when removing exhaust system components.*

b) *Apply penetrating oil to the exhaust system component fasteners* **(see illustration)** *to make them easier to remove.*

c) *While you're waiting for the penetrant to loosen up the exhaust system fasteners, always disconnect the electrical connector for the downstream oxygen sensor and remove the sensor before removing the exhaust pipe section that includes the catalytic converter.*

d) *Use new gaskets, hangers and clamps when installing exhaust systems components.*

e) *Apply anti-seize compound to the threads of all exhaust system fasteners during reassembly.*

f) *Allow sufficient clearance between newly installed parts and all points on the underbody to avoid overheating the floor pan and possibly damaging the interior carpet and insulation. Pay particularly close attention to the catalytic converter and heat shield.*

14   Compressed Natural Gas (CNG)/ gasoline fuel system – general information

Warning: *CNG is very flammable and is stored under very high pressure at all times. Take extra precautions when working around any part of the CNG system. Do not try to service any parts of the system as the fuel line is under high pressure and must be purged by a CNG specialist.*

The Compressed Natural Gas (CNG) system is used in conjunction with gasoline. The engine always starts and runs on CNG and when the engine coolant temperature it at or above 18° C (64.4° F). The engine operates on gasoline when the ambient temperature is below 18° C (64.4° F). The engine will switch back to gasoline when the CNG system falls below the low-fuel threshold in the system.

To ensure that the gasoline does not get stale, the PCM is programmed to ensure that one tank of gasoline is consumed every 8000 miles.

The CNG is contained in two cylindrical tanks in the storage compartment in the pickup bed. Various valves, pressure regulators and fuel lines are within this storage compartment and they are not to be serviced by other than a CNG specialist.

Chapter 4 Part B
Fuel and exhaust systems - diesel engine

Contents

Specifications

Fuel injector washer thickness	0.060 inch

Torque specifications — Ft-lbs (unless otherwise indicated)

Note: *One foot-pound (ft-lb) of torque is equivalent to 12 inch-pounds (in-lbs) of torque. Torque values below approximately 15 ft-lbs are expressed in inch-pounds, since most foot-pound torque wrenches are not accurate at these smaller values.*

Fuel Control Actuator (FCA) mounting bolts	62 in-lbs
Fuel filter housing mounting bolts	24
Fuel injection pump mounting nuts/bolts	18
Fuel injection pump shaft nut (drive gear-to-pump shaft nut)	77
Fuel line banjo bolts	18
Fuel line tube nut fitting	30
Intake air heater manifold mounting bolts	18
Intake manifold mounting bolts	18
Turbocharger	
Turbocharger-to-exhaust manifold stud mounting nuts	32
Turbocharger oil return line bolts	27
Turbocharger oil supply line fitting	27

1 General Information

General information

The fuel system is what most sets diesel engines apart from their gasoline powered cousins. Simply stated, fuel is injected directly into the combustion chambers, and the extremely high pressures produced in the cylinders is what ignites the charge. Unlike gasoline engines, diesels have no throttle plate to limit the entry of air into the intake manifold. The only control is the amount of fuel injected; an unrestricted supply of air is always available through the intake.

There are two major sub-systems in the fuel injection system: the low pressure (also known as the supply or transfer) portion and the high-pressure injection (delivery) portion.

The low pressure system moves fuel from the fuel tank to the injection pump. Fuel is drawn from the fuel tank by the pump which is a part of the fuel pump module. It's then pumped through a fuel filter/water separator and then into the high-pressure injection pump. A bypass system allows excess fuel to return to the tank.

The fuel injection pump supplies high-pressure fuel to a fuel rail, which is attached by high-pressure fuel lines to the fuel injectors. The injectors are electrically actuated, in firing order sequence, by the ECM.

Fuel system problems are by far the most frequent cause of breakdowns and loss of power in diesel-powered vehicles. When-ever a diesel engine quits running or loses power for no apparent reason, check the fuel system first. Begin with the most obvious items, such as the fuel filter and damaged fuel lines.

The fuel system on diesel engines is extremely sensitive to contamination. Because of the very small clearances in the injection pump and the minute orifices in the injection nozzles, fuel contamination can be a serious problem. The injection pump and the injectors can be damaged or ruined by contamination. Water-contaminated diesel fuel is a major problem. If it remains in the fuel system too long, water will cause serious and expensive damage. The fuel lines and the fuel filter can also become plugged with rust particles or clogged with ice in cold weather.

Diesel fuel contamination

Warning: *The pressure in the high-pressure fuel lines can reach extremely high pressure (as much as 23,000 psi), so use extreme caution when working near any part of the fuel injection system with the engine running.*

Before you replace an injection pump or some other expensive component, find out what caused the failure. If water contamination is present, buying a new or rebuilt pump or other component won't do much good. The following procedure will help you determine if water contamination is present:

a) *Remove the engine fuel filter and inspect the contents for the presence of water or gasoline (see Chapter 1).*

b) *If the vehicle has been stalling, performance has been poor or the engine has been knocking loudly, suspect fuel contamination. Gasoline or water must be removed by flushing (see below).*

c) *If you find a lot of water in the fuel filter, remove the injection pump fuel return line and check for water there. If the pump has water in it, flush the system.*

d) *Small quantities of surface rust won't create a problem. If contamination is excessive, the vehicle will probably stall.*

e) *Sometimes contamination in the system becomes severe enough to cause damage to the internal parts of the pump. If the damage reaches this stage, have the damaged parts replaced and the pump rebuilt by an authorized fuel injection shop, or buy a rebuilt pump.*

Storage

Good quality diesel fuel contains inhibitors to stop the formation of rust in the fuel lines and the injectors, so as long as there are no leaks in the fuel system, it's generally safe from water contamination. Diesel fuel is usually contaminated by water as a result of careless storage. There's not much you can do about the storage practices of service stations where you buy diesel fuel, but if you keep a small supply of diesel fuel on hand at home, as many diesel owners do, follow these simple rules:

a) *Diesel fuel ages and goes stale. Don't store containers of diesel fuel for long periods of time. Use it up regularly and replace it with fresh fuel.*

b) *Keep fuel storage containers out of direct sunlight. Variations in heat and humidity promote condensation inside fuel containers.*

c) *Don't store diesel fuel in galvanized containers. It may cause the galvanizing to flake off, contaminating the fuel and clogging filters when the fuel is used.*

d) *Label containers properly as containing diesel fuel.*

Fighting fungi and bacteria with biocides

If there's water in the fuel, fungi and/or bacteria can form in warm or humid weather. Fungi and bacteria plug fuel lines, fuel filters and injection nozzles; they can also cause corrosion in the fuel system.

If you've had problems with water in the fuel system and you live in a warm or humid climate, have a diesel specialist correct the problem. Then, use a diesel fuel biocide to sterilize the fuel system in accordance with the manufacturer's instructions. Biocides are available from your dealer, service stations and auto parts stores. Consult your dealer or a diesel specialist for advice on using biocides in your area and for recommendations on which ones to use.

Cleaning the low-pressure fuel system

Warning: *Diesel fuel is flammable, so take extra precautions when you work on any part of the fuel system. Don't smoke or allow open flames or bare light bulbs near the work area, and don't work in a garage where a gas-type appliance (such as a water heater or a clothes dryer) is present. Since diesel fuel is carcinogenic, wear fuel-resistant gloves when there's a possibility of being exposed to fuel, and, if you spill any fuel on your skin, rinse it off immediately with soap and water. Mop up any spills immediately and do not store diesel fuel-soaked rags where they could ignite. When you perform any kind of work on the fuel system, wear safety glasses and have a Class B type fire extinguisher on hand.*

Water-In-Fuel (WIF) warning system

1   The WIF system detects the presence of water in the fuel filter when it reaches excessive amounts. Water is detected by a probe located in the fuel filter that completes a circuit through a wire to a light in the instrument cluster that shows a gas pump and water drops.

2   The WIF system includes a bulb-check feature: When the ignition is turned to ON, the bulb glows momentarily, then fades away.

3   If the light comes on immediately after you've filled the tank or let the vehicle sit for an extended period of time, drain the water from the system immediately. Do not start the engine. There might be enough water in the system to shut the engine down before you've driven even a short distance. If, however, the light comes on during a cornering or braking maneuver, there's less water in the system; the engine probably won't shut down immediately, but you still should drain the water soon.

4   Water is heavier than diesel fuel, so it sinks to the bottom of the fuel tank. An extended return pipe on the fuel tank sending unit, which reaches down into the bottom of the tank, enables you to siphon most of the water from the tank without having to remove the tank. But siphoning won't remove all of the water; you'll still need to remove the tank and

thoroughly clean it. **Warning:** *Do not start a siphon by mouth - use a siphoning kit (available at most auto parts stores).*

Removing water from the fuel system

5   Disconnect the cables from the negative terminals of the batteries (see Chapter 5).

6   Remove the fuel tank (see Chapter 4A).

7   Remove the fuel pump module, then drain the tank into an approved container. Dispose of the fuel following local regulations.

8   Thoroughly clean the fuel tank. If it's damaged or otherwise unusable, replace it. Clean or replace the fuel pick-up filter.

9   Install the fuel tank and add fuel.

10   Open the fuel filter drain valve and attach an extension hose to it that empties into a container (see Chapter 1).

11   Perform the fuel system priming procedure by turning the key to the Start position momentarily. This will energize the transfer fuel pump for a 30-second interval. Repeat this until you're satisfied that clear fuel is coming out.

12   Replace the fuel filter (see Chapter 1).

Gasoline in the fuel system

Warning: *Gasoline and diesel fuel are flammable, so take extra precautions when you work on any part of the fuel system. Don't smoke or allow open flames or bare light bulbs near the work area, and don't work in a garage where a gas-type appliance (such as a water heater or a clothes dryer) is present. Since diesel fuel is carcinogenic, wear fuel-resistant gloves when there's a possibility of being exposed to fuel, and, if you spill any fuel on your skin, rinse it off immediately with soap and water. Mop up any spills immediately and do not store diesel fuel-soaked rags where they could ignite. When you perform any kind of work on the fuel system, wear safety glasses and have a Class B type fire extinguisher on hand.*

13   If gasoline has been accidentally pumped into the fuel tank, it should be drained immediately. Gasoline in the fuel in small amounts - up to 30 percent - isn't usually noticeable. At higher ratios, the engine may make a knocking noise, which will get louder as the ratio of gasoline increases. Here's how to rid the fuel system of gasoline:

14   Perform the same procedure as for removing water from the fuel system (see Steps 5 through 12).

15   Try to start the engine. If it doesn't start, purge the injection pump and lines: Place rags around the fuel rail. Loosen the fuel line fittings a little, just enough for fuel to leak out. Depress the accelerator pedal to the floor and, holding it there, crank the engine until all gasoline is removed and diesel fuel leaks out of the fittings. Tighten the fittings. Limit cranking to 30 seconds with two or three minute intervals between cranking. **Warning:** *Avoid sources of ignition and have a fire extinguisher handy.*

16   Start the engine and run it at idle for 15 minutes.

2   Fuel system priming

Warning: *The pressure in the high-pressure fuel lines can be extremely high (as much as 23,200 psi), so use extreme caution when working near any part of the fuel injection system with the engine running.*

1   The fuel system is primed by the transfer (lift) pump. When the key is turned to ON (without cranking over the engine), the pump runs for about two seconds, then it shuts off. It will also operate for as much as 25 seconds after the starter has been quickly engaged, then disengaged without allowing the engine to start. **Note:** *The pump shuts off immediately if the key is turned to ON and the engine stops running.*

2   Turn the ignition key to the CRANK position, then quickly release the key to the ON position before the engine starts. This step will energize the transfer pump for about 25 seconds.

3   If the engine doesn't start, turn the key to OFF and repeat the previous step until the engine *does* start.

4   The fuel system is now primed.

5   Try to start the engine again. If it starts but runs erratically and noisily, that's normal. It will clean out in a few minutes. If the engine won't start, proceed to the next step.

6   Perform the previous fuel priming procedure steps. Make sure that there is fuel in the tank.

7   Crank the engine for 30 seconds at a time to allow the fuel system to prime. **Caution:** *Operate the starter for no more than 30 seconds at a time. Let it cool for at least three minutes, then try to start it again.*

3   Fuel transfer pump - check and replacement

Warning: *Diesel fuel is flammable, so take extra precautions when you work on any part of the fuel system. Don't smoke or allow open flames or bare light bulbs near the work area, and don't work in a garage where a gas-type appliance (such as a water heater or a clothes dryer) is present. Since diesel fuel is carcinogenic, wear fuel-resistant gloves when there's a possibility of being exposed to fuel, and, if you spill any fuel on your skin, rinse it off immediately with soap and water. Mop up any spills immediately and do not store diesel fuel-soaked rags where they could ignite. When you perform any kind of work on the fuel system, wear safety glasses and have a Class B type fire extinguisher on hand.*

1   The transfer pump is a part of the fuel pump module. This module contains the pump, a filter and the fuel level sensor.

2   The module is located in the fuel tank; see Chapter 4A for the check and replacement procedures.

4   Fuel lines and fittings - general information

Warning: *Diesel fuel is flammable, so take extra precautions when you work on any part of the fuel system. See the* **Warning** *in Section 3.*

Warning: *The pressure in the high-pressure fuel lines can be extremely high (as much as 23,200 psi), so use extreme caution when inspecting for high-pressure fuel leaks. Do not move your hand near a suspect leak - instead, use a piece of cardboard. High-pressure fuel leaks can cause severe injury.*

Caution: *Do not attempt to weld high-pressure fuel lines or repair lines that are bent, kinked or otherwise damaged. Replace them with factory replacement fuel lines.*

Low-pressure fuel lines

1   The low-pressure side of the fuel system includes the following lines:

> *Fuel supply line from the fuel tank to the fuel transfer pump*
> *Fuel return line back to the fuel tank*
> *Fuel drain manifold line at the rear of the cylinder head*
> *Fuel supply line from the fuel filter to the fuel injection pump*

2   Leaks in the low-pressure fuel lines can cause fuel starvation, which will result in low power. You should be able to smell a leak on the low-pressure side of the fuel system. If you find a leak at a fuel line connection, tighten the fitting and note whether the leak stops. If it doesn't, the fitting might be stripped, and must therefore be replaced.

3   Obstructions in the low-pressure fuel lines can cause starting problems and, because they're restricting the fuel supply to the fuel injection pump, they can prevent the engine from accelerating. The usual symptoms are low power and/or a white fog-like exhaust. Inspect the low-pressure fuel lines for bends, kinks and other damage. If you find a damaged line, replace it; don't try to repair it.

4   The low-pressure lines use various types of quick-connect fittings at connection points. For step-by-step instructions showing how to disconnect and reconnect these fittings, refer to *Fuel lines and fittings - repair and replacement* in Chapter 4A.

5   After tightening the fitting(s) and/or replacing any low-pressure fuel lines, prime the fuel system (see Section 2).

High-pressure fuel lines

6   The high-pressure side of the fuel system includes the following lines:

> *Fuel line from the fuel injection pump to the fuel rail*
> *Six fuel lines from the fuel rail to the fuel injector connector tubes*

7   Leaks in the high-pressure fuel lines are usually pretty obvious, and can be extremely dangerous. Not only as a fire hazard, but a stream of high-pressure fuel can cut right through your skin. Start the engine, put on a pair of safety goggles and move a piece of clean cardboard over and around the high-pressure fuel lines and their connections. If a high-pressure line connection is leaking, it will spray the cardboard. Tighten the fitting, then prime the fuel system (see Section 2). If a line itself is leaking, the line must be replaced - don't attempt to repair a damaged line.

8   Unless you're replacing the injection pump or some other component on the high-pressure side of the fuel system, we don't recommend disassembling the high-pressure side of the fuel system.

5   Fuel Control Actuator (FCA) - replacement

Refer to illustration 5.2

Warning: *Diesel fuel is flammable, so take extra precautions when you work on any part of the fuel system. See the* **Warning** *in Section 3.*

Note: *The FCA is located on the rear of the fuel injection pump.*

1   Thoroughly clean off the area around the FCA.

2   Disconnect the electrical connector from the FCA **(see illustration).**

3   Remove the FCA mounting bolts and remove the FCA.

4   Remove and discard the old FCA O-ring. Always install a new O-ring before installing the FCA.

5   Inspect the FCA for corrosion and damage. Shake the FCA. It should rattle; if it doesn't, replace it.

6   Apply a little clean light grease to the new O-rings and install the FCA on the fuel injection pump. Make sure that the FCA is flush with its mounting surface on the injection pump, then tighten the FCA mounting bolts to the torque listed in this Chapter's Specifications.

7   Installation is otherwise the reverse of removal.

5.2 To detach the FCA from the back of the fuel injection pump, disconnect the electrical connector and remove the two FCA mounting bolts

6   Fuel injection pump - removal and installation

Warning: *Diesel fuel is flammable, so take extra precautions when you work on any part of the fuel system. See the **Warning** in Section 3.*

Removal

Refer to illustrations 6.2, 6.7 and 6.12

1   Disconnect the cables from the negative battery terminals (see Chapter 5).
2   Remove the air inlet tube and its hose from above the injection pump **(see illustration)**.
3   Remove the drivebelt (see Chapter 1).
4   Clean the rear part of the injection pump and the line fittings. Clean the other ends of the line fittings to avoid contamination when they're disassembled.
5   Detach the fuel line quick-connect fitting by pushing its button.
6   Remove the high pressure fuel line that attaches to the fuel rail.
7   Remove the banjo bolt from directly above the quick-connect fitting **(see illustration)**. Remove the old sealing washers from the banjo bolts and discard them. Always use new sealing washers when reconnecting these banjo fittings.
8   Disconnect the wiring from the fuel control actuator on the rear of the pump, then remove the clamp that secures the two fuel lines to each other.
9   Unscrew the fuel pump access cover using a ratchet with no socket on it.
10   Remove the nut and the washer from the fuel pump drive gear.
11   Attach a suitable gear puller to the pump drive gear with two bolts and separate the gear from the pump. Leave it in the housing for now.
12   Remove the three injection pump mounting nuts **(see illustration)** and remove the pump from the engine.
13   Remove the old injection pump O-ring from the machined groove in the pump mounting surface. Discard the O-ring.

Installation

Refer to illustration 6.16

14   Thoroughly clean off the pump mount-

6.2 Loosen the hose clamps on the connecting tubes for the intercooler duct at each end of the duct, then remove the duct

ing surfaces of the pump and the timing gear cover. Also clean off the machined tapers on both the injection pump shaft and on the injection pump gear using lacquer thinner or acetone. These surfaces must be absolutely dry and free of all dirt and oil to ensure correct gear-to-shaft mounting.
15   Apply a little clean engine oil to the new injection pump O-ring, then install a new O-ring into the machined groove in the pump mounting surface.
16   Check the fuel injection pump phasing before installing the pump onto the engine. Locate the numbers on the fuel injection pump shaft. There should be a 0 and 750 or a 754 stamped into the shaft. Set the number "5" digit at the 9 o'clock position **(see illustration)**. Place the pump in position on the rear of the timing gear cover with the digit "5" set in the 9 o'clock position. Make sure the engine is at TDC (see Chapter 2C). **Note:** *The engine can be positioned in TDC number 1 or TDC number 6 for this procedure.*
17   Once the pump is flush with the mounting surface on the timing gear cover, install the three pump mounting nuts and tighten them finger tight. To prevent damage to any components, tighten the injection pump mounting nuts and the drive gear-to-pump shaft nut in

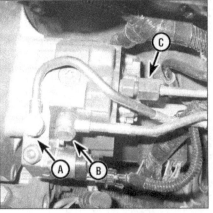

6.7 Fuel injection pump-to-fuel line fittings

A   *Banjo bolt for fuel filter/water separator-to-injection pump line*
B   *Banjo bolt for fuel pressure limiting valve (on fuel rail)-to-injection pump line*
C   *Tube nut type fitting for high-pressure line to fuel rail*

the following sequence:
a) *Install the washer and nut on the injection pump shaft and hand tighten the nut until it's finger tight, then carefully tighten it a bit more.*
b) *Tighten the three injection pump mounting nuts to the torque listed in this Chapter's Specifications.*
c) *Tighten the injection pump shaft-to-gear nut to the torque listed in this Chapter's Specifications.*

18   Using new sealing washers, reconnect the fuel line banjo bolts and tighten them to the torque listed in this Chapter's Specifications. Reconnect the tube nut fitting for the high-pressure fuel line and tighten it to the torque listed in this Chapter's Specifications.

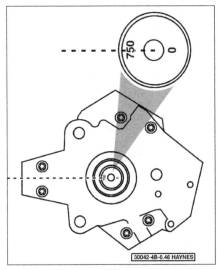

6.16 Set the fuel injection pump phasing with the digit "5" in the 9 o'clock position

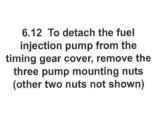

6.12 To detach the fuel injection pump from the timing gear cover, remove the three pump mounting nuts (other two nuts not shown)

19   The remainder of installation is the reverse of removal.
20   Reconnect the cables to the negative battery terminals.
21   Prime the fuel system (see Section 2).
22   Start the engine and check for fuel leaks.

7   Fuel injectors - check and replacement

Check

1   A leaking fuel injector could cause various symptoms, depending on the severity of the leak. Fuel knock, poor acceleration and performance, low fuel economy and rough engine idle can be caused by pintle-valve leaks. Defective needle valve operation might cause the engine to misfire. **Note:** *A leak in the high-pressure fuel line(s) can cause many of the same problems and symptoms. Check for fuel line leaks before proceeding (see Section 4).* Any checks on diesel fuel injectors should be performed by a dealer service department or other qualified diesel repair facility. Injector tests require special high-pressure testing equipment and adapters for accurate results.

Replacement

2   Because of the expensive special tools needed to remove and install the connector tubes to the injectors and the injectors themselves, we don't recommend attempting this job at home. The injectors used by these engines generally provide a long and trouble-free service life. By the time that you round up all the special tools necessary to remove and install the injector connector tubes and the injectors, you will have spent as much, if not more, than it will cost to have this job done by a dealer service department. And even if you're willing to spend the money on the tools you'll need for this job, you'll probably never use them again! So we recommend having this job done by a diesel specialist if the need ever arises.

8   Fuel heater, fuel filter housing and fuel heater relay - replacement

Warning: *Diesel fuel is flammable, so take extra precautions when you work on any part of the fuel system. See the* **Warning** *in Section 3.*

Fuel heater

1   The fuel heater prevents diesel fuel from waxing during cold weather operation. A defective fuel heater can cause wax build-up in the fuel filter/water separator. This clogging effect can make the engine difficult to start and prevent the engine from revving up. This condition can also cause a fog-like blue or white exhaust. If the heater doesn't operate in a cold climate, the engine might not operate at all because of fuel waxing.

2009 models

Refer to illustration 8.3

Note: *The fuel heater is located on the fuel filter/water separator housing.*
2   Drain the water and fuel contaminants from the fuel filter assembly (see Chapter 1).
3   Disconnect the fuel heater element connector **(see illustration)**.
4   Remove the two fuel heater element mounting screws.
5   Remove the fuel heater element from the fuel filter assembly.
6   Replace the O-ring with a new one.
7   Installation is the reverse of removal.

2010 and later models

8   The heater on 2010 and later models is an integral part of the fuel filter housing assembly. It can't be replaced separately; go to Step 9 for the filter housing replacement procedure.

Fuel filter housing

9   Drain the water and fuel contaminants from the fuel filter assembly (see Chapter 1).
10   If necessary, detach the engine oil dipstick tube from the filter housing.
11   Disconnect the wiring from the filter assembly.
12   Disconnect all of the fuel lines from the filter assembly.
13   Remove the mounting bolts and remove the filter unit from the engine.
14   Thoroughly clean all mounting surfaces, then install the housing and the mounting bolts and tighten them to the torque listed in this Chapter's Specifications.
15   Replace the banjo bolt washers with new ones. The remainder of installation is the reverse of removal.

Fuel heater relay

16   There is no serviceable fuel heater relay; it is an integral part of the underhood fuse/relay box, which the manufacturer calls the Totally Integrated Power Module (TIPM).

9   Intake manifold, air heater and air heater relay - replacement

Intake manifold

2009 models

Note: *The two intake air heater elements are housed inside a metal block that's bolted to the top of the intake manifold cover by the four intake air heater manifold bolts. The metal block and intake air heater elements are not serviceable separately. They must be replaced as a single assembly.*
1   Disconnect the cables from both negative battery terminals (see Chapter 5).
2   Disconnect the wiring from the EGR temperature sensor and the EGR valve actuator.3 Remove the fasteners from the oil dip-

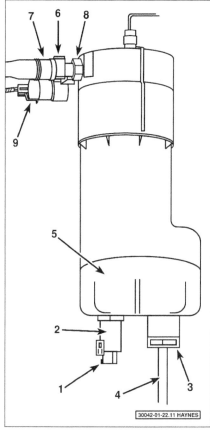

8.3 Fuel filter assembly details

1   WIF harness connector
2   WIF sensor
3   Drain valve
4   Drain tube
5   Fuel filter canister
6   Fuel line connector
7   Fuel line
8   Filter screen
9   Fuel heater element connector

stick tube.
4   Remove the intercooler outlet tube.
5   Remove the air transfer tube and the bolt from the P-clip.
6   Remove the six air inlet bolts and the air inlet.
7   Carefully clean each end of each fuel supply tube, then remove the lines that are accessible. You'll have to loosen and tilt the bracket on the rear of the cylinder head for access to the rear fuel line fittings.
8   Disconnect the wiring from the sensor at the rear of the fuel rail.
9   Remove the high pressure fuel line from the dump valve.
10   Remove the fuel rail.
11   Disconnect the remaining wiring harnesses from the intake manifold.
12   Remove the intake manifold.
13   Remove the front fuel inlet lines that are now accessible, if necessary.
14   Installation is the reverse of removal.

9.24a To access the intake air heater elements, disconnect or remove the following:

1   *Heater grid positive cable nut*
2   *Heater grid positive cable nut*
3   *Heater grid ground cable - it's not necessary to remove this bolt unless you're replacing the ground cable; instead disconnect the other end of the ground cable from its terminal stud*
4   *Intake air heater manifold bolts*
5   *Dipstick tube mounting bracket bolt*

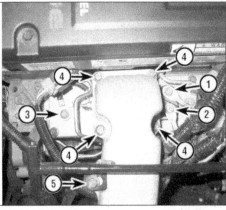

9.24b To disconnect the heater element ground strap from its stud terminal, remove this nut

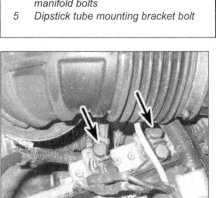

9.28 The two heater relays are located next to the right battery

2010 and later models

15   Disconnect the cables from both negative battery terminals (see Chapter 5).
16   Disconnect the left intercooler tube from the EGR air flow control valve.
17   Remove the EGR crossover tube.
18   Disconnect the interfering wiring harnesses.
19   Remove the bolt and the nut from the oil dipstick tube.
20   Remove the bolts and the intake manifold. Discard the gasket; it must be replaced with a new one.
21   Installation is the reverse of removal.

Air heater

Refer to illustrations 9.24a and 9.24b

22   The two intake air heater grids are housed inside a metal box, which is located on top of the intake manifold cover. The intake air heater elements heat incoming air to make the engine easier to start during cold start-ups and to improve drivability in cool and cold ambient temperatures. The Engine Control Module (ECM) controls the current to the heater elements through two heater relays located in the engine compartment.
23   Remove the intake manifold (see earlier in this Section).
24   Disconnect the heater cable from the heater **(see illustrations)**. Also disconnect and remove the fuel injector wiring harness. Disconnect the pressure sensor wiring har-

ness from the valve cover.
25   Detach the vent tube from the valve cover, then remove the air intake heater mounting bolts. Lift the heater assembly off and discard its gasket.
26   Installation is the reverse of removal. Clean the old gasket material from the air intake housing and intake manifold and from both ends of the heater block.
27   Use new upper and lower heater block gaskets and tighten the air intake housing bolts to the torque listed in this Chapter's Specifications.

Air heater relay

Refer to illustration 9.28

28   The intake air heater relay is located in the engine compartment on a bracket that is bolted to the right battery tray **(see illustration)**. The following replacement procedure applies to all heater relays.
29   Disconnect the cables from both negative battery terminals.
30   Clearly label the trigger wires, then disconnect them.
31   Remove the nuts from the cable terminals and disconnect the cables. Label the wires to prevent crossed wires during reassembly.
32   Remove the relay mounting bracket bolts and remove the relay.
33   Installation is the reverse of removal.

10   Air filter housing - removal and installation

Refer to illustration 10.3

1   Disconnect the wiring connectors from the sensors.
2   Loosen the hose clamp that secures the air intake duct to the upper half of the air filter housing, then detach the duct.
3   Remove the housing mounting bolt **(see illustration)**.
4   Pull the air filter housing straight up to disengage the locator pins from their grommets.
5   While the air filter housing is removed, inspect the rubber locator pin grommets for cracks, tears and deterioration. If they're damaged, replace them.
6   Installation is the reverse of removal.

11   Accelerator cable - removal and installation

These models do not have a conventional accelerator cable between the accelerator pedal and the throttle lever. They have an Accelerator Pedal Position (APP) sensor installed at the pedal. See Chapter 6 for additional information on this sensor.

10.3 Air filter housing details

1   *Intake Air Temperature sensor connector*
2   *Hose clamp*
3   *Mounting bolt*

12 Turbocharger - description and inspection

Description

1   A turbocharger improves engine power, lowers the density of exhaust smoke, improves fuel economy, reduces engine noise and lessens the effects of lower density air at higher altitude. The turbocharger uses an exhaust gas-driven turbine to pressurize the air entering the combustion chambers.

2   The amount of boost (intake manifold pressure) is controlled by a wastegate (exhaust bypass valve). The wastegate is operated by a spring-loaded actuator assembly, which controls the boost by allowing a certain amount of exhaust gas to bypass the turbine.

3   Turbocharged models are equipped with an intercooler - a heat exchanger through which the compressed air intake charge is routed to lower its temperature. Cooler air is denser, which promotes combustion efficiency, increasing power and reducing emissions.

Inspection

4   Though it's a relatively simple device, the turbocharger is a precision component. Special tools are needed to disassemble and overhaul a turbocharger, so servicing should be left to a dealer service department. However, you can inspect some things yourself, such as a cracked turbo mounting flange, a blocked or restricted oil supply line, a worn out or overheated turbine/compressor shaft bearing or a defective wastegate actuator.

5   A turbocharger has its own distinctive sound, so a change in the quality or the quantity of noise can be a sign of potential problems. But before assuming that a funny sound is caused by a defective turbocharger, inspect the exhaust manifold for cracks and loose connections. For example, a high-pitched or whistling sound might indicate an intake air or exhaust gas leak. Inspect the turbocharger mounting flange at the exhaust manifold and make sure that the hose clamp that attaches the air intake duct to the turbocharger is tight.

6   If an unusual sound *is* coming from the turbocharger, turn off the engine and allow it to cool completely. Remove the intake duct between the air cleaner housing and the turbocharger. Turn the compressor wheel to make sure it spins freely. **Warning:** *The turbine or compressor wheels have very sharp blades; do not turn the blades with your fingers. Use a plastic pen.* If it doesn't, it's possible the turbo lubricating oil has sludged or coked-up from overheating. Push in on the turbine wheel and check for binding. The turbine should rotate freely with no binding or rubbing on the housing. If it does the turbine or compressor shaft bearing is worn out. **Warning:** *Inspect the turbocharger with the*

engine off and cool to the touch.

7   The turbocharger is lubricated by engine oil that has been pressurized, cooled and filtered. Oil is delivered to the turbocharger by a supply line that's tapped into the oil filter head. Oil travels to the turbocharger's bearing housing, where it lubricates the shaft and bearings. A return pipe at the bottom of the turbocharger routes the engine oil back to the crankcase. Because the turbine and compressor wheels spin at speeds up to 140,000 rpm, severe damage can result from the interruption or contamination of the oil supply to the turbocharger bearings. Look for leaks in the oil supply line (the one on top). If a fitting is leaking, tighten it and note whether the leak stops. If the supply line itself is leaking, replace it. Remove the oil return line (on the bottom) and inspect it for obstructions. A blocked return line can cause a loss of oil through the turbocharger seals. Burned oil on the turbine housing is a sign of a blocked return line.

Caution: *Whenever a major engine bearing such as a main, connecting rod or camshaft bearing is replaced, flush the turbocharger oil passages with clean oil.*

13.4a To disconnect the air intake duct from the turbocharger, loosen this hose clamp screw

13 Turbocharger - removal and installation

Caution: *The turbocharger is a precision component that has been assembled and balanced to very fine tolerances. Do not disassemble it or try to repair it. Turbochargers should only be overhauled or repaired by authorized turbocharger repair facilities. An incorrectly assembled turbocharger could result in damage to the turbocharger and/or the engine.*

Note: *This procedure requires a special wrench (see Step 15). Don't start work until you have this tool.*

Removal

Refer to illustrations 13.4a, 13.4b, 13.4c, 13.6, 13.13 and 13.16

1   Disconnect the negative battery cables from both batteries (see Chapter 5).

2   Drain sufficient coolant from the radiator (see Chapter 1).

3   Remove the air filter housing (see Section 10).

4   Disconnect the hoses and tubes from the turbocharger **(see illustrations)**.

13.4b To disconnect the turbocharger outlet duct from the turbocharger, loosen this hose clamp screw and pull off the duct

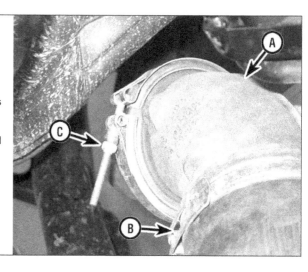

13.4c To disconnect the exhaust discharge elbow (A) from the turbocharger, loosen this clamp (B) and separate the elbow from the turbocharger. If you need more room, loosen the hose clamp (C) at the lower end of the elbow and remove the elbow entirely

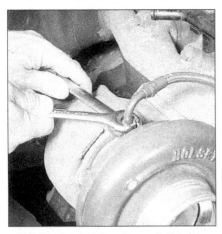

13.6 Disconnect the oil supply line from the top of the turbocharger. Use a back-up wrench to protect the elbow in the line from kinks

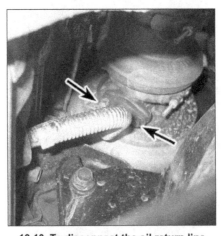

13.13 To disconnect the oil return line from the bottom of the turbocharger, remove these two bolts

13.16 Carefully lift the turbocharger out of the engine compartment; it's heavy, so take care to avoid damaging the wastegate system

5   Disconnect the wiring from the turbocharger.
6   Disconnect the oil supply line from the turbocharger **(see illustration)**. Remove the coolant lines completely.
7   On 2012 and later models, remove the engine-driven viscous fan and the fan shroud (see Chapter 3).
8   Raise the vehicle and support it securely on jackstands. Remove the right inner fender splash shield.
9   Install an engine support fixture or an engine hoist (see Chapter 2D) and use it to take the weight of the engine.
10   Remove the right engine mount.
11   Remove the interfering exhaust bracket from the transmission.
12   Remove the exhaust pipe V-band clamp from the turbocharger.
13   Remove the oil drain tube from the turbocharger and the engine **(see illustration)**, then lower the vehicle.
14   Remove the Grid Heater relay from the battery tray and set it aside.
15   Remove the turbocharger mounting nuts. **Note:** *This requires a special wrench (Miller #9866) which is available at Dodge dealers and specialty automotive tool dealers.*
16   Remove the turbocharger, being careful not to damage the wastegate actuator assem-

bly **(see illustration)**. **Note:** *The wastegate actuator is precisely adjusted. Be careful when laying the complete turbocharger unit on the bench, so as not to disturb wastegate actuator alignment.*

Installation

17   Clean the turbocharger and exhaust manifold sealing surfaces.
18   Use a die to clean the studs in the turbocharger mounting portion of the exhaust manifold and coat them with anti-seize compound. Bolt the turbocharger onto the exhaust manifold using a new gasket.
19   Reinstall the oil drain line using new O-rings. Also replace the seals on the coolant lines.
20   Prime the center bearing of the turbocharger with oil by squirting some clean engine oil into the oil supply hole on top, while turning the compressor wheel, then install the supply line. **Warning:** *The turbine or compressor wheels have very sharp blades; do not turn the blades with your fingers. Use a plastic pen.*
21   The remainder of installation is the reverse of removal. Tighten the turbocharger-to-exhaust manifold nuts to the torque listed in this Chapter's Specifications.

14   Intercooler - removal and installation

Warning: *The air conditioning system is under high pressure - have a dealer service department or service station evacuate the system and recover the refrigerant before disconnecting any of the hoses or fittings.*

1   Have the air conditioning system discharged by an automotive air conditioning technician (see **Warning** above).
2   On 2011 and later models, remove the grille (see Chapter 11).
3   Remove the air conditioning condenser (see Chapter 3).
4   Place a drain pan beneath it, then disconnect the remove the automatic transmission cooler lines (if so equipped). Plug the openings to prevent contamination. Remove the transmission cooler.
5   Disconnect the two air hoses from the intercooler.
6   Remove the intercooler mounting bolts, then rotate it to remove it from the vehicle.
7   Installation is the reverse of removal.
8   Have the air conditioning system evacuated, recharged and leak tested by the shop that discharged it.

Chapter 5
Engine electrical systems

Contents

Specifications

General

Battery voltage
Engine off	12.6 to 13.2 volts
Engine running	13.5 to 15 volts

Firing order
3.6L V6 engine	1-2-3-4-5-6
3.7L V6 engine	1-6-5-4-3-2
V8 engines	1-8-4-3-6-5-7-2
Diesel engine	1-5-3-6-2-4

Ignition system

Ignition coil resistance (at 70 to 80-degrees F)
3.6L, 3.7L V6 and 4.7L V8 engines
Primary resistance	0.6 to 0.9 ohms
Secondary resistance	6 to 9 K-ohms
5.7L and 6.4LHemi engine, primary resistance	0.558 to 0.682 ohms

Spark plug wire resistance
Minimum	250 ohms per inch/3000 ohms per foot
Maximum	1,000 ohms per inch/12,000 ohms per foot

3.2 To test the open circuit voltage of the battery, connect a voltmeter to the battery terminals. A fully charged battery should have at least 12.6 volts

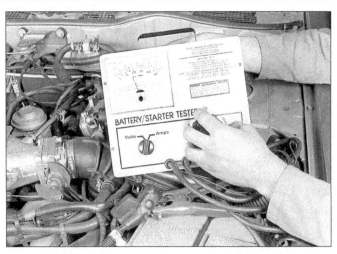

3.3 Connect a battery load tester to the battery and check the battery condition under load following the tool manufacturer's instructions

1   General information

The engine electrical systems include all ignition, charging and starting components. Because of their engine-related functions, these components are discussed separately from chassis electrical devices such as the lights, the instruments, etc. (which are included in Chapter 12).

Always observe the following precautions when working on the electrical systems:

a) *Be extremely careful when servicing engine electrical components. They are easily damaged if checked, connected or handled improperly.*

b) *Never leave the ignition switch on for long periods of time with the engine off.*

c) *Don't disconnect the battery cables while the engine is running.*

d) *Maintain correct polarity when connecting a battery cable from another vehicle during jump-starting.*

e) *Always disconnect the negative cable first and hook it up last or the battery may be shorted by the tool being used to loosen the cable clamps.*

It's also a good idea to review the safety-related information regarding the engine electrical systems located in the "Safety First!" Section near the front of this manual before beginning any operation included in this Chapter.

2   Battery - emergency jump starting

Refer to the "Booster battery (jump) starting" procedure at the front of this manual.

3   Battery - check and replacement

Warning: *Hydrogen gas is produced by the battery, so keep open flames and lighted cigarettes away from it at all times. Always wear eye protection when working around a battery. Rinse off spilled electrolyte immediately with large amounts of water.*
Caution: *Always disconnect the negative cable first and hook it up last or you might accidentally short the battery with the tool that you're using to loosen the cable clamps.*

Check

Refer to illustrations 3.2 and 3.3

1   To check the battery state of charge, look at the indicator eye on the top of the battery (the eye is the top of a hydrometer that's built into the battery). If the indicator eye is green, the battery is 75 to 100 percent charged. If the indicator eye is black, the battery is 0 to 75 percent charged. If the indicator eye is clear (or bright), the battery electrolyte level is low. All factory-installed batteries are the *maintenance-free* type; the cell caps cannot be removed, so no water can be added. If the indicator eye is clear on a maintenance-free battery, replace the battery. If the maintenance-free battery has been replaced by a *low-maintenance* battery with removable cell caps, remove the caps and add enough water to bring it up to the correct level (which should be marked on the outside of the battery case). If there are no MINIMUM and MAXIMUM lines on the battery case, add enough water to each cell so that the plates are fully immersed. Wait a few hours for the electrolyte in the plates to go back into solution, then charge the battery (see Chapter 1). **Note:** *A low-electrolyte/low-water condition is often a symptom of overcharging, so after recharging the battery, check the alternator charging voltage (see Section 9) and, if necessary, replace the alternator. Oth-*

erwise, the same condition will reoccur.
2   Perform an open-voltage circuit test using a voltmeter **(see illustration)**. **Note:** *To obtain an accurate voltage measurement, you must first remove the battery's surface charge. To remove the surface charge, turn on the high beams for ten seconds, then turn them off and let the vehicle stand for two minutes.* With the engine and all accessories off, touch the negative probe of the voltmeter to the negative terminal of the battery and the positive probe to the positive terminal of the battery. The battery voltage should be 12.6 volts or more. If the battery is less than the specified voltage, charge the battery before proceeding to the next test. Do not proceed with the battery load test unless the battery charge is correct.
3   Perform a battery load test. An accurate check of the battery condition can only be performed with a battery load tester (available at most auto parts stores). This test evaluates the ability of the battery to operate the starter and other accessories during periods of heavy amperage draw (load). The tool utilizes a carbon pile to increase the load demand (amperage draw) on the battery. Install a special battery load-testing tool onto the terminals **(see illustration)**. Load test the battery according to the tool manufacturer's instructions. Typically a load of 50-percent of the cold cranking amperage rating is applied during the test. The cold cranking amperage rating can usually be found on the battery label. Maintain the load on the battery for a maximum of 15 seconds. The battery voltage should not drop below 9.6 volts during the test. If the battery condition is weak or defective, the tool will indicate this condition immediately. **Note:** *Cold temperatures will cause the voltage readings to drop slightly. Follow the chart given in the tool manufacturer's instructions to compensate for cold climates. Minimum load voltage for freezing temperatures (32-degrees F) should be approximately 9.1 volts.*

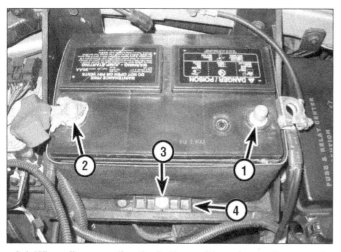

3.4 When disconnecting the battery cables, always disconnect the cable from the negative terminal (1) first, then disconnect the cable from the positive terminal (2). After the battery cables are disconnected, remove the battery hold-down bolt (3) and the hold-down clamp (4)

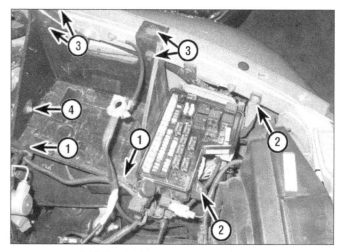

3.12 Top battery tray mounting details

1   Wiring harness clips
2   Fuse/relay box mounting fasteners
3   Tray mounting bolts
4   ABS controller bolt

Replacement

Battery

Refer to illustration 3.4

4   Disconnect the cable from the negative battery terminal **(see illustration)**, then disconnect the cable from the positive terminal.

5   Remove the battery hold-down bolt and hold-down clamp.

6   Lift out the battery. Be careful - it's heavy.

Note: *Battery straps and handlers are available at most auto parts stores for a reasonable price. They make it easier to remove and carry the battery.*

7   While the battery is out, inspect the area underneath the tray for corrosion. Clean the battery tray, then use a baking soda/water solution to neutralize any deposits to prevent further oxidation. If the metal around the tray is corroded, too, clean it as well and spray the area with a rust-inhibiting paint.

8   If corrosion has leaked down past the battery tray, remove the tray (see Steps 12 through 23) for further cleaning.

9   If you are replacing the battery, make sure you get one that's identical, with the same dimensions, amperage rating, cold cranking rating, etc.

10   Installation is the reverse of removal.

Battery tray

Left side

Refer to illustrations 3.12, 3.19 and 3.21

11   Remove the battery (see Steps 4 through 6).

12   Detach the wiring harness clips from the battery tray **(see illustration)**.

13   Remove the engine compartment fuse and relay box mounting bolts.

14   Slide the fuse and relay box towards the center of the engine compartment to disengage the two locator pins from their respec-

3.19 To detach the EVAP purge solenoid from its mounting bracket, depress this locking tang and slide the solenoid off its bracket

tive slots in the front wall of the battery box and lift up the fuse and relay box.

15   Remove the ABS controller mounting bolt, then support the controller with some wire or with a bungee cord.

16   Loosen the lug nuts for the left front wheel. Raise the front of the vehicle and place it securely on jackstands. Remove the left front wheel. Remove the left front wheelhouse splash shield (see Chapter 11).

17   Mark the location of the cruise control servo (if equipped), then remove the servo retaining screws and detach the servo from the battery tray.

18   Unplug the electrical connector from the battery temperature sensor if your vehicle uses one.

19   Detach the Evaporative Emissions (EVAP) system purge solenoid from its

3.21 Battery tray lower details and mounting bolts

mounting bracket **(see illustration)**. It's not necessary to disconnect the electrical connector from the purge solenoid or to detach the solenoid mounting bracket from the battery tray, which is attached to the underside of the tray by a pair of retaining screws (unless you're planning to *replace* the tray).

20   Disconnect the ground cable from the left front fender (on most models).

21   Remove the battery tray mounting bolts and remove the battery tray **(see illustration)**.

22   Installation is the reverse of removal.

Right side

23   Removal of the right-side battery tray, on models so equipped, is similar to the removal procedure of the left side battery tray, except that the air filter housing and the relay bracket must be removed (see Chapter 4), and some of the steps related to left side tray removal will not apply.

4.4a To detach the body ground cable from the left front fender, remove this bolt

4.4b To detach the frame ground cable from the left frame rail, remove this bolt

4.4c To detach the engine ground cable from the front of the engine, remove this bolt (as seen from underneath the vehicle, looking straight up)

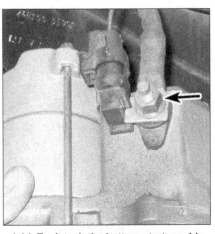

4.4d To detach the battery starter cable from the solenoid terminal, remove this nut

4   Battery cables - check and replacement

Refer to illustrations 4.4a, 4.4b, 4.4c and 4.4d

1   Periodically inspect the entire length of each battery cable for damage, cracked or burned insulation and corrosion. Poor battery cable connections can cause starting problems and decreased engine performance.

2   Check the cable-to-terminal connections at the ends of the cables for cracks, loose wire strands and corrosion. The presence of white, fluffy deposits under the insulation at the cable terminal connection is a sign that the cable is corroded and should be replaced. Check the terminals for distortion, missing mounting bolts and corrosion.

3   When removing the cables, always disconnect the cable from the negative battery terminal first and hook it up last or the battery could be accidentally shorted by the tool you're using to loosen the cable clamps. Even if you're only replacing the cable for the

positive terminal, always disconnect the cable from the negative battery terminal first.

4   After disconnecting the old cables from the battery, trace each of them to their opposite ends and detach them from the starter solenoid and ground terminals **(see illustrations)**. Note the routing of each cable to ensure correct installation. If a cable is bundled with another wiring harness, cut the electrical tape tying them together, remove any protective sheathing and separate them.

5   If you are replacing either or both of the battery cables, take them with you when buying new cables. It is vitally important that you replace the cables with identical parts. Cables have characteristics that make them easy to identify: positive cables are usually red and larger in cross-section; ground cables are usually black and smaller in cross-section.

6   Clean the threads of the solenoid or ground connection with a wire brush to remove rust and corrosion. Apply a light coat of battery terminal corrosion inhibitor or petroleum jelly to the threads to prevent future corrosion.

7   Attach the cable to the solenoid or ground connection and tighten the mounting nut/bolt securely.

8   Before connecting a new cable to the battery, make sure that it reaches the battery post without having to be stretched.

9   Connect the positive cable first, followed by the negative cable.

5   Ignition system - general information

All gasoline engines are equipped with a distributorless ignition system. The ignition system consists of the battery, the ignition coil(s), the spark plug wires (3.7L and 4.7L engines), the spark plugs, a knock sensor on each cylinder head, the Camshaft Position (CMP) sensor, the Crankshaft Position (CKP) sensor, the Manifold Absolute Pressure (MAP) sensor, the Throttle Position (TP) sensor and

the Powertrain Control Module (PCM). For more information about the CMP sensor, CKP sensor, knock sensors, MAP sensor, TP sensor and PCM, refer to Chapter 6. The PCM controls the base ignition timing and the ignition timing advance on all engines. The base ignition timing is not adjustable on any model.

The PCM controls the ignition system by opening and closing the ignition coil ground circuit. The computerized ignition system provides complete control of the ignition timing by determining the optimum timing in response to engine speed, coolant temperature, throttle position and vacuum pressure in the intake manifold. These parameters are relayed to the PCM by the camshaft position sensor, crankshaft position sensor, throttle position sensor, coolant temperature sensor and manifold absolute pressure sensor. The PCM and the crankshaft position sensor are very important components of the ignition system. The ignition system will not operate and the engine will not start if the PCM or the crankshaft position sensor are defective. Refer to Chapter 6 for additional information on the various sensors.

3.7L V6 engine

The 3.7L V6 uses three coils, each of which is mounted on top of a spark plug of the odd-numbered cylinders. Each coil is connected to an even-numbered spark plug by a spark plug wire so that when the coil is energized, it fires both spark plugs at the same time. Only one of the two cylinders has pressure when the spark plug is fired, so it is the only cylinder that draws significant power from the coil at that time - this is a waste-spark system.

4.7L V8 engine

The 4.7L V8 engine uses a coil-over-plug system. On this engine there are eight ignition coils (each with two secondary terminals), eight spark plugs wires and 16 spark plugs. The engine has two spark plugs per cylinder, so each coil is located on top of and connected

6.5a To use a calibrated ignition tester on a 5.7L Hemi V8, remove an ignition coil and connect the tester to the spark plug boot, clip the tester to a convenient ground and crank over the engine. If there's enough power to fire the plug, bright blue sparks will be visible between the electrode tip and the tester body (weak sparks or intermittent sparks are the same as no sparks)

6.5b To use a ignition tester on a 3.7L V6 or a 4.7L V8, you need to test for spark twice at each cylinder. First, disconnect the plug wire from the lower spark plug, connect it to the tester, connect the other end of the tester to the spark plug and crank over the engine. Using this type of tester, if there's enough power to fire the plug, the bulb inside the tester housing will flash. If the plug wire is putting out a healthy spark, remove the tester and reconnect the spark plug wire . . .

directly to one of those plugs with a rubber boot type seal and fires a spark plug at the bottom of the combustion chamber via a second high-tension terminal and a spark plug wire.

3.6L V6 and Hemi engines

The Hemi engines use a coil-over-plug system which consists of individual coils, one above each pair of spark plugs and connected directly to the spark plugs. There are no spark plug wires on these engines. Battery voltage is supplied to the ignition coil primary circuits. The PCM fires the coils in firing order sequence by turning the ground paths for their primary circuits on and off.

---

6   Ignition system - check

Refer to illustrations 6.5a, 6.5b and 6.5c

Warning: *Because of the high voltage generated by the ignition system, extreme care should be taken whenever an operation is performed involving ignition components. This not only includes the ignition coil, but also related components and test equipment.*
Warning: *The following procedure requires the engine to be cranked during testing. When cranking the engine, make sure that no test leads, loose clothing, long hair, etc. comes into contact with any moving parts (drivebelt, cooling fan, etc.).*

1   Before proceeding with the ignition system, check the following items:

a) *Make sure the battery cable clamps, where they connect to the battery, are clean and tight.*
b) *Test the condition of the battery (see Section 3). If it does not pass all the tests, replace it with a new battery.*

c) *Check the ignition system wiring and connections for tightness, damage, corrosion or any other signs of a bad connection.*
d) *Check the related fuses inside the engine compartment fuse and relay box (see Chapter 12). If they're burned, determine the cause and repair the circuit.*

2   If the engine turns over but won't start or has a severe misfire, perform the following steps using a calibrated ignition tester to make sure there is sufficient secondary ignition voltage to fire the spark plugs. There are several different types of calibrated ignition testers available. Either style will work on any of the engines covered in this manual.
3   Disable the fuel system by removing the fuel pump relay, which is located in the engine compartment fuse and relay box (see Chapter 12).
4   On 3.6L V6 and Hemi engines, you'll have to remove each ignition coil (see Section 7) to test for spark to the plug that it fires. On 3.7L V6 and 4.7L V8 engines, you'll also have to remove each ignition coil to verify its output to the plug that it fires directly, but you can simply disconnect the spark plug wire to its companion cylinder and attach an ignition tester to the wire to verify spark to its companion cylinder.
5   Attach a calibrated ignition system tester (available at most auto parts stores) to the spark plug boot **(see illustrations)**. If you're using a conventional style spark tester (like the one shown in the first illustration), ground the clip on the tester to a bolt or metal bracket on the engine. If you're using a newer style tester (like the one shown in the second illustration), it's not necessary to ground the tester because you can simply connect it to the spark plug. Crank the engine while watching the tester. Note whether a bright blue, well-defined spark occurs (conventional style tes-

6.5c . . . then remove the ignition coil (see Section 7), connect the tester to the spark plug boot, connect the other end of the tester to the spark plug and crank over the engine again

ter) or if the bulb within the tester body flashes (newer style tester).
6   If sparks occur (or the tester body flashes) during cranking, sufficient voltage is reaching the plug to fire it. Repeat this test for each spark plug (which means both spark plugs at each cylinder on the 3.6L V6 and Hemi) to verify that all the coils are OK. However, be aware that even if all ignition coils are able to fire the spark tester, the plugs themselves might be fouled, so remove and inspect the plugs too (see Chapter 1).
7   If the spark plug is in good shape, the coil might be defective (see Section 7).
8   If no sparks occur during cranking at one cylinder, inspect the primary wire connection at the coil from which you're not getting any spark. Make sure that it's clean and tight.

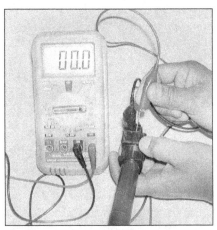

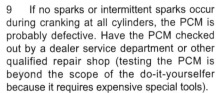

7.3 To check the primary resistance of an ignition coil, measure the resistance across the two coil primary terminals and compare your measurement to the primary resistance listed in this Chapter's Specifications

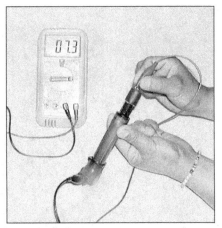

7.4 Measure the resistance between the positive terminal of the primary side and the high-tension terminal and compare your measurement to the secondary resistance listed in this Chapter's Specifications

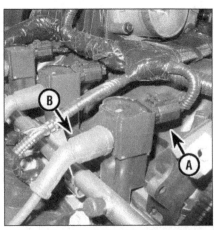

7.7 4.7L V8 ignition coil primary wiring connector (A) and the secondary wire that connects to a lower spark plug (B) - the coil itself is mounted directly on top of the upper spark plug

9   If no sparks or intermittent sparks occur during cranking at all cylinders, the PCM is probably defective. Have the PCM checked out by a dealer service department or other qualified repair shop (testing the PCM is beyond the scope of the do-it-yourselfer because it requires expensive special tools).

10   Any additional testing of the ignition system must be done by a dealer service department or by an independent repair shop with the right tools.

7   Ignition coil - check and replacement

3.7L V6 and 4.7L V8 engines

Check

Refer to illustrations 7.3 and 7.4

1   Remove the ignition coil (see Steps 6 through 9).

2   Clean the outer case and check it for cracks and other damage. Clean the coil primary terminals and check the coil tower terminal for corrosion. Clean it with a wire brush if any corrosion is found. Keeping all coil terminals clean and dry is essential for proper operation of the ignition system.

3   To check the coil primary resistance, attach the leads of an ohmmeter to the two terminals of the connector **(see illustration)**. Compare your measurement to the primary resistance value listed in this Chapter's Specifications. If the measured primary resistance value is not close to the specified range, replace the ignition coil.

4   To check the coil secondary resistance, connect one of the ohmmeter leads to the positive terminal of the connector and the other ohmmeter lead to the high-tension terminal at the coil and then also to the terminal at the end of the spark plug wire **(see illustration)**. Compare your measurements to

the secondary resistance value listed in this Chapter's Specifications. If the measured secondary resistance value is not within the specified range, replace the ignition coil.

5   If the primary and secondary resistance values are within specification or close, but the coil seems to be misfiring or not firing at all, or is causing the PCM to set a Diagnostic Trouble Code (DTC) indicating a misfire, try swapping it with an adjacent coil. If the suspect coil was causing a DTC for one cylinder, it will likely set the same DTC when it's installed above another cylinder. If this is the case, the coil is probably defective because it's unlikely that the harnesses for two adjacent ignition coils would both be defective. At any rate, no further testing is possible at home. If you're not sure whether you should replace the coil at this point, consult a dealer service department or other qualified repair shop.

Replacement

Refer to illustration 7.7

6   Depending on which coil you're planning to check or replace, remove either the air intake duct or the air intake resonator box (see Chapter 4A).

7   Disconnect the electrical connector from the coil **(see Illustration)**. To release the connector, push down on the release lock on top of the connector and pull off the connector.

8   To prevent dirt and debris from falling down into the spark plug well, use compressed air to blow out the area around the base of the ignition coil.

9   Remove the ignition coil mounting nut and detach the coil from the spark plug. Detach the spark plug wire from the coil by pulling on the boot, not the wire.

10   Installation is the reverse of removal.

3.6L V6 and Hemi engines

Check

11   On Hemi engines, if you're checking one

of the front coils on the left cylinder bank (cylinder numbers 1 or 3) or the front ignition coil on the right cylinder bank (cylinder number 2), it's not really necessary to remove them to check coil primary resistance. But if you're going to check one of the two rear coils on the left cylinder bank (cylinder numbers 5 or 7) or any of the three coils on the right cylinder bank (cylinder numbers 4, 6 or 8) besides the one for cylinder number 2, it will be easier to check these coils on the bench. To remove an ignition coil, see Steps 17 through 20.

12   On 3.6L V6 models, it isn't necessary to completely remove the ignition coils to test them, but the upper intake manifold will have to be removed to check the left-side cylinders (2-4-6). To remove the ignition coils, see Steps 22 through 27.

13   Clean the outer case and check it for cracks and other damage. Clean the coil primary terminals and check the coil high-tension terminal for corrosion. If you find any corrosion, clean the high-tension terminal with a wire brush. Keeping the coil terminals and wires clean and dry is essential for proper operation of the ignition system.

14   To check the coil primary resistance, attach the leads of an ohmmeter to the two terminals of the connector **(see illustration 7.3)**.

15   Compare your measurement to the primary resistance value listed in this Chapter's Specifications. If the measured resistance value is not within the specified range, replace the ignition coil.

16   There is no published secondary resistance value available for the ignition coils used on the Hemi engine. If the primary resistance is within specification, but the coil seems to be misfiring or not firing at all, or is causing the PCM to set a Diagnostic Trouble Code (DTC) that indicates a misfire, try swapping it with an adjacent coil (and don't forget to swap the spark plug wire for the companion cylinders of both coils too). If the suspect

coil was causing a DTC for one cylinder, it will likely set the same DTC when it's installed above another cylinder. If this is the case, the coil is probably defective because it's unlikely that the harnesses for two adjacent ignition coils would both be defective. At any rate, no further testing is possible at home. If you're not sure whether you should replace the coil at this point, have it checked by a dealer service department.

Replacement

Hemi engines

17   If you're planning to check or replace the ignition coils for cylinder numbers 2, 4, 6 and/or 8, remove the air intake duct and the resonator box (see Chapter 4A).
18   Disconnect the electrical connector.
19   To prevent dirt and debris from falling into the spark plug well, blow out the area surrounding the base of the ignition coil with compressed air.
20   Remove the two coil mounting bolts. To remove the coil, carefully pull it up with a twisting motion.
21   Installation is the reverse of removal.

3.6L V6 engine

22   Remove the engine cover (see Chapter 1).
23   Disconnect and remove the resonator (see Chapter 4A).
24   Remove the upper intake manifold (see Chapter 2D).
Note: The ignition coils for cylinders 1 and 3 can be removed without having to remove the upper intake manifold.
25   Disconnect the electrical connector from the ignition coil.
26   Remove the ignition coil mounting bolt.
27   Grasp the coil firmly and pull it off the spark plug using a twisting motion.
28   Installation is the reverse of removal. Tighten the ignition coil bolts securely.

8   Charging system - general information and precautions

The charging system consists of the alternator; the Electronic Voltage Regulator (EVR), the ignition switch, the battery, a battery temperature sensor, a voltage gauge, a charge indicator light and the wiring between all the components. The charging system supplies electrical power for the ignition system, the lights, the radio, etc. The alternator is driven by a serpentine drivebelt at the front of the engine.

The ignition switch turns the charging system on and off. The system remains on as long as the engine is running and the Automatic Shut Down (ASD) relay is energized. When the ASD relay is on, voltage is supplied to the ASD relay sense circuit at the PCM (gasoline engines) or the ECM (diesel engines). This voltage is connected through the PCM/ECM and supplied to the alternator field terminal (GEN SOURCE +) on the backside of the alternator. The EVR's

field control circuit inside the PCM/ECM regulates the current output of the alternator (there is no external voltage regulator). The field control circuit is connected in series with the second rotor field terminal and ground.

The battery temperature sensor, which is mounted on the underside of the battery tray, is a Negative Temperature Coefficient (NTC) thermistor that receives a 5-volt signal from the PCM/ECM and is grounded through a return wire. As the temperature of the battery increases, the resistance in the temperature sensor decreases and the detection voltage at the PCM/ECM increases. The PCM/ECM uses the signal from the battery temperature sensor and data from monitored line voltage to vary the battery charging rate. It does so by cycling the alternator's ground path, which controls the strength of the rotor's magnetic field. The PCM/ECM regulates the alternator's current output accordingly. System voltage is higher when the engine is cold and is gradually reduced to a lower voltage as the temperature increases. To replace the battery temperature sensor, refer to Chapter 6.

When the engine is running, the voltage gauge on the instrument cluster indicates electrical system voltage. The indicator needle should be within the normal range if the battery is charged. If the needle moves outside the normal range and stays there during normal driving, inspect the charging system (see Section 9). If the voltage gauge is defective, replace the instrument cluster (see Chapter 12). The voltage gauge cannot be serviced separately from the cluster.

On some models a CHECK GAUGES indicator light, which is located on the instrument cluster, illuminates when the voltage gauge, the oil pressure gauge or the engine coolant gauge indicates a reading that is either too high or too low. If this happens, the system being monitored by the gauge that's out of range requires immediate attention.

The charging system doesn't ordinarily require periodic maintenance. However, you should inspect the drivebelt, the battery, the charging system wiring harness and all connections at the intervals outlined in Chapter 1. Be very careful when making electrical circuit connections to the alternator or the charging system circuit and note the following:

a) When reconnecting wires to the alternator from the battery, be sure to note the polarity.
b) Before using arc-welding equipment to repair any part of the vehicle, disconnect the wires from the alternator and the battery terminals.
c) Never start the engine with a battery charger connected.
d) Always disconnect both battery cables before using a battery charger.
e) The alternator is turned by an engine drivebelt, which could cause serious injury if your hands, hair or clothes become entangled in it with the engine running.

f) Because the alternator is connected directly to the battery, it could arc or cause a fire if overloaded or shorted out.
g) Wrap a plastic bag over the alternator and secure it with rubber bands before steam-cleaning the engine.

9   Charging system - check

Note: These vehicles are equipped with an On-Board Diagnostic-II (OBD-II) system that is useful for detecting charging system problems because it can provide you with the Diagnostic Trouble Code (DTC) that will indicate the general nature of the problem. Refer to Chapter 6 for a list of the DTCs used by the PCM/ECM on these vehicles and for the procedure you'll need to use to obtain DTCs.
1   If a malfunction occurs in the charging circuit, do not immediately assume that the alternator is causing the problem. First check the following items:

a) The battery cables where they connect to the battery. Make sure the connections are clean and tight.
b) The battery electrolyte specific gravity (by observing the charge indicator on the battery). If it is low, charge the battery.
c) Inspect the external alternator wiring and connections.
d) Check the drivebelt condition and tension (see Chapter 1).
e) Check the alternator mounting bolts for tightness.
f) Run the engine and check the alternator for abnormal noise.

2   Using a voltmeter, check the battery voltage with the engine off. It should be at least 12.6 volts with a fully charged battery.
3   Start the engine and check the battery voltage again. It should now be greater than the voltage recorded in Step 2, but should not read more than 15 volts.
4   If the indicated voltage reading is less or more than the specified charging voltage, have the charging system checked at a dealer service department or other properly equipped repair facility. The voltage regulator on these models is contained within the PCM and it cannot be adjusted, removed or tampered with in any way.

10   Alternator - removal and installation

1   Disconnect the cable(s) from the negative battery terminal(s).
2   Remove the drivebelt (see Chapter 1).

3.6L V6, 3.7L V6 and 4.7L V8 engines

Refer to illustration 10.3
3   Disconnect the battery cable from the B+ output terminal and disconnect the field wire

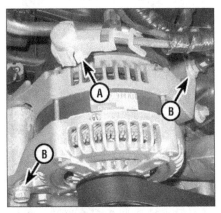

10.3 Remove the nut from the battery output terminal stud (A) and disconnect the battery cable, then disconnect the electrical connector for the field terminal from the rear of the alternator - the mounting bolts are shown at (B) (4.7L V8 engine shown, 3.7L V6 engine similar)

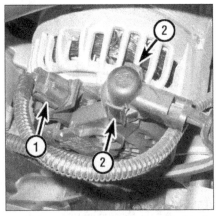

10.8 Disconnect the field wire electrical connector (1) from the field terminal, then depress the two release tabs (2) on the plastic insulator cap, pull off the cap and remove the nut that secures the battery cable to the B+ terminal (Hemi engine)

electrical connector from the field terminal **(see illustration)**.

4   Remove the mounting bolts and remove the alternator.

5   If you are replacing the alternator, take the old one with you when purchasing a replacement unit. Make sure the new/rebuilt unit looks identical to the old alternator. Look at the terminals - they should be the same in number, size and location as the terminals on the old alternator. Finally, look at the identification numbers - they will be stamped into the housing or printed on a tag attached to the housing. Make sure the numbers are the same on both alternators.

6   Many new and remanufactured alternators do not have a pulley installed, so you may have to switch the pulley from the old unit to the new/rebuilt one. When buying an alternator, find out the shop's policy regarding pulleys; some shops will perform this service free of charge.

7   Installation is the reverse of removal.

Hemi engine

Refer to illustration 10.8

8   Disconnect the field wire electrical connector from the field terminal **(see illustration)**.

9   Unsnap the plastic insulator cap from the B+ output terminal, remove the battery cable retaining nut from the B+ output terminal and disconnect the battery cable from the B+ output terminal.

10   Remove the alternator support bracket nuts and bolts and remove the alternator support bracket.

11   Remove the alternator mounting bolts and remove the alternator.

12   Installation is the reverse of removal.

Diesel engine

13   Remove the upper alternator bracket mounting bolt.

14   Remove the lower alternator mounting bolt and nut and remove the alternator.

15   Unsnap the plastic insulator cap from the B+ output terminal, remove the nut that secures the battery cable to the B+ output terminal and disconnect the cable from the B+ terminal. Disconnect the field wire electrical connector from the field terminal.

16   Installation is the reverse of removal.

11   Starting system - general description and precautions

General description

The starting system consists of the starter relay, the starter motor and starter solenoid assembly, the battery, the battery cables, the ignition switch and key lock cylinder, the clutch pedal position switch (manual transmissions), the Transmission Range (TR) sensor (automatic transmissions) and the wiring connecting these components. All starter motors are located on the lower part of the engine, near the transmission bellhousing, where they can engage the ring gear on the flywheel (manual transmissions) or the driveplate (automatic transmissions).

The starting system has two separate circuits: A low-amperage control circuit, which operates on less than 20 amps, and a high-amperage supply circuit that delivers between 150 and 350 amps to the starter motor (700 amps on diesel engines). The low-amp control circuit includes the ignition switch, the clutch pedal position switch or PNP switch, the starter relay, the coil inside the starter solenoid and the wire harness connecting these components. The high-amp supply circuit consists of the battery, the battery cables, the contact disc in the starter solenoid and the starter motor itself.

On vehicles with a manual transmission, the clutch pedal position switch is installed in series between the ignition switch and the bat-

tery terminal on the control side of the starter relay coil. The clutch pedal position switch is normally open to prevent the starter relay from being energized when the ignition switch is turned to START unless the clutch pedal is depressed. This setup prevents the starter motor from operating while the clutch disc and the flywheel are engaged. The starter relay coil ground terminal is always grounded.

On vehicles with an automatic transmission, the TR sensor is installed in series between the starter relay ground terminal and ground. The TR sensor is normally open to prevent the starter relay from being energized unless the shift lever is in the NEUTRAL or PARK position. When the ignition switch is turned to START, battery voltage is supplied through the low-amperage control circuit to the battery terminal of the starter relay coil if the shift lever is in the NEUTRAL or PARK position. If it isn't, the starter circuit remains open and the engine won't start.

When the starter relay coil is energized, the normally-open relay contacts close and connect the relay common supply terminal to the relay's normally-open terminal. The closed relay contacts energize the windings of the starter solenoid pull-in coil, which pulls in the solenoid plunger, which pulls the shift lever in the starter motor, which engages the starter's overrunning clutch and pinion gear with the starter's ring gear. As the solenoid plunger reaches the end of its travel, the solenoid contact disc completes the high-current starter supply circuit and energizes the solenoid plunger hold-in coil. Current flows from the solenoid battery terminal to the starter motor and energizes the starter.

The starter motors used on the vehicles covered in this manual are not rebuildable because no parts are available. They're sold as complete new or remanufactured assemblies. If any part of the starter motor fails, including the starter solenoid, the entire assembly must be replaced.

Precautions

Always observe the following precautions when working on the starting system:

a)   *Excessive cranking of the starter motor can overheat it and cause serious damage. Never operate the starter motor for more than 15 seconds at a time without pausing to allow it to cool for at least two minutes.*

b)   *The starter is connected directly to the battery and could arc or cause a fire if mishandled, overloaded or shorted.*

c)   *Always detach the cable(s) from the negative battery terminal(s) before working on the starting system.*

12   Starter motor and circuit - check

Refer to illustrations 12.3 and 12.4

1   If a malfunction occurs in the starting circuit, do not immediately assume that the

12.3 Use an inductive ammeter to measure starter current draw

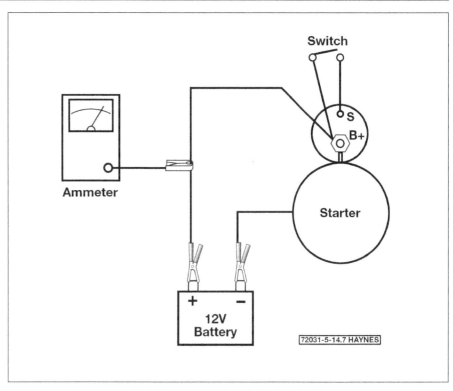

12.4 Starter motor bench-testing details

starter is causing the problem. First, check the following items:

a) *Make sure the battery cable clamps, where they connect to the battery, are clean and tight.*

b) *Check the condition of the battery cables (see Section 4). Replace any defective battery cables with new parts.*

c) *Test the condition of the battery (see Section 3). If it does not pass all the tests, replace it with a new battery.*

d) *Check the starter motor wiring and connections.*

e) *Check the starter motor mounting bolts for tightness.*

f) *Check the related fuses in the engine compartment fuse box (see Chapter 12). If they're blown, determine the cause and repair the circuit.*

g) *Check the ignition switch circuit for correct operation (see Chapter 12).*

h) *Check the starter relay (see Chapter 12).*

i) *Check the operation of the clutch pedal position switch (see Chapter 8) or the Transmission Range (TR) sensor (see Chapter 6). These systems must operate correctly to provide battery voltage to the starter solenoid.*

2   If the starter does not activate when the ignition switch is turned to the start position, check for battery voltage to the starter solenoid. This will determine if the solenoid is receiving the correct voltage from the ignition switch. Connect a 12-volt test light or a voltmeter to the starter solenoid positive terminal. While an assistant turns the ignition switch to the start position, observe the test light or voltmeter. The test light should shine brightly or battery voltage should be indicated on the voltmeter. If voltage is not available to the starter solenoid, refer to the wiring diagrams in Chapter 12 and check the fuses and starter relay in series with the starting system. If voltage is available but there is no movement

from the starter motor, remove the starter from the engine (see Section 13) and bench test the starter (see Step 4).

3   If the starter turns over slowly, check the starter cranking voltage and the current draw from the battery. This test must be performed with the starter assembly on the engine. Crank the engine over (for 10 seconds or less) and observe the battery voltage. It should not drop below 9.6 volts. Also, observe the current draw with an ammeter **(see illustration)**. Typically a starter should not exceed 160 amps. If the starter motor amperage draw is excessive, have it tested by a dealer service department or other qualified repair shop. There are several conditions that may affect the starter cranking potential. The battery must be in good condition and the battery cold-cranking rating must not be under-rated for the particular application. Check the battery specifications carefully. The battery terminals and cables must be clean and not corroded. Also, in cases of extreme cold temperatures, make sure the battery and/or engine block is warmed before performing the tests.

4   If the starter is receiving voltage but does not activate, remove and check the starter motor assembly on the bench. Most likely the solenoid is defective. In some rare cases, the engine may be seized so be sure to try and rotate the crankshaft pulley (see Chapter 2) before proceeding. With the starter assembly mounted in a vise on the bench, install one jumper cable from the positive terminal of a test battery to the B+ terminal on the starter. Install another jumper cable from the nega-

tive terminal of the battery to the body of the starter **(see illustration)**. Install a starter switch and apply battery voltage to the solenoid S terminal (for 10 seconds or less) and observe the solenoid plunger, shift lever and overrunning clutch extend and rotate the pinion drive. If the pinion drive extends but does not rotate, the solenoid is operating but the starter motor is defective. If there is no movement but the solenoid clicks, the solenoid and/or the starter motor is defective. If the solenoid plunger extends and rotates the pinion drive, the starter assembly is operating properly.

13   Starter motor - removal and installation

1   Disconnect the cable(s) from the negative battery terminal(s).
2   Raise the vehicle and support it securely on jackstands.

3.6L V6, 3.7L V6 and 4.7L V8 engines

Refer to illustrations 13.4, 13.5 and 13.6

3   On 4WD models with certain transmissions, there's a support bracket between the front axle and the side of the transmission that blocks access to the lower starter mounting bolt. If your vehicle has this support bracket, remove the two support bracket bolts at the transmission, then pry the support bracket aside enough to gain access to the lower starter mounting bolt.

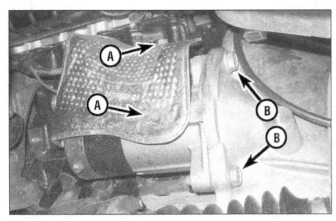

13.4 If the starter motor is equipped with a heat shield, remove these two bolts (A) to detach it from the starter. To remove the starter from an automatic transmission (shown), remove these two bolts (B) (starter motor on 3.7L V6 with automatic and 4WD shown, starters on other 3.7L V6 engines and 4.7L V8 engines similar)

13.5 Disconnect the electrical connector (A) from the spade terminal on the solenoid, then remove the nut and disconnect the battery cable (B) from the terminal stud on the solenoid (starter motor on 3.7L V6 shown, starters on 4.7L V8 similar)

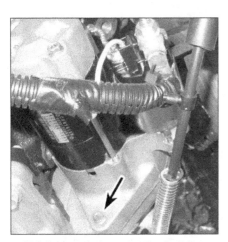

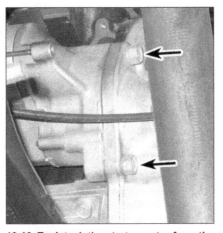

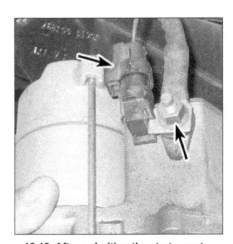

13.6 Lower starter mounting bolt on a 4.7L V8 engine

13.10 To detach the starter motor from the transmission on a Hemi engine, remove these two bolts

13.12 After unbolting the starter motor and moving it forward, remove the nut that secures the battery positive cable to the stud terminal on the solenoid, disconnect the positive cable from the stud, then disconnect the electrical connector from the spade terminal on the solenoid

4   Remove the heat shield, if equipped, from the starter motor **(see illustration)**.
5   Remove the nut that secures the battery cable to the terminal stud on the starter solenoid, disconnect the cable from the terminal stud and disconnect the electrical connector from the spade terminal on the solenoid **(see illustration)**. **Note:** *If you have difficulty disconnecting the battery cable or the electrical connector from the solenoid, leave them connected until you detach the starter from the transmission bellhousing and move it to a position where you can access the wiring connectors more easily.*
6   On models with a manual transmission, remove the starter mounting bolt and nut. On models with an automatic transmission, remove the two starter mounting bolts **(see illustration)**.
7   Move the starter motor toward the front of the vehicle until the nose of the starter pinion housing clears the transmission bellhousing, then tilt the nose down and, if you haven't already done so, lower the starter until you can disconnect the electrical connector and

remove the nut that secures the battery cable to the terminal stud on the starter solenoid. Once everything is disconnected, remove the starter. **Caution:** *The starter motor is fairly heavy, so be sure to support it while removing it. Do NOT allow it to hang by the wiring harness.*
8   Installation is the reverse of removal.

Hemi engines

Refer to illustrations 13.10 and 13.12

9   On some 4WD models with certain transmissions, there's a support bracket between the front axle and the side of the transmission that blocks access to the lower starter mounting bolt. If your vehicle has this support bracket, remove the two support bracket bolts at the transmission, then pry the support bracket aside enough to gain access to the lower starter mounting bolt.
10   Remove the two starter motor mounting bolts **(see illustration)**.
11   Move the starter toward the front of the

vehicle until the starter pinion housing clears the transmission bellhousing. **Caution:** *The starter motor is fairly heavy, so be sure to support it while removing it. Do NOT allow it to hang by the wiring harness.*
12   Disconnect the battery cable and electrical connector from the terminals on the starter motor solenoid **(see illustration)**.
13   Installation is the reverse of removal.

Diesel engine

14   Disconnect the wiring harness and the battery cable from the starter.
15   Remove the three starter mounting bolts.
16   Remove the starter. **Caution:** *This starter is very heavy. Take care to avoid dropping it.*
17   Installation is the reverse of removal.

Chapter 6
Emissions and engine control systems

Contents

Specifications

Torque specifications

Ft-lbs (unless otherwise indicated)

Note: *One foot-pound (ft-lb) of torque is equivalent to 12 inch-pounds (in-lbs) of torque. Torque values below approximately 15 ft-lbs are expressed in inch-pounds, since most foot-pound torque wrenches are not accurate at these smaller values.*

Engine Coolant Temperature (ECT) sensor	
3.6L V6 engine	22
3.7L V6, 4.7L V8 and all Hemi engines	96 in-lbs
Diesel engine	156 in-lbs
Knock sensor mounting bolts (gasoline engines)	
3.6L V6 engine	16
3.7L V6, 4.7L V8 and all Hemi engines	15
Oxygen sensors	22
Crankshaft position sensor (CMP) bolt	
Gasoline engine	106 in-lbs
Diesel engine	89 in-lbs
Camshaft position sensor mounting bolts (gasoline engines)	
3.6L V6 engine	80 in-lbs
3.7L V6, 4.7L V8 and all Hemi engines	106 in-lbs

1   General information

Refer to illustration 1.7

To prevent pollution of the atmosphere from incompletely burned and evaporating gases, and to maintain good drivability and fuel economy, a number of emission control systems are incorporated on the vehicles covered in this manual. These emission control systems and their components are an integral part of the engine management system. The engine management system also includes all the government mandated diagnostic features of the second generation of on-board diagnostics, which is known as On-Board Diagnostics II (OBD-II).

At the center of the engine management and OBD-II systems is the on-board computer, which is known as the Powertrain Control Module (PCM). Using a variety of information sensors, the PCM monitors all of the important engine operating parameters (temperature, speed, load, etc.). It also uses an array of output actuators - such as the ignition coils, the fuel injectors, the Idle Air Control (IAC) motor, the Torque Converter Clutch (TCC) and various solenoids and relays - to respond to and alter these parameters as necessary to maintain optimal performance, economy and emissions. The principal emission control systems used on the vehicles covered in this manual include the:

Catalytic converters
Evaporative Emission Control (EVAP) system
Exhaust Gas Recirculation (EGR) system
Positive Crankcase Ventilation (PCV) system
Torque Converter Clutch (TCC) system

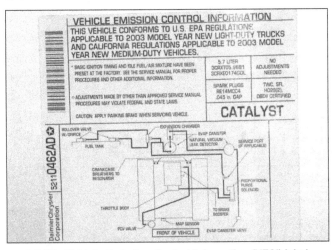

1.7 The Vehicle Emission Control Information (VECI) label, located in the engine compartment, contains information on the emission devices installed on your vehicle and a vacuum hose routing schematic

2.2 Hand-held scan tools like these can extract computer codes and also perform diagnostics

The Sections in this Chapter include general descriptions and component replacement procedures for most of the information sensors and output actuators, as well as the important components that are part of the systems listed above. Refer to Chapter 4 for more information on the air induction, fuel delivery and injection systems and exhaust systems, and to Chapter 5 for information on the ignition system. Refer to Chapter 1 for any scheduled maintenance for emission-related systems and components.

The procedures in this Chapter are intended to be practical, affordable and within the capabilities of the home mechanic. The diagnosis of most engine and emission control functions and drivability problems requires specialized tools, equipment and training. When servicing emission devices or systems becomes too difficult or requires special test equipment, consult a dealer service department.

Although engine and emission control systems are very sophisticated on late-model vehicles, you can do most of the regular maintenance and some servicing at home with common tune-up and hand tools and relatively inexpensive digital multimeters. Because of the Federally mandated extended warranty that covers the emission control system, check with a dealer about warranty coverage before working on any emission-related systems. After the warranty has expired, you might want to perform some of the component replacement procedures in this Chapter to save money. Remember that the most frequent cause of emission and drivability problems is a loose electrical connector or a broken wire or vacuum hose, so before jumping to conclusions the first thing you should always do is to inspect all electrical connections, electrical wiring and vacuum hoses related to a system.

Pay close attention to any special precautions given in this Chapter. Remember

that illustrations of various system components might not exactly match the component installed on the vehicle on which you're working because of changes made by the manufacturer during production or from year to year.

A Vehicle Emission Control Information (VECI) label **(see illustration)** is located in the engine compartment. This label contains emission-control and engine tune-up specifications and adjustment information. It also includes a vacuum hose routing diagram for emission-control components. When servicing the engine or emission systems, always check the VECI label in your vehicle. If any information in this manual contradicts what you read on the VECI label on your vehicle, always defer to the information on the VECI label.

2   On-Board Diagnostic (OBD) system and Diagnostic Trouble Codes (DTCs)

Scan tool information

Refer to illustration 2.2

1   Hand-held scanners are handy for analyzing the engine management systems used on late-model vehicles. Because extracting the Diagnostic Trouble Codes (DTCs) from an engine management system is now the first step in troubleshooting many computer-controlled systems and components, even the most basic generic scan tools are capable of accessing a computer's DTCs. More powerful scan tools can also perform many of the diagnostics once associated with expensive factory scan tools. If you're planning to obtain a generic scan tool for your vehicle, make sure that it's compatible with the year, make and model of the vehicle(s) on which you plan to use it. Some of the more versa-

tile scan tools accept removable cartridges, each of which contains the diagnostics for a particular manufacturer. An aftermarket generic scanner should work with any model covered by this manual. But before purchasing a scan tool, contact the manufacturer of the scanner you're planning to buy and verify that it will work properly with the system you want to scan. If you don't plan to purchase a scan tool and don't have access to one, you can have the codes extracted by a dealer service department or by an independent repair shop.

2   With the advent of the Federally mandated emission control system known as On-Board Diagnostics-II (OBD-II), specially designed scanners were developed. Several tool manufacturers have released OBD-II scan tools for the home mechanic **(see illustration)**.

OBD-II system general description

3   All vehicles covered by this manual are equipped with the OBD-II system. This system consists of the on-board computer, known as the Powertrain Control Module (PCM), and information sensors that monitor various functions of the engine and send a constant stream of data to the PCM during engine operation. Unlike earlier on-board diagnostics systems, the OBD-II system doesn't just monitor everything, store Diagnostic Trouble Codes (DTCs) and illuminate a Check Engine light or Malfunction Indicator Light (MIL) when there's a problem. This warning light was referred to as the "Check Engine" light prior to OBD-II, and many do-it-yourselfers and professional technicians still use this term. However, its name was changed to "Malfunction Indicator Light," or simply "MIL," as part of the Society of Automotive Engineers' standard terminology that was introduced in 1996 to encourage all manufacturers to use the same terms when

referring to the same components. So in this manual we will refer to this warning light as the Malfunction Indicator Light, or MIL.

4   The PCM is the brain of the electronically controlled OBD-II system. It receives data from a number of information sensors and switches. Based on the data that it receives from the sensors, the PCM constantly alters engine operating conditions to optimize drivability, performance, emissions and fuel economy. It does so by turning on and off and by controlling various output actuators such as relays, solenoids, valves and other devices. The PCM can only be accessed with an OBD-II scan tool plugged into the 16-pin Data Link Connector (DLC), which is located underneath the driver's end of the dashboard, near the steering column.

5   If your vehicle is still under warranty, virtually every fuel, ignition and emission control component in the OBD-II system is covered by a Federally mandated emissions warranty that is longer than the warranty covering the rest of the vehicle. Vehicles sold in California and in some other states have even longer emissions warranties than other states. Read your owner's manual for the terms of the warranty protecting the emission-control systems on your vehicle. It isn't a good idea to do-it-yourself at home while the vehicle emission systems are still under warranty because owner-induced damage to the PCM, the sensors and/or the control devices might VOID this warranty. So as long as the emission systems are still under warranty, take the vehicle to a dealer service department if there's a problem.

Information sensors

6   **Accelerator Pedal Position (APP) sensor** - The APP sensor is located on the accelerator pedal. It provides the PCM with information about the angle of the accelerator pedal at all times.

7   **Camshaft Position (CMP) sensor** - The CMP sensor produces a signal that the PCM uses to monitor the position of the camshaft. The CMP sensor is positioned next to a tone wheel mounted on the front of the camshaft. On diesel engines, the tone wheel is located on the timing gear. The tone wheel has notches machined into it. When the engine is operating, the CMP sensor receives a 5-volt signal from the PCM (ECM on diesels), then sends a 0.3-volt signal back to the PCM/ECM that switches to 5 volts every time one of the notches passes by it. This data enables the PCM to determine the position of the camshaft (and therefore the valve train) so that it can time the firing sequence of the fuel injectors. The PCM also uses the signal from the CMP sensor and the signal from the Crankshaft Position (CKP) sensor to distinguish between fuel injection and spark timing. On diesel engines, the CMP sensor's signal is used primarily for starting; once the engine is running, the Powertrain Control Module (PCM) uses the CMP as a back-up sensor for engine speed (the CKP sensor is the primary

engine speed indicator for a running diesel engine).

On 3.7L V6 and 4.7L V8 engines, the CMP sensor is located on the outer side of the front end of the right cylinder head, just ahead of the exhaust manifold. On Hemi engines, the CMP sensor is located on the upper right side of the timing chain cover. On diesel engines, the CMP sensor is located on the left rear of the timing gear cover.

8   **Crankshaft Position (CKP) sensor** - Like the CMP sensor, the CKP sensor is also a Hall effect device. The CKP sensor uses a tone wheel with notches machined into it to flip the CKP sensor's output from 0.3 volt to 5 volts every time a notch passes by it. On gasoline engines, the tone wheel for the CKP sensor is mounted at the rear end of the crankshaft. The tone wheel for the CKP sensor on diesel engines is located on the front end of the crankshaft, behind the harmonic balancer. The PCM uses data from the CKP sensor to calculate engine speed and crankshaft position, which enables it to synchronize ignition timing with fuel injector timing, to control spark knock and to detect misfires.

On gasoline engines, the CKP sensor is located on the right rear side of the block. On diesel engines, the CKP sensor is located on the front left side of the engine, next to the harmonic balancer.

9   **Engine Coolant Temperature (ECT) sensor** - The ECT sensor is a Negative Temperature Coefficient thermistor (temperature-sensitive variable resistor). The resistance of the thermistor decreases as the coolant temperature increases, so the voltage output of the ECT sensor increases. The PCM uses this variable voltage signal to calculate the temperature of the engine coolant. The ECT sensor tells the PCM when the engine is sufficiently warmed up to go into closed-loop operation and helps the PCM control the air/fuel mixture ratio and ignition timing. On 3.7L V6 and 4.7L V8 engines, the ECT sensor is located at the front of the intake manifold. On Hemi engines, the ECT sensor is located at the front of the block, under the air conditioning compressor. On diesel engines, the ECT sensor is located at the right front corner of the cylinder head, near the thermostat housing.

10   **Fuel temperature sensor** - The fuel temperature sensor is an integral part of the fuel heater and can be replaced only by replacing the fuel heater assembly (see Chapter 4B).

11   **Inlet air temperature/pressure sensor** - This dual-function sensor is used only on diesel engines. The inlet air temperature part of the sensor monitors ambient (outside) air temperature and the pressure part of the sensor monitors barometric pressure. The inlet air temperature/pressure sensor is located on the air filter housing cover.

12   **Input Shaft Speed (ISS) sensor** - The ISS sensor is a magnetic pick-up coil that generates a signal to the Transmission Control Module (TCM) that's proportional to the speed of rotation of the input shaft. The TCM

and PCM use this signal to determine the correct transmission gear ratio and to detect problems with the Vehicle Speed Sensor (VSS) signal. The TCM and PCM also compare the ISS signal to the VSS signal to determine whether the Torque Converter Clutch (TCC) is slipping, or is starting to slip. There are five transmissions used on the models in this book; the locations of the ISS sensor vary.

13   **Intake Air Temperature (IAT) sensor** - The IAT sensor is a Negative Temperature Coefficient (NTC) thermistor (temperature-sensitive variable resistor) that monitors the temperature of the air entering the engine and sends a variable voltage signal to the PCM (see the explanation for how an NTC-type thermistor works in the ECT sensor description above). The voltage signal from the IAT sensor is one of the inputs used by the PCM to determine injector pulse-width (the duration of each injector's on-time) and to adjust spark timing. On 3.7L V6 and 4.7L V8 engines, the IAT sensor is located on the left side of the intake manifold. On Hemi engines, the IAT sensor is located on the front of the air resonator box. On diesel engines, the IAT sensor (which is one-half of the dual-function IAT/MAP sensor) is located on top of the intake manifold.

14   **Knock Sensor (KS)** - The Knock Sensor (KS) is a piezoelectric crystal that oscillates in proportion to engine vibration. The oscillation of the crystal produces a voltage output that is monitored by the PCM, which retards the ignition timing when the oscillation exceeds a certain threshold. When the engine is operating normally, the Knock Sensor (KS) oscillates consistently and its voltage signal is steady. When detonation occurs, engine vibration increases, and the oscillation of the Knock Sensor (KS) exceeds a design threshold. Detonation is an uncontrolled explosion, after the spark occurs at the spark plug, which spontaneously combusts the remaining air/fuel mixture, resulting in a pinging or slapping sound. If allowed to continue, the engine could be damaged. Knock sensors are used only on gasoline engines. These engines are equipped with two knock sensors. On 3.7L V6 and 4.7L V8 engines, the knock sensors are located on top of the block, under the intake manifold. On Hemi engines, the knock sensors are located on each side of the block, under the exhaust manifolds.

15   **Manifold Absolute Pressure (MAP) sensor** - Air density changes with altitude and the PCM needs to know the air density at all times. The MAP sensor monitors the pressure or vacuum downstream from the throttle plate, inside the intake manifold. The MAP sensor measures intake manifold pressure and vacuum on the absolute scale, from zero psi, not from sea-level atmospheric pressure (14.7 psi). The MAP sensor converts the absolute pressure into a variable voltage signal that changes with the pressure or vacuum. The PCM uses this signal to calculate intake manifold pressure or vacuum, engine load, injector

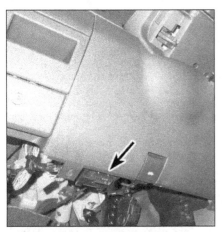

2.26 The Data Link Connector (DLC), or diagnostic connector, is located under the instrument panel

pulse-width, spark advance, shift points, idle speed and deceleration fuel shut-off. On 3.7L V6 and 4.7L V8 engines, the MAP sensor is located on the front of the intake manifold. On Hemi engines, the MAP sensor is located on the front of the air resonator box. On diesel engines, the MAP sensor (which is one-half of the dual-function IAT/MAP sensor) is located on top of the intake manifold.

16 **Output Shaft Speed (OSS) sensor** - The OSS sensor is a magnetic pick-up coil that generates an output to the Transmission Control Module (TCM) that's proportional to the speed of rotation of the output shaft. The TCM and PCM use this signal to determine the correct transmission gear ratio, and to detect problems with the Vehicle Speed Sensor (VSS) signal. The TCM and PCM also compare the OSS signal to the VSS signal to determine whether the Torque Converter Clutch (TCC) is slipping or is starting to slip. The OSS is located on the rear of the transmission on all models.

17 **Oxygen sensors** - The PCM uses the voltage signal from the upstream oxygen sensor to maintain an air/fuel ratio of 14.7:1 by constantly adjusting the on-time of the fuel injectors. There are two oxygen sensors: the upstream sensor is located on the exhaust manifold and a downstream oxygen sensor is located behind the catalyst.

18 **Throttle Position (TP) sensor** - The TP sensor is a potentiometer that receives constant voltage input from the PCM and sends back a voltage signal that varies in relation to the opening angle of the throttle plate inside the throttle body. The PCM uses this, along with information from other sensors, to calculate injector pulse width (the interval of time during which an injector solenoid is energized by the PCM). This sensor is an integral part of the throttle body's ETC motor assembly and cannot be serviced separately. Diesel engines don't have a TP sensor because they don't have a throttle body (or a throttle plate).

19 **Transmission Range (TR) sensor** - The TR sensor is located at the end of the manual shaft, on the side of the automatic

transmission. The TR sensor performs the same functions as a Park/Neutral Position switch: it prevents the engine from starting in any gear other than Park or Neutral, and it closes the circuit for the back-up lights when the shift lever is moved to Reverse. The TR sensor is also connected to the PCM, which sends a voltage signal to the TR sensor. The PCM monitors the voltage output signal from the switch, which corresponds to the position of the manual lever. Thus the PCM is able to determine the gear selected and is able to determine the correct pressure for the electronic pressure control system of the transmission. On most transmissions, the TR sensor is an integral part of the shift solenoid module, which is on top of the valve body, inside the transmission. Replacing the TR sensors on these transmissions is beyond the scope of the home mechanic.

20 **Vehicle Speed Sensor (VSS)** - The speed sensor is a magnetic pick-up coil that monitors output shaft rotating speed on automatic transmissions. The sensor is positioned over the park gear. When the vehicle is moving the park gear lugs rotate past the speed sensor, which generates a pulsing voltage. The frequency of this voltage increases as the speed goes up. This signal is processed by the Transmission Control Module (TCM), which shares this data with the PCM. The TCM and PCM use the speed sensor signal to calculate vehicle speed. The speed sensor is located on the overdrive gear case.

Output actuators

21 **EVAP canister purge solenoid** - The EVAP canister purge solenoid is a PCM-controlled solenoid that controls the purging of evaporative emissions from the EVAP canister to the intake manifold. The EVAP purge solenoid is never turned on during cold start warm-ups or during hot start time delays. But once the engine reaches a specified temperature and enters closed-loop operation, the PCM will energize the canister purge solenoid between 5 and 10 times a second. When the solenoid is energized by the PCM, it allows the fuel vapors that are stored in the EVAP canister to be drawn into the intake manifold, where they're mixed with intake air, then burned along with the normal air/fuel mixture, under certain operating conditions. The PCM regulates the flow rate of the vapors by controlling the pulse-width of the solenoid (the length of time during which the solenoid is turned on) in accordance with operating conditions. The EVAP canister purge solenoid is located under the engine compartment fuse and relay box and the (left) battery tray.

22 **EVAP Natural Vacuum Leak Detection (NVLD) system** - These vehicles use a Natural Vacuum Leak Detection (NVLD) system. The NVLD system is based on the principle that the pressure inside a sealed container (such as the EVAP system) will change if the temperature changes; if the temperature goes up, so does the pressure. If during a leak test the PCM notes that the pressure inside the

EVAP system isn't proportional to the temperature, it sets a Diagnostic Trouble Code (DTC) for a leak. The NVLD unit is located in front of the EVAP canisters, which are located under the vehicle in front of the fuel tank.

23 **Fuel injectors** - The fuel injectors, which spray fuel into the intake ports, are coils under PCM control. The injectors are installed between the fuel rail and the intake ports. For more information about the injectors, see Chapter 4.

24 **Ignition coils** - The ignition coils are under the control of the Powertrain Control Module (PCM). For more information about the ignition coils, refer to Chapter 5.

Obtaining and clearing Diagnostic Trouble Codes (DTCs)

25 All models covered by this manual are equipped with on-board diagnostics. When the PCM recognizes a malfunction in a monitored emission control system, component or circuit, it turns on the Malfunction Indicator Light (MIL) on the dash. The PCM will continue to display the MIL until the problem is fixed and the Diagnostic Trouble Code (DTC) is cleared from the PCM's memory. You'll need a scan tool to access any DTCs stored in the PCM. Before outputting any DTCs stored in the PCM, thoroughly inspect ALL electrical connectors and hoses. Make sure that all electrical connections are tight, clean and free of corrosion. And make sure that all hoses are correctly connected, fit tightly and are in good condition (no cracks or tears). Also, make sure that the engine is tuned up. A poorly running engine is probably one of the biggest causes of emission-related malfunctions. Often, simply giving the engine a good tune-up will correct the problem.

Accessing the DTCs

Refer to illustration 2.26

26 On the vehicles covered in this manual, the Diagnostic Trouble Codes (DTCs) can only be accessed with a scan tool. Plug the connector of the scan tool into the Data Link Connector (DLC) or diagnostic connector, which is located under the lower edge of the dash, just to the left of the hood release handle **(see illustration)**. Then follow the instructions included with the scan tool to extract the DTCs.

27 Once you have outputted all of the stored DTCs, look them up on the accompanying DTC chart.

28 After troubleshooting the source of each DTC, make any necessary repairs or replace the defective component(s).

Clearing the DTCs

29 Clear the DTCs with the scan tool in accordance with the instructions provided by the scan tool's manufacturer.

Diagnostic Trouble Codes

30 The accompanying tables are a list of the Diagnostic Trouble Codes (DTCs) that

can be accessed by a do-it-yourselfer working at home (there are many more DTCs available to professional service technicians with proprietary scan tools and software, but those codes cannot be accessed by a generic scan tool). If, after you have checked and repaired the connectors, wire harness and vacuum hoses (if applicable) for an emission-related system, component or circuit, the problem persists, have the vehicle checked by a dealer service department.

OBD-II Diagnostic Trouble Codes (DTCs)

Note: *Not all trouble codes apply to all models.*

Code	Possible cause
P000F	Fuel system over pressure relief valve activated
P0020	Bank 2, camshaft 1 position actuator circuit - open
P0023	Bank 2, camshaft 2 position actuator circuit - open
P0008	Engine position system performance (bank 1)
P0010	Bank 1, camshaft 1 position actuator circuit - open
P0013	"B" Camshaft position - actuator circuit malfunction (bank 1)
P0016	Crankshaft/camshaft timing misalignment
P0020	Bank 2, camshaft 1 position actuator circuit - open
P0023	Bank 2, camshaft 2 position actuator circuit - open
P003A	Turbocharger boost control module position exceeded learning limit
P0030	HO_2S heater control circuit (bank 1, sensor 1)
P0031	Upstream oxygen sensor (left cylinder bank), heater circuit low voltage
P0032	Upstream oxygen sensor (left cylinder bank), heater circuit high voltage
P0036	HO_2S heater control circuit (bank 1 sensor 2)
P0037	Downstream oxygen sensor (left cylinder bank), heater circuit low voltage
P0038	Downstream oxygen sensor (left cylinder bank), heater circuit high voltage
P0049	Turbo/supercharger turbine - over-speed
P0051	Upstream oxygen sensor (right cylinder bank), heater circuit low voltage
P0052	Upstream oxygen sensor (right cylinder bank), heater circuit high voltage
P0053	HO_2S heater resistance (bank 1, sensor 1)
P0054	HO_2S heater resistance (bank 1, sensor 2)
P0057	Downstream oxygen sensor (right cylinder bank), heater circuit low voltage
P0058	Downstream oxygen sensor (right cylinder bank), heater circuit high voltage
P006E	Turbocharger boost control supply voltage circuit low
P0068	MAP sensor/TP sensor correlation, vacuum leak detected
P007B	Charge air cooler temperature sensor circuit performance
P007C	Charge air cooler temperature sensor circuit low
P007D	Charge air cooler temperature sensor circuit high

OBD-II Diagnostic Trouble Codes (DTCs) (continued)

Note: *Not all trouble codes apply to all models.*

Code	Possible cause
P0070	Ambient temperature sensor stuck
P0071	Ambient temperature sensor performance
P0072	Ambient temperature sensor circuit, low voltage
P0073	Ambient temperature sensor circuit, high voltage
P0087	Fuel rail/system pressure - too low
P0088	Fuel rail/system pressure - too high
P00AF	Turbocharger boost control module performance
P0102	Mass Air Flow (MAF) or volume air flow circuit, low input
P0103	Mass Air Flow (MAF) or volume air flow circuit, high input
P0106	Manifold Absolute Pressure (MAP) sensor performance
P0107	Manifold Absolute Pressure (MAP) sensor circuit, voltage too low
P0108	Manifold Absolute Pressure (MAP) sensor circuit, voltage too high
P0110	Intake Air Temperature (IAT) sensor stuck
P0111	Intake Air Temperature (IAT) sensor performance
P0112	Intake Air Temperature (IAT) sensor circuit, voltage too low
P0113	Intake Air Temperature (IAT) sensor circuit, voltage too high
P0116	Engine Coolant Temperature (ECT) sensor circuit performance
P0117	Engine Coolant Temperature (ECT) sensor circuit, voltage too low
P0118	Engine Coolant Temperature (ECT) sensor circuit, voltage too high
P0121	Throttle Position (TP) sensor No. 1, performance
P0122	Throttle Position (TP) sensor circuit, voltage too low
P0123	Throttle Position (TP) sensor circuit, voltage too high
P0124	Throttle Position (TP) sensor intermittent
P0125	Coolant temperature insufficient to go into closed-loop operation
P0128	Thermostat rationality
P0129	Barometric pressure out-of-range
P013A	O2 sensor half slow response - rich to lean
P013B	O2 sensor half slow response - lean to rich
P0131	Upstream oxygen sensor (left cylinder bank), circuit voltage too low
P0132	Upstream oxygen sensor (left cylinder bank), circuit voltage too high

Code	Possible cause
P0133	Upstream oxygen sensor (left cylinder bank), slow response
P0135	Upstream oxygen sensor (left cylinder bank), heater failure
P0136	Downstream oxygen sensor (left cylinder bank), heater circuit malfunction
P0137	Downstream oxygen sensor (left cylinder bank), circuit voltage too low
P0138	Downstream oxygen sensor (left cylinder bank), circuit voltage too high
P0139	Downstream oxygen sensor (left cylinder bank), slow response
P014C	O2 sensor half slow response - rich to lean
P014D	O2 sensor half slow response - lean to rich
P0141	Downstream oxygen sensor (left cylinder bank), heater failure
P0151	Upstream oxygen sensor (right cylinder bank), heater circuit low voltage
P0152	Upstream oxygen sensor (right cylinder bank), circuit voltage too high
P0153	Upstream oxygen sensor (right cylinder bank), slow response
P0155	Upstream oxygen sensor (right cylinder bank), heater failure
P0157	Downstream oxygen sensor (right cylinder bank), circuit voltage too low
P0158	Downstream oxygen sensor (right cylinder bank), circuit voltage too high
P0159	Downstream oxygen sensor (right cylinder bank), slow response
P0160	O_2 sensor circuit - no activity detected (bank 2, sensor 2)
P0161	Downstream oxygen sensor (right cylinder bank), heater failure
P0169	Incorrect fuel composition
P0171	Fuel system lean at upstream oxygen sensor (left cylinder bank)
P0172	Fuel system rich at upstream oxygen sensor (left cylinder bank)
P0174	Fuel system lean at upstream oxygen sensor (right cylinder bank)
P0175	Fuel system rich at upstream oxygen sensor (right cylinder bank)
P0191	Fuel rail pressure sensor circuit, range or performance problem
P0192	Fuel rail pressure sensor circuit, low input
P0193	Fuel rail pressure sensor circuit, high input
P0196	Fuel rail pressure sensor circuit, range or performance problem
P0197	Fuel rail pressure sensor circuit, low input
P0198	Fuel rail pressure sensor circuit, high input
P020A	Fuel injector 1 performance
P020B	Fuel injector 2 performance
P020C	Fuel injector 3 performance

OBD-II Diagnostic Trouble Codes (DTCs) (continued)

Note: *Not all trouble codes apply to all models.*

Code	Possible cause
P020D	Fuel injector 4 performance
P020E	Fuel injector 5 performance
P020F	Fuel injector 6 performance
P0201	Fuel injector No. 1 control circuit
P0202	Fuel injector No. 2 control circuit
P0203	Fuel injector No. 3 control circuit
P0204	Fuel injector No. 4 control circuit
P0205	Fuel injector No. 5 control circuit
P0206	Fuel injector No. 6 control circuit
P0207	Fuel injector No. 7 control circuit
P0208	Fuel injector No. 8 control circuit
P0217	Engine overheating condition
P0218	High-temperature operation activated
P0219	Engine overspeed condition
P0221	Throttle Position (TP) sensor No. 2, performance
P0222	Throttle Position (TP) sensor No. 2, low voltage
P0223	Throttle Position (TP) sensor No. 2, high voltage
P0234	Engine overboost condition
P0236	Turbocharger boost sensor A circuit, range or performance problem
P0237	Turbocharger boost sensor A circuit, low
P0238	Turbocharger boost sensor A circuit, high
P0253	Injection pump fuel metering control A, low (cam/rotor/injector)
P0254	Injection pump fuel metering control A, high (cam/rotor/injector)
P0255	Injection pump fuel metering control A, intermittent (cam/rotor/injector)
P026A	Charge air cooler efficiency below limit
P0298	Engine oil over temperature
P0299	Manifold pressure sensor out of range low
P02E1	Diesel intake air flow control performance
P02E2	Diesel intake air flow control circuit low
P02E3	Diesel intake air flow control circuit high

Code	Possible cause
P02E7	Diesel intake air flow position sensor performance
P02E8	Diesel intake air flow position sensor circuit low
P02E9	Diesel intake air flow position sensor high
P0300	Multiple cylinder misfire
P0301	Cylinder No. 1 misfire
P0302	Cylinder No. 2 misfire
P0303	Cylinder No. 3 misfire
P0304	Cylinder No. 4 misfire
P0305	Cylinder No. 5 misfire
P0306	Cylinder No. 6 misfire
P0307	Cylinder No. 7 misfire
P0308	Cylinder No. 8 misfire
P0315	Crankshaft position system - variation not learned
P0315	No crank sensor learned
P0320	No crank reference signal from the CKP sensor at the PCM
P0325	Knock sensor No. 1 circuit
P0330	Knock sensor No. 2 circuit
P0335	Crankshaft Position (CKP) sensor circuit
P0336	Crankshaft Position (CKP) sensor A circuit - range or performance problem
P0337	Crankshaft Position (CKP) sensor A circuit - low input
P0338	Crankshaft Position (CKP) sensor A circuit - high input
P0339	Crankshaft Position (CKP) sensor intermittent
P0340	No cam reference signal from the CMP sensor at the PCM
P0341	Crankshaft Position (CKP) sensor "A", circuit - range or performance problem
P0344	Camshaft position (CMP) sensor intermittent
P0345	Camshaft position sensor circuit - bank 2, sensor 1
P0349	Camshaft position sensor intermittent - bank 2, sensor 1
P0344	Camshaft Position (CMP) sensor intermittent
P0351	Ignition coil No. 1 primary circuit
P0352	Ignition coil No. 2 primary circuit
P0353	Ignition coil No. 3 primary circuit
P0354	Ignition coil No. 4 primary circuit

OBD-II Diagnostic Trouble Codes (DTCs) (continued)

Note: Not all trouble codes apply to all models.

Code	Possible cause
P0355	Ignition coil No. 5 primary circuit
P0356	Ignition coil No. 6 primary circuit
P0357	Ignition coil No. 7 primary circuit
P0358	Ignition coil No. 8 primary circuit
P0365	Camshaft position sensor circuit - bank 1, sensor 2
P0369	Camshaft position sensor intermittent - bank 1, sensor 2
P0390	Camshaft position sensor circuit – bank 2, sensor 2
P0394	Camshaft position sensor intermittent - bank 2, sensor 2
P040B	Exhaust gas recirculation temperature sensor A circuit performance
P040C	Exhaust gas recirculation temperature sensor A circuit low
P040D	Exhaust gas recirculation temperature sensor circuit high
P0401	Exhaust gas recirculation, insufficient flow detected
P0403	Exhaust gas recirculation circuit malfunction
P0404	Exhaust gas recirculation circuit, range or performance problem
P0405	Exhaust gas recirculation valve position sensor A circuit low
P0406	Exhaust gas recirculation valve position sensor A circuit high
P0420	Upstream catalytic converter (left cylinder bank) efficiency
P0432	Upstream catalytic converter (right cylinder bank) efficiency
P0440	General EVAP system failure
P0441	EVAP system purge system performance
P0442	EVAP system leak monitor, medium (0.40) leak detected
P0443	EVAP system purge solenoid circuit
P0452	Natural Vacuum Leak Detection (NVLD) pressure switch stuck closed
P0453	Natural Vacuum Leak Detection (NVLD) pressure switch stuck open
P0455	EVAP system leak monitor, large leak detected
P0456	EVAP system leak monitor, small (0.20) leak detected
P0457	Evaporative emission control system leak detected (fuel cap loose/off)
P046C	Exhaust gas recirculation position sensor performance
P0460	Fuel level sending unit, no change over miles
P0461	Fuel level sending unit, no change over time
P0462	Fuel level sending unit, voltage too low

Code	Possible cause
P0463	Fuel level sending unit, voltage too high
P0471	Exhaust pressure sensor, range or performance problem
P0472	Exhaust pressure sensor, low
P0473	Exhaust pressure sensor, high
P0477	Exhaust pressure control valve, low
P0478	Exhaust pressure sensor, high
P0480	Low-speed fan control relay circuit
P0481	Cooling fan no. 2, control circuit malfunction
P0483	Cooling fan rationality check malfunction
P0489	Exhaust gas recirculation (EGR) system, circuit low
P049D	EGR control position exceeded learning limit
P0498	Natural Vacuum Leak Detection (NVLD) canister vent valve solenoid circuit low
P0499	Natural Vacuum Leak Detection (NVLD) canister vent valve solenoid circuit high
P04DB	Crankcase ventilation system disconnected
P050B	Cold start ignition timing problem
P050D	Cold start rough idle
P0500	No vehicle speed signal
P0501	Vehicle Speed Sensor (VSS) signal performance
P0503	Vehicle Speed Sensor (VSS), erratic signal
P0505	Idle air control motor circuit
P0506	Idle speed performance lower than expected
P0507	Idle speed performance higher than expected
P0508	Idle Air Control (IAC) valve signal, circuit low voltage
P0509	Idle Air Control (IAC) valve signal, circuit high voltage
P051B	Crankcase pressure circuit performance
P051C	Crankcase pressure sensor circuit low
P051D	Crankcase pressure sensor circuit high
P0513	Invalid Sentry Key Immobilizer Module (SKIM) key
P0514	Battery temperature sensor circuit range/performance
P0516	Battery temperature sensor, low voltage
P0517	Battery temperature sensor, high voltage
P0520	Engine oil pressure sensor circuit
P0521	Engine oil pressure sensor/switch circuit, range or performance problem

OBD-II Diagnostic Trouble Codes (DTCs) (continued)

Note: Not all trouble codes apply to all models.

Code	Possible cause
P0522	Oil pressure circuit, low voltage
P0524	Engine oil pressure too low
P0523	Oil pressure circuit, high voltage
P0532	Air conditioning pressure switch circuit, low voltage
P0533	Air conditioning pressure switch circuit, high voltage
P0541	Intake air heater "A" circuit low
P0542	Intake air heater "A" circuit high
P0545	Exhaust gas temperature sensor circuit low (bank 1)
P0546	Exhaust gas temperature sensor circuit high (bank 1)
P0551	Power steering switch performance
P0562	Low battery voltage
P0563	High battery voltage
P0571	Brake switch performance
P0572	Brake switch, low circuit voltage
P0573	Brake switch, high circuit voltage
P0579	Speed control switch No. 1 performance
P0580	Speed control switch No. 1, low voltage
P0581	Speed control switch No. 1, high voltage
P0582	Speed control vacuum solenoid circuit
P0585	Speed control switch No. 1 and No. 2 correlation
P0586	Speed control vent solenoid circuit
P0591	Speed control switch No. 2, performance
P0592	Speed control switch No. 2, low voltage
P0593	Speed control switch No. 2, high voltage
P0594	Speed control servo power circuit
P060B	ETC A/D ground performance
P060D	ETC level 2 APP performance
P060E	ETC level 2 TPS performance
P060F	ETC level 2 ECT performance
P0600	Serial communication link
P0601	PCM Internal controller failure (internal memory checksum invalid)
P0602	Control module, programming error

Code	Possible cause
P0604	Internal Transmission Control Module (TCM)
P0605	Internal Transmission Control Module (TCM)
P0606	PCM processor failure
P0607	Control module performance
P061A	ETC level 2 torque performance
P061C	ETC level 2 RPM performance
P0613	Internal Transmission Control Module (TCM)
P062C	ETC level 2 MPH performance
P0622	Alternator field control circuit not switching correctly
P0627	Fuel pump relay circuit
P063A	Alternator voltage sensor performance
P063C	Alternator voltage sense low
P063D	Alternator voltage sense high
P0630	Vehicle Identification Number (VIN) not programmed into PCM
P0632	Odometer not programmed into PCM
P0633	Sentry Key Immobilizer Module (SKIM) not programmed into PCM
P064C	Glow plug module internal performance
P064D	O2 sensor module internal performance
P0642	Primary 5-volt supply low
P0643	Primary 5-volt supply high
P0645	Air conditioning clutch relay circuit
P0646	A/C clutch relay control circuit low
P0647	A/C clutch relay control circuit high
P0652	Auxiliary 5-volt supply low
P0653	Auxiliary 5-volt supply high
P0658	Actuator supply voltage - circuit low
P0685	Automatic Shut Down (ASD) relay control circuit
P0686	ECM power relay control - circuit low
P0688	Automatic Shut Down (ASD) sense circuit, low voltage
P0691	Engine coolant blower motor 1 - short to ground
P0692	Engine coolant blower motor 1 - short to positive
P0698	Sensor reference voltage C - circuit low
P0699	Sensor reference voltage C - circuit high

OBD-II Diagnostic Trouble Codes (DTCs) (continued)

Note: *Not all trouble codes apply to all models.*

Code	Possible cause
P06A4	Sensor voltage reference 4 circuit low
P06A5	Sensor voltage reference 4 circuit high
P0700	Transmission control system (MIL request)
P0703	Torque converter/brake switch B, circuit malfunction
P0706	Check shifter signal
P0711	Transmission temperature sensor performance, no temperature increase after start-up
P0712	Transmission temperature sensor, low voltage
P0713	Transmission temperature sensor, high voltage
P0714	Transmission temperature sensor, intermittent voltage
P0715	Input Speed Sensor (ISS) error
P0720	Output Speed Sensor (OSS) error, low rpm above 15 mph
P0725	Engine speed sensor circuit
P0729	Gear ratio error in sixth gear
P0731	Gear ratio error in first gear
P0732	Gear ratio error in second gear
P0733	Gear ratio error in third gear
P0734	Gear ratio error in fourth gear
P0735	Gear ratio error fifth gear
P0736	Gear ratio error in REVERSE
P0740	Torque Converter Clutch (TCC) control circuit
P0743	Torque Converter Clutch (TCC) solenoid/transmission relay circuits
P0748	Pressure SOL control/transmission relay circuits
P0750	LR solenoid circuit
P0751	Overdrive switch pressed LO for more than 5 minutes
P0753	Transmission 3-4 shift SOL/transmission relay circuits
P0755	2C solenoid circuit
P0760	OD solenoid circuit
P0765	UD solenoid circuit
P0770	4C solenoid circuit
P0836	4WD MUX switch stuck
P0837	4WD MUX switch performance

Code	Possible cause
P0838	4WD mode sensor circuit, low voltage
P0839	4WD mode sensor circuit, high voltage
P0841	LR pressure switch sense circuit
P0845	2C hydraulic pressure test failure
P0846	2C hydraulic pressure switch sense circuit
P0850	Transmission Range (TR) sensor performance
P0868	Line pressure low
P0869	Line pressure high
P0870	Overdrive hydraulic pressure test failure
P0871	Overdrive pressure switch sense circuit
P0875	UD hydraulic pressure test failure
P0876	UD hydraulic switch sense circuit
P0884	Power up at speed
P0888	Relay output always off
P0890	Switched battery
P0891	Transmission relay always on
P0932	Line pressure sensor circuit fault
P0934	Line pressure sensor circuit, low voltage
P0935	Line pressure sensor circuit, high voltage
P0944	Loss of prime
P0987	4C hydraulic pressure test failure
P0988	4C pressure switch sense circuit
P1239	Engine oil temperature too low
P1521	Incorrect engine oil type
P1715	Restricted manual valve in T3 range
P2263	Turbo boost performance
P2302	Ignition coil 1 secondary circuit—insufficient ionization
P2305	Ignition coil 2 secondary circuit—insufficient ionization
P2308	Ignition coil 3 secondary circuit—insufficient ionization
P2314	Ignition coil 4 secondary circuit—insufficient ionization
P2700	Inadequate element volume, low/reverse clutch

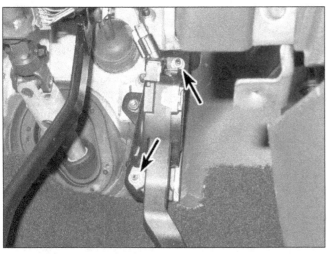

3.2 Accelerator Pedal Position (APP) sensor mounting nuts

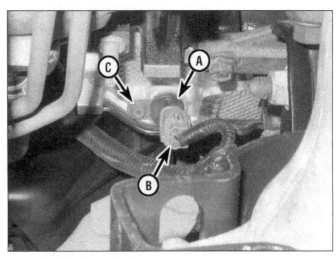

4.7a Disconnect the CMP sensor (A) electrical connector (B) and remove the sensor mounting bolt (C) (as seen from underneath the engine) (3.7L V6 engine)

3   Accelerator Pedal Position (APP) sensor - replacement

Refer to illustration 3.2

1   Disconnect the wiring from the sensor.
2   Remove the two mounting nuts and remove the entire pedal/sensor assembly **(see illustration). Note:** *Don't try to disassemble it. It's replaced as a complete assembly.*
3   Installation is the reverse of removal.

4   Camshaft Position (CMP) sensor - replacement

1   Disconnect the cable(s) from the negative battery terminal(s) (see Chapter 5).

3.6L V6 engine

Note: There are two CMP sensors, located at the rear of the valve covers and are bolted through to the cylinder head.
2   Remove the engine cover, if equipped.
3   Disconnect the electrical connector from the sensor(s), then remove the sensor(s) mounting bolt.
4   Remove the sensor(s) from the valve cover. Inspect the O-ring; if it is not damaged it can be used again.
5   Installation is the reverse of removal. Tighten the sensor mounting bolt(s) to the torque listed in this Chapter's Specifications

3.7L V6 and 4.7L V8 engines

Refer to illustrations 4.7a and 4.7b

Note: *The CMP sensor is located on the top of the front end of the right cylinder head.*
6   Raise the front end of the vehicle and place it securely on jackstands.

7   Disconnect the electrical connector from the CMP sensor **(see illustrations)**.
8   Remove the sensor mounting bolt and remove the CMP sensor.
9   Inspect the CMP sensor O-ring for cracks, tears and other deterioration. If it's damaged, replace it.
10   When installing the CMP sensor, apply a small dab of clean engine oil to the sensor O-ring, then use a slight rocking motion to work the O-ring into the sensor mounting bore. Do NOT use a twisting motion or you will damage the O-ring.
11   Make sure that the CMP sensor mounting flange is fully seated flat against the mounting surface around the sensor mounting hole. **Caution:** *If the CMP sensor is not fully seated against its mounting surface, the sensor mounting tang will be damaged when the sensor mounting bolt is tightened to the specified torque.*

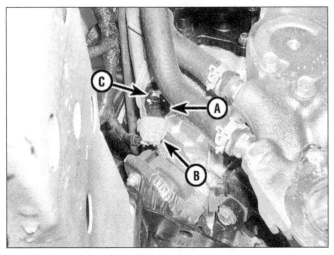

4.7b Disconnect the CMP sensor (A) electrical connector (B) and remove the sensor mounting bolt (C) (as seen from underneath the engine) (4.7L V8 engine)

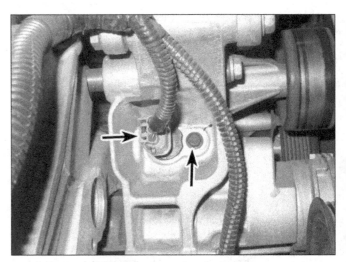

4.13 Disconnect the CMP sensor electrical connector and remove the sensor mounting bolt (Hemi engine)

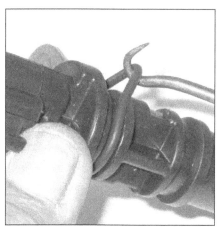

4.15 Remove the old CMP sensor O-ring and inspect it for cracks, tears and deterioration. If it's damaged, replace it

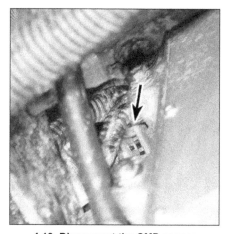

4.19 Disconnect the CMP sensor electrical connector, remove the sensor mounting bolt and pull the sensor out of the timing cover (diesel engines)

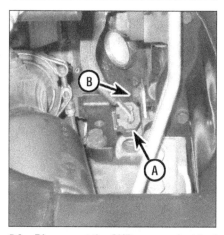

5.3a Disconnect the CKP sensor electrical connector (A) and remove the sensor mounting bolt (B) (3.7L V6 engine)

12   Installation is otherwise the reverse of removal. Tighten the CMP sensor mounting bolt to the torque listed in this Chapter's Specifications.

Hemi engines

Refer to illustrations 4.13 and 4.15

Note: *The CMP sensor is located on the upper right side of the timing chain cover, under the alternator.*

13   Disconnect the electrical connector from the CMP sensor **(see illustration)**.

14   Remove the CMP sensor mounting bolt and remove the CMP sensor from the timing chain cover.

15   Remove the old CMP sensor O-ring **(see illustration)** and inspect it for cracks, tears and deterioration. If the old O-ring is damaged, replace it.

16   When installing the CMP sensor, apply a small dab of clean engine oil to the sensor O-ring, then use a slight rocking motion to work the O-ring into the sensor mounting

bore. Do NOT use a twisting motion or you will damage the O-ring.

17   Make sure that the CMP sensor mounting flange is fully seated flat against the mounting surface around the sensor mounting hole. **Caution:** *If the CMP sensor is not fully seated against its mounting surface, the sensor mounting tang will be damaged when the sensor mounting bolt is tightened to the specified torque.*

18   Installation is otherwise the reverse of removal. Tighten the CMP sensor mounting bolt to the torque listed in this Chapter's Specifications.

Diesel engine

Refer to illustration 4.19

Note: *The CMP sensor is located on the left rear of the timing gear cover, below the injection pump.*

19   Disconnect the electrical connector from the CMP sensor **(see illustration)**. **Note:** *The CMP sensor is very difficult to reach. You*

might want to unbolt the power steering pump mounting bracket for better access.

20   Remove the CMP sensor mounting bolt and remove the CMP sensor from its mounting hole.

21   Remove the CMP sensor O-ring and inspect it for cracks, tears and deterioration. If it's damaged, replace it.

22   Clean out the CMP sensor mounting hole.

23   Apply a small dab of clean engine oil on the CMP sensor O-ring, then install the sensor using a slight rocking (side-to-side) motion. Do NOT twist the sensor into position or you will damage the O-ring.

24   The remainder of installation is the reverse of removal.

---

5   Crankshaft Position (CKP) sensor - replacement

---

1   Disconnect the cable(s) from the negative battery terminal(s) (see Chapter 5).

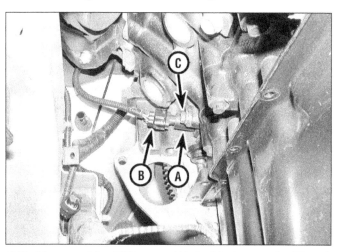

5.3b Disconnect the CKP sensor (A) electrical connector (B) and remove the sensor mounting bolt (C) (4.7L V8 engine)

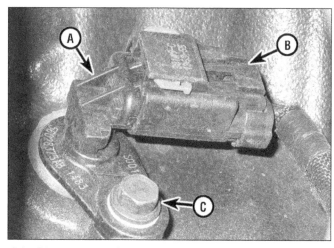

5.3c Disconnect the CKP sensor (A) electrical connector (B) and remove the sensor mounting bolt (C) (Hemi engine, right-rear side shown)

5.8 Disconnect the CKP sensor electrical connector and remove the sensor mounting bolt (diesel engine)

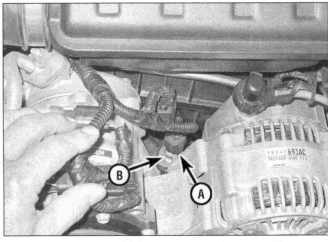

6.3a Disconnect the ECT sensor electrical connector (A) and unscrew the sensor (B) from the manifold (3.7L V6 engine)

2   Raise the front of the vehicle and support it securely on jackstands.

Gasoline engines

Refer to illustrations 5.3a, 5.3b and 5.3c

Note: *The CKP sensor is located on the right rear side of the engine block.*

3   Disconnect the electrical connector from the CKP sensor **(see illustrations)**.

4   Remove the CKP sensor mounting bolt. Remove the sensor from the engine.

5   Remove the CKP sensor O-ring and inspect it for cracks, tears and deterioration. If it's damaged, replace it.

6   Apply a small amount of engine oil on the O-ring and, using a slight rocking motion, push the sensor into its mounting hole in the engine block until the sensor is fully seated.

7   Installation is otherwise the reverse of removal.

Diesel engine

Refer to illustration 5.8

Note: *The CKP sensor is located on the front left side of the engine, next to the harmonic balancer.*

8   Disconnect the electrical connector from the CKP sensor **(see illustration)**.

9   Remove the CKP sensor mounting bolt and remove the CKP sensor.

10   Remove and inspect the old sensor O-ring. If it's cracked, torn or deteriorated, replace it.

11   Before installing the CKP sensor, clean out the sensor's mounting hole and apply a little clean engine oil to the sensor O-ring.

12   Install the CKP sensor using a slight rocking motion. Make sure that the sensor is flush with its mounting surface before installing and tightening the sensor mounting bolt.

13   Installation is otherwise the reverse of removal.

---

6   Engine Coolant Temperature (ECT) sensor - replacement

Warning: *Wait until the engine is completely cool before beginning this procedure.*

1   Partially drain the cooling system (see Chapter 1).

2   Disconnect the cable(s) from the negative battery terminal(s) (see Chapter 5).

36L V6, 3.7L V6 and 4.7L V8 engines

Refer to illustrations 6.3a, 6.3b and 6.5

Note: *On the 3.6L V6 engine, the ECT sensor is located in the water jacket at rear of the cylinder head on the left side of the engine. On 3.7L V6 and 4.7L V8 engines, the sensor is located at the front of the intake manifold.*

3   Disconnect the electrical connector from the ECT sensor **(see illustrations)**.

4   Using a deep socket, carefully unscrew the ECT sensor from the cylinder head or the intake manifold.

5   To prevent leakage and thread corrosion, wrap the threads of the ECT sensor with Teflon sealing tape **(see illustration)** before installing the sensor. **Note:** *Seal the sensor threads whether you're installing the old sensor or a new unit.*

6   Installation is otherwise the reverse of removal. Tighten the ECT sensor to the torque listed in this Chapter's Specifications.

7   Refill the cooling system (see Chapter 1).

6.3b Disconnect the ECT sensor electrical connector and unscrew the sensor from the manifold (4.7L V8 engine)

6.5 Before installing the ECT sensor, wrap the threads of the sensor with Teflon tape to prevent leaks

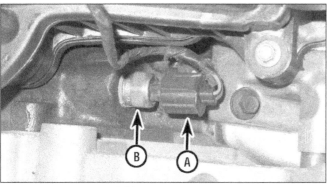

6.10 Disconnect the ECT sensor electrical connector (A) and unscrew the sensor (B) from the manifold (Hemi engine)

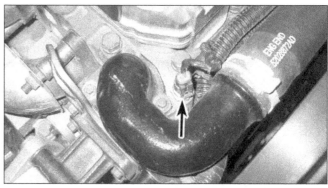

6.17 ECT sensor location - diesel engine

Hemi engines

Refer to illustration 6.10

Note: *The ECT sensor is located at the front of the engine block in the water pump assembly.*

8   Remove the drivebelt (see Chapter 1).

9   To access the ECT sensor, you'll need to unbolt the air conditioning compressor from the engine (see Chapter 3). It is NOT necessary to disconnect any air conditioning hoses from the compressor. Set the compressor aside and support it with a bungee cord or a piece of wire.

10   Disconnect the electrical connector from the ECT sensor **(see illustration)**.

11   Using a deep socket, carefully unscrew the ECT sensor from the intake manifold.

12   To prevent leakage and thread corrosion, wrap the threads of the ECT sensor with Teflon sealing tape **(see illustration 6.5)** before installing the sensor. **Note:** *Seal the sensor threads whether you're installing the old sensor or a new unit.*

13   Installation is otherwise the reverse of removal. Tighten the ECT sensor to the torque listed in this Chapter's Specifications.

14   Refill the cooling system (see Chapter 1).

Diesel engine

Refer to illustration 6.17

Note: *The ECT sensor is located at the right front corner of the cylinder head, near the*

thermostat housing.

15   Remove the interfering upper EGR tube.

16   Remove the heat shield.

17   Disconnect the electrical connector from the ECT sensor **(see illustration)**.

18   Unscrew the ECT sensor from the cylinder head.

19   To prevent leakage and thread corrosion, wrap the threads of the ECT sensor with Teflon sealing tape **(see illustration 6.5)** before installing the sensor. **Note:** *Seal the sensor threads whether you're installing the old sensor or a new one.*

20   Installation is otherwise the reverse of removal. Tighten the ECT sensor to the torque listed in this Chapter's Specifications.

21   Refill the cooling system (see Chapter 1).

---

7   Input Shaft Speed (ISS) and Output Shaft Speed (OSS) sensors - replacement

---

Refer to illustrations 7.3 and 7.4

Note: *On all models except 8-speed transmissions, the ISS and OSS sensors are located on the left side of the transmission. On 8-speed transmissions, the ISS and OSS sensors are located inside the transmission and are not serviceable by the home mechanic.*

1   Disconnect the cable(s) from the nega-

tive battery terminal(s) (see Chapter 5).

2   Raise the vehicle and place it securely on jackstands.

3   The ISS and OSS sensors **(see illustration)** are identical in appearance (and are replaced exactly the same way), so make sure that you've correctly identified the sensor that you wish to replace.

4   Disconnect the electrical connector from the ISS or OSS sensor **(see illustration)**.

5   Place a drain pan underneath the sensor. Remove the sensor mounting bolt and pull out the sensor.

6   Installation is the reverse of removal.

7   Check the transmission fluid level and add fluid as necessary (see Chapter 1).

---

8   Intake Air Temperature (IAT) sensor - replacement

---

1   Disconnect the cable(s) from the negative battery terminal(s) (see Chapter 5).

Gasoline engines

Refer to illustrations 8.2a, 8.2b, 8.2c, 8.3 and 8.4

Note: *On 3.7L V6 and 4.7L V8 engines, the IAT sensor is slightly upstream of the throttle body in the air inlet duct. On Hemi engines, the IAT sensor is in the front of the intake manifold air box plenum.*

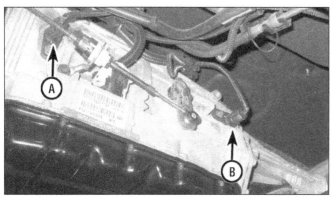

7.3 The ISS sensor (A) and OSS sensor (B) are located on the left side of 45RFE and 545RFE automatic transmissions

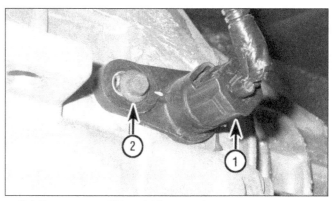

7.4 To remove an ISS or OSS sensor from the transmission, disconnect the electrical connector (1), then remove the sensor mounting bolt (2)

8.2a On 3.7L V6 and 4.7L V8 engines, the IAT sensor is mounted in the air inlet duct

8.2b On Hemi engines, the IAT sensor is located at the right front corner of the air resonator box

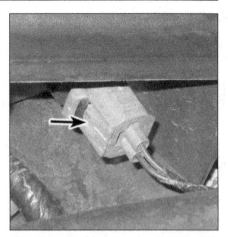

8.2c Depress this tab and pull off the IAT sensor connector (Hemi engine shown, 3.7L V6 and 4.7L V8 similar)

8.3 Lift up the release tab slightly and rotate the IAT sensor counterclockwise 1/4-turn and pull it out (Hemi engine shown, 3.7L V6 and 4.7L V8 similar)

2   Disconnect the electrical connector from the IAT sensor **(see illustrations)**.
3   Lift up the release tab **(see illustration)** slightly, then turn the sensor 1/4-turn counterclockwise and pull it out.
4   Remove the old O-ring from the IAT sensor **(see illustration)** and inspect it for cracks, tears and deterioration. If the O-ring is damaged, replace it.
5   To install the IAT sensor, insert it into its mounting hole and rotate it clockwise 1/4-turn. Make sure that the release tab locks the sensor into place. Installation is otherwise the reverse of removal.

Diesel engine

Refer to illustration 8.6

Note: *This sensor is a dual-function unit that serves as the IAT sensor and as the Manifold Absolute Pressure (MAP) sensor. It monitors the air temperature and the turbocharger boost pressure inside the intake manifold. It's located on top of the intake manifold, below and behind the EGR valve.*

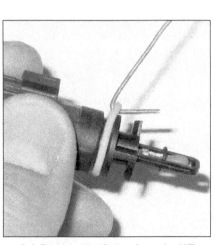

8.4 Remove the O-ring from the IAT sensor and inspect it for cracks, tears and deterioration. If the O-ring is damaged, replace it

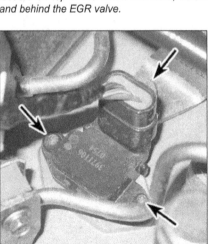

8.6 Disconnect the IAT/MAP sensor electrical connector and remove the two Torx mounting screws (diesel engine)

6   Disconnect the electrical connector from the IAT/MAP sensor **(see illustration)**.
7   Remove the two sensor mounting screws.
8   Remove the IAT/MAP sensor from the intake manifold.
9   Remove the sensor O-ring and inspect it for cracks **(see illustration 8.4)**, tears and other deterioration. If it's damaged, replace it.
10   Installation is the reverse of removal. Tighten the mounting screws securely.

---

9   Knock sensor(s) - replacement

---

1   Disconnect the cable(s) from the negative battery terminal(s) (see Chapter 5).

3.6L V6 engine

Note: There are two knock sensors, both located below the lower intake manifold area. The sensor closest to the front of the engine is Sensor 1 and the other is Sensor 2.
2   Remove the intake manifolds (see Chapter 2D).
3   Disconnect the electrical connector(s), then remove the sensor(s) mounting bolt(s).
4   Remove the sensor(s) from the valley between the cylinder banks.
5   Installation is the reverse of removal. Tighten the knock sensor mounting bolt(s) to the torque listed in this Chapter's Specifications.

3.7L V6 and 4.7L V8 engines

6   Remove the intake manifold (see Chapter 2A).
7   Disconnect the knock sensor electrical connector, which is located at the rear of the engine, behind the intake manifold. **Note:** *If you can't find the knock sensor electrical connector, wait until you have removed the intake manifold. Then locate the two knock sensors and trace their leads back to this common connector.*
8   Remove the knock sensor mounting

9.12 Disconnect the knock sensor electrical connector (A) and remove the knock sensor mounting bolt (B) (Hemi engine)

10.2 On 4.7L V8 engines, the MAP sensor is located on the front of the intake manifold (3.7L V6 similar)

10.6 Slide the release lock (1) away from the MAP sensor (toward the harness), then depress the release tab (2) and pull off the connector (Hemi engine)

bolt(s) and remove the knock sensor(s). **Note:** *Any foam strips on the knock sensor mounting bolt threads are used to retain the bolts during vehicle assembly at the manufacturing plant. They have no other purpose. They are not some form of adhesive, thread sealant or locking compound. Do NOT use any type of adhesive, thread sealant or locking compound when installing these bolts again.*

9   Make sure that the holes for the knock sensor mounting bolts are thoroughly cleaned before installing the bolts. **Note:** *On 3.7L V6 engines, the left knock sensor is identified by an identification tag (LEFT); it also has a larger mounting bolt. Do NOT switch the knock sensors - do not install the left knock sensor in the right sensor mounting position and vice versa. The PCM assumes that the knock sensors are installed in their correct locations. Switching the sensor locations will confuse the PCM and cause it to set a Diagnostic Trouble Code.*

10   Installation is otherwise the reverse of removal. Tighten the knock sensor mounting bolts to the torque listed in this Chapter's Specifications. **Caution:** *Over- or under-tightening the knock sensor mounting bolts will affect knock sensor performance, which might affect the PCM's spark control ability.*

Hemi engines

Refer to illustration 9.12

Note: *There are two knock sensors. One is located on the left side of the block, below the exhaust manifold, and the other is located in the same place on the right side of the block.*

11   Raise the front of the vehicle and place it securely on jackstands.

12   Disconnect the electrical connector from the knock sensor **(see illustration)**.

13   Remove the knock sensor mounting bolt and remove the knock sensor from the engine. **Note:** *Any foam strips on the knock sensor mounting bolt threads are used to retain the bolts during vehicle assembly at the manufacturing plant. They have no other purpose. They are not some form of adhesive, thread sealant or locking compound. Do NOT use any type of*

adhesive, thread sealant or locking compound when installing these bolts again.

14   Installation is the reverse of removal. Tighten the knock sensor mounting bolts to the torque listed in this Chapter's Specifications. **Caution:** *Over- or under-tightening the knock sensor mounting bolts will affect knock sensor performance, which might affect the PCM's spark control ability.*

---

10   Manifold Absolute Pressure (MAP) sensor - replacement

1   Disconnect the cable(s) from the negative battery terminal(s) (see Chapter 5).

3.6L V6, 3.7L V6 and 4.7L V8 engines

Refer to illustration 10.2

Note: On the 3.6L V6 engine, the MAP sensor is located on the top of the intake manifold. On 3.7L V6 and 4.7L V8 engines, the sensor is located in the front of the intake manifold.

2   Disconnect the electrical connector from the MAP sensor **(see illustration)**.

3   Remove the MAP sensor mounting screw and remove the MAP sensor from the intake manifold.

4   Remove the MAP sensor O-ring and inspect it for cracks, tears and deterioration. If the O-ring is damaged, replace it.

5   Installation is the reverse of removal.

Hemi engines

Refer to illustrations 10.6 and 10.7

Note: *The MAP sensor is located on the front of the intake manifold air plenum box.*

6   Disconnect the electrical connector from the MAP sensor **(see illustration)**.

7   Remove the MAP sensor from the intake manifold **(see illustration)**.

8   Remove the old MAP sensor O-ring and inspect it for cracks, tears and deterioration. If it's damaged, replace it.

10.7 Rotate the MAP sensor 1/4-turn counterclockwise and pull it out of its mounting hole. Remove and inspect the MAP sensor O-ring

9   To install the MAP sensor, place it in position, insert it into the mounting hole in the intake manifold and turn it 1/4-turn clockwise.

10   Installation is otherwise the reverse of removal.

Diesel engine

Note: *The MAP sensor is located on top of the intake manifold.*

11   The MAP sensor is one-half of the dual-function Intake Air Temperature/Manifold Absolute Pressure (IAT/MAP) sensor. The replacement procedure is in Section 8.

---

11   Oxygen sensors - general description and replacement

General description

1   The oxygen in the exhaust reacts with the elements inside the oxygen sensor to produce a voltage output that varies from 0.1 volt (high oxygen, lean mixture) to 0.9 volt (low oxygen, rich mixture). The pre-converter oxygen sensor (mounted in the exhaust system before the catalytic converter) provides a

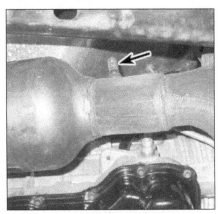

11.8a The upstream oxygen sensor is located ahead of the catalytic converter. To find the electrical connector, trace the electrical lead from the sensor to the connector and disconnect it (Hemi shown, others similar)

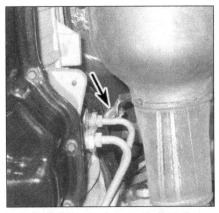

11.8b The downstream oxygen sensor (not visible in this photo, but shown in the next photo) is located behind the catalyst. First, disconnect the electrical connector, then trace the lead up to the sensor

11.9 Remove an oxygen sensor with a wrench if you can (this is the downstream sensor). If there's not enough room for a wrench you might have to use an oxygen sensor socket (available at most auto parts stores)

feedback signal to the PCM that indicates the amount of leftover oxygen in the exhaust. The PCM monitors this variable voltage continuously to determine the required fuel injector pulse width and to control the engine air/fuel ratio. Based on oxygen sensor signals, the PCM tries to maintain the ideal air/fuel ratio at all times.

2   The post-converter oxygen sensor (mounted in the exhaust system after the catalytic converter) has no effect on PCM control of the air/fuel ratio. However, the post-converter sensor is identical to the pre-converter sensor and operates in the same way. The PCM uses the post-converter signal to monitor the efficiency of the catalytic converter. A post-converter oxygen sensor will produce a slower fluctuating voltage signal that reflects the lower oxygen content in the post-catalyst exhaust.

3   Oxygen sensor configuration varies depending on the model and on where it is sold; Federal (49-State) model or California model.

4   An oxygen sensor produces no voltage when it is below its normal operating temperature of about 600-degrees F. During this warm-up period, the PCM operates in an open-loop fuel control mode. It does not use the oxygen sensor signal as a feedback indication of residual oxygen in the exhaust. Instead, the PCM controls fuel metering based on the inputs of other sensors and its own programs.

5   An oxygen sensor depends on four conditions in order to operate correctly:

a) *Electrical* - *The low voltage generated by the sensor requires good, clean connections. Always check the connectors whenever an oxygen sensor problem is suspected or indicated.*

b) *Outside air supply* - *The sensor needs air circulation to the internal portion of the sensor. Whenever the sensor is installed, make sure that the air passages are not restricted.*

c) *Correct operating temperature* - *The PCM will not react to the sensor signal until the sensor reaches approximately 600-degrees F. This factor must be considered when evaluating the performance of the sensor.*

d) *Unleaded fuel* - *Unleaded fuel is essential for correct sensor operation.*

6   The PCM can detect several different oxygen sensor problems and set Diagnostic Trouble Codes (DTCs) to indicate the specific fault (see Section 2). When an oxygen sensor DTC occurs, the PCM disregards the oxygen sensor signal voltage and reverts to open-loop fuel control as described previously.

Replacement

Refer to illustrations 11.8a, 11.8b and 11.9

Warning: *Be careful not to burn yourself during the following procedure.*

Note: *Since the exhaust pipe contracts when cool, the oxygen sensor may be hard to loosen. To make sensor removal easier, start the engine and let it run for a minute or two, then turn it off.*

7   Raise the vehicle and place it securely on jackstands.

8   Locate the upstream or downstream oxygen sensor **(see illustrations)**, trace the sensor's electrical lead to the sensor electrical connector and disconnect it.

9   Remove the upstream or downstream oxygen sensor. On some models you can remove the sensor with a wrench **(see illustration)**. On others, you will have to use an oxygen sensor socket (available at most auto parts stores).

10   Clean the threads inside the sensor mounting hole in the exhaust pipe with an appropriate tap.

11   If you're installing the old sensor, clean off the threads, then apply a coat of anti-seize compound to the threads before installing the sensor. If you're installing a new sensor, do NOT apply anti-seize compound; new sen-

sors are already coated with anti-seize.

12   Installation is otherwise the reverse of removal. Tighten the oxygen sensor to the torque listed in this Chapter's Specifications.

---

12   Powertrain Control Module (PCM) - removal and installation

---

Caution: *Avoid static electricity damage to the Powertrain Control Module (PCM) by grounding yourself to the body of the vehicle before touching the PCM, and using a special anti-static pad on which to store the PCM once it's removed.*

Note: *Any time the PCM is replaced with a new unit, it must be reprogrammed with a scan tool by a dealership service department or other qualified repair shop.*

Note: *Any time the battery is disconnected, stored operating parameters may be lost from the PCM, causing the engine to run rough for a period of time while the PCM relearns the information.*

Gasoline engines

Refer to illustration 12.3

1   The PCM is located in the right rear corner of the engine compartment on the firewall.

2   Disconnect the cable from the negative terminal of the battery (see Chapter 5).

3   Carefully disconnect the large PCM electrical connectors **(see illustration)**.

4   Remove the PCM mounting bolts and remove the PCM.

5   Installation is the reverse of removal.

Diesel engine

Refer to illustration 12.8

6   Disconnect the cables from the negative battery terminals (see Chapter 5). **Note:** *Wait 10 minutes (automatic transmission vehicles) or 75 seconds (manual transmission vehicles) after turning the key Off before*

12.3 To disconnect each of the electrical connectors from the Powertrain Control Module (PCM), release the orange lock, then depress the release tab and pull off the connector

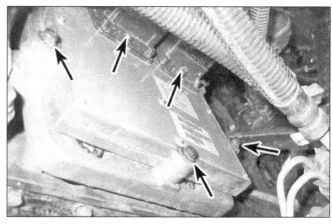

12.8 Disconnect the PCM electrical connectors and remove the mounting bolts (some models use three mounting bolts, while others use five) (diesel engine)

disconnecting the battery. Failure to wait for the PCM to fully power down may result in a diagnostic trouble code (DTC).

7   The PCM is located on the left side of the engine block. For access, raise the front of the vehicle and support it securely on jackstands.

8   Disconnect the electrical connectors from the PCM **(see illustration)**.

9   Remove the mounting bolts and remove the PCM from the engine.

10   Installation is the reverse of removal.

---

13   Catalytic converter - general description, check and replacement

---

Note: *Because of a Federally-mandated extended warranty which covers emission-related components such as the catalytic converter, check with a dealer service department before replacing the converter at your own expense.*

General description

1   A catalytic converter (or catalyst) is an emission control device in the exhaust system that reduces certain pollutants in the exhaust gas stream. There are two types of converters. An oxidation catalyst reduces hydrocarbons (HC) and carbon monoxide (CO). A reduction catalyst reduces oxides of nitrogen (NOx). A catalyst that can reduce *all three pollutants* is known as a "Three-Way Catalyst" (TWC). All models covered by this manual are equipped with TWCs.

Check

2   The test equipment for a catalytic converter (a loaded-mode dynamometer and a 5-gas analyzer) is expensive. If you suspect that the converter on your vehicle is malfunctioning, take it to a dealer or authorized emission inspection facility for diagnosis and repair.

3   Whenever you raise the vehicle to service underbody components, inspect the con-

verter for leaks, corrosion, dents and other damage. Carefully inspect the welds and/or flange bolts and nuts that attach the front and rear ends of the converter to the exhaust system. If you note any damage, replace the converter.

4   Although catalytic converters don't break too often, they can become clogged or even plugged up. The easiest way to check for a restricted converter is to use a vacuum gauge to diagnose the effect of a blocked exhaust on intake vacuum.

a) *Connect a vacuum gauge to an intake manifold vacuum source (see Chapter 2).*

b) *Warm the engine to operating temperature, place the transaxle in Park (automatic models) or Neutral (manual models) and apply the parking brake.*

c) *Note the vacuum reading at idle and jot it down.*

d) *Quickly open the throttle to near its wide-open position and then quickly get off the throttle and allow it to close. Note the vacuum reading and jot it down.*

e) *Do this test three more times, recording your measurement after each test.*

f) *If your fourth reading is more than one in-Hg lower than the reading that you noted at idle, the exhaust system might be restricted (the catalytic converter could be plugged, OR an exhaust pipe or muffler could be restricted).*

Replacement

Refer to illustrations 13.6a and 13.6b

Warning: *Make sure that the exhaust system is completely cooled down before proceeding. If the vehicle has just been driven, the catalytic converter can be hot enough to cause serious burns.*

5   Raise the vehicle and place it securely on jackstands.

6   Spray a liberal amount of penetrant onto the threads of the exhaust pipe-to-exhaust manifold bolts and the clamp bolt behind the catalytic converter **(see illustrations)** and

13.6a To disconnect the exhaust pipe ahead of the catalyst from the exhaust manifold, remove these two bolts. If they're difficult to loosen, spray some penetrant onto the threads, wait awhile and try again. Discard these bolts and use new ones when reconnecting the pipe to the manifold

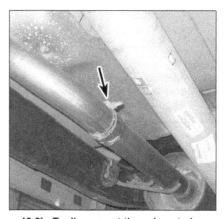

13.6b To disconnect the exhaust pipe behind the catalyst from the exhaust pipe going to the muffler, back off this nut to loosen the clamp. If it's hard to loosen, spray some penetrant onto the threads and wait awhile. If the clamp is in bad shape, replace it with a new one

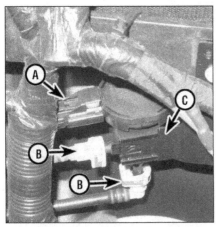

14.9 Disconnect the EVAP canister purge solenoid electrical connector (A), disconnect the vacuum hoses (B) and disengage the solenoid from its mounting bracket (C)

wait awhile for the penetrant to loosen things up.

7   While you're waiting for the penetrant to do its work, disconnect the electrical connectors for the upstream and downstream oxygen sensor and remove both oxygen sensors (see Section 11).

8   Unscrew the upper exhaust pipe-to-exhaust manifold flange bolts. If they're still difficult to loosen, spray the threads with some more penetrant, wait awhile and try again.

9   To loosen the clamp that secures the slip joint between the exhaust pipe behind the catalytic converter and the pipe ahead of the muffler, back off the nut. If it's still difficult to loosen, spray the threads with some more penetrant, wait awhile and try again.

10   Remove the catalytic converter assembly. Remove and discard the old flange gasket.

11   Installation is the reverse of removal. Use a new flange gasket at the exhaust manifold mounting flange. Use new bolts at the front flange. Although the slip joint clamp doesn't get as overheated as the exhaust manifold flange bolts, it's still a good idea to use a new clamp. Coat the threads of the clamp and the exhaust manifold bolts with anti-seize compound to facilitate future removal. Tighten the fasteners securely.

---

14   Evaporative emissions control (EVAP) system - general description and component replacement

---

General description

1   The **Evaporative emissions control (EVAP) system** prevents fuel system vapors (which contain unburned hydrocarbons) from escaping into the atmosphere. On warm days, vapors trapped inside the fuel tank expand until the pressure reaches a certain threshold,

at which point the fuel vapors are routed from the fuel tank through the fuel vapor vent valve and the fuel vapor control valve to the EVAP canister, where they're stored temporarily, until they can be consumed by the engine during normal operation. When the conditions are right (engine warmed up, vehicle up to speed, moderate or heavy load on the engine, etc.) the Powertrain Control Module (PCM) opens the canister purge solenoid, which allows the fuel vapors to be drawn from the canister into the intake manifold, where they mix with the air/fuel mixture before being consumed in the combustion chambers. This system is complex and virtually impossible to troubleshoot without the right tools and training. However, the following description should give you a good idea of how the system works and where the components are located:

2   The **EVAP canister**, which contain activated charcoal, is the container for storing the fuel vapors. Some models use two canisters. You'll have to raise the vehicle to inspect or replace the canister but the canister is designed to be maintenance-free and should last the life of the vehicle. The EVAP canister is underneath the vehicle, just in front of the fuel tank.

3   The **Evaporative System Integrity Monitor (ESIM) switch** is mounted on one of the EVAP canisters in a vertical position. It can't be replaced by itself; you'll have to replace the canister assembly if it's faulty.

4   The **EVAP canister purge solenoid**, which is under the control of the Powertrain Control Module (PCM), regulates the flow of vapors being purged from the EVAP canister into the intake manifold. The canister purge solenoid is normally closed. It opens only when directed to do so by the PCM, which uses the availability of intake manifold vacuum and data from various information sensor inputs to determine when and how long to open the valve. The interval of time during which the purge valve is opened by the PCM is known as its duty cycle. The canister purge solenoid valve is located in the engine compartment, where it is attached to the side of the fuse and relay box.

General system checks

5   The most common symptom of a faulty EVAP system is a strong fuel odor (particularly during hot weather). If you smell fuel while driving or (more likely) right after you park the vehicle and turn off the engine, check the fuel filler cap first. Make sure that it's screwed onto the fuel filler neck all the way. If the odor persists, inspect all EVAP hose connections, both in the engine compartment and under the vehicle. You'll have to raise the vehicle and place it securely on jackstands to inspect most of the EVAP system, since it's located under the vehicle. Inspect each hose attached to the canister for damage and leakage along its entire length. Repair or replace as necessary. Inspect the canister for damage and look for fuel leaking from the bottom. If fuel is leaking or the canister is otherwise damaged, replace it.

6   Poor idle, stalling, and poor drivability

can be caused by a defective fuel vapor vent valve or canister purge solenoid, a damaged canister, cracked hoses, or hoses connected to the wrong tubes. Fuel loss or fuel odor can be caused by fuel leaking from fuel lines or hoses, a cracked or damaged canister, or a defective vapor valve.

7   To check for excessive fuel vapor pressure in the fuel tank, remove the gas cap and listen for the sound of pressure release. If the fuel tank emits a whooshing sound when you open the filler cap, fuel tank vapor pressure is excessive. Inspect the canister vapor hoses and the canister inlet port for blockage or collapsed hoses. Also inspect the hose for the EVAP canister vent valve. A complete test can only be done with a proprietary OBD-II scan tool (see Section 2), which will run a series of checks to detect excessive pressure. You'll have to take the vehicle to a dealer service department to have the EVAP system professionally diagnosed.

Component replacement

EVAP canister purge solenoid

Refer to illustration 14.9

8   Disengage the EVAP canister purge solenoid from its mounting bracket by pulling it upward carefully.

9   Disconnect the electrical connector from the EVAP canister purge solenoid **(see illustration)**.

10   Detach the quick-connect fittings and remove the vapor hoses from the solenoid.

11   Installation is the reverse of removal.

EVAP canisters

12   Raise the vehicle and place it securely on jackstands.

13   Disconnect the vapor hoses from the EVAP canisters.

14   Disconnect the wiring from the ESIM switch (if so equipped).

15   Remove the lower support bracket bolts and remove the lower support bracket.

16   Remove the EVAP canister mounting nuts from the top of the upper support bracket and remove the EVAP canisters.

17   Installation is the reverse of removal.

---

15   Positive Crankcase Ventilation (PCV) system - general description and check

---

General information

Refer to illustrations 15.2a, 15.2b and 15.3

Note: *For specific information on how to replace the PCV valves on the engines covered by this manual, see Chapter 1.*

1   The Positive Crankcase Ventilation (PCV) system reduces hydrocarbon emissions by scavenging crankcase vapors, which are rich in unburned hydrocarbons. A PCV valve regulates the flow of gases into the intake manifold in proportion to the amount of intake vacuum available. At idle, when intake vacuum is very

high, the PCV valve restricts the flow of vapors so that the engine doesn't run poorly. As the throttle plate opens and intake vacuum begins to diminish, the PCV valve opens more to allow vapors to flow more freely.

2   On the 3.6L V6 engine, the PCV valve is installed in the rear of the right-side valve cover. On 3.7L V6 and 4.7L V8 engines, the PCV system consists of a PCV valve, which is installed at the rear of the left cylinder head/valve cover, and the hose that connects it to the intake system **(see illustrations)**.

3   On Hemi engines, the PCV system consists of a fresh air inlet hose and the PCV valve, which is located on the intake manifold, to the right of and behind the throttle body **(see illustration)**. There are also internal passages connecting the PCV to the intake manifold (there is no external crankcase ventilation hose on these models).

Check

4   An engine that is operated without a properly functioning crankcase ventilation system can be damaged. So anytime you're servicing the engine, inspect the PCV system hose(s) for cracks, tears, deterioration and other damage. Disconnect the hose(s) and inspect it/them for damage and obstructions. If a hose is clogged, clean it out. If you're unable to clean it satisfactorily, replace it. On Hemi engines, remove the PCV valve and inspect the O-rings for damage.

5   A plugged PCV hose might cause any or all of the following conditions: A rough idle, stalling or a slow idle speed, oil leaks or sludge in the engine. So if the engine is running roughly, stalling and idling at a lower than normal speed, or is losing oil, or has oil in the throttle body or air intake manifold plenum, or has a build-up of sludge, a PCV system hose might be clogged. Repair or replace the hose(s) as necessary.

6   A leaking PCV hose might cause any or all of the following conditions: a rough idle, stalling or a high idle speed. So if the engine is running roughly, stalling or idling at a higher than normal speed, a PCV system hose might be leaking. Repair or replace the hose(s) as necessary.

7   Here's an easy functional check of the PCV system on a vehicle with a fresh air inlet hose and a crankcase ventilation hose with a PCV valve in it:

1) *Disconnect the crankcase ventilation (PCV) hose (the hose that connects the PCV valve to the intake manifold).*

2) *Start the engine and let it warm up to its normal idle.*

3) *Verify that there is vacuum at the PCV hose. If there is no vacuum, look for a plugged hose or a clogged port or pipe on the intake manifold. Also look for a hose that collapses when it's blocked (when vacuum is applied). Replace clogged or deteriorated hoses.*

4) *Remove the engine oil dipstick and install a vacuum gauge on the upper end of the dipstick tube.*

15.2a PCV valve and hose (4.7L V8 engine shown)

15.2b This hose supplies fresh air to the PCV system (3.7L V6 and 4.7L V8 engines)

5) *Pinch off or plug the PCV system's fresh air inlet hose.*

6) *Run the engine at 1500 rpm for 30 seconds, then read the vacuum gauge while the engine is running at 1500 rpm.*

7) *If there's vacuum present, the crankcase ventilation system is operating correctly.*

8) *If there's NO vacuum present, the engine might be drawing in outside air. The PCV system won't function correctly unless the engine is a sealed system. Inspect the valve cover(s), oil pan gasket or other sealing areas for leaks.*

9) *If the vacuum gauge indicates positive pressure, look for a plugged hose or engine blow-by.*

8   If the PCV system is functioning correctly, but there's evidence of engine oil in the throttle body or air filter housing, it could be caused by excessive crankcase pressure. Have the crankcase pressure tested by a dealer service department.

9   In the PCV system, excessive blow-by (caused by worn rings, pistons and/or cylinders, or by constant heavy loads) is discharged into the intake manifold and consumed. If you discover heavy sludge deposits or a dilution of the engine oil, even though the PCV system is functioning correctly, look for other causes

(see *Troubleshooting* and Chapter 2E) and correct them as soon as possible.

16   Exhaust Gas Recirculation (EGR) system - description and component replacement

Description

1   Oxides of nitrogen, or simply NOx, are compounds that are formed in the engine when the oxygen and nitrogen in the incoming air mix together. When NOx is emitted from the tailpipe, it mixes with other compounds and sunlight to form ozone and smog. The EGR system reduces NOx by recirculating exhaust gases from the exhaust manifold through the EGR valve and intake manifold, then back to the combustion chambers, where it mixes with the incoming air/fuel mixture. These recirculated exhaust gases dilute the incoming air/fuel mixture, which cools the combustion chambers, thereby reducing NOx emissions.

2   The EGR system consists of the Powertrain Control Module (PCM), the EGR valve and various information sensors that the PCM uses to determine when to open the EGR

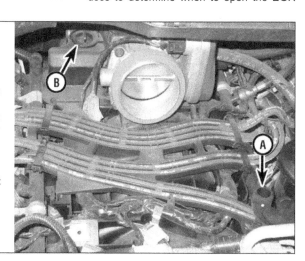

15.3 On Hemi engines, the PCV system consists of a fresh air inlet hose (A) that connects the air resonator box (removed here) to the oil filler tube and the PCV valve (B), which is on the intake manifold to the right of the throttle body

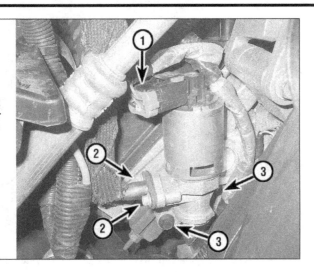

16.5 Disconnect the EGR valve electrical connector (1), remove the EGR pipe flange (2), then remove the EGR valve bolts (3) (typical V6 or V8 engine)

valve. When the PCM chooses, a solenoid inside the EGR valve is energized. This lifts the pintle valve off its seat. The exhaust gas then flows from the exhaust manifold port to the intake manifold.

3   Once activated by the PCM, the EGR valve uses a feedback circuit to control the position of the pintle valve. This enables the PCM to control the position of the pintle with a high degree of precision. A pintle position sensor monitors the position of the pintle.

4   If there is too much EGR flow at idle, cruise or during cold running conditions, the engine will stop after a cold start, stop at idle after deceleration, surge during cruising speeds or idle roughly. If there is too little EGR flow, combustion chamber temperature can become too high during acceleration or under a heavy load, which can cause spark knock (in gasoline engines) and/or engine overheating.

Component replacement
Gasoline engines
EGR valve
Refer to illustration 16.5

Note: *The EGR valve is at the rear of the left cylinder head on 3.7L V6 and 4.7L V8 models. It is at the front of the right cylinder head on Hemi engines.*

3.7L V6 engine
5   Disconnect the electrical connector from the EGR valve **(see illustration)**.
6   Remove the EGR pipe mounting flange bolts.
7   Separate the pipe from the EGR valve, then pull it out of the intake manifold. Remove the gasket and discard it.
8   Remove the EGR valve mounting bolts and remove the EGR valve. Remove and discard the old gasket.

9   Clean the EGR valve mounting surfaces on the cylinder head and on the valve itself.
10   Installation is the reverse of removal. Use new gaskets between the EGR valve and the cylinder head and between the EGR pipe mounting flange and the EGR valve. Tighten the EGR valve and EGR pipe mounting bolts securely.

4.7L V8 engine
11   Remove the throttle body resonator.
12   Disconnect the EVAP system purge hose and the brake vacuum hose from the intake manifold.
13   Disconnect the wiring from the EGR valve **(see illustration 16.5)**.
14   Remove the EGR valve mounting bolts and remove the valve along with the attached EGR tube from the cylinder head. Discard the gasket.
15   Remove the EGR tube from the EGR valve and discard the gasket.
16   Installation is the reverse of removal. Use new gaskets between the EGR valve and the cylinder head and between the EGR pipe mounting flange and the EGR valve. Tighten the EGR valve and EGR pipe mounting bolts securely.

Hemi engines
17   Disconnect the wiring from the EGR valve.
18   Unbolt the EGR tube from the EGR valve, separate them and discard the gasket.
19   Remove the EGR valve mounting bolts and remove the valve from the cylinder head. Discard the gasket.
20   Installation is the reverse of removal. Use new gaskets between the EGR valve and the cylinder head and between the EGR pipe mounting flange and the EGR valve. Tighten the EGR valve and EGR pipe mounting bolts securely.

EGR pipe
Note: *The EGR pipe connects the EGR valve to the intake manifold.*
21   Remove the engine cover.
22   Remove the EGR pipe mounting flange bolts. Remove and discard the old EGR pipe mounting flange gasket.
23   Disconnect the EGR pipe from the intake manifold.
24   Remove the old gasket from the EGR pipe flange and replace it with a new one. Remove the old seal(s) and gaskets from the EGR pipe. To ensure a positive seal, you should always replace them whenever you disconnect the EGR pipe.
25   Clean the EGR pipe mounting surfaces on the EGR valve and on the EGR pipe mounting flange.
26   Installation is the reverse of removal. Tighten the EGR pipe mounting flange bolts securely.

Diesel engines
EGR valve
Note: *The EGR valve is at the front of the engine, on top of the intake manifold.*
27   Remove the cover from the EGR crossover tube.
28   Loosen the left clamp and remove the right clamp from the EGR crossover tube.
29   Remove the bolt from the center bottom part of the crossover tube.
30   Disconnect the wiring from the EGR valve assembly.
31   Remove the four EGR valve mounting bolts.
32   Pry off the EGR valve assembly and remove it.
33   Clean all sealing surfaces of old gasket material and replace all gaskets.
34   Clean the EGR valve as well as the pipe and all other accessible parts (see Chapter 1).
35   Installation is the reverse of removal.

EGR cooler 6.4L Hemi and Diesel
36   Diesel models incorporate an EGR cooler that uses engine coolant to cool the hot exhaust gases to a temperature where they will better do their job of reducing combustion temperature when they are routed into the engine.

6.4L Hemi models
37   The EGR cooler is located under the intake manifold.

Diesel models
38   The EGR cooler/bypass valve assembly is mounted on the top of the exhaust manifold on the passenger's side of the engine.

All models
39   Refer to Chapter 1 for the removal and cleaning procedure.

Chapter 7 Part A
Manual transmission

Contents

Specifications

General
Transmission lubricant type.. See Chapter 1

Torque specifications
	Ft-lbs
Back-up light switch	20 to 25
Drain and fill plugs	42
Shift tower-to-transmission bolts	17
Structural dust cover bolts, Hemi engine	40
Transmission-to-engine mounting bolts	
Transmission-to-engine bolts (no washers)	30
Engine-to-transmission bolts (washers)	50

1   General information

The vehicles covered by this manual are equipped with a manual or an automatic transmission. Information on the manual transmission is included in this Part of Chapter 7. Information on the automatic transmission can be found in Part B of this Chapter. Information on the transfer case is in Chapter 7C.

All vehicles equipped with a manual transmission use a G56 six-speed unit.

Depending on the cost of having a transmission overhauled, it might be a better idea to replace it with a used or rebuilt unit. Your local auto parts store, dealer or transmission shop should be able to supply information concerning cost, availability and exchange policy. Regardless of how you decide to remedy a transmission problem, you can still save a lot of money by removing and installing the unit yourself.

2.4 Use a hammer and chisel to dislodge the rear seal from the extension housing on 2WD models (shown) or, on 4WD models, from the transfer case

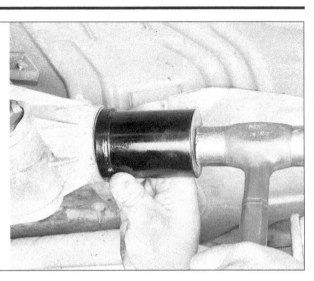

2.6 Use a large socket (shown) or a seal driver to drive the new seal into the extension housing (shown) or transfer case. The outside diameter of the socket must be slightly smaller than the outside diameter of the seal, just enough to clear the edges of the seal bore

2 Extension housing oil seal - replacement

Refer to illustrations 2.4 and 2.6

Note: *This procedure also applies to the transfer case extension housing seal on 4WD models.*

1   Oil leaks at the extension housing oil seal are usually caused by a worn seal lip. Replacing this seal is relatively easy, since you can do so without removing the transmission from the vehicle. The extension housing oil seal is located at the rear tip of the transmission extension housing, where the driveshaft is attached. **Note:** *If the vehicle is a 4WD model, there is no extension housing. Instead, the transfer case is bolted to the rear of the transmission housing. But the following procedure applies to replacing the rear seal on 4WD models as well.*

2   If you suspect that the extension housing seal is leaking (because you've seen puddles below the extension housing seal and/or because the transmission seems to be losing lubricant), raise the vehicle and support it securely on jackstands. If the seal *is* leaking, gear lube will be oozing or dripping from the rear of the transmission, coating the forward end of the driveshaft and leaking onto the ground.

3   Remove the driveshaft (see Chapter 8).

4   Some seals have a metal casing with a lip. With this type of seal you can use a chisel and hammer to carefully pry the oil seal out of the rear of the transmission **(see illustration)**. Do not damage the splines on the transmission output shaft.

5   If the seal doesn't have a metal lip, obtain a special oil seal removal tool (available at most auto parts stores) to do the job.

6   Using a seal driver or a very large socket

as a drift, install the new oil seal **(see illustration)**. Drive it into the bore squarely and make sure that it's completely seated.

7   Lubricate the splines of the transmission output shaft and the outside of the driveshaft yoke with lightweight grease, then install the driveshaft (see Chapter 8). Be careful not to damage the lip of the new seal.

3 Shift lever - removal and installation

1   If equipped, remove the center console (see Chapter 11). With the parking brake on and the wheels blocked, place the shift lever in Neutral.

2   Remove the retaining screws for the shift lever boot and slide the shift lever boot up the shift lever.

3   Remove the four shift tower mounting bolts and remove the shift lever and tower from the transmission as a single assembly.

4   Disassemble the shift lever assembly and clean the parts in solvent. Blow everything dry with compressed air, coat all friction surfaces with clean multi-purpose grease, then reassemble.

5   Installation is the reverse of removal. Tighten the shift tower mounting bolts to the torque listed in this Chapter's Specifications.

4 Back-up light switch - check and replacement

Check

1   The back-up light switch is located on the left side of the transmission case.

2   Turn the ignition key to the ON position, move the shift lever to the REVERSE position and verify that the back-up lights come on.

3   If the back-up lights don't go on, check the back-up light fuse, which is located in the engine compartment fuse and relay box (see Chapter 12).

4   If the fuse is blown, troubleshoot the back-up light circuit for a short circuit.

5   If the fuse is okay, put the shift lever in REVERSE, then raise the vehicle and support it securely on jackstands.

6   Working under the vehicle, disconnect the electrical connector from the back-up light switch. Using an ohmmeter, check for continuity across the terminals of the *switch* (not the connector). There should be continuity.

7   If there is no continuity between the switch terminals with the shift lever in REVERSE, replace the switch.

8   If there is continuity at the switch, check for voltage at the electrical connector. One of the two terminals should have battery voltage present with the ignition key in the ON position.

9   If there is no voltage at the switch electrical connector, troubleshoot the circuit between the engine compartment fuse and relay box and the back-up light switch connector for an open circuit condition.

10   If there is voltage at the switch electrical connector, trace the back-up light circuit between the electrical connector and the back-up light bulbs for an open circuit condition. **Note:** *Although not very likely, the back-up light bulbs could both be burned out, but don't rule out this possibility.*

Replacement

11   Raise the vehicle and support it securely on jackstands, if not already done.

12   Disconnect the electrical connector from the back-up light switch.

13   Unscrew the back-up light switch from the transmission case.

14   Apply RTV sealant or Teflon tape to the threads of the new switch to prevent leakage. Install the switch in the transmission case and tighten it to the torque listed in this Chapter's Specifications.

15   The remainder of installation is the reverse of removal.

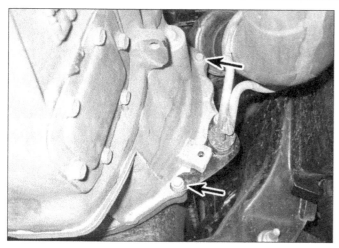

5.20a Right-side transmission mounting bolts (not all bolts are visible in this photo)

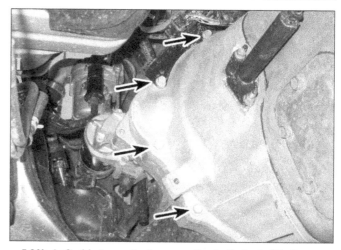

5.20b Left-side transmission mounting bolts (not all bolts are visible in this photo; the two center bolts are the starter mounting bolts, which will already have been removed)

5   Transmission - removal and installation

Removal

Refer to illustrations 5.20a and 5.20b

1   Disconnect the cable(s) from the negative battery terminal(s) (see Chapter 5).

2   Put the transmission shift lever into the NEUTRAL position.

3   Remove the shift tower and lever assembly (see Section 3).

4   Raise the vehicle and support it securely on jackstands.

5   Remove the skid plate, if equipped.

6   Disconnect the electrical connector from the back-up light switch and detach the wiring harness from any clips on the transmission

7   Drain the transmission lubricant (see Chapter 1).

8   Remove the driveshaft (see Chapter 8). Mark the flange and the driveshaft so it can be installed in the same position. Repeat the procedure for the front driveshaft on 4WD models.

9   On 4WD models, remove the transfer case shift linkage, then remove the transfer case (see Chapter 7C). **Caution:** *Don't allow the release lever plunger to tilt or become angled in its bore. This can cause oil leakage.*

10   Detach the clutch release cylinder from the clutch housing (see Chapter 8), then move the cylinder aside for clearance. If you're careful, you should be able to set the release cylinder aside without actually disconnecting the clutch hydraulic line from the release cylinder.

11   On Hemi engines, remove the starter motor (see Chapter 5).

12   On Hemi engines, remove the structural dust cover. **Note:** *Don't remove the dust cover from the engine block to avoid having to align it and the clutch housing with the engine.*

13   On diesel models, remove the interfering exhaust bracket.

14   Remove the exhaust pipe if it interferes with removal.

15   Support the engine from above with an engine hoist, or place a jack (with a block of wood as an insulator) under the engine oil pan. The engine must remain supported at all times while the transmission is out of the vehicle.

16   Support the transmission with a transmission jack (available at auto parts stores and at tool rental yards) or with a large heavy-duty floor jack. If you're going to use a floor jack to support the transmission, make sure that you use a transmission jack adapter head (also available at auto parts stores and at tool rental yards). These transmissions are very heavy, so it's a good idea to have someone help you lower the transmission to make sure that it doesn't fall off the jack.

17   Raise the transmission slightly, then disconnect and remove the transmission mount between the extension housing and the crossmember.

18   Remove the bolts and nuts attaching the crossmember to the frame rails, then remove the crossmember.

19   After the crossmember has been removed, lower the jack slightly so that the transmission is still supported but neither raised nor lowered by the jack.

20   Remove the bolts attaching the transmission to the engine **(see illustrations)**.

21   Make a final inspection for any wiring harness or hoses that might still be connected to the transmission.

22   Keeping the transmission level, roll the jack toward the rear of the vehicle until the transmission input shaft clears the splined hub in the clutch disc.

23   Once the input shaft is clear, lower the transmission to the floor and remove it from under the vehicle.

24   While the transmission is removed, be sure to remove and inspect all clutch components (see Chapter 8). Always install new clutch components anytime that you have to remove the transmission.

Installation

25   Apply a light coat of high-temperature bearing grease (or a suitable equivalent) to the following components:

> *Pilot bearing (in the rear end of the crankshaft)*
> *Driveshaft slip yoke*
> *Input shaft splines*
> *Release bearing bore*
> *Release bearing sliding surface*
> *Release fork ballstud*

After lubricating the above components, install the clutch components (see Chapter 8).

26   Raise the transmission into position and carefully slide it forward, engaging the input shaft with the clutch plate hub. Do not use excessive force to install the transmission - if the input shaft does not slide into place, readjust the angle of the transmission so it is level and/or turn the input shaft so the splines engage properly with the clutch.

27   Install the transmission-to-engine bolts and tighten them to the torque listed in this Chapter's Specifications. **Caution:** *Don't use the bolts to draw the transmission to the engine. If the transmission doesn't slide forward easily and mate with the engine block, find out why before proceeding.*

28   Raise the transmission extension housing just high enough to clear the crossmember, install the crossmember and attach it to the frame rails. Install the transmission mount between the extension housing and the crossmember. Carefully lower the transmission extension housing onto the mount and the crossmember. When everything is properly aligned, tighten all nuts and bolts securely.

29   Remove the jacks supporting the transmission and the engine.

30   On 4WD models, install the transfer case and shift linkage (see Chapter 7C).

31   Install the various components previously removed:

   Clutch release cylinder (see Chapter 8)
   Driveshaft (see Chapter 8)
   Exhaust system components (see
      Chapter 4)
   Starter motor (see Chapter 5)

32   Install the Crankshaft Position (CKP) sensor (see Chapter 6).

33   Reconnect the electrical connector for the back-up light switch. Also reconnect any other wiring harness connectors that you might have disconnected. Make sure that any harnesses that were attached to clips or cable guides on the transmission housing are reattached.

34   Remove the jackstands and lower the vehicle.

35   Install the shift tower and shift lever assembly (see Section 3).

36   Fill the transmission to the correct level with the specified lubricant (see Chapter 1).

37   Reconnect the cable(s) to the negative battery terminal(s).

38   Road test the vehicle for proper operation and check for leakage.

---

6   Transmission overhaul - general information

Overhauling a manual transmission is a difficult job for the do-it-yourselfer. It involves the disassembly and reassembly of many small parts. Numerous clearances must be precisely measured and, if necessary, changed with select fit spacers and snap-rings. As a result, if transmission problems arise, it can be removed and installed by a competent do-it-yourselfer, but overhaul should be left to a transmission repair shop. Rebuilt transmissions might be available. Check with your dealer parts department and auto parts stores. At any rate, the time and money involved in an overhaul is almost sure to exceed the cost of a rebuilt unit.

Nevertheless, it's not impossible for an inexperienced mechanic to rebuild a transmission if the special tools are available and the job is done in a deliberate step-by-step manner so nothing is overlooked.

The tools necessary for an overhaul include internal and external snap-ring pliers, a bearing puller, a slide hammer, a set of pin punches, a dial indicator and possibly a hydraulic press. In addition, a large, sturdy workbench and a vise or transmission stand will be required.

During disassembly of the transmission, make careful notes of how each piece comes off, where it fits in relation to other pieces and what holds it in place. If you note how each part is installed before removing it, getting the transmission back together again will be much easier.

Before taking the transmission apart for repair, it will help if you have some idea what area of the transmission is malfunctioning. Certain problems can be closely tied to specific areas in the transmission, which can make component examination and replacement easier. Refer to the *Troubleshooting* Section at the front of this manual for information regarding possible sources of trouble.

Chapter 7 Part B
Automatic transmission

Contents

Specifications

General
Transmission fluid type .... See Chapter 1

Torque specifications

	Ft-lbs
Torque converter-to-driveplate bolts	
All except AS68RC and 8HP45/845RE	65
AS68RC	37
8HP45/845RE	31
Transmission-to-engine bolts	
42RLE, 68RFE	50
AS68RC, AS69RC	47
8HP45/845RE	31
545RFE, 65RFE, 66RFE	
Upper bolts	30
Lower bolts	40
8HP45/845RE and 8HP70	41

1   General information

The vehicles covered in this manual are equipped with a six-speed manual transmission or a four, five or eight-speed automatic transmission. Information on manual transmissions is in Part A of this Chapter. You'll also find extension housing oil seal replacement, which is a virtually identical procedure for both automatic and manual transmissions, in Part A. Information on automatic transmissions is included in this Part of Chapter 7. Information on the transfer case is in Part C of this Chapter.

All automatic transmissions are equipped with a Torque Converter Clutch (TCC) system that engages in the higher gears. The TCC system provides a direct connection between the engine and the drive wheels, which improves fuel economy. The TCC system consists of a solenoid, controlled by the Powertrain Control Module (PCM), that locks the torque converter when appropriate.

All automatic transmissions covered in this Chapter are equipped with an external air-to-oil transmission oil cooler, which is located in front of the radiator. Diesel engines are also equipped a water-to-oil transmission cooler, which is located on the left side of the engine. On diesels, transmission fluid is routed through this water-to-oil transmission cooler first, then it's routed through the external air-to-oil cooler.

The air-to-oil coolers used on diesel engines are equipped with an internal thermostat that controls the flow of transmission fluid through the cooler. When the transmission fluid is below its operating temperature, it's routed through a cooler bypass. When the fluid reaches operating temperature, the thermostat closes off the bypass, allowing transmission fluid to flow through the cooler. The thermostat can be serviced separately from the cooler (see Section 5). If you're going to back-flush the oil cooler, the thermostat MUST be removed.

All vehicles with an automatic transmission are equipped with a Brake Transmission Shift Interlock (BTSI) system that locks the shift lever in the PARK position and prevents the driver from shifting out of PARK unless the brake pedal is depressed. The BTSI system also prevents the ignition key from being turned to the LOCK or ACCESSORY position unless the shift lever is fully locked into the PARK position.

Due to the complexity of the automatic transmissions covered in this manual and the need for specialized equipment to perform most service operations, this Chapter is limited to general diagnosis, routine maintenance, adjustments and removal and installation procedures.

If the transmission requires major repair work, leave it to a dealer service department or a transmission repair shop. However, even if a transmission shop does the repairs, you can save some money by removing and installing the transmission yourself.

2   Diagnosis - general

Note: Automatic transmission malfunctions may be caused by five general conditions: poor engine performance, incorrect adjustments, hydraulic malfunctions, mechanical malfunctions or malfunctions in the computer or its signal network. Diagnosis of these problems should always begin with a check of the easily repaired items: fluid level and condition (see Chapter 1), shift cable adjustment and, if equipped, Throttle Valve (TV) cable adjustment. Next, perform a road test to determine if the problem has been corrected or if more diagnosis is necessary. If the problem persists after the preliminary tests and corrections are completed, additional diagnosis should be done by a dealer service department or transmission repair shop. Refer to the Troubleshooting *section at the front of this manual for information on symptoms of transmission problems.*

Preliminary checks

1   Drive the vehicle to warm up the transmission to its normal operating temperature.
2   Check the fluid level as described in Chapter 1:

 a) *If the fluid level is unusually low, add enough fluid to bring the level within the area between the high and low marks on the dipstick (see Section 4 in Chapter 1), then check for external leaks (see below).*

 b) *If the fluid level is abnormally high, it might have been overfilled. Drain off the excess. On models with diesel engines, which use a main water-to-oil cooler in addition to an auxiliary air-to-oil cooler, the presence of engine coolant in the automatic transmission fluid could indicate a leak in the water-to-oil cooler, so check the drained fluid for coolant contamination. Only diesel engines use a water-to-oil cooler in addition to the external air-to-oil cooler.*

 c) *If the fluid is foaming, drain it and refill the transmission, then check for coolant in the fluid or a high fluid level.*

3   Check the engine idle speed. **Note:** *If the engine is malfunctioning, do not proceed with the preliminary checks until it has been repaired and runs normally.*
4   Inspect the shift cable (see Section 3). Make sure it's properly adjusted and that it operates smoothly.

Fluid leak diagnosis

5   Most fluid leaks are usually easy to locate because they leave a visible stain and/or wet spot. Most repairs are simply a matter of replacing a seal or gasket. If a leak is more difficult to find, the following procedure will help.
6   Identify the fluid. Make sure that it's *transmission* fluid, *not* engine oil or brake fluid. One way to positively identify Automatic Transmission Fluid (ATF) is by its deep red color.

7   Try to pinpoint the source of the leak. Drive the vehicle several miles, then park it over a large sheet of cardboard. After a minute or two, you should be able to locate the leak by determining the source of the fluid dripping onto the cardboard.
8   Make a careful visual inspection of the suspected component and the area immediately around it. Pay particular attention to gasket mating surfaces. A flashlight and mirror are often helpful for finding leaks in areas that are hard to see.
9   If you still can't find the leak, thoroughly clean the suspected area with a degreaser or solvent, then dry it off.
10   Drive the vehicle for several miles at normal operating temperature and varying speeds. After driving the vehicle, visually inspect the suspected component again.
11   Once you have located the leak, you must determine the source before you can repair it properly. For example, if you replace a pan gasket but the sealing flange is warped or bent, the new gasket won't stop the leak. The flange must first be straightened.
12   Before attempting to repair a leak verify that the following conditions are corrected or they might cause another leak. **Note:** *Some of the following conditions cannot be fixed without highly specialized tools and expertise. Such problems must be referred to a transmission repair shop or a dealer service department.*

Gasket leaks

13   Inspect the pan periodically. Make sure that the bolts are tight, that no bolts are missing, that the gasket is in good condition and that the pan is flat (dents in the pan might indicate damage to the valve body inside).
14   If the pan gasket is leaking, the fluid level or the fluid pressure might be too high, the vent might be plugged, the pan bolts might be too tight, the pan sealing flange might be warped, the sealing surface of the transmission housing might be damaged, the gasket might be damaged or the transmission casting might be cracked or porous. If sealant instead of gasket material has been used to form a seal between the pan and the transmission housing, it may be the wrong type sealant.

Seal leaks

15   If a transmission seal is leaking, the fluid level or pressure might be too high, the vent might be plugged, the seal bore might be damaged, the seal itself might be damaged or incorrectly installed, the surface of the shaft protruding through the seal might be damaged or a loose bearing might be causing excessive shaft movement.
16   Make sure that the dipstick tube seal is in good condition and that the tube is correctly seated.

Case leaks

17   If the case itself appears to be leaking, the casting is porous. A porous casting must be repaired or replaced.
18   Make sure that the oil cooler hose fittings are tight and in good condition.

3.4 Use a screwdriver to pry the shift cable off the shift lever at the transmission

3.7 Slide this clip loose to adjust the length of the transmission shift cable

Fluid comes out vent pipe or fill tube

19 If this condition occurs, the transmission is overfilled, there is coolant in the fluid, the case is porous, the dipstick is incorrect, the vent is plugged or the drain-back holes are plugged.

---

3 Shift cable - check, adjustment and replacement

Note: On 8HP45/845RE 8-speed models, no shift cable is used. The transmission uses an electronic shifter (E-shifter) which is an electronic switch/module mounted to the instrument panel, using electronic messaging to communicate the transmission gear changes.

Check

1 Firmly apply the parking brake and try to momentarily operate the starter in each shift lever position. The starter should only operate when the shift lever is in the PARK

or NEUTRAL positions. If the starter operates in any position other than PARK or NEUTRAL, adjust the shift cable (see below). If, after adjustment, the starter still operates in positions other than PARK or NEUTRAL, the Transmission Range (TR) sensor is defective (see Chapter 6).

Adjustment

Refer to illustrations 3.4 and 3.7

2 Place the shift lever in the PARK position.
3 Raise the vehicle and support it securely on jackstands. **Note:** *The rear of the vehicle must also be raised, so the driveshaft can be turned in Step 6 (to verify that the transmission is completely engaged in PARK).*
4 Working at the transmission end of the cable, pry the cable end off the manual shift lever **(see illustration)**.
5 Verify that the manual shift lever on the transmission is all the way to the rear, in the last detent. This is the PARK position.
6 Verify that the park lock pawl inside the

transmission is engaged by trying to rotate the driveshaft. The driveshaft will not rotate if the transmission is correctly engaged in PARK.
7 Disengage the locking clip on the cable end by pushing it up from the rear side **(see illustration). Note:** *This adjustment mechanism is at the transmission end of the cable on most models. On some diesel models, it's at the opposite end of the cable, so instrument panel trim will have to be removed for access (see Chapter 11).*
8 Move the end of the cable in or out as required so that it snaps over the stud on the shift lever.
9 Snap the locking clip into place by pushing it down firmly.
10 Lower the vehicle and replace any trim panels removed. With the parking brake firmly applied, make sure the engine starts (don't move the shift lever from PARK yet).
11 If the linkage appears to be adjusted correctly, but the starter still operates in any other position(s) besides PARK and NEUTRAL, replace the Transmission Range (TR) sensor (see Chapter 6).

Replacement

Refer to illustrations 3.18, 3.19 and 3.20

12 Make sure the shift lever is in the PARK position.
13 Raise the vehicle and place it securely on jackstands.
14 Working at the transmission end of the cable, remove the cable end from the shift lever, then pull it out of its mounting bracket.
15 Lower the vehicle.
16 Remove the instrument panel trim panels necessary to reach the upper end of the shift cable.
17 Remove the shift cable grommet from the firewall after pulling back the insulation.
18 Disconnect the electrical connector from the Brake Transmission Shift Interlock (BTSI) solenoid **(see illustration)**
19 Pry the cable end off the lever on the steering column **(see illustration)**

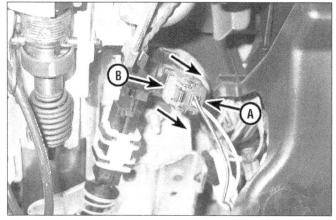

3.18 To disconnect the electrical connector (A) from the Brake Transmission Shift Interlock (BTSI) solenoid, slide the sliding lock toward you, then depress the button (B) on the slide lock and pull off the connector

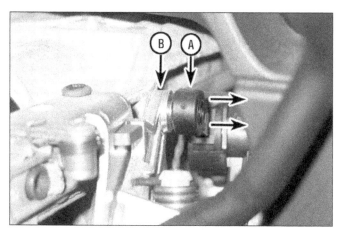

3.19 Insert a small screwdriver between the shift cable (A) and the shift lever pin (B) to disconnect them

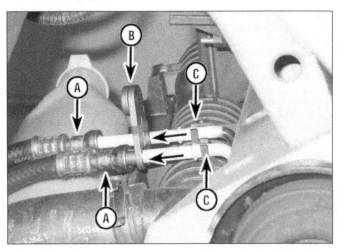

3.20 To disengage the shift cable assembly from its mounting bracket, squeeze the tangs on the cable housing together, then pull the cable assembly straight down. To remove the cable assembly, disengage the grommet from the cable hole in the firewall, then pull the cable through the hole into the cab

5.4 To disconnect each transmission oil cooler line fitting (A), insert the special quick-connect release tool (B) into each fitting, push the tool into the fitting until it releases the locking fingers inside the fitting, then pull the two sides of the fitting apart. Once both fittings are disconnected, detach both lines from the clips (C) on the radiator side tank (typical assembly shown)

20   To disengage the cable assembly from the steering column bracket, squeeze the two tangs on the cable housing together, then pull the cable assembly straight down **(see illustration)**.

21   Installation is the reverse of removal. Adjust the cable (see Steps 2 through 11).

---

4   Brake Transmission Shift Interlock (BTSI) system - description, check and adjustment

---

Description

1   The Brake Transmission Shift Interlock (BTSI) system is a solenoid-operated device, located on the shift cable, that locks the shift lever into the PARK position when the ignition key is in the LOCK or ACCESSORY position. When the ignition key is in the RUN position, a magnetic holding device, inline with the park lock cable, is energized. When the system is functioning correctly, the only way to unlock the shift lever and move it out of PARK is to depress the brake pedal. The BTSI system

also prevents the ignition key from being turned to the LOCK or ACCESSORY position unless the shift lever is fully locked into the PARK position.

Check

2   Verify that the ignition key can be removed *only* in the PARK position.

3   When the shift lever is in the PARK position and the shift lever Overdrive Off (O/D OFF) button is not activated, you *should* be able to rotate the ignition key from OFF to LOCK. But when the shift lever is in any gear position other than PARK (including NEUTRAL), you should not be able to rotate the ignition key to the LOCK position.

4   You should *not* be able to move the shift lever out of the PARK position when the ignition key is turned to the OFF position.

5   You should *not* be able to move the shift lever out of the PARK position when the ignition key is turned to the RUN or START position until you depress the brake pedal.

6   You should *not* be able to move the shift lever out of the PARK position when the ignition key is turned to the ACC or LOCK position.

7   Once in gear, with the ignition key in the RUN position, you *should* be able to move the shift lever between gears, or put it into NEUTRAL or PARK, without depressing the brake pedal.

8   If the BTSI system doesn't operate as described, have it checked by a dealer service department.

Adjustment

9   The BTSI is not adjustable on the models covered by this manual. It is a part of the shift lever assembly.

---

5   Transmission oil cooler - removal and installation

Air-to-oil cooler

Refer to illustrations 5.4 and 5.6

Note: *All vehicles with an automatic transmission are equipped with an external air-to-oil cooler as standard equipment. All coolers are mounted in front of the radiator in similar fashions, although they vary in size and in the location and number of mounting bolts.*

1   Disconnect the cable(s) from the negative battery terminal(s).

2   Put a drain pan underneath the oil cooler line fittings to catch any spilled transmission fluid.

3   Raise the front of the vehicle and place it securely on jackstands.

4   Using a quick-connect release tool (available at most auto parts stores), disconnect the transmission oil cooler line fittings **(see illustration)**. Plug the lines to prevent fluid spills.

5   Detach the transmission oil cooler lines from the radiator side tank.

6   Remove the transmission oil cooler mounting bolts and/or nuts **(see illustration)**

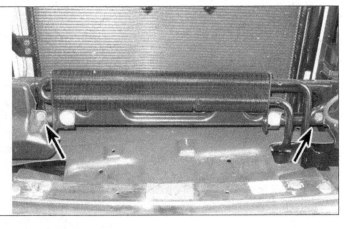

5.6 Typical transmission oil cooler mounting bolt locations

6.2 To check the transmission mount, insert a large screwdriver between the extension housing and the crossmember and try to lever the transmission up

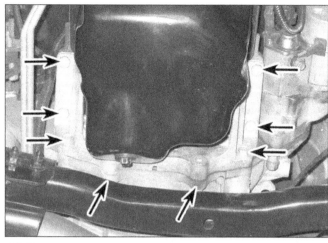

7.8 To remove the transmission brace on a vehicle with 45RFE or 545RFE transmission, remove these bolts (Hemi engine shown, others similar)

and remove the cooler. Be careful not to damage the oil cooler tubes or the radiator cooling fins.

7   On diesel models, remove the oil cooler thermostat (see Steps 11 through 13) from the cooler, clean it, inspect it, then install the old thermostat or a new one.

8   If you removed the cooler in order to flush it after a transmission failure, have it flushed by a dealer service department or by a transmission shop. A number of special tools are needed to flush the cooler correctly.

9   Installation is the reverse of removal. Check the transmission fluid level and add some if necessary (see Chapter 1).

Thermostat

Note: *The air-to-oil coolers on diesel engines are equipped with a serviceable internal thermostat. The thermostat is located inside a housing at the upper left corner of the cooler, at the left end of the cooler bypass tube. Gasoline engines are not equipped with this thermostat.*

10   Remove the transmission oil cooler (see Steps 1 through 6).

11   Remove the snap-ring that retains the thermostat end-plug.

12   Remove the end-plug, the thermostat and the spring from the thermostat housing.

13   Thoroughly clean out the thermostat bore in the housing and wash the end-plug, thermostat and spring in clean solvent. Inspect the parts for wear and damage. If anything is worn or damaged, replace it.

14   Installation is the reverse of removal.

Water-to-oil cooler

Warning: *Wait until the engine is completely cool before beginning this procedure.*
Note: *Only diesel engines are equipped with this cooler.*

15   Disconnect the cables from the negative battery terminals.

16   Remove the starter motor (see Chapter 5).

17   Drain the cooling system (see Chapter 1).

18   Disconnect the coolant lines from the oil cooler. Use a back-up wrench on the transmission cooler boss to protect the cooler.

19   Disconnect the transmission fluid lines from the oil cooler. Plug the cooler lines to prevent dirt and moisture from entering the lines.

20   Remove the cooler mounting bracket-to-transmission adapter bolt.

21   On 2009 models, detach the wiring harness from the bottom of the coolant bracket.

22   On 2010 and later models, detach the ground cable from its stud, then remove the upper nut.

23   Remove the cooler assembly.

24   Installation is the reverse of removal. Check the transmission fluid level and add some if necessary (see Chapter 1).

6   Transmission mount - check and replacement

Check

Refer to illustration 6.2

1   Raise the vehicle and support it securely on jackstands.

2   Insert a large screwdriver or prybar into the space between the transmission extension housing and the crossmember and try to pry the transmission up slightly **(see illustration)**.

3   The transmission should not move much at all and the rubber in the center of the mount should fully insulate the center of the mount from the mount bracket around it.

Replacement

4   To replace the mount, remove the bolts or nuts attaching the mount to the crossmember and the bolts attaching the mount to the transmission.

5   Raise the transmission slightly with a jack and remove the mount.

6   Installation is the reverse of the removal procedure. Tighten all fasteners securely.

7   Automatic transmission - removal and installation

Removal

Refer to illustrations 7.8, 7.9, 7.12, 7.13, 7.20 and 7.23

Caution: *The transmission and torque converter must be removed as a single assembly. If you try to leave the torque converter attached to the driveplate, the converter driveplate, pump bushing and oil seal will be damaged. The driveplate is not designed to support the load, so none of the weight of the transmission should be allowed to rest on the plate during removal.*

1   Disconnect the cable(s) from the negative battery terminal(s) (see Chapter 5).

2   Raise the vehicle and support it securely on jackstands.

3   Remove the skid plate, if equipped.

4   Remove all exhaust components that interfere with transmission removal (see Chapter 4). This includes the particulate filter on diesel engines.

5   If the transmission is being removed for overhaul, drain the transmission fluid (see Chapter 1).

6   On gasoline engines, remove the starter motor (see Chapter 5).

7   Mark the yokes and remove the driveshaft (see Chapter 8). On 4WD models, remove both driveshafts.

8   On 3.7L V6 and 4.7L V8 models, remove the engine-to-transmission brace **(see illustration)**

9   Remove the torque converter access

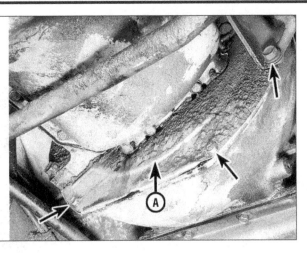

7.9 To remove the torque converter access cover (A), remove all the retaining bolts and pull it off. The number and location of the cover bolts varies with the engine-transmission combination

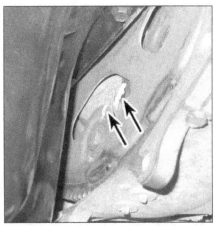

7.12 If you're going to re-use the old torque converter, mark the relationship of the torque converter to the driveplate to ensure that their dynamic balance is preserved when the torque converter is reattached

cover, if equipped **(see illustration)**. On Hemi engines, this is a structural component.

10　On Hemi engines, remove the transmission-to-differential bracket and the exhaust support.

11　On diesel engines, remove the bracket from the rear of the transmission, if so equipped.

12　Mark the relationship of the torque converter to the driveplate **(see illustration)** to ensure that their dynamic balance is maintained when the converter is reattached to the driveplate.

13　Remove the torque converter-to-driveplate bolts **(see illustration)**. Turn the crankshaft for access to each bolt.

14　Disconnect all electrical connectors from the transmission. The connectors to various electrical and/or electronic devices on the transmission are different in shape, color and the number of terminals, so there's little danger of accidentally reconnecting a connector to the wrong device. The wiring harness is also designed so that each connector will only reach the device to which it's supposed to be connected. However, if any of the connectors look identical or look like they could be accidentally reconnected to the wrong device, be

sure to mark them to prevent mix-ups.

15　Disconnect the shift cable from the transmission (see Section 3).

16　Support the rear end of the transmission with a floor jack, then raise the transmission slightly to take the weight off the crossmember.

17　Unbolt the transmission mounting bracket from the transmission and from the crossmember and remove the bracket.

18　Unbolt and remove the transmission crossmember.

19　On 4WD models, remove the transfer case (see Chapter 7C). **Note:** *If you are not planning to replace the transmission, but are removing it in order to gain access to other components such as the torque converter, it isn't really necessary to remove the transfer case. However, the two components are awkward and heavy when removed and installed as a single assembly. They're much easier to maneuver off and on as separate units. If you decide to leave the transfer case attached, disconnect the shift rod from the transfer case shift lever, or remove the shift lever from the transfer case and tie the rod and lever to the chassis (see Chapter 7C).* **Warning:** *If you decide to remove the transfer case and*

transmission as a single assembly, use safety chains to help stabilize them and to prevent them from falling off the jack head, which could cause serious damage to the transmission and/or transfer case and serious bodily injury to you.

20　Disconnect the transmission cooler lines from the transmission **(see illustration)**. Plug the ends of the lines to prevent fluid from leaking out after you disconnect them. **Caution:** *The lines on some models are equipped with quick-connect fittings. Use the proper tool to separate these fittings to avoid damage.*

21　Remove the oil filler tube bracket bolts and withdraw the tube from the transmission. Don't lose the filler tube seal (unless it's damaged, in which case you should replace it). On 4WD models, remove the bolt that attaches the transfer case vent tube to the converter housing (unless you already did so when removing the transfer case).

7.13 To remove the torque converter-to-driveplate bolts, turn the crankshaft to access each bolt

7.20 To prevent damage to the lines, use a back-up wrench on the stationary fittings when unscrewing the transmission cooler line fittings

22   Support the transmission with a transmission jack (available at most equipment rental facilities) and secure the transmission to the jack with safety chains. Support the engine with a jack. Use a block of wood under the oil pan to spread the load.
23   Remove the bolts securing the transmission to the engine **(see illustration)**. A long extension and a U-joint socket will greatly simplify this step. **Note:** *The upper bolts are easier to remove after the transmission has been lowered (see the next Step). Also, on some models, you might have to remove the oil filter (see Chapter 1) before you can remove the lower right (passenger's side) bolt.*
24   Lower the engine and transmission slightly and clamp a pair of locking pliers onto the lower portion of the transmission case, just in front of the torque converter. The pliers will prevent the torque converter from falling out while you're removing the transmission.
25   Move the transmission to the rear to disengage it from the engine block dowel pins and make sure the torque converter is detached from the driveplate. Lower the transmission with the jack.

Installation

26   Prior to installation, make sure the torque converter is securely engaged in the pump. If you've removed the converter, spread transmission fluid on the torque converter rear hub, where the transmission front seal rides. With the front of the transmission facing up, rotate the converter back and forth. It should drop down into the transmission front pump in stages. To make sure the converter is fully engaged, lay a straightedge across the transmission-to-engine mating surface and make sure the converter lugs are at least 1/2-inch below the straightedge. Reinstall the locking pliers to hold the converter in this position.
27   With the transmission secured to the jack, raise it into position. Connect the transmission fluid cooler lines.
28   Turn the torque converter to line up the holes with the holes in the driveplate. The marks on the torque converter and driveplate made during removal must line up.
29   Move the transmission forward carefully until the dowel pins and the transmission are engaged. Make sure the transmission mates with the engine with no gap. If there's a gap, make sure there are no wires or other objects pinched between the engine and transmission and also make sure the torque converter is completely engaged in the transmission front pump. Try to rotate the converter - if it doesn't rotate easily, it's probably not fully engaged in the pump. If necessary, lower the transmis-

sion and install the converter fully.
30   Install the transmission-to-engine bolts and tighten them to the torque listed in this Chapter's Specifications. As you're tightening the bolts, make *sure* that the engine and transmission mate completely at all points. If not, find out why. Never try to force the engine and transmission together with the transmission-to-engine bolts or you'll break the transmission case!
31   Install the torque converter-to-driveplate bolts. Tighten them to the torque listed in this Chapter's Specifications. **Caution:** *Using the correct length bolts for bolting the converter to the driveplate is critical. A number of different converters are used on the vehicles covered by this manual. If the bolts are too long, they will damage the converter. If you're planning to use new bolts, make sure you obtain original equipment replacement bolts of the same length.* **Note:** *Install all of the bolts before tightening any of them.*
32   The remainder of installation is the reverse of removal.
33   Refill the transmission with the specified fluid (see Chapter 1), run the engine and check for fluid leaks.

8   Automatic transmission overhaul - general information

In the event of a fault occurring, it will be necessary to establish whether the fault is electrical, mechanical or hydraulic in nature, before repair work can be contemplated. Diagnosis requires detailed knowledge of the transmission's operation and construction, as well as access to specialized test equipment, and so is beyond the scope of this manual. It is essential that problems with the automatic transmission are referred to a dealer service department or other qualified repair facility for assessment.

Note that a faulty transmission should not be removed before the vehicle has been assessed by a knowledgeable technician equipped with the proper tools, as troubleshooting must be performed with the transmission installed in the vehicle.

9   Shifter (8HP45/845RE 8-speed models) – removal and installation

1   Disconnect the cable(s) from the negative battery terminal(s) (see Chapter 5).
2   Remove the center instrument panel bezel (see Chapter 11).

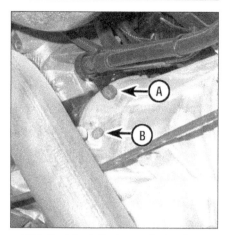

7.23 To detach the transmission from the engine, remove all transmission-to-engine bolts (A) and engine-to-transmission bolts (B) (left side shown, right side similar; not all bolts visible in this photo)

3   Disconnect the electrical connector from the shifter.
4   Remove the shifter-to-instrument panel bezel mounting screws and shifter from the back side of the panel.
5   Installation is the reverse of removal.

10   Shifter (65RFE 6-speed models) – removal and installation

1   Disconnect the cable(s) from the negative battery terminal(s) (see Chapter 5).
2   Remove the center console and storage tray (see Chapter 11).
3   Disconnect the power outlet harness electrical connector and the shifter harness electrical connector.
4   Twist the lower trim piece on the base of the shifter counterclockwise and push down to disengage the trim piece.
5   Remove the shifter.
6   Disconnect the electrical connector lamp from PRNDL bezel.
7   Release the clips on the PRNDL bezel and remove the bezel.
8   Disconnect the shift cable retainer from the bracket.
9   Disconnect the shift cable end from the shifter lever.
10   Remove the shifter lever assembly.
11   Installation is the reverse of removal.

Notes

Chapter 7 Part C
Transfer case

Contents

Specifications

Torque specifications

Ft-lbs (unless otherwise indicated)

Note: *One foot-pound (ft-lb) of torque is equivalent to 12 inch-pounds (in-lbs) of torque. Torque values below approximately 15 ft-lbs are expressed in inch-pounds, since most foot-pound torque wrenches are not accurate at these smaller values.*

Electric shift motor mounting bolts
    New Venture cases .... 12 to 18
    Borg Warner cases .... 72 to 96 in-lbs
Shift-rod lock bolt .... 90 in-lbs
Transfer case-to-transmission bolts/nuts
    New Venture cases .... 20 to 25
    Borg Warner cases 30
Companion flange nut, if equipped
    New Venture cases* .... 130 to 200
    Borg Warner cases 178

Use a new nut

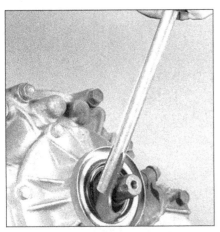

5.9 Use a seal removal tool to pry the transfer case seal out of the housing

5.11 The new seal can be driven into place using a seal installer tool or a large socket

1   General information

The transfer case is a device which transmits power from the transmission to the front and rear driveshafts. The models covered by this manual may be equipped with any one of the following transfer cases, all of them manufactured by New Venture (NV) and Borg Warner (BW):

a) *The NV 243, 246, 271 and 273 transfer cases*

b) *The BW 44-44, 44-45, 44-46 and 44-47 transfer cases*

We don't recommend trying to rebuild any of these transfer cases at home. They're difficult to overhaul without special tools, and rebuilt units may even be available (on an exchange basis) for less than it would cost to rebuild your own. However, there are a number of components that you can check, adjust and/or replace, and those items are covered in this Chapter.

2   Shift linkage adjustment (manual shift models)

1   Move the shift lever into the 2H position.
2   Raise the vehicle and support it securely on jackstands.
3   Loosen the shift-rod lock bolt at the trunnion.
4   Check the fit of the shift rod in the trunnion. Make sure it doesn't bind in the trunnion.
5   Verify that the transfer case range lever is in the 2H position. The 2H position on the transfer case shift arm is the second position from full forward.
6   Align the adjustment locating hole on the lower shifter lever with the adjustment channel located on the shifter bracket assembly.
7   Insert a drill bit with the correct diameter through the adjustment channel and into the locating hole to lock the shifter in position.
8   Tighten the shift-rod lock bolt to the

torque listed in this Chapter's Specifications.
9   Remove the jackstands and lower the vehicle.

3   Shift range selector switch (electric shift models) - replacement

1   Disconnect the cable(s) from the negative battery terminal(s) (see Chapter 5).
2   Raise the vehicle and support it securely on jackstands.
3   Remove the center instrument panel bezel (see Chapter 11).
4   Disconnect the switch assembly electrical connectors.
5   Separate the shift range selector switch from the panel bezel.
6   Installation is the reverse of the removal.

4   Electric shift motor - replacement

Note: *New Venture motors: New shift motors are packaged with the shift motor positioned in the 2WD/AWD mode. The transfer case must be selected for 2WD/AWD before the shift motor can be installed.*
Note: *Borg Warner motors: Mode sensors can't be separated from the shift motors. They are an integral part of the sensor.*
1   Disconnect the cable(s) from the negative battery terminal(s) (see Chapter 5).
2   Raise the vehicle and support it securely on jackstands.
3   Disconnect the shift motor and the mode sensor connectors.
4   Remove the shift motor and mode sensor assembly mounting bolts and separate the unit from the transfer case.
5   Installation is the reverse of removal. Replace the shift sector O-ring with a new part. If the sector shaft does not align with the shift motor, manually shift the transfer case to the correct position.
6   Tighten the electric shift motor mounting bolts to the torque listed in this Chapter's Specifications.

5   Oil seals - replacement

1   Disconnect the cable(s) from the negative battery terminal(s) (see Chapter 5).
2   Raise the vehicle and support it securely on jackstands.
3   Remove the skid plate, if equipped.
4   Drain the transfer case lubricant (see Chapter 1).

Front (output shaft) seal

Refer to illustrations 5.9 and 5.11
5   Remove the front driveshaft (see Chapter 8).
6   On models equipped with a companion flange, remove the companion flange nut. On New Venture transfer cases, discard the nut; it's not reusable.
7   If equipped, tap the companion flange off the front output shaft with a brass or plastic hammer.
8   On models without a companion flange, remove the driveshaft seal boot from the seal slinger around the output shaft, then remove the seal slinger by bending its ears outward.
9   Carefully pry out the old seal with a screwdriver or a seal removal tool **(see illustration)**. Make sure you don't scratch or gouge the seal bore.
10   Lubricate the lips and the outer diameter of the new seal with multi-purpose grease. Place the seal in position, square to the bore, making sure the garter spring faces toward the inside of the transfer case.
11   Use a seal driver or a suitable equivalent to drive the seal into place. A large deep socket **(see illustration)** with an outside circumference slightly smaller than the circumference of the new seal will work fine. Start the seal in the bore with light hammer taps. Continue tapping the seal into place until it is recessed the correct amount.
12   Install the seal slinger and driveshaft seal boot. Secure the boot with a new clamp.
13   Install the companion flange, if equipped, on the front output shaft, then install a NEW flange nut and tighten it to the torque listed in this Chapter's Specifications.

14   Install the front driveshaft (see Chapter 8).
15   Remove the jackstands and lower the vehicle.

Extension housing seal

16   This procedure is identical to the extension housing seal replacement procedure for the transmission (see Chapter 7A).

6   Transfer case - removal and installation

Note: *Before starting this procedure, shift the transfer case into 2WD or AWD.*

Removal

Refer to illustration 6.10

1   Disconnect the cable(s) from the negative battery terminal(s) (see Chapter 5).
2   Raise the vehicle and support it securely on jackstands.
3   Remove the skid plate, if equipped. Drain the transfer case lubricant.
4   Detach all vacuum/vent lines, if equipped, and electrical connectors from the transfer case.
5   On manual shift models, disconnect the shift lever rod from the grommet in the transfer case shift lever or from the shift lever arm on the floor, whichever provides easier access. Press the rod out of the grommet with adjustable pliers.
6   On electric shift models, disconnect the transfer case shift motor and transfer case sensor.
7   Remove the front and rear driveshafts (see Chapter 8).
8   Support the transmission with a transmission jack.
9   Support the transfer case with a transmission jack. Secure the transfer case to the transmission jack with chains or tie-down straps.
10   Remove the transfer case-to-transmission bolts/nuts **(see illustration)**.
11   Make a final check that all wires and hoses have been disconnected from the transfer case, then move the transfer case and jack toward the rear of the vehicle until the transfer case is clear of the transmission. Keep the transfer case level as this is done.
12   Once the input shaft is clear, lower the transfer case and remove it from under the vehicle.

Installation

13   Remove all gasket material from the rear of the transmission. Apply RTV sealant to both sides of the transfer case-to-transmission gasket and position the gasket on the mating surface of the transmission.
14   With the transfer case secured to the jack as on removal, raise it into position behind the transmission and carefully slide it forward, engaging the input shaft with the transmission output shaft. Do not use excessive force to install the transfer case - if the input shaft does not slide into place, readjust the angle so it is level and/or turn the input shaft so the splines engage properly with the transmission.
15   Install the transfer case-to-transmission bolts/nuts, tightening them to the torque listed in this Chapter's Specifications.
16   Remove the safety chains and remove the jack supporting the transfer case.
17   Install the rear crossmember, if removed.
18   Remove the transmission jack from under the transmission.
19   Install the driveshafts (see Chapter 8).
20   Reattach all vacuum and/or vent lines. Plug in all electrical connectors.
21   On manual shift models, connect the shift rod to the transfer case shift lever or to the floor-mounted shift lever arm, and adjust the shift linkage (see Section 2).
22   On electric shift models, connect the transfer case shift motor and transfer case sensor electrical connectors.
23   Refill the transfer case with lubricant (see Chapter 1). If the vehicle has a manual transmission, this is also a good time to check the lubricant level for the transmission (see Chapter 1).
24   Install the skid plate, if equipped.
25   Remove the jackstands and lower the vehicle.
26   Connect the negative battery cable(s).
27   Road test the vehicle for proper operation and check for leakage.

7   Transfer case overhaul - general information

Overhauling a transfer case is a difficult job for the do-it-yourselfer. It involves the disassembly and reassembly of many small parts. Numerous clearances must be precisely measured and, if necessary, changed with select fit spacers and snap-rings. As a result, if transfer case problems arise, it can be removed and

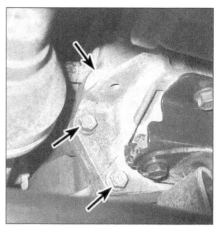

6.10 Remove the transfer case bolts/nuts from the transmission - remaining bolts/ nuts hidden from view

installed by a competent do-it-yourselfer, but overhaul should be left to a transmission repair shop. Rebuilt transfer cases may be available - check with your dealer parts department and auto parts stores. At any rate, the time and money involved in an overhaul is almost sure to exceed the cost of a rebuilt unit.

Nevertheless, it's not impossible for an inexperienced mechanic to rebuild a transfer case if the special tools are available and the job is done in a deliberate step-by-step manner so nothing is overlooked.

The tools necessary for an overhaul include internal and external snap-ring pliers, a bearing puller, a slide hammer, a set of pin punches, a dial indicator and possibly a hydraulic press. In addition, a large, sturdy workbench and a vise or transmission stand will be required.

During disassembly of the transfer case, make careful notes of how each piece comes off, where it fits in relation to other pieces and what holds it in place. Note how parts are installed when you remove them; this will make it much easier to get the transfer case back together.

Before taking the transfer case apart for repair, it will help if you have some idea what area of the transfer case is malfunctioning. Certain problems can be closely tied to specific areas in the transfer case, which can make component examination and replacement easier. Refer to the *Troubleshooting* section in the introductory pages of this manual for information regarding possible sources of trouble.

Notes

Chapter 8
Clutch and driveline

Contents

Specifications

Torque specifications

Ft-lbs (unless otherwise indicated)

Note: *One foot-pound (ft-lb) of torque is equivalent to 12 inch-pounds (in-lbs) of torque. Torque values below approximately 15 ft-lbs are expressed in inch-pounds, since most foot-pound torque wrenches are not accurate at these smaller values.*

Clutch master cylinder mounting nuts	21
Clutch release cylinder mounting nuts	17
Clutch pressure plate-to-flywheel bolts	
Gasoline engines	37
Diesel engine	22
Driveaxle hub/nut (independent front suspension)	185
Driveshaft center support bearing mounting bolts	40
Driveshaft-to-front axle bolts (4WD models)	
2009 and 2010 models	85
2011 and later models	
Light-duty	85
Heavy-duty (DP)	55
Heavy-duty (DJ, D2, DD)	21

Torque specifications

Ft-lbs (unless otherwise indicated)

Note: *One foot-pound (ft-lb) of torque is equivalent to 12 inch-pounds (in-lbs) of torque. Torque values below approximately 15 ft-lbs are expressed in inch-pounds, since most foot-pound torque wrenches are not accurate at these smaller values.*

Driveshaft-to-rear axle bolts
  2009 and 2010 models.. 85
  2011 and later models
    Light-duty... 85
    Heavy-duty (DP)... 55
    Heavy-duty (DJ, D2, DD)... 85
Driveshaft-to-transfer case bolts (4WD models, front and rear)............. 65
Front hub bearing-to-steering knuckle nuts (solid axle)......................... 148
Front axle hub/nut (solid axle)
  9-1/4-inch axle... 263
  275FBI axle... 243
Rear axle shaft flange bolts
  All except 12.8-inch ring gear axle 95
  12.8-inch ring gear axle... 98
Rear axle shaft hub nut
  10.5-inch and 11.5-inch ring gear axles
    Step 1 (while turning hub)... 22
    Step 2*.. Turn 30 degrees counterclockwise and align slots
  12.8-inch (302RBI) ring gear axles
    Step 1 (while turning hub)... 70
    Step 2 .... Turn counterclockwise 90 degrees
    Step 3**.. 30
Rear differential pinion mate shaft lock bolt (semi-floating axle) .... 96 in-lbs
Wheel lug nuts.. See Chapter 1

*Endplay should be 0.001 to 0.010 inch
**Endplay should be near zero

1   General information

The information in this Chapter deals with the components from the rear of the engine to the rear wheels, except for the transmission (and transfer case, if equipped), which is dealt with in the previous Chapter. For the purposes of this Chapter, these components are grouped into three categories: clutch, driveshaft and axles. Separate Sections within this Chapter offer general descriptions and checking procedures for components in each of the three groups.

Since nearly all the procedures covered in this Chapter involve working under the vehicle, make sure it's securely supported on sturdy jackstands or on a hoist where the vehicle can be easily raised and lowered.

2   Clutch - description and check

1   All vehicles with a manual transmission have a single dry plate, diaphragm spring-type clutch. The clutch disc has a splined hub which allows it to slide along the splines of the transmission input shaft. A sleeve-type release bearing is operated by a release fork in the clutch housing. The fork pivots on a ballstud mounted inside the housing.

2   The clutch release system is operated by hydraulic pressure. The hydraulic release system consists of the clutch pedal, a master cylinder and fluid reservoir, the hydraulic line, a release cylinder, and a release fork mounted inside the clutch housing.

3   When force is applied to the clutch pedal to release the clutch, hydraulic pressure is exerted against the release fork by the release cylinder pushrod. When the release fork is moved, it pushes against the release bearing. The bearing pushes against the fingers of the diaphragm spring of the pressure plate assembly, which in turn releases the clutch plate.

4   Terminology can be a problem when discussing the clutch components because common names are in some cases different from those used by the manufacturer. For example, the driven plate is also called the clutch plate or disc, the clutch release bearing is sometimes called a throwout bearing, and the release cylinder is sometimes called the slave cylinder.

5   Unless you're replacing components with obvious damage, perform these preliminary checks to diagnose clutch problems:

a) *The first check should be of the fluid level in the master cylinder. If the fluid level is low, add fluid as necessary and inspect the hydraulic system for leaks.*

b) *To check clutch spin-down time, run the engine at normal idle speed with the transmission in Neutral (clutch pedal up - engaged). Disengage the clutch (pedal down), wait several seconds and shift the transmission into Reverse. No grinding noise should be heard. A grinding*

noise would most likely indicate a bad pressure plate or clutch disc.

c) *To check for complete clutch release, run the engine (with the parking brake applied to prevent vehicle movement) and hold the clutch pedal approximately 1/2-inch from the floor. Shift the transmission between 1st gear and Reverse several times. If the shift is rough, component failure is indicated.*

d) *Visually inspect the pivot bushing at the top of the clutch pedal to make sure there's no binding or excessive play.*

3   Clutch hydraulic release system - removal and installation

Removal

Note: *The clutch hydraulic release system is serviced as an assembly. The components cannot be serviced separately. They're filled with fluid during manufacture, then sealed. They must not be disassembled or disconnected. If the system is leaking or has air in it, replace the entire assembly.*

1   Raise the vehicle and support it securely on jackstands.

2   Remove the fasteners attaching the release cylinder to the clutch housing, then separate the release cylinder from the clutch housing.

3   Disengage the clutch hydraulic fluid line from the retaining clip and stud on the firewall.

4   Lower the vehicle.

5   Pry off the retainer clip from the pushrod-to-clutch pedal pin, then slide the pushrod off the pin. Remove the pin bushing and inspect it. If it's worn, replace it.

6   Unplug the electrical connector for the clutch pedal position switch, then remove the clutch master cylinder mounting fasteners.

7   To avoid spillage, make sure that the cap on the master cylinder's remote reservoir is tight, then remove the reservoir mounting fasteners. **Caution:** *Brake fluid will damage paint. If any is accidentally spilled, wash it off with water.*

8   Detach the master cylinder from the firewall, then carefully lift the system from the vehicle.

Installation

Caution: *If you're installing a new clutch hydraulic release system, you'll notice a shipping stop on the master cylinder pushrod. DO NOT remove this shipping stop until after the clutch release cylinder has been installed, but DO remove it BEFORE the clutch is operated.*

Note: *The new clutch hydraulic system comes as two pieces to ease installation. After connecting the hydraulic line to the release cylinder, the line should not be disconnected.*

9   Carefully maneuver the clutch hydraulic release system assembly into position.

10   Insert the clutch master cylinder in the

firewall. Install the nuts and tighten them to the torque listed in this Chapter's Specifications.

11   Install the remote reservoir and tighten the fasteners securely.

12   Install the bushing on the pushrod-to-clutch pedal pivot pin. Use a new bushing, if necessary.

13   Install the master cylinder pushrod on the pin. Secure the rod with the retainer clip. Use a new clip if the old one is weak.

14   Plug in the electrical connector for the clutch pedal position switch.

15   Raise the vehicle and support it securely on jackstands.

16   Attach the clutch fluid hydraulic line to the retaining clip and stud on the firewall.

17   Install the release cylinder. Make sure the cap at the end of the cylinder rod is seated properly in the release fork.

18   Install the release cylinder mounting fasteners and tighten them securely. If a new clutch hydraulic release system has been installed, connect the hydraulic line to the release cylinder.

19   Remove the jackstands and lower the vehicle.

20   If a new clutch hydraulic release system has been installed, remove the plastic shipping stop from the master cylinder pushrod before operating the clutch.

21   Test drive the vehicle and make sure the clutch hydraulic release system is operating properly.

4   Clutch components - removal, inspection and installation

Warning: *Dust produced by clutch wear and deposited on clutch components is hazardous to your health. DO NOT blow it out with compressed air and DO NOT inhale it. DO NOT use gasoline or petroleum-based solvents to remove the dust. Brake system cleaner should be used to flush the dust into a drain pan. After the clutch components are wiped clean with a rag, dispose of the contaminated rags and cleaner in a covered, marked container.*

Removal

Refer to illustration 4.7

1   Access to the clutch components is normally accomplished by removing the transmission and clutch housing, leaving the engine in the vehicle. If, of course, the engine is being removed for major overhaul, then check the clutch for wear and replace worn components as necessary. However, the relatively low cost of the clutch components compared to the time and trouble spent gaining access to them warrants their replacement anytime the engine or transmission is removed, unless they are new or in near-perfect condition. The following procedures are based on the assumption the engine will stay in place.

2   Raise the vehicle and support it securely on jackstands.

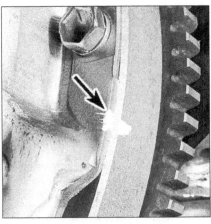

4.7 Mark the relationship of the pressure plate to the flywheel (if you're planning to re-use the old pressure plate)

4.11 The clutch disc

1 **Lining** - *this will wear down in use*
2 **Springs or dampers** - *check for cracking and deformation*
3 **Splined hub** - *the splines must not be worn and should slide smoothly on the transmission input shaft splines*
4 **Rivets** - *these secure the lining and will damage the flywheel or pressure plate if allowed to contact the surfaces*

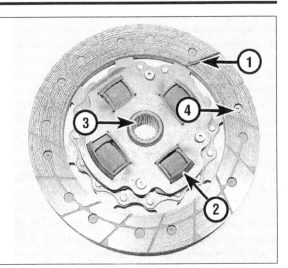

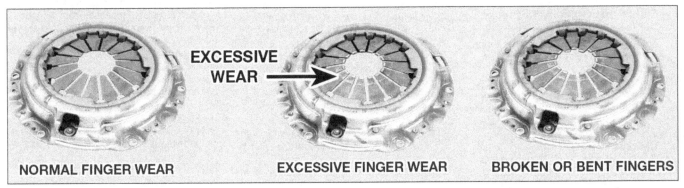

EXCESSIVE WEAR

NORMAL FINGER WEAR   EXCESSIVE FINGER WEAR   BROKEN OR BENT FINGERS

4.13a Replace the pressure plate if excessive wear or damage is noted

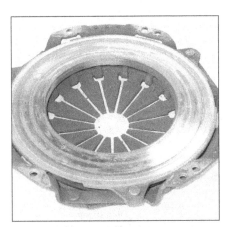

4.13b Inspect the pressure plate surface for excessive score marks, cracks and signs of overheating

3   Detach the clutch release cylinder from the clutch housing (see Section 3). **Caution:** *Do not disconnect the clutch hydraulic fluid line from the release cylinder. The hydraulic release system cannot be bled if air is allowed to enter the system.*
4   Remove the transmission (see Chapter 7A).
5   The clutch fork and release bearing can remain attached to the clutch housing for the time being.

6   To support the clutch disc during removal, install a clutch alignment tool through the clutch disc hub.
7   Carefully inspect the flywheel and pressure plate for indexing marks. The marks are usually an X, an O or a white letter. If they cannot be found, scribe marks yourself so the pressure plate and the flywheel will be in the same alignment during installation **(see illustration)**.
8   Turning each bolt only 1/4-turn at a time, loosen the pressure plate-to-flywheel bolts. Work in a criss-cross pattern until all spring pressure is relieved. Then hold the pressure plate securely and completely remove the bolts, followed by the pressure plate and clutch disc.

Inspection

Refer to illustrations 4.11, 4.13a and 4.13b

9   Ordinarily, when a problem occurs in the clutch, it can be attributed to wear of the clutch driven plate assembly (clutch disc). However, all components should be inspected at this time.
10   Inspect the flywheel for cracks, heat checking, grooves and other obvious defects. If the imperfections are slight, a machine shop can machine the surface flat and smooth, which is highly recommended regardless of the surface appearance. Refer to Chapter 2

for the flywheel removal and installation procedure.
11   Inspect the lining on the clutch disc. There should be at least 1/16-inch of lining above the rivet heads. Check for loose rivets, distortion, cracks, broken springs and other obvious damage **(see illustration)**. As mentioned above, ordinarily the clutch disc is routinely replaced, so if in doubt about the condition, replace it with a new one.
12   The release bearing should also be replaced along with the clutch disc (see Section 5).
13   Check the machined surfaces and the diaphragm spring fingers of the pressure plate **(see illustrations)**. If the surface is grooved or otherwise damaged, replace the pressure plate. Also check for obvious damage, distortion, cracking, etc. Light glazing can be removed with emery cloth or sandpaper. If a new pressure plate is required, new and remanufactured units are available.
14   Check the pilot bearing in the end of the crankshaft for excessive wear, scoring, dryness, roughness and any other obvious damage. If any of these conditions are noted, replace the bearing (see Section 6). **Note:** *Considering the amount of work you've done to get to this point, it's a good idea to go ahead and replace the pilot bearing at this time.*

4.16 Center the clutch disc in the pressure plate with an alignment tool before the bolts are tightened

5.5 To check the clutch release bearing, hold the hub (the center) of the bearing and rotate the outer portion while applying pressure; if the bearing doesn't turn smoothly or if it's noisy or rough, replace it

6.5 To remove the pilot bearing, use a special puller designed for the job

6.6 Tap the bearing into place with a bushing driver or a socket slightly smaller than the outside diameter of the bearing

Installation

Refer to illustration 4.16

15   Clean the machined surfaces with lacquer thinner or acetone. It's important that no oil or grease is on these surfaces or the lining of the clutch disc. Handle the parts only with clean hands. Install the flywheel (see Chapter 2).

16   Position the clutch disc and pressure plate against the flywheel. If installing the original pressure plate, align the index marks. Hold the clutch in place with an alignment tool **(see illustration)**. Make sure it's installed properly (most replacement clutch plates will be marked "flywheel side" or something similar - if it's not marked, install the clutch disc with the damper springs toward the transmission).

17   Tighten the pressure plate-to-flywheel bolts only finger-tight, working around the pressure plate.

18   Center the clutch disc by ensuring the alignment tool extends through the splined hub and into the pilot bearing in the crankshaft. Wiggle the tool up, down or from side-to-side as needed to bottom the tool in the pilot bearing. Tighten the pressure plate-to-flywheel bolts a little at a time, working in a criss-cross pattern, to prevent distorting the cover. After all the bolts are snug, tighten them to the torque listed in this Chapter's Specifications. Remove the alignment tool.

19   Using high-temperature grease, lubricate the inner groove of the release bearing (see Section 5). Also place grease on the release lever contact areas and the transmission input shaft bearing retainer.

20   Install the clutch release bearing, if removed (see Section 5).

21   Install the transmission (see Chapter 7A) and all components removed previously.

22   Remove the jackstands and lower the vehicle.

5   Clutch release bearing - removal, inspection and installation

Warning: *Dust produced by clutch wear is hazardous to your health. DO NOT blow it out with*

compressed air and DO NOT inhale it. DO NOT use gasoline or petroleum-based solvents to remove the dust. Brake system cleaner should be used to flush the dust into a drain pan. After the clutch components are wiped clean with a rag, dispose of the contaminated rags and cleaner in a covered, marked container.*

Removal

1   Raise the vehicle and support it securely on jackstands.

2   Remove the transmission (see Chapter 7A).

3   Detach the clutch release lever from the ballstud and remove the bearing and lever from the input shaft.

Inspection

Refer to illustration 5.5

4   Wipe off the bearing with a clean rag and inspect it for damage, wear and cracks. Don't immerse the bearing in solvent - it's sealed for life and immersion in solvent will ruin it.

5   Hold the center of the bearing and rotate the outer portion while applying pressure **(see illustration)**. If the bearing doesn't turn smoothly or if it's noisy or rough, replace it. **Note:** *Considering the difficulty involved with replacing the release bearing, we recommend replacing the release bearing whenever the clutch components are replaced.*

Installation

6   Lightly lubricate the friction surfaces of the release bearing, ballstud and the input shaft with high-temperature grease.

7   Install the release lever and bearing onto the input shaft.

8   The remainder of installation is the reverse of removal.

6   Pilot bearing - inspection and replacement

Refer to illustrations 6.5 and 6.6

1   The clutch pilot bearing is a needle roller

type bearing which is pressed into the rear of the crankshaft. It's greased at the factory and does not require additional lubrication. Its primary purpose is to support the front of the transmission input shaft. The pilot bearing should be inspected whenever the clutch components are removed from the engine. Because of its inaccessibility, replace it with a new one if you have any doubt about its condition. **Note:** *If the engine has been removed from the vehicle, disregard the following Steps which don't apply.*

2   Remove the transmission (see Chapter 7A).

3   Remove the clutch components (see Section 4).

4   Using a flashlight, inspect the bearing for excessive wear, scoring, dryness, roughness and any other obvious damage. If any of these conditions are noted, replace the bearing.

5   Removal can be accomplished with a special puller available at most auto parts stores **(see illustration)**.

6   To install the new bearing, lightly lubricate the outside surface with grease, then with the letter side of the bearing facing the transmission, drive it into the recess with a hammer and socket **(see illustration)** or a bushing driver.

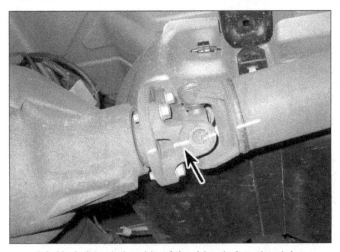

9.2 Mark the relationship of the driveshaft to the pinion flange or yoke

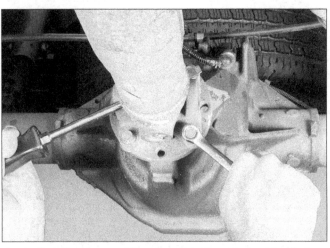

9.3 Insert a large screwdriver or prybar through the driveshaft or pinion yoke as shown, to prevent the shaft from turning when you loosen the fasteners

7   Install the clutch components, transmission and all other components removed previously. Tighten all fasteners to the recommended torque.

7   Clutch pedal position switch - check and replacement

Check

1   The clutch pedal position switch, which is part of the starter relay circuit, is mounted on the clutch master cylinder pushrod. The switch closes the starter relay circuit only when the clutch pedal is fully depressed. The switch is an integral part of the clutch master cylinder pushrod and cannot be serviced separately. If the switch must be replaced, so must the clutch hydraulic release system (see Section 3).
2   To test the switch, verify that the engine will not crank over when the clutch pedal is in the released position, and that it does crank over with the pedal depressed.
3   If the engine starts without depressing the clutch pedal, replace the switch.

Replacement

4   The switch is an integral part of the clutch master cylinder pushrod and cannot be serviced separately. Replace the clutch hydraulic release system (see Section 3).

8   Driveshaft(s) - general information

1   A driveshaft is a tube that transmits power between the transmission or transfer case and the differential(s). Universal joints are located at either end of the driveshaft and allow the driveshaft to operate at different angles as the suspension moves.

2   Three different types of universal joints are used: single-cardan, double-cardan and constant velocity.
3   Some models have a two-piece driveshaft. The two driveshafts are connected at a center support bearing.
4   The rear driveshaft employs a splined yoke at the front, which slips into the extension housing of the transmission. The front driveshaft on 4WD models (and the rear half of the driveshaft on models with a two-piece driveshaft) incorporates a slip yoke as part of the shaft. This arrangement allows the driveshaft to alter its length during vehicle operation. An oil seal prevents fluid from leaking out of the extension housing and keeps dirt from entering the transmission or transfer case. If leakage is evident at the front of the driveshaft, replace the extension housing oil seal (see Chapter 7).
5   The driveshaft assembly requires very little service. Factory U-joints are lubricated for life and must be replaced if problems develop. The driveshaft must be removed from the vehicle for this procedure. **Note:** *Some aftermarket universal joints have grease fittings to allow periodic lubrication.*
6   Since the driveshaft is a balanced unit, it's important that no undercoating, mud, etc. be allowed to accumulate on it. When the vehicle is raised for service it's a good idea to clean the driveshaft and inspect it for any obvious damage. Also, make sure the small weights used to originally balance the driveshaft are in place and securely attached. Whenever the driveshaft is removed it must be reinstalled in the same relative position to preserve the balance.
7   Problems with the driveshaft are usually indicated by a noise or vibration while driving the vehicle. A road test should verify if the problem is the driveshaft or another vehicle component. Refer to the *Troubleshooting* section at the front of this manual.

9   Driveshaft(s) - removal and installation

Note: *The manufacturer recommends replacing driveshaft fasteners with new ones when installing the driveshaft.*

Rear driveshaft
Removal
Refer to illustrations 9.2 and 9.3
Note: *Where a two-piece driveshaft is involved, the rear shaft must be removed before the front shaft.*
1   Raise the vehicle and support it securely on jackstands.
2   Use chalk or a scribe to mark the relationship of the driveshaft to the differential axle assembly mating flange or yoke. This ensures correct alignment when the driveshaft is reinstalled **(see illustration).**
3   Remove the bolts securing the driveshaft flange or universal joint clamps to the differential pinion flange or yoke **(see illustration).** Turn the driveshaft (or wheels) as necessary to bring the bolts into the most accessible position.
4   Pry the universal joint away from its mating flange or yoke and remove the shaft from the flange. Be careful not to let the caps fall off of the universal joint (which would cause contamination and loss of the needle bearings).
5   Lower the rear of the driveshaft. If the driveshaft is a one-piece unit, slide the front end of the driveshaft out of the transmission extension housing; if it's a two-piece driveshaft, mark the relationship of the center support bearing to the support bracket, then unbolt the center support bearing and slide the front end of the front driveshaft out of the extension housing.
6   Wrap a plastic bag over the extension housing and hold it in place with a rubber

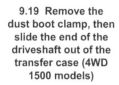

9.19 Remove the dust boot clamp, then slide the end of the driveshaft out of the transfer case (4WD 1500 models)

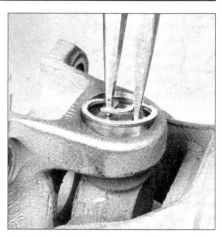

12.2 A pair of needle-nose pliers can be used to remove the universal joint snap-rings

band. This will prevent loss of fluid and protect against contamination while the driveshaft is out.

Installation

7   Remove the plastic bag from the transmission or transfer case extension housing and wipe the area clean. Inspect the oil seal carefully. If it's leaking, now is the time to replace it (see Chapter 7A).
8   Inspect the center support bearing, if equipped. If it's rough or noisy, replace it (see Section 10).
9   Insert the front end of the driveshaft assembly into the transmission or transfer case extension housing.
10   If the driveshaft is a one-piece unit, raise the rear of the driveshaft into position, checking to be sure the marks are in alignment. If not, turn the pinion flange until the marks line up.
11   If the driveshaft is a two-piece unit, raise the center support bearing and bolt it loosely into place. Raise the rear end of the rear shaft into position and make sure the alignment marks are in alignment; if not, turn the pinion flange until they are.
12   Remove the tape securing the bearing caps and install the clamps, if equipped, and fasteners. Tighten the center support bearing bolts to the torque listed in this Chapter's Specifications.

Front driveshaft (4WD models)

Refer to illustration 9.19

13   Raise the vehicle and support it securely on jackstands.
14   If equipped, remove the skid plate.
15   If necessary, remove the exhaust crossover pipe.
16   Use chalk or a scribe to mark the relationship of the driveshaft to the differential axle assembly mating flange or yoke **(see illustration 9.2)**. This ensures correct alignment when the driveshaft is reinstalled.
17   Remove the bolts and straps that secure the front end of the driveshaft to the differential mating flange or yoke.
18   On models equipped with a transfer case flange, make alignment marks on the

driveshaft and transfer case flanges. Unbolt the flange that secures the driveshaft universal joint to the transfer case and differential, then remove the driveshaft.
19   On 1500 models, remove the dust boot clamp, then slide the end of the driveshaft out of the transfer case **(see illustration)**.
20   Installation is the reverse of removal. If the shaft cannot be lined up due to the components of the differential or transfer case having been rotated, put the vehicle in Neutral or rotate one wheel to allow the original alignment to be achieved. Make sure the universal joint caps are properly placed in the flange seat. Tighten the fasteners to the torque listed in this Chapter's Specifications.

10   Driveshaft center support bearing - removal and installation

1   Raise the vehicle and support it securely on jackstands.
2   Remove the driveshaft assembly (see Section 9). Mark the relationship of the front portion of the driveshaft to the rear portion of the driveshaft.
3   Loosen the slip-joint boot clamp and pull back the boot on the front of the rear driveshaft.
4   Pull the rear driveshaft out of the center bearing.
5   Take the front driveshaft and center bearing to an automotive machine shop and have the old bearing pressed off and a new bearing pressed on.
6   Installation is the reverse of removal. Be sure the match marks line up so the driveshaft is properly phased.

11   Universal joints - general information and check

1   Universal joints are mechanical couplings which connect two rotating components that meet each other at different angles.
2   These joints are composed of a yoke on each side connected by a crosspiece called a trunnion. Cups at each end of the trunnion

contain needle bearings which provide smooth transfer of the torque load. Snap-rings, either inside or outside of the bearing cups, hold the assembly together.
3   Wear in the needle roller bearings is characterized by vibration in the driveline, noise during acceleration, and in extreme cases of lack of lubrication, metallic squeaking and ultimately grating and shrieking sounds as the bearings disintegrate.
4   It is easy to check if the needle bearings are worn with the driveshaft in position, by trying to turn the shaft with one hand, the other hand holding the rear axle flange when the rear universal joint is being checked, and the front half coupling when the front universal joint is being checked. Any movement between the driveshaft and the front half couplings, and around the rear half couplings, is indicative of considerable wear. Another method of checking for universal joint wear is to use a prybar inserted into the gap between the universal joint and the driveshaft or flange. Leave the vehicle in gear and try to pry the joint both radially and axially. Any looseness should be apparent with this method. A final test for wear is to attempt to lift the shaft and note any movement between the yokes of the joints.
5   If any of the above conditions exist, replace the universal joints with new ones.

12   Universal joints - replacement

Single-cardan U-joints

Refer to illustrations 12.2, 12.4 and 12.9

Note: *A press or large vise will be required for this procedure. It may be advisable to take the driveshaft to a local dealer service department, service station or machine shop where the universal joints can be replaced for you, normally at a reasonable charge.*

1   Remove the driveshaft (see Section 9).
2   On U-joints with external snap-rings, use a small pair of pliers to remove the snap-rings from the spider **(see illustration)**.

12.4 To press the universal joint out of the driveshaft yoke, set it up in a vise with the small socket pushing the joint and bearing cap into the large socket

12.9 If the snap-ring will not seat in the groove, strike the yoke with a brass hammer - this will relieve the tension that has set up in the yoke and slightly spring the yoke ears (this should also be done if the joint feels tight when assembled)

3   Supporting the driveshaft, place it in position on a workbench equipped with a vise.
4   Place a piece of pipe or a large socket, having an inside diameter slightly larger than the outside diameter of the bearing caps, over one of the bearing caps. Position a socket with an outside diameter slightly smaller than that of the opposite bearing cap against the cap **(see illustration)** and use the vise or press to force the bearing cap out (inside the pipe or large socket). Use the vise or large pliers to work the bearing cap the rest of the way out.
5   Transfer the sockets to the other side and press the opposite bearing cap out in the same manner.
6   Pack the new universal joint bearings with grease. Ordinarily, specific instructions for lubrication will be included with the universal joint servicing kit and should be followed carefully.
7   Position the spider in the yoke and partially install one bearing cap in the yoke.
8   Start the spider into the bearing cap, then partially install the other cap. Align the spider and press the bearing caps into position, being careful not to damage the dust seals.
9   Install the snap-rings. If difficulty is encountered in seating the snap-rings, strike the driveshaft yoke sharply with a hammer. This should spring the yoke ears slightly and allow the snap-rings to seat in the groove **(see illustration).**
10   Install the grease fitting and fill the joint with grease. Do not overfill the joint, as this could blow out the grease seals.
11   Install the driveshaft (see Section 9).

Double-cardan U-joints

12   Use the above procedure, but note that it will have to be repeated because the double-cardan joint is made up of two single-cardan joints. Also pay attention to how the spring,

centering ball and bearing are arranged. **Note:** *Both U-joints in the double-cardan assembly must be replaced at the same time, even if only half of it is worn out.*

13 Driveaxle (4WD models with independent front suspension) - removal and installation

1   Loosen the wheel lug nuts, raise the vehicle and support it securely on jackstands. Remove the wheel(s).
2   Remove the caliper and brake disc as outlined in Chapter 9, then disconnect the ABS sensor and secure the sensor and wire harness aside.
3   Remove the driveaxle/hub nut. Place a prybar between two of the wheel studs to prevent the hub from turning while loosening the nut.
4   Support the lower control arm with a floor jack, then slightly raise it to take the force of the spring off the upper control arm. Remove the shock absorber lower mounting bolt, then remove the upper control arm balljoint nut and separate the control arm from the steering knuckle (see Chapter 10).
5   Using a prybar or slide hammer with CV joint adapter, pry the inner CV joint assembly from the front differential. Be careful not to damage the front differential. Suspend the driveaxle with a piece of wire - don't let it hang, or damage to the outer CV joint may occur.
6   Pull the steering knuckle out and away from the outer CV joint of the driveaxle. If the driveaxle proves difficult to remove, tap the end of the driveaxle with a soft-faced hammer or a hammer and a brass-punch. If the driveaxle is stuck in the hub splines and won't

move, it may be necessary to push it from the hub with a two-jaw puller.
7   Once the driveaxle is loose from the hub splines, pull out on the hub/knuckle assembly. Remove the support wire and guide the driveaxle out from under the vehicle.
8   Installation is the reverse of removal, noting the following points:
   a) *Thoroughly clean the splines and bearing shield on the outer CV joint. This is very important, as the bearing shield protects the wheel bearings from water and contamination. Also clean the wheel bearing area of the steering knuckle.*
   b) *Thoroughly clean the splines and oil seal sealing surface on the inner tripod CV joint. Apply a film of multi-purpose grease around the oil seal contact surface of the inner CV joint.*
   c) *When installing the driveaxle, hold the driveaxle straight out, then push it in sharply to seat the set-ring on the splines of the inner CV joint. To make sure the set-ring is properly seated, attempt to pull the inner CV joint housing out of the differential by hand. If the set-ring is properly seated, the inner joint will not move out.*
   d) *Tighten the driveaxle/hub nut to the torque listed in this Chapter's Specifications.*
   e) *Install the wheel and lug nuts, lower the vehicle and tighten the lug nuts to the torque listed in the Chapter 1 Specifications.*
   f) *The upper balljoint-to-steering knuckle nut should not be reused. A new one should always be used. Tighten it to the torque listed in the Chapter 10 Specifications.*

14 Driveaxle boot - replacement

Note: *If the CV joints exhibit wear, indicating the need for an overhaul (usually due to torn boots), explore all options before beginning the job. Complete rebuilt driveaxles may be available on an exchange basis, which eliminates a lot of time and work. Whatever is decided, check on the cost and availability of parts before disassembling the joints.*
1   Loosen the wheel lug nuts. Raise the vehicle and support it securely on jackstands, then remove the wheel.
2   Remove the driveaxle (see Section 13).

Inner CV joint
Disassembly

Refer to illustration 14.6

3   Mount the driveaxle in a bench vise with wood blocks to protect it. **Caution:** *Do not overtighten the vise.*
4   Remove the boot retaining clamps and slide the inner boot back onto the shaft.
5   Pull the inner CV joint housing off the shaft and tripod.

14.6 Remove the snap-ring from the groove in the end of the axleshaft

14.12 Wrap the driveshaft splines with tape to prevent damaging the boot as it's slid onto the shaft

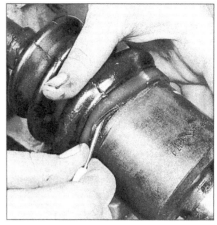

14.19 After positioning the joint mid-way through its travel, equalize the pressure inside the boot by inserting a small, dull screwdriver between the boot and the CV joint housing

6   Use a pair of snap-ring pliers and remove the snap-ring from the end of the shaft **(see illustration)**.
7   Mark the end of the shaft and the tripod, then remove the tripod from the shaft.
8   Remove the boot from the shaft.
9   Clean the housing and the tripod with solvent.
10   Check the tripod components and the housing for excessive wear and/or damage.
11   If any components are worn or damaged, the entire joint must be replaced.

Assembly

Refer to illustrations 14.12, 14.19 and 14.20

12   Wrap the splines of the shaft with tape to prevent damage to the boot, then install the small boot clamp and boot onto the shaft **(see illustration)**.
13   Install the tripod onto the end of the shaft, with the mark you made facing the end of the shaft (and aligned with the mark on the shaft).
14   Install the snap-ring, making sure it is completely seated in its groove.
15   Apply CV joint grease to the tripod and interior of the housing. **Note:** *If grease was not included with the new boot, obtain some CV joint grease - don't use any other type of grease.*
16   Apply the remainder of the grease into the boot, then insert the shaft and tripod into the housing.
17   Position the large-diameter end of the boot over the edge of the housing and seat the lip of the boot into the locating groove at the edge of the housing. Insert the lip of the small-diameter end of the boot into the locating groove on the shaft.
18   Adjust the length of the joint by positioning it mid-way through its travel.
19   Insert a small screwdriver between the boot and the housing to equalize the pressure inside the boot **(see illustration)**
20   Tighten the boot clamps **(see illustration)**.

Outer CV joint
Disassembly

Refer to illustrations 14.23, 14.27, 14.28, 14.30, 14.31, 14.34a and 14.34b

21   Mount the axleshaft in a vise with wood blocks to protect it, remove the boot clamps and push the boot back.
22   Wipe the grease from the joint.
23   Using a pair of snap-ring pliers, expand the snap-ring retaining the outer joint to the shaft, then remove the joint **(see illustration)**.
24   Slide the boot off the driveaxle.
25   Clean the axle spline area and inspect for wear, damage, corrosion and broken splines.
26   Clean the outer CV joint bearing assembly with a clean cloth to remove excess grease.
27   Mark the relative position of the bearing cage, inner race and housing **(see illustration)**.

14.20 Depending on the type of clamps furnished with the replacement boot, you'll most likely need a special pair of clamp tightening pliers (most auto parts stores carry these)

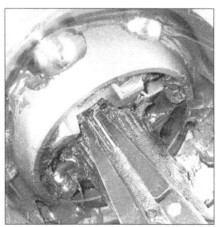

14.23 After expanding the snap-ring, the outer joint assembly can be removed

14.27 Mark the bearing cage, inner race and housing relationship after removing the grease

14.28 With the cage and inner race tilted, the balls can be removed one at a time

14.30 Align one of the elongated windows in the cage with one of the lands on the housing (outer race), then rock the cage and inner race out of the housing

14.31 Tilt the inner race 90-degrees, align the race lands with the windows in the cage, then separate the two components

28   Mount the CV joint in the vise with wood blocks to protect the stub shaft. Push down one side of the cage and remove the ball bearing from the opposite side **(see illustration)**. The balls may have to be pried out.

29   Repeat this procedure until all of the balls are removed. If the joint is tight, tap on the inner race (not the cage) with a hammer and brass drift.

30   Remove the bearing assembly from the housing by tilting it vertically and aligning two opposing cage windows in the area between the ball grooves **(see illustration)**.

31   Turn the inner race 90-degrees to the cage and align one of the spherical lands with an elongated cage window. Raise the land into the window and swivel the inner race out of the cage **(see illustration)**.

32   Clean all of the parts with solvent and dry them with compressed air (if available).

33   Inspect the housing, splines, balls and races for damage, corrosion, wear and cracks.

34   Check the inner race for wear and scoring. If any of the components are not serviceable, the entire CV joint assembly must be replaced with a new one **(see illustrations)**.

Assembly

Refer to illustration 14.40

35   Apply a thin film of oil to all CV joint components before beginning reassembly.

36   Align the marks and install the inner race in the cage so one of the race lands fits into the elongated window **(see illustration 14.31)**.

37   Rotate the inner race into position in the cage and install the assembly in the CV joint housing, again using the elongated window for clearance **(see illustration 14.30)**.

38   Rotate the inner race into position in the housing. Be sure the large counterbore of the inner race faces out. The marks made during disassembly should face out and be aligned.

39   Pack the lubricant from the kit into the ball races and grooves.

40   Install the balls into the holes, one at a time, until they are all in position. Fill the joint with grease through the splined hole, then insert a wooden dowel into the splined hole to force the grease into the joint **(see illustration)**.

41   Place the driveaxle in the vise and slide the inner clamp and boot over it (wrap the shaft splines with tape to prevent damaging the boot) **(see illustration 14.12)**.

42   Place the CV joint housing in position on the axle, align the splines and push it into place. If necessary, tap it on with a soft-face hammer. Make sure it is seated on the snapring by attempting to pull it from the shaft.

43   Make sure the boot is not distorted, then install and tighten the clamps **(see illustration 14.20)**.

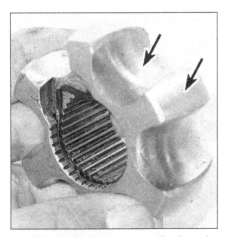

14.34a Check the inner race lands and grooves for pitting and score marks

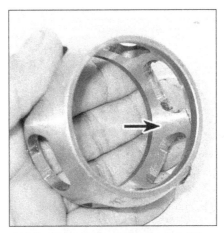

14.34b Check the cage for cracks, pitting and score marks (shiny spots are normal and don't affect operation)

14.40 Apply grease through the splined hole, then insert a wooden dowel into the hole and push down - the dowel will force the grease into the joint

16.10 The rear axle is retained to each leaf spring by two U-bolts and four nuts

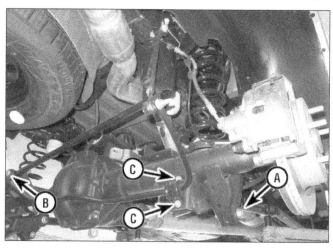

16.17 All suspension components must be detached from the rear axle including shock absorbers (A), track bar (B) and stabilizer bar (C); the control arms mount at the front of the axle

15   Rear axle - general information

The rear axle assembly consists of a straight, hollow housing enclosing a differential assembly and axleshafts. These assemblies support the vehicle's sprung weight components through leaf springs attached between the axle housings and the vehicle's frame rails.

There are two different rear axle assemblies: light-duty axles are a semi-floating design, in which the axle supports the weight of the vehicle on the axleshaft in addition to transmitting driving forces to the rear wheels.

Heavy-duty models use a full-floating axle. A full floating axleshaft doesn't carry any of the vehicle's weight; this is supported on the axle housing itself by roller bearings. Full-floating axles can be identified by the large hub projecting from the center of the wheel. The axle flange is secured by bolts on the end of the hub.

Due to the need for special tools and equipment, it is recommended that operations on these models be limited to those described in this Chapter. Where repair or overhaul is required, remove the axle assembly and take it to a rebuilder, or exchange it for a new or reconditioned unit. Always make sure that an axle unit is exchanged for one of identical type and gear ratio.

16   Rear axle assembly - removal and installation

Removal

1   Raise the rear of the vehicle and support it with jackstands placed under the frame rails.
2   Remove the rear wheels.
3   Disconnect the driveshaft from the rear axle (see Section 9).

Leaf spring models

Refer to illustration 16.10

4   Disconnect the ABS sensor(s).
5   Disconnect the parking brake cable from the parking brake lever (see Chapter 9).
6   Unscrew the vent hose fitting to detach the brake line junction block from the axle tube.
7   Disconnect the brake lines from the clips and brackets on the axle housing. Remove the rear brake calipers (see Chapter 9). **Caution:** *Tie the calipers up with wire to keep any stress off the flexible brake lines.*
8   Support the rear axle with a floor jack. If the rear differential is offset to one side, you'll have to use two jacks - one placed under each axle tube.
9   Remove the lower mounting bolts securing the rear shocks to the axle (see Chapter 10).
10   With the jack(s) supporting the axle, remove the nuts and U-bolts securing the axle to the springs **(see illustration)**.
11   Lower the axle assembly and remove it from under the vehicle.

Coil spring models

Refer to illustration 16.17

12   Disconnect the wiring from the rear wheel speed sensors, then unbolt the sensors and remove them.
13   Remove the mounting bolts from the rear brake calipers and hang them out of the way using wire. Take care to avoid putting tension on the brake hoses.
14   Disconnect the vent hose from the axle tube.
15   Disconnect the parking brake cables from the rear brake assemblies (see Chapter 9).
16   Detach the parking brake cable clamps, then move the cable clear of the rear axle.
17   Remove the bolts from the stabilizer bar clamps **(see illustration)**. Support the bar with wire.

18   Support the center of the axle with a floor jack. Raise the axle enough to avoid fully extending the shock absorbers.
19   Remove the bolts from the lower shock absorber mounts.
20   Unbolt the track bar from the rear axle and secure it out of the way.
21   Remove the bolts that secure the four control arms to the rear axle. **Note:** *Discard all suspension mounting bolts. They should be replaced with new ones (see Chapter 10).*
22   With the help of an assistant, slowly lower the rear axle until the coil springs can be removed.
23   Fully lower the axle and remove it from under the vehicle.

Installation

24   Installation is the reverse of removal.
25   Tighten all fasteners to the torques listed in the Chapter 10 Specifications. Tighten the caliper mounting bolts to the torque listed in the Chapter 9 Specifications. If necessary, check and fill the axle with the specified lubricant (see Chapter 1).

17   Rear axleshaft - removal and installation

Semi-floating axleshaft

Refer to illustrations 17.3a, 17.3b and 17.4

Warning: *The dust created by the brake system is harmful to your health. Never blow it out with compressed air and don't inhale any of it. An approved filtering mask should be worn when working on the brakes. Do not, under any circumstances, use petroleum-based solvents to clean brake parts. Use brake system cleaner only!*

1   The axleshaft is usually removed only when the bearing is worn or the seal is leaking. To check the bearing, raise the vehicle, support it securely on jackstands and remove

17.3a Remove the pinion mate shaft lock bolt . . .

17.3b . . . then carefully remove the pinion mate shaft from the differential carrier (don't turn the wheels or the carrier after the shaft has been removed, or the pinion gears may fall out)

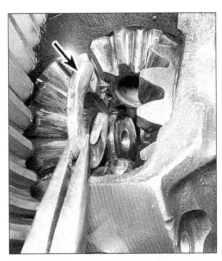

17.4 Push the axle flange in, then remove the C-lock from the inner end of the axleshaft

the wheel and brake drum (see Chapter 9). Try to move the axle flange up and down. If the axleshaft moves up and down, bearing wear is excessive. Also check for differential lubricant leaking out from below the axleshaft - this indicates the seal is leaking.

2   Remove the cover from the differential carrier and allow the lubricant to drain into a suitable container.

3   Remove the lock bolt from the differential pinion mate shaft and remove the shaft (see illustrations).

4   Push in the outer (flanged) end of the axleshaft and remove the C-lock from the inner end of the shaft (see illustration).

5   Withdraw the axleshaft, taking care not to damage the oil seal in the end of the axle housing as the splined end of the axleshaft passes through it.

6   Installation is the reverse of removal. Apply thread-locking compound to the threads and tighten the pinion shaft lock bolt to the torque listed in this Chapter's Specifications.

7   Always use a new cover gasket (or clean off the old RTV sealant and apply new sealant [see Chapter 1]) and tighten the cover bolts to the torque listed in the Chapter 1 Specifications.

8   Refill the axle with the correct quantity and grade of lubricant (see Chapter 1).

Full-floating axleshaft

Refer to illustrations 17.10, 17.11, 17.14a and 17.14b

9   Loosen the rear wheel lug nuts, raise the rear of the vehicle and support it securely on jackstands. Remove the wheel. **Note:** *If you're just removing the axleshaft, it isn't necessary to remove the wheel or raise the rear of the vehicle. If you are going to replace the hub seals and/or bearings, the rear of the vehicle must be supported on jackstands and the wheel will have to be removed.*

10   Remove the axleshaft flange bolts (see illustration).

11   Pull out the axleshaft (see illustration).

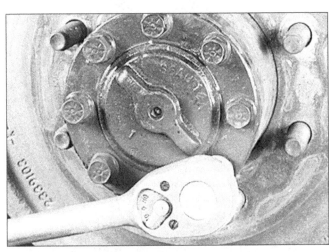

17.10 Remove the axleshaft flange bolts . . .

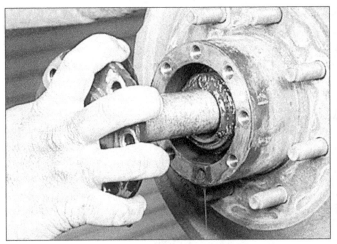

17.11 . . . and pull out the axleshaft

If the same axleshaft is going to be installed, clean the flange mating surface.

12   Clean the flange mating surface on the hub. While the axleshaft is removed, inspect and, if necessary, replace the hub and bearing seals and bearings (see Section 18). This is also a good time to inspect the rear brake assembly (see Chapter 9).

13   Slip a new gasket over the end of the axleshaft and slide the axleshaft into the axle housing.

14   Before engaging the axleshaft splines with the differential, apply RTV sealant to the gasket mating surface of the hub and place the new gasket in position **(see illustrations)**. Push the axleshaft into the axle housing until the axleshaft splines are fully engaged with the differential. If you have difficulty engaging the splines with the differential, have an assistant turn the other wheel slightly while you push on the axleshaft.

15   Install the flange bolts and tighten them to the torque listed in this Chapter's Specifications. **Caution:** *The manufacturer recommends using new bolts, but the old bolts can be used if the threads are cleaned and a thread locking compound is used when they are installed.*

16   If the wheel was removed, install it and the lug nuts. Lower the vehicle and tighten the lug nuts to the torque listed in the Chapter 1 Specifications.

18   Rear wheel hub bearing and grease seal (full-floating axle) - removal, installation and adjustment

10.5-inch and 11.5-inch ring gear axles

Removal

1   Remove the axleshaft (see Section 17).

2   Remove the retaining ring and key from the end of the axle housing.

3   Remove the hub nut, using a special socket available at most auto parts stores.

4   Pull the hub and bearing assembly straight off the axle tube.

5   Remove and discard the oil seal from the back of the hub.

6   To further disassemble the hub, use a hammer and a long bar or drift punch to knock out the inner bearing, cup (race) and oil seal.

7   Remove the outer retaining ring, then knock the outer bearing and cup from the hub.

Installation

8   Clean the old sealing compound from the seal bore in the hub.

9   Use solvent to clean the bearings, hub and axle tube. A small brush may prove useful; make sure no bristles from the brush embed themselves in the bearing rollers. Spray the bearings with brake system cleaner, which will remove the solvent and allow the bearings to

17.14a Apply a coat of RTV sealant to the mating surface of the hub . . .

dry much more rapidly.

10   Carefully inspect the bearings for cracks, wear and damage. Check the axle tube flange, studs and hub splines for damage and corrosion. Check the bearing cups (races) for pitting or scoring. Worn or damaged components must be replaced with new ones.

11   Lubricate the bearings, races and the axle tube contact areas with wheel bearing grease. Work the grease completely into the bearings, forcing it between the rollers, cone and cage.

12   Reassemble the hub by reversing the disassembly procedure. Use only the proper size bearing driver when installing the new bearing cups (races), and make sure they are driven in straight and completely.

13   Make sure the axle housing oil deflector is in position. Place the hub assembly on the axle tube, taking care not to damage the oil seals.

14   Install the hub nut and adjust the bearings as described below.

Adjustment

15   Rotate the hub, making sure it turns freely.

16   While rotating the hub in the normal direction of rotation (forward), tighten the hub nut to 22 ft-lbs with a torque wrench. Again, this will require the special socket, which is available at most auto parts stores.

17   Back the nut off 1/4-turn, then tighten the nut hand-tight (just enough to remove the freeplay in the bearing) with the special socket.

18   Turn the nut to align the closest slot in the nut with the keyway in the spindle, then install the key.

19   Install the retaining ring in the end of the spindle.

20   Try to wiggle the hub assembly; you shouldn't be able to detect any play, but the hub should turn freely (there shouldn't be any preload on the bearings, but there shouldn't be any freeplay, either). If necessary, repeat the adjustment procedure.

21   Install the axleshaft (see Section 17).

17.14b . . . then place the new gasket in position on the hub

12.8-inch ring gear (302RBI) axles

Removal

22   Remove the axleshaft (see Section 17).

23   Remove the hub nut, using a special socket available at most auto parts stores.

24   Remove the hub and brake disc assembly. **Caution:** *This assembly is heavy. Take care to avoid dropping it.*

25   Remove the outer bearing from the hub, then drive out the oil seal using a hammer and a long punch or a seal removal tool.

26   Remove the inner bearing.

27   Drive out the bearing races using a hammer and a brass drift to avoid damaging them.

28   Refer to Steps 9 through 12 to service and install the bearings.

Installation

29   Carefully install the hub and rotor assembly. **Note:** *It may be easiest to have an assistant help you with this. The assembly is heavy and the seal is easily damaged during installation.*

30   Install the hub nut with its attached washer, lining up the washer's tab with the groove in the housing.

31   Using the special socket and a torque wrench, tighten the nut to the torque listed in this Chapter's Specifications.

32   Refer to Section 17 to complete the installation procedure.

19   Rear axleshaft oil seal (semi-floating axle) - replacement

1   Remove the axleshaft (see Section 17).

2   Pry the oil seal out of the end of the axle housing with a seal removal tool.

3   Apply a film of multi-purpose grease to the oil seal recess and tap the new seal evenly into place with a hammer and seal installation tool, large socket or piece of pipe, so the lips are facing in and the metal face is visible

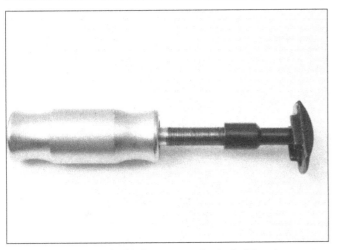

20.2 A typical slide hammer and axleshaft bearing remover attachment

21.3 Use an inch-pound torque wrench to check the torque necessary to rotate the pinion shaft

from the end of the axle housing. When correctly installed, the face of the oil seal should be flush with the end of the axle housing.

4   Install the axleshaft (see Section 17).

20   Rear axleshaft bearing (semi-floating axle) - replacement

Refer to illustration 20.2

1   Remove the axleshaft (see Section 17) and the oil seal (see Section 19).

2   A bearing puller which grips the bearing from behind will be required for this job **(see illustration).** Extract the bearing from the axle housing.

3   Clean out the bearing recess and drive in the new bearing with a bearing driver. Make sure the bearing is tapped in to the full depth of the recess.

4   Install a new oil seal (see Section 19), then install the axleshaft (see Section 17).

21   Pinion oil seal - replacement

Refer to illustrations 21.3, 21.5, 21.8 and 21.9

Note: *This procedure applies to the front and rear pinion oil seals.*

1   Loosen the wheel lug nuts. Raise the front (for front differential) or rear (for rear differential) of the vehicle and support it securely on jackstands. Block the opposite set of wheels to keep the vehicle from rolling off the stands. Remove the wheels.

2   Disconnect the driveshaft from the differential pinion flange and support it out of the way with a piece of wire or rope (see Section 9).

3   Rotate the pinion a few times by hand. Use a beam-type or dial-type inch-pound torque wrench to check the torque required to rotate the pinion **(see illustration).** Record it for use later.

4   Mark the relationship of the pinion flange to the shaft then count and write down the number of exposed threads on the shaft.

5   A special tool, available at most auto parts stores, can be used to keep the companion flange from moving while the self-locking pinion nut is loosened. A chain wrench can also be used to immobilize the flange **(see illustration).**

6   Remove the pinion nut.

7   Withdraw the flange. It may be necessary to use a two-jaw puller engaged behind the flange to draw it off. Do not attempt to pry or hammer behind the flange or hammer on the end of the pinion shaft.

8   Pry out the old seal and discard it **(see illustration).**

9   Lubricate the lips of the new seal and fill the space between the seal lips with wheel bearing grease, then tap it evenly into position with a seal installation tool or a large socket **(see illustration).** Make sure it enters the housing squarely and is tapped in to its full depth.

10   Install the pinion flange, lining up the marks made in Step 4. If necessary, tighten the pinion nut to draw the flange into place.

21.5 A chain wrench can be used to hold the pinion flange while the nut is loosened

21.8 Use a seal removal tool or a large screwdriver to remove the pinion seal (be careful not to disturb the pinion while doing this)

21.9 A large socket with a diameter the same as that of the new seal can be used to drive the pinion seal into the differential housing

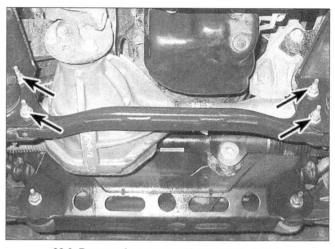

22.6 Remove the crossmember mounting bolts

Do not try to hammer the flange into position.
11   Apply a bead of RTV sealant to the ends of the splines visible in the center of the flange so oil will be sealed in.
12   Install the washer and a new pinion nut. Tighten the nut until the number of threads recorded in Step 4 are exposed.
13   Measure the torque required to rotate the pinion and tighten the nut in small increments (no more than 5 ft-lbs) until it matches the figure recorded in Step 3. To compensate for the drag of the new oil seal, the nut should be tightened a little more until the rotational torque of the pinion exceeds the earlier recording by 5 in-lbs.
14   Reinstall all components removed previously by reversing the removal Steps, tightening all fasteners to their specified torque values.

22   Front axle assembly (4WD models) - removal and installation

Independent front suspension

Refer to illustrations 22.6 and 22.10

1   Loosen the front wheel lug nuts. Raise the front of the vehicle and support it with jackstands placed under the frame rails.
2   Detach the vent tube from the axle and seal it. **Caution:** *Make sure that you perform this Step now and not while the axle assembly is being removed. If you wait, and the vent gets filled with oil, it may be ruined and damage the entire axle.*
3   Remove the driveaxles (see Section 13).
4   Disconnect the driveshaft from the differential pinion flange and support it out of the way with a piece of wire or rope (see Section 9). Also remove the companion flange from the differential. **Note:** *Mark the positions of the flange and the nut so they can be installed in their original positions.*

5   Remove the skid plate.
6   Remove the suspension crossmember **(see illustration)**.
7   Disconnect the wiring from the stabilizer bar disconnect device.
8   Support the axle assembly with a floor jack under the differential.
9   Remove any remaining interfering wiring harness clips.
10   Remove the axle assembly pinion mounting bolts **(see illustration)**. Also remove the bolts from the axle shaft tube and the differential housing-to-engine mounting bolts.
11   Lower the jack slowly, then roll the axle assembly from under the vehicle.
12   Installation is the reverse of removal.

Solid axle

13   Loosen the front wheel lug nuts. Raise the vehicle and support it securely with jackstands positioned under the frame rails. Remove the front wheels.
14   Remove the brake calipers and discs (see Chapter 9) and disconnect the ABS wheel speed sensors.
15   Disconnect the vent hose and the brake hose bracket.
16   Mark the driveshaft to the companion flange, then disconnect it from the differential pinion flange. Support the driveshaft out of the way with a piece of wire or cord (see Section 9).
17   Disconnect the stabilizer bar links from the axle brackets (see Chapter 10).
18   Disconnect the shock absorbers from the axle brackets.
19   Disconnect the tie-rod and drag link from the steering knuckle (see Chapter 10). Support the axle assembly with two floor jacks, one positioned under the differential and one positioned under the long axle tube.
20   Disconnect the track bar from the axle bracket (see Chapter 10).
21   On heavy-duty models, disconnect the

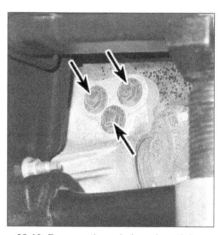

22.10 Remove the axle housing pinion mounting bolts (shown), then the axle tube mounting bolts and the differential housing-to-engine bolts

vibration damper from the axle bracket.
22   Mark the relationship of the alignment cams to ensure proper reassembly, then remove the bolts and disconnect the control arms from the axle bracket.
23   Lower the jacks and remove the axle from under the vehicle.
24   Installation is the reverse of removal. Tighten all bolts to the torque listed in this Chapter's Specifications and in the Specifications in Chapters 9 and 10.

23   Front driveaxle oil seal (independent front suspension) - removal and installation

1   Remove the driveaxle (see Section 13).
2   Pry the oil seal out of the end of the axle housing with a seal removal tool.
3   Apply high-temperature grease to the oil

seal recess and tap the new seal evenly into place with a hammer and seal driver, so the lips are facing in and the metal face is visible from the end of the axle housing. When correctly installed, the face of the oil seal should be flush with the end of the axle housing.

4   Lubricate the lips of the seal with multi-purpose grease, then install the driveaxle (see Section 13).

5   Check the front differential lubricant level, adding as necessary (see Chapter 1).

24   Front hub bearing assembly and axleshaft (solid axle) - removal and installation

Removal

Note: *If you're removing the axleshaft in order to replace the axleshaft seal or bearing, we recommend having the job done by a dealer service department or a qualified independent garage. Replacing the seal or the bearing requires special tools.*

1   Loosen the wheel lug nuts. Raise the vehicle and support it securely on jackstands. Remove the wheel.

2   Remove the brake caliper and support it out of the way with wire (see Chapter 9). Remove the brake disc.

3   On light-duty axles, remove the cotter pin and axle hub nut. On heavy-duty models, bend the nut's collar out of the slot in the axle.

4   Remove the hub-to-knuckle bolts.

5   Remove the hub bearing from the steering knuckle and axleshaft. If the axleshaft splines stick in the hub, a two-jaw puller may be required to push the axle out.

6   Carefully pull the axleshaft from the axle housing. **Note:** *On heavy-duty models, there is a large shim on the outer end of the axleshaft that must retained.* If the U-joint is worn, it can be replaced (see Section 12).

Installation

7   Clean the axleshaft and apply a thin film of wheel bearing grease to the shaft splines, seal contact surface and hub bore. Install the axleshaft (with its shim, on heavy-duty models) engaging the splines with the differential side gears. Be very careful not to damage the axleshaft oil seals.

8   Install the hub bearing. Install the hub bearing-to-steering knuckle bolts and tighten them to the torque listed in this Chapter's Specifications.

9   Install the axleshaft washer and nut, tighten the nut to the torque listed in this Chapter's Specifications. Line-up the nut with the next cotter pin hole and install a new cotter pin.

10   Install the brake disc and caliper (see Chapter 9).

11   Install the wheel and hand tighten the wheel lug nuts.

12   Remove the jackstands, lower the vehicle and tighten the wheel lug nuts to the torque listed in the Chapter 1 Specifications.

Chapter 9 Brakes

Contents

Specifications

General
Brake fluid type .... See Chapter 1

Disc brakes
Brake pad minimum thickness	See Chapter 1
Brake disc-to-hub bolts (models equipped with dual rear wheels)	114
Disc lateral runout limit	0.005 inch
Disc minimum thickness	Cast into disc

Torque specifications
	Ft-lbs
Brake hose-to-caliper banjo bolts	20
Caliper mounting bolts (pins)	
Front	
2014 and earlier	24
2015 and later	31
Rear	
2014 and earlier	22
2015 and later	24
Caliper mounting bracket bolts	
Front	
1500 models	130
2500/3500 models	
2012 and earlier	275
2013 and later	260
Rear	
1500 models	
2012 and earlier	120
2013 and 2014	142
2015 and later	132
2500/3500 models	
2012 and earlier	
Upper bolts	163
Lower bolts	190
2013 and later	190
Hydraulic booster pressure lines	30
Master cylinder mounting nuts	18
Power brake booster mounting nuts	21
Vacuum pump bracket mounting bolts	41

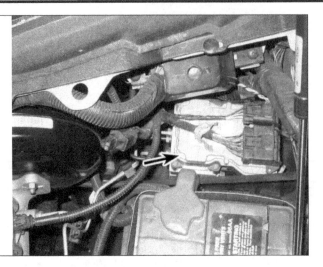

2.2 The ABS hydraulic unit is mounted in the rear left corner of the engine compartment

1   General information

The vehicles covered by this manual are equipped with hydraulically operated front and rear brake systems. The front and rear brakes are disc type. Both the front and rear brakes are self adjusting; disc brakes automatically compensate for pad wear.

Hydraulic system

The hydraulic system consists of two separate circuits. The master cylinder has separate reservoirs for the two circuits, and, in the event of a leak or failure in one hydraulic circuit, the other circuit will remain operative.

Power brake booster

The power brake booster on models with a gasoline engine uses engine manifold vacuum to provide assistance to the brakes. It is mounted on the firewall in the engine compartment, directly behind the master cylinder.

Diesel engines, which have no manifold vacuum, are equipped with a hydraulic brake booster that relies on pressure provided by the power steering pump.

Parking brake

The parking brake mechanically operates the rear brakes only. The parking brake cables actuate a pair of parking brake shoes mounted inside the drum (hub) portion of each rear brake disc.

Service

After completing any operation involving disassembly of any part of the brake system, always test drive the vehicle to check for proper braking performance before resuming normal driving. When testing the brakes, perform the tests on a clean, dry, flat surface. Conditions other than these can lead to inaccurate test results.

Test the brakes at various speeds with both light and heavy pedal pressure. The vehicle should stop evenly without pulling to one side or the other.

Tires, vehicle load and wheel alignment are factors which also affect braking performance.

Precautions

There are some general cautions and warnings involving the brake system on this vehicle:

a) *Use only brake fluid conforming to DOT 3 or 4 specifications.*

b) *The brake pads and linings contain fibers which are hazardous to your health if inhaled. Whenever you work on brake system components, clean all parts with brake system cleaner. Do not allow the fine dust to become airborne. Also, wear an approved filtering mask.*

c) *Safety should be paramount whenever any servicing of the brake components is performed. Do not use parts or fasteners which are not in perfect condition, and be sure that all clearances and torque specifications are adhered to. If you are at all unsure about a certain procedure, seek professional advice. Upon completion of any brake system work, test the brakes carefully in a controlled area before putting the vehicle into normal service. If a problem is suspected in the brake system, don't drive the vehicle until it's fixed.*

2   Anti-lock Brake System (ABS) - general information

General information

Refer to illustration 2.2

1   The anti-lock brake system is designed to maintain vehicle steerability, directional stability and optimum deceleration under severe braking conditions on most road surfaces. It does so by monitoring the rotational speed of each wheel and controlling the brake line pressure to each wheel during braking. This prevents the wheels from locking up.

2   The ABS system has three main components - the wheel speed sensors, the anti-lock brake control module and the hydraulic control unit **(see illustration)**. Wheel speed sensors - one at each front wheel and another located on the rear differential - send a variable voltage signal to the control unit, which monitors these signals, compares them to its program and determines whether a wheel is about to lock up. When a wheel is about to lock up, the control unit signals the hydraulic unit to reduce hydraulic pressure (or not increase it further) at that wheel's brake caliper. Pressure modulation is handled by electrically operated solenoid valves.

3   If a problem develops within the system, an ABS warning light will glow on the dashboard. Sometimes, a visual inspection of the ABS system can help you locate the problem. Carefully inspect the ABS wiring harness. Pay particularly close attention to the harness and connections near each wheel. Look for signs of chafing and other damage caused by incorrectly routed wires. If a wheel sensor harness is damaged, the sensor must be replaced. **Warning:** *Do NOT try to repair an ABS wiring harness. The ABS system is sensitive to even the smallest changes in resistance. Repairing the harness could alter resistance values and cause the system to malfunction. If the ABS wiring harness is damaged in any way, it must be replaced.* **Caution:** *Make sure the ignition is turned off before unplugging or reattaching any electrical connections.*

4   If a dashboard warning light comes on and stays on while the vehicle is in operation, the ABS system requires attention. Although special electronic ABS diagnostic testing tools are necessary to properly diagnose the system, you can perform a few preliminary checks before taking the vehicle to a dealer service department.

a) *Check the brake fluid level in the reservoir.*

b) *Verify that the computer electrical connectors are securely connected.*

c) *Check the electrical connectors at the hydraulic control unit.*

d) *Check the fuses.*

e) *Follow the wiring harness to each wheel and verify that all connections are secure and that the wiring is undamaged.*

5   If the above preliminary checks do not rectify the problem, the vehicle should be diagnosed by a dealer service department or other qualified repair shop. Due to the complex nature of this system, all actual repair work must be done by a qualified automotive technician.

Wheel speed sensor - removal and installation

Refer to illustrations 2.9a and 2.9b

6   Loosen the wheel lug nuts, raise the vehicle and support it securely on jackstands. Remove the wheel.

7   Make sure the ignition key is turned to the Off position.

8   Trace the wiring back from the sensor,

2.9a Front wheel speed sensor

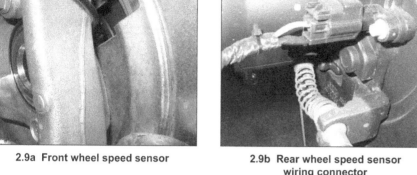

2.9b Rear wheel speed sensor
wiring connector

3.5 Before removing the caliper, depress
the piston into the bottom of its bore in
the caliper with a large C-clamp to make
room for the new pads

detaching all brackets and clips while noting
its correct routing, then disconnect the electrical connector.
9   Remove the mounting fastener and carefully pull the sensor out from the knuckle or
rear differential (see illustrations).
10   Installation is the reverse of the removal
procedure. Tighten the mounting fastener
securely.
11   Install the wheel and lug nuts, tightening
them securely. Lower the vehicle and tighten
the lug nuts to the torque listed in the Chapter 1 Specifications.

3   Disc brake pads - replacement

Refer to illustrations 3.5 and 3.6a through 3.6o

Warning: *Disc brake pads must be replaced
on both front or rear wheels at the same time
- never replace the pads on only one wheel.
Also, the dust created by the brake system is
harmful to your health. Never blow it out with
compressed air and don't inhale any of it. An
approved filtering mask should be worn when
working on the brakes. Do not, under any cir-*
cumstances, use petroleum-based solvents to
clean brake parts. Use brake system cleaner
only!
Note: *This procedure applies to the front and
rear brake pads.*
1   Remove the cap from the brake fluid reservoir.
2   Loosen the wheel lug nuts, raise the end
of the vehicle you're working on and support
it securely on jackstands. Block the wheels at
the opposite end.
3   Remove the wheels. Work on one brake
assembly at a time, using the assembled
brake for reference if necessary.
4   Inspect the brake disc carefully (see
Section 5). If machining is necessary, follow
the information in that Section to remove the
disc, at which time the pads can be removed
as well.
5   Push the piston back into its bore to provide room for the new brake pads. A C-clamp
can be used to accomplish this (see illustration). As the piston is depressed to the bottom of the caliper bore, the fluid in the master
cylinder will rise. Make sure that it doesn't
overflow. If necessary, siphon off some of the
fluid.

6   Follow the accompanying photos (illustrations 3.6a through 3.6o) for the actual
pad replacement procedure. Be sure to stay
in order and read the caption under each illustration.

3.6a Always wash the brakes with brake
cleaner before disassembling anything

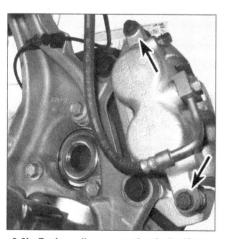

3.6b Brake caliper mounting bolts (front
caliper shown, rear caliper similar)

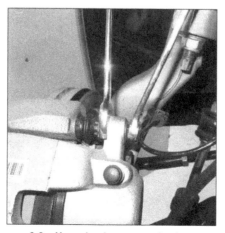

3.6c Use a back-up wrench when
removing the caliper bolts

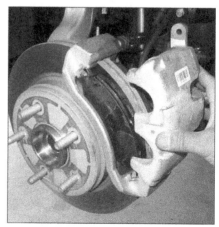

3.6d Remove the caliper . . .

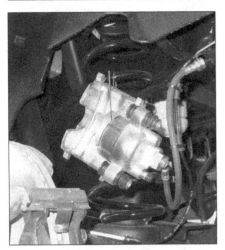

3.6e . . . and use a piece of wire to tie it to the control arm - never let the caliper hang by the brake hose

3.6f Remove the inner pad . . .

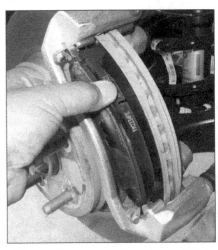

3.6g . . . then remove the outer pad from the caliper mounting bracket

3.6h Remove the anti-rattle clips, paying close attention to how they're installed in the mounting bracket after cleaning and lubricating them

3.6i Rear anti-rattle clips have different design . . .

3.6j . . . and are installed like this when cleaned and lubricated

3.6k Install the inner pad . . .

3.6l . . . and outer pad into the mounting bracket - make sure both pads are fully seated . . .

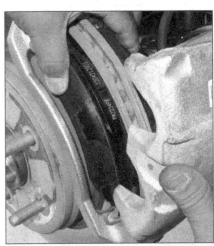

3.6m . . . then hold them in place and install the caliper

7   After the job has been completed, firmly depress the brake pedal a few times to bring the pads into contact with the disc. Check the level of the brake fluid, adding some if necessary. Check the operation of the brakes carefully before placing the vehicle into normal service.

4   Disc brake caliper - removal and installation

Warning: *The dust created by the brake system is harmful to your health. Never blow it out with compressed air and don't inhale any of it. An approved filtering mask should be worn when working on the brakes. Do not, under any circumstances, use petroleum-based solvents to clean brake parts. Use brake system cleaner only!*
Note: *This procedure applies to the front and rear disc brake calipers.*

Removal
Refer to illustration 4.2

1   Loosen the front or rear wheel lug nuts, raise the front or rear of the vehicle and place it securely on jackstands. Block the wheels at the opposite end. Remove the front or rear wheel.
2   Remove the brake hose-to-caliper banjo bolt and disconnect the brake hose from the caliper **(see illustration)**. Discard the old sealing washers. Plug the brake hose immediately to keep contaminants and air out of the brake system and to prevent losing any more brake fluid than is necessary. **Note:** *If you are simply removing the caliper for access to other components, leave the brake hose connected and suspend the caliper with a length of wire - don't let it hang by the hose* **(see illustration 3.6e).**
3   Remove the caliper mounting bolts and detach the caliper from the mounting bracket **(see illustration 3.6b).**

3.6n Clean, inspect and lubricate the bushings and boots . . .

Installation
4   Installation is the reverse of removal. Use new sealing washers on each side of the brake hose fitting and tighten the banjo bolt and the caliper mounting bolts to the torque listed in this Chapter's Specifications.
5   Bleed the brake system (see Section 9).
Note: *If the brake hose was not disconnected, bleeding won't be required.* Make sure there are no leaks from the hose connections. Test the brakes carefully before returning the vehicle to normal service.

5   Brake disc - inspection, removal and installation

Inspection
Refer to illustrations 5.2, 5.3, 5.4a and 5.4b

1   Loosen the wheel lug nuts, raise the vehicle and support it securely on jackstands. Apply the parking brake. Remove the wheels.

3.6o . . . then install the bushings and mounting bolts. Tighten the bolts to the torque listed in this Chapter's Specifications

2   Visually inspect the disc surface for score marks and other damage **(see illustration)**. Light scratches and shallow grooves are normal after use and won't affect brake operation. Deep grooves require disc removal and refinishing by an automotive machine shop. Be sure to check both sides of the disc.
3   To check disc runout, place a dial indicator at a point about 1/2-inch from the outer edge of the disc **(see illustration)**. If you're checking a front disc, or rear disc on a vehicle with single rear wheels, install the lug nuts, with the flat sides facing in, and tighten them securely to hold the disc in place. Set the indicator to zero and turn the disc. The indicator reading should not exceed the runout limit listed in this Chapter's Specifications. If it does, the disc should be refinished by an automotive machine shop. If you elect not to have the discs resurfaced, deglaze them with sandpaper or emery cloth.
4   The disc must not be machined to a thickness less than the specified minimum thick-

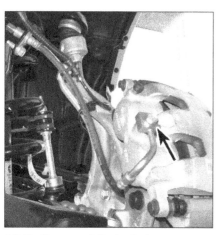
4.2 To disconnect the brake hose, remove the brake hose-to-caliper banjo bolt (front caliper shown)

5.2 The brake pads on this vehicle were obviously neglected - they wore down completely and cut deep grooves into the disc (wear this severe means the disc must be replaced)

5.3 Measure the brake disc runout with a dial indicator

5.4a The minimum (discard) thickness of the brake disc is cast into the disc

5.4b Measure the brake disc thickness at several points with a micrometer

5.6 The caliper mounting bracket is retained by two bolts (rear shown, front similar)

ness, which is cast into the disc **(see illustration)**. The disc thickness can be checked with a micrometer **(see illustration)**.

Removal and installation

Refer to illustration 5.6

5   Remove the brake caliper (don't disconnect the brake hose) and hang it out of the way (see Section 4).

6   Remove the caliper mounting bracket **(see illustration)**.

7   If equipped, remove the hub extender from the hub.

8   Remove the lug nuts installed in Step 3 and pull the disc off the hub.

9   If you're removing a rear disc on a model equipped with dual rear wheels, remove the rear axleshaft, remove the hub and bearing assembly (see Chapter 8), then unbolt the disc from the hub.

10   Installation is the reverse of removal. On

models so equipped, tighten the disc-to-hub bolts securely.

11   Lower the vehicle and tighten the wheel lug nuts to the torque listed in the Chapter 1 Specifications.

---

6   Master cylinder - removal and installation

Removal

Refer to illustration 6.4

1   The master cylinder is located in the engine compartment, mounted to the power brake booster.

2   Remove as much fluid as you can from the reservoir with a syringe, such as an old turkey baster. **Warning:** *If a baster is used, never again use it for the preparation of food.*

3   Place rags under the fluid fittings and prepare caps or plastic bags to cover the

ends of the lines once they are disconnected. **Caution:** *Brake fluid will damage paint. Cover all pained surfaces around the work area and be careful not to spill fluid during this procedure.*

4   Loosen the fittings at the ends of the brake lines where they enter the master cylinder **(see illustration)**. To prevent rounding off the corners on these nuts, the use of a flare-nut wrench, which wraps around the nut, is preferred. Pull the brake lines slightly away from the master cylinder and plug the ends to prevent contamination.

5   Disconnect the electrical connector at the brake fluid level switch on the master cylinder reservoir, then remove the nuts attaching the master cylinder to the power booster. Pull the master cylinder off the studs and out of the engine compartment. Again, be careful not to spill the fluid as this is done.

6   If a new master cylinder is being installed, remove the fastener securing the reservoir to the master cylinder then pull up on the reservoir to remove it from the master cylinder. Transfer the reservoir to the new master cylinder. **Note:** *Install new seals when transferring the reservoir.*

Installation

Refer to illustration 6.8

7   Bench bleed the new master cylinder before installing it. Mount the master cylinder in a vise, with the jaws of the vise clamping on the mounting flange.

8   Attach a pair of master cylinder bleeder tubes to the outlet ports of the master cylinder **(see illustration)**.

9   Fill the reservoir with brake fluid of the recommended type (see Chapter 1).

10   Slowly push the pistons into the master cylinder (a large Phillips screwdriver can be used for this) - air will be expelled from the pressure chambers and into the reservoir. Because the tubes are submerged in fluid, air can't be drawn back into the master cylinder when you release the pistons.

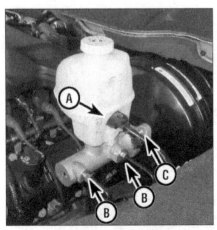

6.4 Brake master cylinder mounting details:

A   *Brake fluid switch electrical connector*
B   *Brake lines*
C   *Mounting nuts (other nut not visible)*

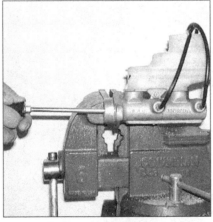

6.8 The best way to bleed air from the master cylinder before installing it on the vehicle is with a pair of bleeder tubes that direct brake fluid into the reservoir during bleeding

7.9 Remove the retaining clip, then detach the pushrod from the pedal

7.10 Remove the brake booster mounting fasteners

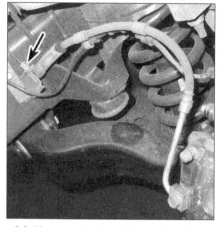

8.3 Unscrew the line fitting with a flare-nut wrench

11   Repeat the procedure until no more air bubbles are present.

12   Remove the bleed tubes, one at a time, and install plugs in the open ports to prevent fluid leakage and air from entering. Install the reservoir cap.

13   Install the master cylinder over the studs on the power brake booster and tighten the attaching nuts only finger tight at this time.

14   Thread the brake line fittings into the master cylinder. Since the master cylinder is still a bit loose, it can be moved slightly in order for the fittings to thread easily. Do not strip the threads as the fittings are tightened.

15   Fully tighten the mounting nuts, then the brake line fittings. Tighten the nuts to the torque listed in this Chapter's Specifications.

16   Connect the brake fluid switch electrical connector.

17   Fill the master cylinder reservoir with fluid, then bleed the master cylinder and the brake system (see Section 9). To bleed the cylinder on the vehicle, have an assistant depress the brake pedal and hold the pedal to the floor. Loosen the fitting to allow air and fluid to escape. Repeat this procedure on both fittings until the fluid is clear of air bubbles. **Caution:** *Have plenty of rags on hand to catch the fluid - brake fluid will ruin painted surfaces. After the bleeding procedure is completed, rinse the area under the master cylinder with clean water.*

18   Test the operation of the brake system carefully before placing the vehicle into normal service. **Warning:** *Do not operate the vehicle if you are in doubt about the effectiveness of the brake system. It is possible for air to become trapped in the anti-lock brake system hydraulic control unit, so, if the pedal continues to feel spongy after repeated bleedings or the BRAKE or ANTI-LOCK light stays on, have the vehicle towed to a dealer service department or other qualified shop to be bled with the aid of a scan tool.*

7   Power brake booster (gasoline engine models) - check, removal and installation

1   The power brake booster unit requires no special maintenance apart from periodic inspection of the vacuum hose and the case.

2   Disassembly of the power unit requires special tools and is not ordinarily performed by the home mechanic. If a problem develops, it's recommended that a new or factory rebuilt unit be installed.

Operating check

3   Depress the brake pedal several times with the engine off and make sure that there is no change in the pedal reserve distance.

4   Depress the pedal and start the engine. If the pedal goes down slightly, operation is normal.

Airtightness check

5   Start the engine and turn it off after one or two minutes. Depress the brake pedal several times slowly. If the pedal goes down farther the first time but gradually rises after the second or third depression, the booster is airtight.

6   Depress the brake pedal while the engine is running, then stop the engine with the pedal depressed. If there is no change in the pedal reserve travel after holding the pedal for 30 seconds, the booster is airtight.

Removal

Refer to illustrations 7.9 and 7.10

7   Remove the master cylinder (see Section 6).

8   Disconnect the vacuum hose from the power brake booster.

9   Working under the dash, disconnect the power brake pushrod from the top of the brake pedal by prying off the clip **(see illustration)**.

10   Remove the nuts attaching the booster

to the firewall **(see illustration)**.

11   Carefully lift the booster unit away from the firewall and out of the engine compartment.

Installation

12   To install the booster, place it into position and tighten the retaining nuts to the torque listed in this Chapter's Specifications.

13   Connect the booster pushrod to the brake pedal.

14   Install the master cylinder (see Section 6).

15   Connect the brake lines to the master cylinder and tighten all fitting nuts securely.

16   Connect the vacuum hose to the brake booster assembly.

17   Bleed the brake system (see Section 9).

18   Carefully test the operation of the brakes before placing the vehicle in normal operation.

8   Brake hoses and lines - inspection and replacement

Inspection

1   Whenever the vehicle is raised and supported securely on jackstands, the rubber hoses which connect the steel brake lines with the front and rear brake assemblies should be inspected for cracks, chafing of the outer cover, leaks, blisters and other damage. These are important and vulnerable parts of the brake system and inspection should be thorough. A light and mirror will be helpful for a complete check. If a hose exhibits any of the above conditions, replace it immediately.

Flexible hose replacement

Refer to illustration 8.3

2   Clean all dirt away from the hose fittings.

3   Using a flare-nut wrench, disconnect the metal brake line from the hose fitting **(see illustration)**. Be careful not to bend the frame

9.8 When bleeding the brakes, a hose is connected to the bleed screw at the caliper and submerged in brake fluid - air will be seen as bubbles in the tube and container (all air must be expelled before moving to the next wheel)

bracket or line. If the threaded fitting is corroded, spray it with penetrating oil and allow it to soak in for about 10 minutes, then try again. If you try to break loose a fitting nut that's frozen, you will kink the metal line, which will then have to be replaced.

4   Disconnect the brake hose bracket from the frame. Immediately plug the metal line to prevent excessive leakage and contamination.

5   Unscrew the banjo bolt at the caliper and remove the hose, discarding the sealing washers on either side of the fitting.

6   Attach the new brake hose to the caliper. **Note:** *When connecting a brake hose to a caliper, always use new sealing washers.* Tighten the banjo bolt to the torque listed this Chapter's Specifications.

7   Connect the other end of the new hose, making sure the hose isn't kinked or twisted. Fit the metal line to the hose (or hose fitting), then tighten the hose bracket and the brake tube fitting nut securely.

8   Carefully check to make sure the suspension or steering components don't make contact with the hose. Have an assistant push down on the vehicle while you watch to see whether the hose interferes with suspension operation. If you're replacing a front hose, have your assistant turn the steering wheel lock-to-lock while you make sure the hose doesn't interfere with the steering linkage or the steering knuckle.

9   After installation, bleed the brakes (see Section 9). Check the master cylinder fluid level and add fluid as necessary. Carefully test brake operation before returning the vehicle to normal service.

Metal brake lines

10   When replacing brake lines, be sure to use the correct parts. Do not use copper tubing for any brake system components. Purchase steel brake lines from a dealer parts

department or auto parts store.

11   Prefabricated brake line, with the tube ends already flared and fittings installed, is available at auto parts stores and dealer parts departments. If it is necessary to bend a line, use a tubing bender to prevent kinking the line.

12   When installing the new line, make sure it's well supported in the brackets and has plenty of clearance between moving or hot components. Make sure you tighten the fittings securely.

13   After installation, check the master cylinder fluid level and add fluid as necessary. Bleed the brakes (see Section 9). Carefully test brake operation before resuming normal operation.

9   Brake hydraulic system - bleeding

Refer to illustration 9.8

Warning: *The following procedure is a manual bleeding procedure. This is the only bleeding procedure which can be performed at home without special tools. However, if air has found its way into the hydraulic control unit, the entire system must be bled manually, then with a scan tool, then manually a second time. If the brake pedal feels spongy even after bleeding the brakes, or the ABS light on the instrument panel does not go off, or if you have any doubts whatsoever about the effectiveness of the brake system, have the vehicle towed to a dealer service department or other repair shop equipped with the necessary tools for bleeding the system.*

Warning: *Wear eye protection when bleeding the brake system. If the fluid comes in contact with your eyes, immediately rinse them with water and seek medical attention.*

Note: *Bleeding the hydraulic system is necessary to remove any air that manages to find its way into the system when it's been opened during removal and installation of a hydraulic component.*

1   It will be necessary to bleed the complete system if air has entered the system due to low fluid level, or if the brake lines have been disconnected at the master cylinder.

2   If a brake line was disconnected only at a wheel, then only that caliper must be bled.

3   If a brake line is disconnected at a fitting located between the master cylinder and any of the brakes, that part of the system served by the disconnected line must be bled. The following procedure describes bleeding the entire system, however.

4   Remove any residual vacuum (or hydraulic pressure) from the brake power booster by applying the brake several times with the engine off.

5   Remove the cap from the master cylinder reservoir and fill the reservoir with brake fluid. Reinstall the cap. **Note:** *Check the fluid level often during the bleeding operation and add fluid as necessary to prevent the fluid level*

from falling low enough to allow air bubbles into the master cylinder.

6   Have an assistant on hand, as well as a supply of new brake fluid, a clear container partially filled with clean brake fluid, a length of clear tubing to fit over the bleeder valve and a wrench to open and close the bleeder valve.

7   Beginning at the right rear wheel, loosen the bleeder screw slightly, then tighten it to a point where it's snug but can still be loosened quickly and easily.

8   Place one end of the tubing over the bleeder screw fitting and submerge the other end in brake fluid in the container **(see illustration).**

9   Have the assistant slowly depress the brake pedal and hold it in the depressed position.

10   While the pedal is held depressed, open the bleeder screw just enough to allow a flow of fluid to leave the valve. Watch for air bubbles to exit the submerged end of the tube. When the fluid flow slows after a couple of seconds, tighten the screw and have your assistant release the pedal.

11   Repeat Steps 9 and 10 until no more air is seen leaving the tube, then tighten the bleeder screw and proceed to the left rear wheel, the right front wheel and the front left wheel, in that order, and perform the same procedure. Be sure to check the fluid in the master cylinder reservoir frequently.

12   Never use old brake fluid. It contains moisture which can boil, rendering the brake system inoperative.

13   Refill the master cylinder with fluid at the end of the operation.

14   Check the operation of the brakes. The pedal should feel solid when depressed, with no sponginess. If necessary, repeat the entire process. **Warning:** *Do not operate the vehicle if you are in doubt about the effectiveness of the brake system. It is possible for air to become trapped in the anti-lock brake system hydraulic control unit, so, if the pedal continues to feel spongy after repeated bleedings or the BRAKE or ANTI-LOCK light stays on, have the vehicle towed to a dealer service department or other qualified shop to be bled with the aid of a scan tool.*

10   Parking brake shoes - replacement

Refer to illustrations 10.3 and 10.6a through 10.6t

Warning: *Dust created by the brake system is harmful to your health. Never blow it out with compressed air and don't inhale any of it. An approved filtering mask should be worn when working on the brakes. Do not, under any circumstances, use petroleum-based solvents to clean brake parts. Use brake system cleaner only!*

Note: *Although the typical procedure illustrated here is shown with the axle still in place,*

10.3 Remove the plug from the hole in the brake backing plate, insert a brake adjuster tool through the hole and rotate the star wheel, moving the parking brake shoes away from the disc hub (which is already removed in this photo for the sake of clarity)

it takes some dexterity to work behind the axle flange while replacing the parking brake shoes. If this proves difficult, you can remove the axle for better access (see Chapter 8).
1   Loosen the wheel lug nuts, release the parking brake, raise the rear of the vehicle and support it securely on jackstands. Block the front wheels to keep the vehicle from rolling. Remove the rear wheels.
2   Remove the rear brake caliper (see Section 4) and hang it with a length of wire, then remove the caliper mounting bracket (see Section 5).
3   Remove the brake disc (see Section 5).
Note: *If the brake disc cannot be easily pulled off the axle and shoe assembly, make sure that the parking brake is completely released, then apply some penetrating oil at the hub-to-disc joint. Allow the oil to soak in and try to pull the disc off. If the disc still cannot be pulled off, the parking brake shoes will have to be retracted. This is accomplished by first remov-*

10.6a Pull down on the parking brake cable, then clamp a pair of locking pliers on the cable to retain slack . . .

ing the plug from the backing plate. With the plug removed, turn the adjusting wheel with a narrow screwdriver or brake adjusting tool, moving the shoes away from the braking sur-

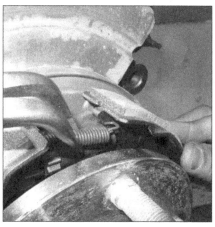

10.6c Use needle nose pliers to detach the upper return spring from both shoes . . .

10.6b . . . then release the cable from the lever behind the brake backing plate

face **(see illustration).** *The disc should now come off.*
4   Working behind the brake assembly, detach the parking brake cable. **Note:** *This will make disassembly of the brakes easier.*

10.6d . . . then remove the spring

10.6e Spread the shoes enough to remove the adjuster

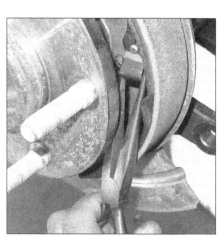

10.6f Remove the clip from the rear hold-down pin

10.6g Disengage the lower return spring from the shoe and remove the shoe

10.6h Remove the clip that retains the front shoe

10.6i Note how the shoe engages the lower support . . .

10.6j . . . then remove the shoe

10.6k Clean the backing plates thoroughly, then apply high-temperature brake grease to the points where the shoes contact them

10.6l Place the lower return spring into position . . .

10.6m . . . then engage it with the front shoe

10.6n Put the rear shoe into position

10.6o Pull the spring across . . .

10.6p . . . and attach it to the rear shoe

10.6q Clean the adjuster, then apply high-temperature brake grease to the threads

10.6r Install the adjuster so the starwheel is centered over the adjustment hole

10.6s Secure both shoes to the hold-down pins with the clips

10.6t Install the upper return spring, then check that all parts are correctly engaged

5   Remove the axle for access (see Chapter 8). **Note:** *Removing the axles makes the job easier, but the photos illustrate the job with the axle in place. Both methods are correct.*
6   Clean the parking brake shoe assembly with brake system cleaner, then follow the accompanying photos **(illustrations 10.6a through 10.6t)** for the parking brake shoe replacement procedure. Be sure to stay in order and read the caption under each illustration. **Note:** *All four parking brake shoes must be replaced at the same time, but to avoid mixing up parts, work on only one brake assembly at a time.*
7   Before reinstalling the disc, check the parking brake surfaces of the disc hub for cracks, score marks, deep scratches and hard spots, which will appear as small discolored areas. If hard spots or any of the other conditions listed above cannot be removed with sandpaper or emery cloth, the disc must be replaced.
8   Once all of the new parking brake shoes are in place, install the brake discs (see Section 5) and the brake calipers (see Section 4).

9   Remove the rubber plugs from the brake backing plates, insert a narrow screwdriver or brake adjusting tool through the adjustment hole and turn the star wheel until the shoes drag slightly as the disc is turned. Turn the star wheel in the opposite direction until the disc turns freely. Install the backing plate plugs.
10   Install the rear wheels and lug nuts, lower the vehicle and tighten the lug nuts to the torque listed in the Chapter 1 Specifications.
11   Operate the parking brake lever several times to release the lockout spring and adjust the cables.
12   Carefully check the operation of the brakes before placing the vehicle in normal service.

---

11   Hydraulic brake booster (diesel engine models) - removal and installation

---

Warning: *The accumulator portion of the booster contains high-pressure gas. Do not carry the booster by the accumulator or drop*

the unit on the accumulator.
Note: *If the brake booster is being replaced due to power steering fluid contamination, flush the power steering system prior to installing the new brake booster (see Chapter 10).*
1   Disconnect the cable(s) from the negative battery terminal(s) (see Chapter 5).
2   With the engine off, depress the brake pedal several times to discharge the accumulator.
3   Remove the master cylinder (see Section 6).
4   Remove the return hose and pressure line(s) from the brake booster.
5   Working inside the vehicle, remove the clip and washer, then disengage the brake booster pushrod from the brake pedal **(see illustration 7.9)**. Remove the brake booster mounting nuts **(see illustration 7.10)**.
6   Working inside the engine compartment, remove the brake booster from the firewall.
7   Installation is the reverse of removal. Tighten the booster mounting nuts and the pressure lines to the torque listed in this Chapter's Specifications.

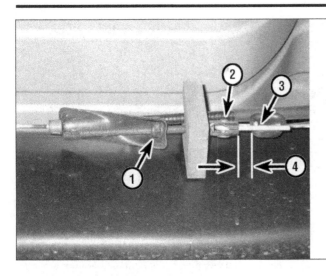

12.2 Parking brake cable adjustment details:

1. Adjuster nut
2. Tensioner
3. Tensioner rod
4. 1/4 inch mark

8   Bleed the brake hydraulic system (see Section 9), then bleed the power steering system (see Chapter 10).

12   Parking brake - adjustment

Refer to illustration 12.2

1   Raise the rear of the vehicle and support it securely on jackstands. Block the front wheels to prevent the vehicle from rolling.
2   Adjust the parking brake shoes (see Section 10). Loosen the adjusting nut at the equalizer enough to create slack in the cables **(see illustration).**
3   Verify that the drums rotate freely without drag, then fully apply the parking brake.
4   Mark the tensioner rod about 1/4-inch from the tensioner.
5   Tighten the adjusting nut at the tensioner bracket until the mark on the tensioner rod moves into alignment with the tensioner. **Caution:** *Do NOT loosen or tighten the tensioner adjusting nut after completing this adjustment.*
6   Release the parking brake pedal and verify that the rear wheels rotate freely without any drag.
7   Remove the jackstands and lower the vehicle.
8   Test the operation of the parking brake on an incline (be sure to remain in the vehicle for this check).

13   Brake light switch - removal and installation

Removal

1   Disconnect the cable(s) from the negative battery terminal(s) (see Chapter 5).
2   The switch is mounted at the brake pedal. Disconnect the wiring from the switch.
3   Hold the brake pedal down, then turn the switch slightly clockwise, then pull it from its bracket. **Note:** *On early 2009 models, the switch must be replaced with a new one whenever it's removed.*
4   Let up the brake pedal.

Installation

Early 2009 models

5   Hold the brake pedal down.

6   Insert the new switch into its bracket until it's properly seated, then turn it counterclockwise to lock it in place.
7   Let go of the brake pedal, but don't pull it up.
8   Turn the plunger release lever clockwise until it snaps into place. When done correctly, the lever will be in line with the wiring harness socket. **Note:** *Make sure the switch is seated properly before moving the lever. If an error is made, the switch will have to be discarded and replaced with a new one.*
9   Connect the wiring harness and the battery cable(s).

Late 2009 and later models

10   Verify that the switch plunger is completely extended.
11   Hold the brake pedal down.
12   Install the switch into its bracket and seat it fully.
13   Turn the switch clockwise to lock it in place.
14   Let go of the brake pedal to allow the switch to adjust itself.
15   Connect the wiring harness and the battery cable(s).

14   Vacuum pump (2013 and later 1500 models) – removal and installation

1   Remove the air filter housing (see Chapter 4A).
2   Disconnect the electrical connector from the pump, then remove the harness from the bracket.
3   Depress the locking tab on the quick-connect fitting and separate the vacuum hose from the pump.
4   Remove the mounting bolts and detach the pump from the vehicle.
5   Installation is the reverse of removal.

Chapter 10
Suspension and steering systems

Contents

Specifications

Torque specifications

Ft-lbs (unless otherwise indicated)

Front suspension
Hub bearing mounting bolts
    2WD models
        1500 120
        2500/3500 130
    4WD models
        1500 120
        2500/3500 149
Shock absorber
    Upper mounting nuts
        2WD models
            1500 40
            2500/3500 45
        4WD models
            1500 45
            2500/3500 40
    Lower mounting bolts/nuts
        2WD models
            1500 19
            2500/3500 21
        4WD models
            1500 155
            2500/3500 100
    Upper mounting bracket nuts (2500/3500 4WD models) 55

Torque specifications

Ft-lbs (unless otherwise indicated)

Front suspension (continued)

Stabilizer bar
  1500
    Bracket bolts 44
    Link-to-control arm nuts 75
    Link-to-stabilizer bar nuts.. 20
  2500/3500
    Independent suspension
      Bracket bolts.. 47
      Link-to-control arm nuts... 82
      Link-to-stabilizer bar nuts 20
    Solid axle
      Bracket bolts.. 45
      Link-to-stabilizer bar nuts 27
      Link-to-axle bracket fasteners 110

Independent front suspension

Upper control arm
  Arm-to-frame pivot bolts
    1500... 130
    2500/3500... 110
  Balljoint-to-steering knuckle nut
    1500
      Step 1... 40
      Step 2... Tighten an additional 200 degrees
    2500/3500
      2009 and 2010 models... 50
      2011 and later models... 92
Lower control arm
  Arm-to-frame pivot bolts
    1500... 155
    2500/3500... 190
  Balljoint-to-steering knuckle nut
    1500
      Step 1... 37
      Step 2... Tighten an additional 90 degrees
    2500/3500
      2009 models.. 110
      2010 models.. 100
      2011 and later models... 116

Link/coil front suspension (4WD w/solid front axle)

Upper suspension arm-to-axle bracket nut... 120
Upper suspension arm-to-frame nut... 120
Lower suspension arm-to-axle bracket nut... 200
Lower suspension arm-to-frame nut... 200
Steering knuckle/balljoints
  Step 1 - lower balljoint nut... 35
  Step 2 - upper balljoint nut... 70
  Step 3 - lower balljoint nut... 148
Track bar mounting nuts/bolts 200

Rear suspension, coil spring (1500)

Shock absorber mounting bolts/nuts 100
Control arm bolts/nuts 225
Track bar bolts.. 129
Stabilizer bar link-to-frame bolts 56
Stabilizer bar link-to-bar bolts 74
Stabilizer bar bracket bolts 37

Rear suspension, leaf spring (2500/3500)

Shock absorber mounting bolts/nuts 100
Leaf spring-to-axle U-bolt nuts 110
Leaf spring-to-front and rear spring hanger nut and bolt........................ 255
Leaf spring rear shackle-to-frame bracket nut and bolt.......................... 160

Torque specifications

Ft-lbs (unless otherwise indicated)

Steering

Steering wheel bolt	45
Steering column nuts	20
Intermediate shaft pinch bolts	
Upper (to steering column shaft) and lower	28
Coupler (to rack-and-pinion steering gear)	
Non-EPS steering models	42
EPS steering models	36
Coupler (to recirculating ball steering gear)	36
Recirculating ball steering gear	
Pitman arm nut	225
Steering gear-to-frame mounting bolts	145
Tie-rod end adjuster clamp bolts	45
Tie-rod-to-steering knuckle nut	78
Steering damper-to-axle nut/bolt	75
Steering damper-to-tie-rod nut/bolt	60
Drag link nuts	100
Rack-and-pinion steering gear	
Steering gear-to-crossmember mounting bolts	
1500 models	
2012 and earlier models	235
2013 and later models	
Step 1	133
Step 2	Tighten an additional 90-degrees
2500/3500 2WD models	185
Tie-rod-to-steering gear	125
Tie-rod jam nut	94
Tie-rod end-to steering knuckle nut	
Step 1	45
Step 2	Tighten an additional 90 degrees
Wheel lug nuts	See Chapter 1

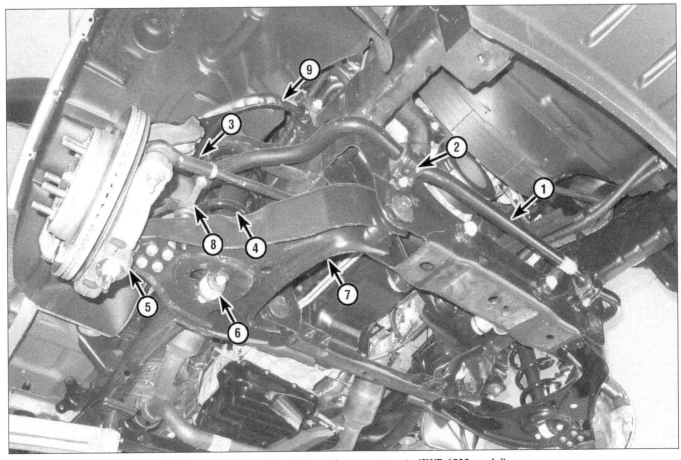

1.1 Front suspension and steering components (2WD 1500 model)

1	Stabilizer bar	4	Coil spring	7	Lower control arm
2	Stabilizer bar bushing clamps	5	Lower balljoint	8	Stabilizer bar link
3	Tie-rod end	6	Shock absorber lower mount	9	Upper control arm

1 General information

Front suspension

Refer to illustration 1.1

2WD models and are equipped with an independent front suspension system with upper and lower control arms, coil springs and shock absorbers. A stabilizer bar controls body roll. Each steering knuckle is positioned by a pair of balljoints in the ends of the upper and lower control arms **(see illustration)**. Light-duty 4WD models are equipped with an independent front suspension system with upper and lower control arms, and coil-over shock absorbers. A stabilizer bar controls body roll. Each steering knuckle is positioned by a pair of balljoints in the ends of the upper and lower control arms. Heavy-duty 4WD models utilize a solid front axle located by a pair of longitudinal suspension arms on each side. The front axle is suspended by a pair of coil springs and shock absorbers, and located laterally by a track bar.

Rear suspension

Refer to illustrations 1.2 and 1.3

The rear suspension on 1500 models uses coil springs, shock absorbers and four control arms. A stabilizer bar is used to control body roll and a lateral track bar keeps the rear axle centered **(see illustration)**.

1.2 Rear suspension components (2WD 1500 model)

1	Stabilizer bar	3	Shock absorber lower mount	5	Upper control arm
2	Track bar	4	Lower control arm	6	Coil spring

1.3 Rear suspension components (2WD)

1	Leaf springs	3	Spring plates	5	Axle
2	Shock absorbers	4	Spring plate U-bolts		

The rear suspension on 2500 and 3500 models consists of two shock absorbers and two leaf springs **(see illustration).**

Steering

The steering system on heavy-duty 4WD models consists of a recirculating-ball steering gearbox, Pitman arm, drag link, tie-rod and a steering damper. When the steering wheel is turned, the gear rotates the Pitman arm, which forces the drag link to one side. The drag link is connected to the right steering knuckle; the tie-rod is connected to the drag link and the left steering knuckle. The steering damper is attached to a bracket on the axle and to the tie-rod.

On non-heavy-duty 4WD models, the

steering system consists of a rack-and-pinion steering gear and two adjustable tie-rods. All non-heavy-duty 4WD models, except 2013 and later 1500 models have hydraulic power assist steering. 2013 and later 1500 models have an Electric Power Steering (EPS) system.

.Frequently, when working on the suspension or steering system components, you may come across fasteners which seem impossible to loosen. These fasteners on the underside of the vehicle are continually subjected to water, road grime, mud, etc., and can become rusted or frozen, making them extremely difficult to remove. In order to unscrew these stubborn fasteners without damaging them (or other components), use lots of penetrating oil and allow it to soak in for a while. Using a wire brush to clean exposed

threads will also ease removal of the nut or bolt and prevent damage to the threads. Sometimes a sharp blow with a hammer and punch is effective in breaking the bond between a nut and bolt threads, but care must be taken to prevent the punch from slipping off the fastener and ruining the threads. Heating the stuck fastener and surrounding area with a torch sometimes helps too, but isn't recommended because of the obvious dangers associated with fire. Long breaker bars and extension, or cheater, pipes will increase leverage, but never use an extension pipe on a ratchet - the ratcheting mechanism could be damaged. Sometimes, turning the nut or bolt in the tightening (clockwise) direction first will help to break it loose. Fasteners that require drastic measures to unscrew should always

2.3 Hold the shock absorber stem (A) with a wrench to prevent it from turning while loosening the upper mounting nut (B)

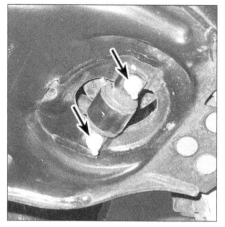

2.5 Shock absorber lower mounting nuts (2WD model)

3.2 Unscrew the stabilizer bar link nut while holding the link with a wrench

be replaced with new ones.

Since most of the procedures that are dealt with in this Chapter involve jacking up the vehicle and working underneath it, a good pair of jackstands will be needed. A hydraulic floor jack is the preferred type of jack to lift the vehicle, and it can also be used to support certain components during various operations. **Warning:** *Never, under any circumstances, rely on a jack to support the vehicle while working on it.*

Whenever any of the suspension or steering fasteners are loosened or removed they must be inspected and, if necessary, be replaced with new ones of the same part number or of original equipment quality and design. Torque specifications must be followed for proper reassembly and component retention. Never attempt to heat or straighten any suspension or steering components. Instead, replace any bent or damaged part with a new one.

2   Shock absorber (front) - removal and installation

1   Loosen the wheel lug nuts. If you're working on a 4WD model with independent front suspension, also loosen the driveaxle/hub nut (see Chapter 8).
2   Support the outer end of the lower control arm with a floor jack (the shock absorber serves as the down-stop for the suspension). The jack must remain in this position throughout the entire procedure.

2WD models

Refer to illustrations 2.3 and 2.5

3   Using a back-up wrench on the stem to prevent it from turning, remove the shock absorber upper mounting nut **(see illustration).**
4   Remove the retainer (metal washer) and grommet (rubber washer).
5   Working underneath the vehicle, remove the two nuts that attach the lower end of the

shock absorber to the lower control arm **(see illustration)** and pull the shock out from below.
6   Remove the lower grommet and retainer from the stem.
7   Before installing the shock absorber upper mounting nut, apply a thread locking compound to the threads of the shock absorber stud, then tighten the upper mounting nut and the lower mounting nuts to the torque listed in this Chapter's Specifications.

2500/3500 4WD models

8   Using a back-up wrench on the stem, remove the shock absorber upper mounting nut. On 4WD models with a solid front axle, remove the three nuts and take off the shock absorber upper mounting bracket.
9   Remove the shock absorber lower mounting bolt.
10   Remove the shock absorber.

1500 4WD models

11   Loosen the driveaxle/hub nut. Loosen the front wheel lug nuts. Raise the front of the vehicle and support it securely on jackstands, then remove the wheel.
12   Remove the three upper shock absorber mounting nuts.
13   Remove the lower shock absorber mounting nut and bolt.
14   Remove the brake caliper and disc (see Chapter 9).
15   Detach the wheel speed sensor from the steering knuckle and upper control arm (if equipped).
16   Detach the upper balljoint from the steering knuckle (see Section 8).
17   Detach the stabilizer link from the lower control arm (see Section 3).
18   Remove the driveaxle/hub nut, then pull the top of the knuckle out. Remove the shock absorber/coil spring assembly. **Warning:** *A special tool is necessary to compress the coil spring to remove it from the shock absorber. Because of this and the safety concerns involved with this procedure, we recommend that a professional repair shop remove the coil spring from the shock absorber and install*

it onto the replacement shock absorber.

All models

19   Installation is the reverse of removal. Tighten all fasteners to the torque listed in this Chapter's Specifications. On 1500 4WD models, tighten the brake caliper mounting bolts to the torque listed in the Chapter 9 Specifications. Tighten the driveaxle/hub nut to the torque listed in the Chapter 8 Specifications.
20   Install the wheels and lug nuts, lower the vehicle and tighten the lug nuts to the torque listed in the Chapter 1 Specifications.

3   Stabilizer bar (front) - removal and installation

Refer to illustrations 3.2 and 3.3

1   Raise the vehicle and support it securely on jackstands.
2   Remove the nuts from the link bolts and remove the link bolts **(see illustration). Note:** *Keep the parts for the left and right sides separate.*
3   On conventional units, remove the stabilizer bar bracket bolts and remove the brackets **(see illustration).**
4   On disconnecting-type front stabilizer

3.3 Remove the stabilizer bar bracket bolts

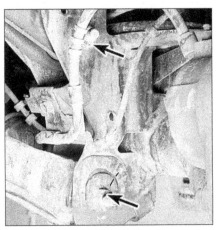

4.4 To detach a front suspension arm from the front axle, remove the nut(s) and knock out the bolt(s) with a punch

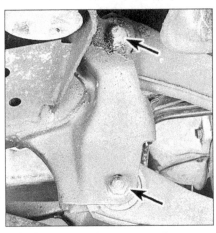

4.5 To detach a front suspension arm from the frame bracket, remove the nut and knock out the pivot bolt

bars, disconnect the wiring harness from the disconnection unit, then unbolt the unit and the support brackets. **Note:** *The disconnection unit is a part of the stabilizer bar and can't be replaced separately. If you're replacing the whole unit, you'll have to have the replacement stabilizer bar programmed to the vehicle by a dealer or other qualified shop with a scan tool.*

5   Remove the stabilizer bar. Remove the rubber bushings from the stabilizer bar.
6   Inspect all rubber bushings for wear and damage. If any of the rubber parts are cracked, torn or generally deteriorated, replace them.
7   Installation is the reverse of removal. Tighten all the fasteners to the torque listed in this Chapter's Specifications. Tighten the wheel lug nuts to the torque listed in the Chapter 1 Specifications.

---

4   Suspension arms (front) (link/coil suspension) - removal and installation

Note: *This procedure applies to vehicles equipped with a solid front axle.*

Upper and lower control arms

Refer to illustrations 4.4 and 4.5

1   Loosen the front wheel lug nuts. Raise the front of the vehicle and support it securely on jackstands. Remove the wheel.
2   If you're removing a lower suspension arm, paint or scribe alignment marks on the adjuster cams to ensure that the arm is installed correctly.
3   If you're removing an upper right side suspension arm, disconnect the exhaust system from the manifolds (see Chapter 4).
4   Remove the suspension arm nut, cams and cam bolt from the axle **(see illustration).**
Note: *The manufacturer recommends that you replace all control arm fasteners with new*

ones whenever they are removed.
5   Remove the suspension arm nut and pivot bolt from the frame bracket **(see illustration).**
6   Installation is the reverse of removal. If you're installing a lower arm, align the marks made in Step 2. Tighten all suspension fasteners to the torque listed in this Chapter's Specifications after the vehicle is on the ground.
7   Have the front end alignment checked and, if necessary, adjusted.

Track bar

8   Loosen the wheel lug nuts, raise the front of the vehicle and support it securely on jackstands placed under the frame rails. Remove the wheels and support the axle with a floor jack placed under the axle tube. Raise the axle enough to support its weight.
9   Remove the fasteners securing the track bar to the frame and axle. Remove the track bar.
10   Installation is the reverse of removal. Tighten the track bar mounting fasteners to the torque values listed in this Chapter's Specifications after the vehicle has been lowered.

---

5   Coil spring (front) - removal and installation

Warning: *Removing a coil spring is potentially dangerous and utmost attention must be directed to the job or serious injury may result. Use only a high-quality spring compressor and carefully follow the manufacturer's instructions furnished with it.*

Models with independent front suspension (except 1500 4WD models)

Warning: *On coil-over shock absorber assemblies found on 4WD models with independent front suspension, a special tool is necessary to compress the coil spring to remove it from the shock absorber (with the shock absorber assembly removed from the vehicle). Because of this and the safety concerns involved with this procedure, we recommend that a professional repair shop remove the coil spring from the shock absorber and install it onto the replacement shock absorber. See Section 2 to remove the shock absorber/coil spring assembly.*

Removal

Refer to illustration 5.3

1   Loosen the wheel lug nuts, raise the vehicle and support it securely on jackstands placed under the frame rails. Remove the wheel.
2   Support the outer end of the lower control arm with a floor jack. Remove the shock absorber (see Section 2).
3   Install a suitable internal-type spring compressor in accordance with the tool manufacturer's instructions **(see illustration).** Compress the spring sufficiently to relieve all pressure from the upper spring seat. This can be verified by wiggling the spring.
4   Disconnect the stabilizer bar link from the lower control arm (see Section 3).
5   Separate the lower control arm from the steering knuckle; loosen the lower balljoint nut a few turns (don't remove it), install a balljoint separator and break the balljoint loose from the knuckle. Now remove the nut. **Note:** *If you*

5.3 A typical aftermarket internal spring compressor tool: the hooked arms grip the upper coils of the spring, the plate is inserted below the lower coil, and when the threaded rod is turned, the spring is compressed

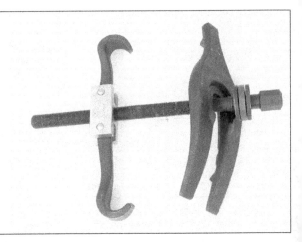

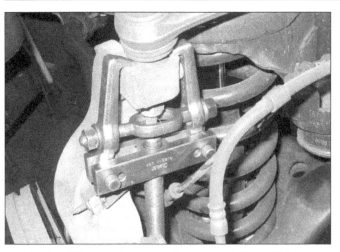

6.2 Back off the nut a few turns, then separate the balljoint from the steering knuckle (leaving the nut on the ballstud will prevent the balljoint from separating violently)

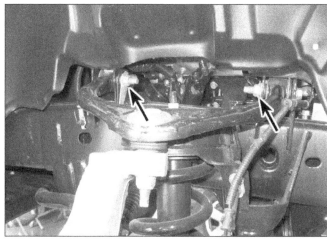

6.5 Remove the nuts and the upper control arm pivot bolts; turn the bolts, not the nuts

don't have the proper balljoint removal tool, a picklefork type balljoint separator can be used, but keep in mind that this type of tool will probably destroy the balljoint boot.
6　Lift the knuckle and hub assembly up, then place a block of wood between the upper control arm and the frame to support the assembly out of the way.
7　Lower the floor jack, then guide the compressed coil spring out.

Installation
8　Place the insulator on top of the coil spring (the upper end of the spring is the end with the more tightly wound coils).
9　Install the top of the spring into the spring pocket and the bottom in the lower control arm.
10　Connect the lower control arm to the steering knuckle (see Section 7). Remove the spring compressor.
11　The remainder of installation is the reverse of removal. Tighten all fasteners to the proper torque values. Tighten the wheel lug nuts to the torque listed in the Chapter 1 Specifications.
12　Have the front end alignment checked and, if necessary, adjusted.

Models with a solid front axle
13　Loosen the front wheel lug nuts, raise the front of the vehicle and support it securely on jackstands placed under the frame rails. Remove the wheels.
14　Use floor jacks to support the end of the axle from which you're removing the spring.
15　Install a coil spring compressor on the coil spring and compress the spring slightly (this is to make sure it won't slip, fall off or fly out).
16　Remove the upper suspension arm and loosen the lower suspension arm nuts and pivot bolts (see Section 4).

17　Mark and disconnect the front driveshaft from the front axle (see Chapter 8).
18　Disconnect the track bar from the frame rail bracket.
19　Disconnect the drag link from the Pitman arm (see Section 18).
20　Disconnect the stabilizer bar link from the axle (see Section 3).
21　Disconnect the shock absorber from the axle (see Section 2).
22　Lower the axle until the coil spring is free and remove the spring.
23　Installation is the reverse of removal.

6　Upper control arm (front) - removal and installation

Note: *This procedure applies to models equipped with independent front suspension.*

Removal
Refer to illustrations 6.2 and 6.5
1　Loosen the wheel lug nuts, raise the vehicle and support it securely on jackstands placed under the frame rails. Remove the wheel.
2　Loosen (but don't remove) the nut on the upper balljoint stud, then disconnect the balljoint from the steering knuckle with a balljoint removal tool **(see illustration). Note:** *If you don't have the proper balljoint removal tool, a picklefork type balljoint separator can be used, but keep in mind that this type of tool will probably destroy the balljoint boot.*
3　If necessary for clearance, remove the brake disc (see Chapter 9).
4　Detach the wiring harness from the control arm.
5　Remove the nuts and pivot bolts that attach the control arm to the frame **(see illustration).** Pull the upper arm from its frame brackets.

Installation
6　Installation is the reverse of removal. Tighten all fasteners to the torque values listed in this Chapter's Specifications, but don't tighten the pivot bolt nuts until the vehicle is sitting at normal ride height. If it's too hard to get to the nuts with the wheel on, normal ride height can be simulated by raising the outer end of the lower control arm with a floor jack.
7　Install the wheel and lug nuts. Lower the vehicle and tighten the lug nuts to the torque listed in the Chapter 1 Specifications. Have the front end alignment checked and, if necessary, adjusted.

7　Lower control arm (front) - removal and installation

Note: *This procedure applies to models equipped with independent front suspension.*

2WD models
Removal
Refer to illustrations 7.5a, 7.5b, 7.9a and 7.9b
1　Loosen the wheel lug nuts, raise the vehicle and support it securely on jackstands placed under the frame rails. Remove the wheel.
2　Support the outer end of the lower control arm with a floor jack. Remove the shock absorber (see Section 2).
3　Remove the brake caliper, caliper bracket and disc (see Chapter 9).
4　Install a suitable internal type spring compressor in accordance with the tool manufacturer's instructions **(see illustration 5.3).** Compress the spring sufficiently to relieve all force from the upper spring seat. This can be verified by wiggling the spring.

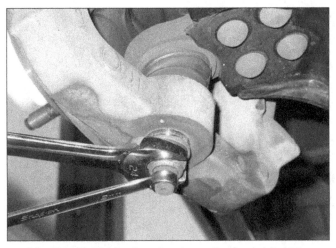

7.5a If the balljoint spins when loosening the nut, hold the exposed stud with a wrench

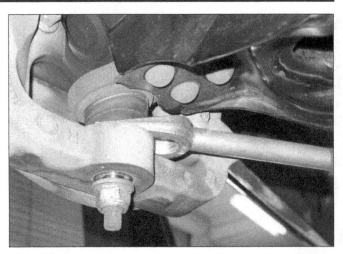

7.5b Separating the lower control arm balljoint with a picklefork type balljoint separator

7.9a Mark the positions of these alignment washers, as they must be installed in the same positions (unless you're going to have the front aligned later)

7.9b Lower control arm pivot bolts (2WD 1500 model)

5   Loosen (but don't remove) the nut on the lower balljoint stud (see illustration), then disconnect the balljoint from the steering knuckle with a balljoint removal tool. Note: *If you don't have the proper balljoint removal tool, a picklefork type balljoint separator can be used, but keep in mind that this type of tool will probably destroy the balljoint boot* (see illustration).
6   Disconnect the link from the stabilizer bar.
7   Lift the knuckle and hub assembly up, then place a block of wood between the upper control arm and the frame to support the assembly out of the way.
8   Pull the lower control arm down, then guide the compressed coil spring out.
9   Mark the positions of the eccentric washers so they can be installed in the same positions (see illustration). Remove the bolts that attach the control arm to the frame (see illustration). Pull the lower arm from its frame brackets.

Installation

10   Installation is the reverse of removal. Tighten all fasteners to the torque values listed in this Chapter's Specifications, but don't tighten the pivot bolt nuts until the vehicle is sitting at normal ride height. If it's too hard to get to the nuts with the wheel on, normal ride height can be simulated by raising the outer end of the lower control arm with a floor jack.
11   Install the wheel and lug nuts. Lower the vehicle and tighten the lug nuts to the torque listed in the Chapter 1 Specifications. Have the front end alignment checked and, if necessary, adjusted.

4WD models

Removal

12   Loosen the wheel lug nuts, raise the vehicle and support it securely on jackstands placed under the frame rails. Remove the wheel.
13   Remove the driveaxle (see Chapter 8), then reinstall the upper balljoint back into the steering knuckle and install the nut temporarily.

14   Disconnect the stabilizer bar from the lower control arm (see Section 3).
15   Remove the lower shock absorber bolt (see Section 2).
16   Loosen (but don't remove) the nut on the lower balljoint stud a few turns, then disconnect the balljoint from the steering knuckle arm with a balljoint removal tool. Now remove the nut. Note: *If you don't have the proper balljoint removal tool, a picklefork type balljoint separator can be used, but keep in mind that this type of tool will probably destroy the balljoint boot* (see illustration 7.5a).
17   Remove the lower control arm pivot bolts and pull the lower arm from its frame brackets.

Installation

18   Installation is the reverse of removal. Tighten all fasteners to the torque values listed in this Chapter's Specifications, but don't tighten the pivot bolt nuts until the vehicle is sitting at normal ride height. If it's too hard to get to the nuts with the wheel on, normal ride height can be simulated by raising the outer end of the lower control arm with a floor jack.

8   Balljoints - replacement

1   Inspect the control arm balljoints for looseness any time either of them is separated from the steering knuckle. See if you can turn the ballstud in its socket with your fingers. If the balljoint is loose, or if the ballstud can be turned, replace the balljoint. You can also check the balljoints with the suspension assembled as follows.

Upper balljoints

Independent front suspension

2   Raise the lower control arm with a floor jack until the tire just barely touches the ground.
3   Using a large prybar inserted between the upper control arm and the steering knuckle, pry upwards on the upper control arm. The manu-

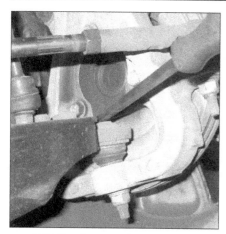

8.15 Pry down on the balljoint to check for wear

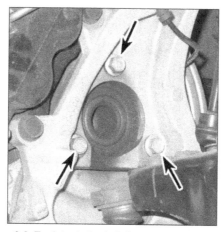

9.6 To detach the hub from the steering knuckle, remove these bolts (2WD shown, 4WD similar)

facturer specifies up to 0.020-inch movement is allowed; a dial indicator can be used to check for play. If you are unable to accurately measure balljoint play, have the balljoint checked at an automotive repair shop.
4   If replacement is indicated, loosen the wheel lug nuts, raise the front of the vehicle and support it securely on jackstands. Place a floor jack under the lower control arm, near the outer end, and raise it until the upper control arm lifts off its rebound bumper. **Warning:** *The jack must remain in this position throughout the entire procedure.*
5   The balljoint is press fit in the lower control arm, which necessitates the use of a special press tool and receiver cup to remove and install the balljoint. Equipment rental yards and some auto parts stores have these tools available for rent. If you don't have access to this tool, remove the control arm (see Section 6) and take it to an automotive machine shop or other qualified repair facility to have the balljoint replaced. **Note:** *Some balljoint replacement tools are similar to a heavy-duty C-clamp. With this type of tool it is only necessary to separate the balljoint from the steering knuckle.*

Link/coil suspension
6   Raise the vehicle and support it securely on jackstands. Grasp the top of the tire and rock the tire in and out. If there is significant movement at the balljoint, have the balljoint checked by an automotive repair shop.
7   If replacement is indicated, the balljoints on 4WD solid axle models are a press fit in the axle tube yoke, which necessitates the use of a special press tool and receiver cup to remove and install them. Equipment rental yards and some auto parts stores have these tools available for rent. If you don't have access to this tool, take the vehicle to an automotive machine shop or other qualified repair facility to have the balljoint replaced.
8   If you have access to this tool, remove the hub/bearing assembly (see Chapter 8).
9   Remove the knuckle (see Section 10).
10 Remove the axleshaft (see Chapter 8).
11 Remove the snap-ring from lower balljoints, then use the balljoint removal tool to

press out the balljoint.
12   Installation is the reverse of removal.

Lower balljoints
Independent front suspension
Refer to illustration 8.15
13   Raise the vehicle and support it securely on jackstands.
14   Place a floor jack under the lower control arm, near the outer end, and raise it until the upper control arm lifts off its rebound bumper.
15   Insert a prybar between the top of the balljoint and the steering knuckle and pry down **(see illustration)**. The manufacturer specifies up to 0.020-inch movement is allowed; a dial indicator can be used to check for play. If you are unable to accurately measure balljoint play, have the balljoint checked at an automotive repair shop.
16   If replacement is indicated, remove the lower control arm (see Section 7); the balljoint is press fit in the lower control arm, which necessitates the use of a special press tool and receiver cup to remove and install the balljoint. Equipment rental yards and some auto parts stores have these tools available for rent. If you don't have access to this tool, take the vehicle to an automotive machine shop or other qualified repair facility to have the balljoint replaced. **Note:** *Some balljoint replacement tools are similar to a heavy-duty C-clamp. With this type of tool it is only necessary to separate the balljoint from the steering knuckle (see Section 10).* **Note:** *The lower balljoints on some models are replaceable, but on others they're not. Check with your local auto parts store to check on the availability of replacement parts before disassembling your vehicle.*

Link/coil suspension
17   Raise the vehicle and support it securely on jackstands.
18   Using a large prybar, pry upwards on the bottom of the steering knuckle, checking for significant movement at the balljoint. If there is significant movement at the balljoint, have the

balljoint checked by an automotive repair shop.
19   If replacement is indicated, the balljoints on 4WD solid axle models are a press fit in the axle tube yoke, which necessitates the use of a special press tool and receiver cup to remove and install them. Equipment rental yards and some auto parts stores have these tools available for rent. If you don't have access to this tool, take the vehicle to an automotive machine shop or other qualified repair facility to have the balljoint replaced.
20   If you have the tool, refer to Steps 8 through 12 for the replacement procedure.

9   Hub and bearing assembly - removal and installation

Independent front suspension models
Removal
Refer to illustration 9.6
Warning: *The dust created by the brake system is harmful to your health. Never blow it out with compressed air and don't inhale any of it. Do not, under any circumstances, use petroleum-based solvents to clean brake parts. Use brake system cleaner only.*
Note: *The hub and bearing assembly is sealed-for-life. If worn or damaged, it must be replaced as a unit.*
1   On 4WD models, loosen the driveaxle/hub nut. Loosen the front wheel lug nuts, raise the vehicle and support it securely on jackstands. Remove the wheel.
2   Remove the brake caliper and hang it out of the way with a piece of wire, then remove the caliper mounting bracket (see Chapter 9). Pull the disc off the hub.
3   Remove the wheel speed sensor.
4   On 4WD models, separate the tie-rod and the upper balljoint from the steering knuckle
5   On 4WD models, pull the top of the knuckle out to separate it from the axleshaft.
6   Working from the rear of the steering knuckle, remove the hub retaining bolts from the steering knuckle **(see illustration).**
7   Remove the hub from the steering knuckle. Remove the disc shield.

Installation
8   Clean the mating surfaces on the steering knuckle, bearing flange and knuckle bore.
9   Position the disc shield, insert the hub and bearing assembly into the steering knuckle and install the bolts, tightening them to the torque listed in this Chapter's Specifications.
10   Installation is the reverse of removal, noting the following points:
a) *On 4WD models, tighten the driveaxle/hub nut to the torque listed in the Chapter 8 Specifications.*
b) *Tighten the brake caliper mounting bracket and brake caliper mounting bolts to the torque listed in the Chapter 9 Specifications.*

11.2 Rear shock absorber mounting fasteners (2WD 1500 model)

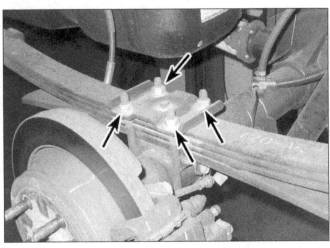

12.2 To remove the spring plate and U-bolts, remove these four nuts

c) *Install the wheel, lower the vehicle and tighten the lug nuts to the torque listed in the Chapter 1 Specifications.*

Solid front axle models

11   See Chapter 8 for this procedure.

10 Steering knuckle - removal and installation

Independent front suspension

1   Loosen the wheel lug nuts, raise the vehicle and support it securely on jackstands. Remove the wheel.
2   On 4WD models, unscrew the driveaxle/hub nut with a socket and large breaker bar (see Chapter 8). Brace a large prybar across two of the wheel studs or insert a large screwdriver through the center of the brake caliper and into the disc cooling vanes to prevent the hub from turning as the nut is loosened.
3   Remove the disc brake caliper and disc (see Chapter 9). If equipped, disconnect the electrical connector from the wheel speed sensor and remove the sensor.
4   Disconnect the tie-rod from the steering knuckle (see Section 18).
5   Remove the driveaxle/hub nut.
6   Support the lower control arm with a floor jack. **Warning:** *The jack must remain in this position throughout the entire procedure.*
7   Separate the lower control arm from the steering knuckle (see Section 7).
8   Separate the upper control arm from the steering knuckle (see Section 6).
9   Remove the knuckle.
10   Carefully inspect the steering knuckle for cracks, especially around the steering arm and spindle mounting area. Also inspect the balljoint stud holes. If they're elongated, or if you find any cracks in the knuckle, replace the steering knuckle.
11   Installation is the reverse of removal. Tighten all suspension fasteners to the torque listed in this Chapter's Specifications.

Link/coil suspension

12   Loosen the wheel lug nuts, raise the vehicle and support it securely on jackstands. Remove the wheel.
13   Remove the hub bearing and axle shaft (see Chapter 8).
14   Disconnect the tie-rod end or drag link from the steering knuckle (see Section 18).
15   Remove the wheel speed sensor.
16   Loosen the steering knuckle upper ballstud nut a few turns, but don't remove it.
17   Remove the lower balljoint nut. Use a balljoint separation tool to detach the balljoint from the knuckle. **Note:** *If you don't have the proper balljoint removal tool, a picklefork type balljoint separator can be used, but keep in mind that this type of tool will probably ruin the balljoint boot* **(see illustration 7.5b).**
18   Remove the upper balljoint nut.
19   Separate the steering knuckle. Use a brass hammer to knock it loose if necessary.
20   Installation is the reverse of removal. Tighten all suspension fasteners to the torque listed in this Chapter's Specifications.

11 Shock absorber (rear) - removal and installation

Refer to illustration 11.2

1   Raise the rear of the vehicle and support it securely on jackstands placed under the frame rails. Support the rear axle with a floor jack placed under the axle tube on the side being worked on. Don't raise the axle - just support its weight.
2   Remove the nut and bolt that attach the upper end of the shock absorber to the frame **(see illustration).** If the nut won't loosen because of rust, apply some penetrating oil and allow it to soak in for awhile.
3   Remove the nut and bolt that attach the lower end of the shock to the axle bracket. Again, if the nut is frozen, apply some penetrating oil, wait awhile and try again.
4   Extend the new shock absorber as far as

possible. Install new rubber grommets into the shock absorber eyes (if they are not already present).
5   Installation is the reverse of removal. Tighten the shock absorber mounting fasteners to the torque listed in this Chapter's Specifications.

12 Leaf spring - removal and installation

Removal

Refer to illustrations 12.2, 12.3 and 12.4

1   Loosen the wheel lug nuts, raise the rear of the vehicle and support it securely on jackstands placed under the frame rails. Remove the wheel and support the rear axle with a floor jack placed under the axle tube. Don't raise the axle - just support its weight.
2   Remove the nuts, U-bolts, spring plate, and spring seat that clamp the leaf spring to the axle **(see illustration).**
3   Remove the leaf spring shackle bolts and the shackle **(see illustration).**

12.3 To detach the rear end of the leaf spring, remove the shackle nuts and bolts

12.4 To detach the front end of the leaf spring from the forward bracket, remove this nut and bolt

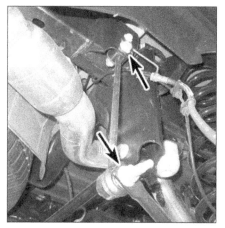

14.2 Rear stabilizer bar link bolt locations

14.3 Rear stabilizer bar clamp bolts - check the clamp bushings for wear and replace them if necessary

4   Unscrew the front bolt **(see illustration)** and remove the leaf spring from the vehicle.

Installation

5   To install the leaf spring, position the spring on the axle tube so that the spring center bolt enters the locating hole on the axle tube.
6   Line up the spring front eye with the mounting bracket and install the front bolt and nut.
7   Install the rear shackle assembly.
8   Tighten the shackle bolt and the front bolt until all slack is taken up.
9   Install the U-bolts and nuts. Tighten the nuts until they force the spring plate against the axle, but don't torque them yet.
10   The auxiliary spring, if equipped, must align with the main spring.
11   Install the wheel and lug nuts. Remove the jackstands and lower the vehicle. Tighten the lug nuts to the torque listed in the Chapter 1 Specifications.
12   Tighten the U-bolt nuts, front bolt/nut and shackle bolts/nuts to the torque listed in this Chapter's Specifications.

13 Coil spring (rear) - removal and installation

1   Raise the vehicle and support it securely on jackstands.
2   Place a jack under the rear axle to support its weight.
3   Remove the bolt from the lower shock absorber mount.
4   Slowly lower the jack until the spring and insulator can be pulled out. **Caution:** *Don't put tension on the rubber brake hoses or allow the rear axle to fall from the jack.*
5   Installation is the reverse of removal. Tighten all fasteners to the torque listed in this Chapter's Specifications.

14 Stabilizer bar (rear) - removal and installation

Refer to illustrations 14.2 and 14.3
Note: *This procedure applies only to models with coil spring rear suspension.*
1   Raise the vehicle and support it securely on jackstands.
2   Remove the nuts from the stabilizer bar

ends **(see illustration)**.
3   Remove the stabilizer bar brackets from the rear axle assembly **(see illustration)**.
4   Remove the stabilizer bar.
5   Inspect the bushings for wear and damage and replace them as required.
6   Installation is the reverse of removal.

15 Suspension arms (rear) - removal and installation

Note: *These procedures apply only to models with coil spring rear suspension.*

Control arms

Refer to illustration 15.4
1   Raise the vehicle and support it securely on jackstands.
2   Place a jack under the rear axle to support its weight.
3   If you're working on the left lower control arm, remove the cable guide bolt.
4   Remove the control arm bolts and nuts and remove the control arm **(see illustration)**. **Note:** *The manufacturer states that the nut from the axle end of the control arm must be replaced with a new one whenever it's removed.*
5   Installation is the reverse of removal. Don't tighten the nuts and bolts at this time.
6   Lower the vehicle and allow its full weight to load the rear suspension, then tighten the control arm bolts to the torque listed in this Chapter's Specifications.

Track bar

Note: *The track bar is the one that is parallel with the rear axle* **(see illustration 1.2)**.
7   Raise the vehicle and support it securely on jackstands.
8   Place a jack under the rear axle to stabilize it.
9   Remove the track bar bolts and nuts and remove the track bar.

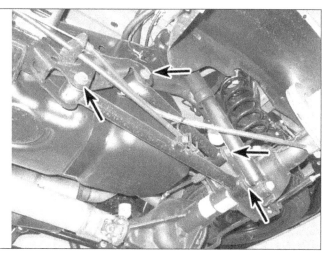

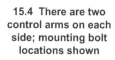

15.4 There are two control arms on each side; mounting bolt locations shown

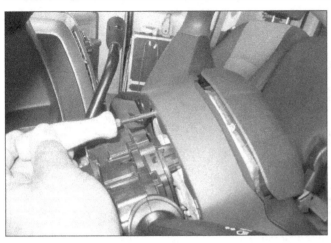

16.6a You can reach the airbag retaining clips when the upper steering column cover is removed

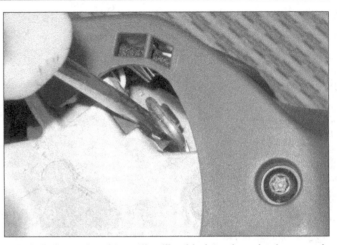

16.6b Release the airbag clips like this (steering wheel removed for clarity)

10   Installation is the reverse of removal. Don't tighten the nuts and bolts at this time.

11   Lower the vehicle and allow its full weight to load the rear suspension, then tighten the track bar bolts to the torque listed in this Chapter's Specifications.

---

16   Steering wheel - removal and installation

Warning: *These models are equipped with airbags. Always disable the airbag system whenever working in the vicinity of any airbag system component to avoid the possibility of accidental airbag deployment, which could cause personal injury (see Chapter 12).*

Removal

Refer to illustrations 16.6a, 16.6b, 16.7, 16.9, 16.12a and 16.12b

1   Park the vehicle with the front wheels in the straight-ahead position. **Warning:** *Do*

NOT turn the steering shaft before, during or after steering wheel removal. If the shaft is turned while the steering wheel is removed, the clockspring can be damaged. The clockspring, which maintains a continuous electrical circuit between the wiring harness and the airbag module, consists of a flat, ribbon-like electrically conductive tape which winds and unwinds as the steering wheel is turned.

2   Disconnect the cable(s) from the negative battery terminal(s) (see Chapter 5). Wait two minutes for the electrical system to discharge before working in the area of the airbag (see Chapter 12).

3   Remove the upper steering column cover (see Chapter 11).

4   Lower the steering column fully.

5   Turn the steering wheel to bring each of the three airbag retaining clips to the top position one at a time.

6   Insert a small screwdriver under each clip and pry to release it **(see illustrations)**. **Note:** *Use a small mirror to make sure that you get the screwdriver under the clips properly.*

7   Pull the airbag module out far enough to reach the wiring harnesses at its rear **(see illustration)**. Carefully disconnect the wiring harnesses without pulling on the wires.

8   Lift off the airbag module. **Warning:** *Carry the airbag module with the trim cover (upholstered side) facing away from you, and set the airbag module in a safe location with the trim cover facing up.*

9   Remove the steering wheel bolt **(see illustration)**. **Note:** *The manufacturer recommends that this bolt be replaced with a new one whenever it's removed.*

10   Mark the relationship of the steering wheel to the steering shaft.

11   Use a jaw-type puller and remove the steering wheel. The hooks of the jaws should face inward. Detach the wires and feed them through the steering wheel as you remove it. **Caution:** *Any attempt to remove the steering wheel without using a puller can damage the steering column.*

12   If it is necessary to remove the clockspring, unplug the electrical connector for the

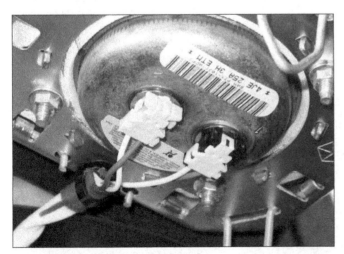

16.7 Carefully disconnect these wires from the rear of the airbag after you lift it off

16.9 The steering wheel bolt should be replaced with a new one whenever it's removed; insert the hooks of the steering wheel puller into these holes, with the hooks facing toward the hub

16.12a Clockspring upper screw

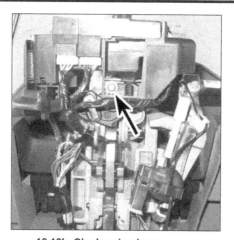

16.12b Clockspring lower screw

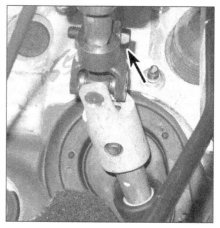

17.7 Remove the steering shaft coupler bolt

clockspring, then remove the screws and lift the clockspring from the steering column **(see illustrations)**.

Installation

13 When installing the clockspring, make absolutely sure that the airbag clockspring is centered, with the arrows on the clockspring rotor and case lined up. This shouldn't be a problem as long as you have not turned the steering shaft while the wheel was removed. If for some reason the shaft was turned, center the clockspring as follows:

a) *Rotate the clockspring clockwise until it stops (don't apply too much force, though).*

b) *Rotate the clockspring counterclockwise about 2-1/2 turns until the arrows on the clockspring rotor and case line up.*

14 Installation is the reverse of removal, noting the following points:

a) *Make sure the airbag clockspring is centered before installing the steering wheel.*

b) *When installing the steering wheel, align the marks on the shaft and the steering wheel hub.*

c) *Install a NEW steering wheel bolt and tighten it to the torque listed in this Chapter's Specifications.*

d) *Enable the airbag system (see Chapter 12).*

17 Steering column - removal and installation

Warning: *These models are equipped with airbags. Always disable the airbag system whenever working in the vicinity of any airbag system component to avoid the possibility of accidental airbag deployment, which could cause personal injury (see Chapter 12).*

Removal

Refer to illustration 17.7

1 Park the vehicle with the wheels pointing straight ahead. Disconnect the cable(s) from the negative battery terminal(s) (see Chapter 5). Wait at least two minutes before proceeding (to allow the backup power supply for the airbag system to become depleted).

2 Remove the steering column covers (see Chapter 11).

3 Remove the steering wheel and the clockspring (see Section 16), then turn the ignition key to the LOCK position to prevent the steering shaft from turning. **Caution:** *If this is not done, the airbag clockspring could be damaged.* **Note:** *Some automatic transmission models may not be equipped with an internal locking shaft. Remove the airbag clockspring to prevent accidental damage (see Section 16).*

4 On models with a column-mounted shifter, detach the shift cable from the shift lever on the column. Also detach the electrical connector (see Chapter 7B). On models with a floor-mounted shifter, release the shift cable from its clip.

5 Unplug any other electrical connectors that would interfere with column removal.

6 Remove the brake light switch.

7 Remove the shaft coupler bolt (securing the steering shaft to the intermediate shaft) **(see illustration)**. Separate the intermediate shaft from the steering shaft. **Note:** *The manufacturer recommends that this bolt be replaced with a new one whenever it's removed.*

8 Remove the steering column mounting nuts, lower the column and pull it to the rear, making sure nothing is still connected.

Installation

9 Guide the steering column into position and install the mounting nuts, but don't tighten them yet.

10 Connect the intermediate shaft to the steering column. Install a new coupler bolt,

then tighten the nut to the torque listed in this Chapter's Specifications.

11 Tighten the column mounting nuts to the torque listed in this Chapter's Specifications.

12 The remainder of installation is the reverse of removal. Adjust the shift cable following the procedures described in Chapter 7B.

18 Steering linkage - removal and installation

Independent front suspension
Tie-rod end
Removal

Refer to illustrations 18.2, 18.3, 18.4a, and 18.4b

1 Loosen the wheel lug nuts, raise the vehicle and support it securely on jackstands. Apply the parking brake. Remove the wheel.

2 Loosen the tie-rod end jam nut **(see illustration).**

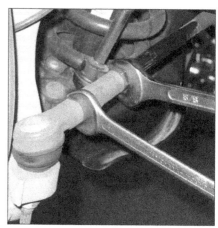

18.2 Hold the tie-rod end with a wrench while loosening the jam nut

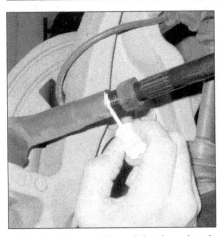

18.3 Mark the position of the tie-rod end in relation to the threads

18.4a If the ballstud turns while attempting to loosen the nut, hold it stationary with a wrench

18.4b Back-off the ballstud nut a few turns, then separate the tie-rod end from the steering knuckle with a puller (leaving the nut on the ballstud will prevent the tie-rod end from separating violently)

3   Mark the relationship of the tie-rod end to the threaded portion of the tie-rod. This will ensure the toe-in setting is restored when reassembled **(see illustration).**
4   Loosen (but don't remove) the nut on the tie-rod end ballstud and disconnect the tie-rod end from the steering knuckle arm with a puller **(see illustrations).**
5   If you're replacing the tie-rod end, unscrew the tie-rod end from the tie-rod, then thread the new tie-rod end onto the tie-rod to the marked position.

Installation
6   Connect the tie-rod end to the steering knuckle arm. Install the nut on the ballstud and tighten it to the torque listed in this Chapter's Specifications. Install the wheel. Lower the vehicle and tighten the lug nuts to the torque listed in the Chapter 1 Specifications.
7   Have the front end alignment checked and, if necessary, adjusted.

Link/coil suspension

Tie-rod end

Removal
8   Loosen the wheel lug nuts, raise the vehicle and support it securely on jackstands. Apply the parking brake. Remove the wheel.
9   Measure the distance between the adjuster tube and the center line of the ballstud. This will ensure the toe-in setting is restored when reassembled.
10   Loosen (but don't remove) the nut on the tie-rod end ballstud and disconnect the tie-rod end from the steering knuckle arm with a puller.
11   If you're replacing the tie-rod end, lubricate the threaded portion of the tie-rod end with chassis grease. Screw the new tie-rod end into the adjuster tube and adjust the distance from the tube to the ballstud by threading the tie-rod end into the adjuster tube until

the same number of threads are showing as before. Or use the measurement you made before removing the tie-rod end. Don't tighten the adjuster tube clamps yet.

Installation
12   Insert the tie-rod ballstud into the steering knuckle. Make sure the ballstud is fully seated. Install the nut and tighten it to the torque listed in this Chapter's Specifications. If the ballstud spins when attempting to tighten the nut, hold it stationary with a small wrench.
13   Tighten the tie-rod adjuster clamp bolts to the torque listed in this Chapter's Specifications. Have the front end alignment checked and, if necessary, adjusted.

Drag link

Removal
14   Loosen the wheel lug nuts, raise the vehicle and support it securely on jackstands. Apply the parking brake. Remove the wheels.
15   Remove the nut from the drag link ballstud.
16   Using a puller, separate the drag link from the Pitman arm.
17   Loosen (but don't remove) the nuts on the tie-rod end ballstuds and disconnect the tie-rod ends from the steering knuckle and drag link with a puller.

Installation
18   Installation is the reverse of removal. Tighten all of the fasteners to the torque listed in this Chapter's Specifications.
19   Have the front end alignment checked and, if necessary, adjusted.

Pitman arm

Removal
20   Loosen the wheel lug nuts, raise the vehicle and support it securely on jackstands. Apply the parking brake. Remove the right wheel.

21   Remove the drag link nut from the Pitman arm ballstud.
22   Working at the right side steering knuckle, loosen (but don't remove) the nut on the end of the drag link ballstud, then disconnect the end of the drag link from the steering knuckle arm with a puller.
23   Remove the Pitman arm nut and washer. Mark the Pitman arm and the steering gear shaft to ensure proper alignment at reassembly (only if the same Pitman arm is going to be used).

Installation
24   Installation is the reverse of removal. Tighten all of the fasteners to the torque listed in this Chapter's Specifications. Have the front end alignment checked and, if necessary, adjusted.

Steering damper
25   Inspect the steering damper for fluid leakage. A slight film of fluid near the shaft seal is normal, but if there's excessive fluid present and it's obviously coming from the steering damper, replace the damper.
26   Inspect the steering damper bushing for excessive wear. If it's in bad shape, replace the damper.
27   To test the damper itself, disconnect it from the tie-rod. Using as much travel as possible, extend and compress the damper. The resistance should be smooth and constant for each stroke. If any binding, dead spots or unusual noises are present, replace the damper.
28   Remove the damper-to-tie-rod mounting bolt and nut. Separate the damper from the tie-rod.
29   Remove the damper-to-axle mounting bolt and nut. Separate the damper from the axle.
30   Installation is the reverse of removal. Tighten the fasteners to the torque values listed in this Chapter's Specifications.

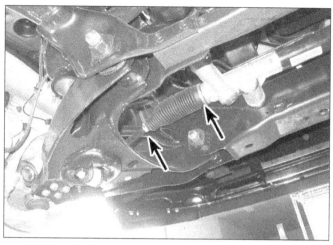

19.3 The steering gear boots are secured by these clamps

20.3 Remove the pinch bolt from the lower end of the intermediate shaft coupler (B), then unscrew the pressure and return line fittings from the power steering gear (A)

19 Steering gear boots - replacement

Refer to illustration 19.3

Note: *This procedure applies to models with independent front suspension only.*

1   Loosen the wheel lug nuts, raise the vehicle and support it securely on jackstands. Remove the wheel.

2   Remove the tie-rod end and jam nut (see Section 18).

3   Remove the steering gear boot clamps **(see illustration)** and slide the boot off. **Note:** *Check for the presence of power steering fluid in the boot. If there is a substantial amount, it means the rack seals are leaking and the power steering gear should be replaced with a new or rebuilt unit.*

4   Before installing the new boot, wrap the threads and serrations on the end of the steering rod with a layer of tape so the small end of the new boot isn't damaged.

5   Slide the new boot into position on the steering gear until it seats in the grooves, then install new clamps.

6   Remove the tape and install the tie-rod end (see Section 18).

7   Install the wheel and lug nuts. Lower the vehicle and tighten the lug nuts to the torque listed in the Chapter 1 Specifications.

8   Have the front end alignment checked and, if necessary, adjusted.

20 Steering gear - removal and installation

Warning: *DO NOT allow the steering column shaft to rotate with the steering gear removed or damage to the airbag system could occur.*

1   Loosen the front wheel lug nuts, raise the front of the vehicle and support it securely on jackstands. Apply the parking brake. Remove the wheels.

2   Remove the skid plate, if equipped.

Models with hydraulic rack-and-pinion steering gear

Refer to illustrations 20.3 and 20.6

3   Remove the intermediate shaft coupler pinch bolt and discard it **(see illustration)**. It should be replaced with a new one according to the manufacturer.

4   Position a drain pan under the steering gear. Using a flare-nut wrench, unscrew the power steering pressure and return lines from the steering gear. Cap the lines to prevent leakage.

5   Detach the tie-rod ends from the steering knuckles (see Section 18).

6   Unscrew the mounting bolts, then lower the steering gear from the vehicle **(see illustration).**

7   Installation is the reverse of removal, noting the following points:

   a) *Tighten the steering gear mounting bolts and the intermediate shaft coupler pinch bolt to the torque values listed in this Chapter's Specifications.*

   b) *Tighten the wheel lug nuts to the torque listed in the Chapter 1 Specifications.*

   c) *Check the power steering fluid level and add some, if necessary (see Chapter 1), then bleed the system (see Section 22).*

   d) *Re-check the power steering fluid level.*

Models with Electric Power Steering (EPS) rack-and-pinion steering gear

Note: *The steering column on these vehicles may not be equipped with an internal locking shaft that allows the ignition key cylinder to be locked with the key. You will have to secure and prevent the steering wheel from turning during servicing of the steering gear.*

8   Disconnect the cable from the negative battery terminal (see Chapter 5).

9   Remove the intermediate shaft coupler pinch bolt and discard it. It should be replaced with a new one according to the manufacturer.

10   Disconnect the EPS electrical connector at the steering gear.

11   Detach the tie-rod ends from the steering knuckles (see Section 18).

12   Remove the steering gear mounting bolts and slide the gear out from under the stabilizer bar.

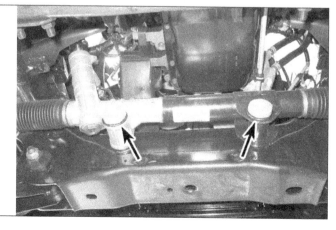

20.6 Location of the rack-and-pinion steering gear mounting bolts

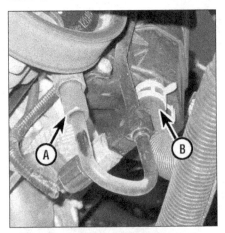

21.3 Detach the pressure (A) and return (B) lines from the power steering pump (Hemi model shown)

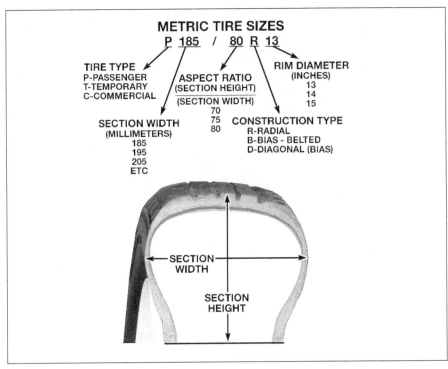

23.1 Metric tire size code

13   Installation is the reverse of removal.

14   Once the unit is installed, reconnect the battery (see Chapter 5), then insert the ignition key and turn the key to the Run position. Wait approximately 8 to 10 seconds then turn the key to the Off position. Complete the calibration procedure by turning the key to the Run position again and check for trouble codes. If no codes are present, the calibration is complete.

Models with recirculating ball steering gear

15   Raise the front of the vehicle and support it securely on jackstands. Apply the parking brake.

16   Lock the steering wheel in place, using the internal mechanism or by tying it with a cord.

17   Remove the intermediate shaft coupler pinch bolt and discard it. According to the manufacturer, it should be replaced with a new one.

18   Position a drain pan under the steering gear, then unscrew the power steering lines from the steering gear. Cap the lines to prevent leakage.

19   Separate the drag link from the Pitman arm (see Section 18).

20   Remove the steering gear retaining bolts from the frame rail, then detach the steering gear from the frame and remove it.

21   If you're installing a new steering gear or a new Pitman arm, remove the Pitman arm from the steering gear sector shaft (see Section 18).

22   Installation is the reverse of removal, noting the following points:

a) Tighten the steering gear mounting bolts and the intermediate shaft coupler pinch bolt to the torque values listed in this Chapter's Specifications.

b) Tighten the wheel lug nuts to the torque listed in the Chapter 1 Specifications.

c) Check the power steering fluid level and add some, if necessary (see Chapter 1), then bleed the system (see Section 22).

d) Re-check the power steering fluid.

21   Power steering pump - removal and installation

Refer to illustration 21.3

1   Disconnect the cable(s) from the negative battery terminal(s) (see Chapter 5).

2   Remove the drivebelt (see Chapter 1).

3   Using a large syringe or suction gun, suck as much fluid out of the power steering fluid reservoir. Position a drain pan under the power steering pump, then disconnect the hoses from the pump **(see illustration)**. Plug the hoses to prevent excessive fluid loss and the entry of contaminants. **Caution:** *Use a back-up wrench on the fitting into which the high-pressure hose attaches. If this fitting becomes loose, tighten it to 50 ft-lbs. If the fitting is allowed to come all the way out, the pump will have to be replaced with a new one, as the internal components won't be able to be installed correctly.*

4   Remove any interfering engine components.

5   Remove the pump mounting bolts. The bolts can be accessed through the holes in the power steering pump pulley.

6   Lift the pump from the engine compartment, being careful not to let any power steering fluid drip on the vehicle's paint.

7   Installation is the reverse of removal. Tighten all fasteners securely. Fill the power

steering reservoir with the recommended fluid (see Chapter 1), then bleed the system (see Section 22). Re-check the power steering fluid level.

22   Power steering system - bleeding

1   The power steering system must be bled whenever a line is disconnected. Bubbles can be seen in power steering fluid that has air in it and the fluid will often have a milky appearance. Low fluid level can cause air to mix with the fluid, resulting in a noisy pump as well as foaming of the fluid.

2   Open the hood and check the fluid level in the reservoir, adding the specified fluid necessary to bring it up to the proper level (see Chapter 1).

3   Start the engine and slowly turn the steering wheel several times from left-to-right and back again. Do not turn the wheel completely from lock-to-lock. Check the fluid level, topping it up as necessary until it remains steady and no more bubbles are visible.

23   Wheels and tires - general information

Refer to illustration 23.1

Most models covered by this manual are equipped with radial tires **(see illustration)**, or inch-pattern light truck tires. Use of other size or type of tires may affect the ride and handling of the vehicle. Don't mix different types of tires, such as radials and bias belted tires, on the same vehicle - handling may be seriously affected. It's recommended that tires be replaced in pairs on the same axle, but if only one tire is being replaced, be sure it's the same size, structure and tread design as the other tire on the same axle.

Because tire pressure has a substantial effect on handling and wear, the pressure of all tires should be checked at least once a month or before any extended trips are taken (see Chapter 1).

Wheels must be replaced if they are bent, dented, leak air, have elongated bolt holes, are heavily rusted, out of vertical symmetry or if the lug nuts won't stay tight. Wheel repairs that use welding or peening are not recommended.

Tire and wheel balance are important to the overall handling, braking and performance of the vehicle. Unbalanced wheels can adversely affect handling and ride characteristics as well as tire life. Whenever a tire is installed on a wheel, the tire and wheel should be balanced by a shop with the proper equipment and expertise.

24 Wheel alignment - general information

Refer to illustration 24.1

Note: *Since wheel alignment requires special equipment and techniques it is beyond the scope of this manual. This section is intended only to familiarize the reader with the basic terms used and procedures followed during a typical wheel alignment.*

The three basic checks made when aligning a vehicle's front wheels are camber, caster and toe-in **(see illustration)**.

Camber and caster are the angles at which the wheels and suspension are inclined in relation to a vertical centerline. Camber is the angle of the wheel in the lateral, or side-to-side plane, while caster is the tilt between the steering axis and the vertical plane, as viewed from the side. Camber angle affects the amount of tire tread which contacts the road and compensates for changes in suspension geometry as the vehicle travels around curves and over bumps. Caster angle affects the self-centering action of the steering, which governs straight-line stability.

Toe-in is the amount the front wheels are angled in relationship to the center line of the vehicle. For example, in a vehicle with zero toe-in, the distance measured between the front edges of the wheels and the dis-

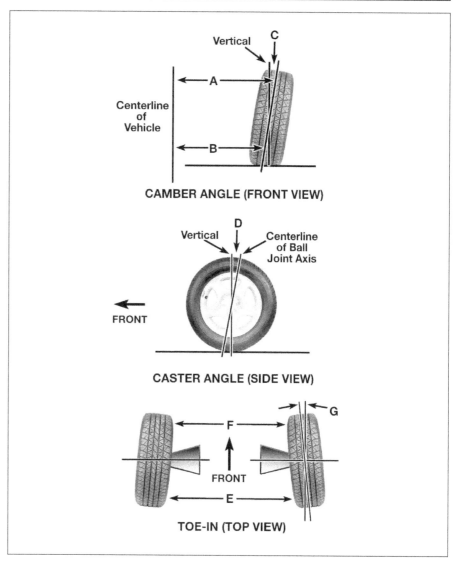

CAMBER ANGLE (FRONT VIEW)

CASTER ANGLE (SIDE VIEW)

TOE-IN (TOP VIEW)

24.1 Front wheel alignment details

A minus B = C (degrees camber)
D = degrees caster

E minus F = toe-in (measured in inches)
G = toe-in (expressed in degrees)

tance measured between the rear edges of the wheels are the same. In other words, the wheels are running parallel with the centerline of the vehicle. Toe-in is adjusted by lengthening or shortening the tie-rods. Incorrect toe-in will cause the tires to wear improperly by allowing them to scrub against the road surface.

Proper wheel alignment is essential for safe steering and even tire wear. Symptoms of alignment problems are pulling of the steering to one side or the other and uneven tire wear. If these symptoms are present,

check for the following before having the alignment adjusted:

a) *Loose steering gear mounting bolts*
b) *Damaged or worn steering gear mounts*
c) *Worn or damaged wheel bearings*
d) *Bent tie-rods*
e) *Worn balljoints*
f) *Improper tire pressures*
g) *Mixing tires of different construction*

Front wheel alignment should be left to an alignment shop with the proper equipment and experienced personnel.

Notes

Chapter 11 Body

Contents

Specifications

Torque specifications
Seat mounting bolts .... 35 ft-lbs

1 General information

These models feature a body-on-frame construction. The frame is a ladder-type, consisting of two box sectioned steel side rails joined by crossmembers. The crossmembers are welded to the side rails, with the exception of the transmission crossmember, which is bolted in place for easy removal.

Certain components are particularly vulnerable to accident damage and can be unbolted and repaired or replaced. Among these parts are the body moldings, bumpers, hood, fenders, doors, tailgate and all glass.

Only general body maintenance practices and body panel repair procedures within the scope of the do-it-yourselfer are included in this Chapter.

2 Body - maintenance

1   The condition of your vehicle's body is very important, because the resale value depends a great deal on it. It's much more difficult to repair a neglected or damaged body than it is to repair mechanical components. The hidden areas of the body, such as the wheel wells, the frame and the engine compartment, are equally important, although they don't require as frequent attention as the rest of the body.
2   Once a year, or every 12,000 miles, it's a good idea to have the underside of the body steam cleaned. All traces of dirt and oil will be removed and the area can then be inspected carefully for rust, damaged brake lines, frayed

electrical wires, damaged cables and other problems. The front suspension components should be greased after completion of this job.

3   At the same time, clean the engine and the engine compartment with a steam cleaner or water soluble degreaser.

4   The wheel wells should be given close attention, since undercoating can peel away and stones and dirt thrown up by the tires can cause the paint to chip and flake, allowing rust to set in. If rust is found, clean down to the bare metal and apply an anti-rust paint.

5   The body should be washed about once a week. Wet the vehicle thoroughly to soften the dirt, then wash it down with a soft sponge and plenty of clean soapy water. If the surplus dirt is not washed off very carefully, it can wear down the paint.

6   Spots of tar or asphalt thrown up from the road should be removed with a cloth soaked in solvent.

7   Once every six months, wax the body and chrome trim. If a chrome cleaner is used to remove rust from any of the vehicle's plated parts, remember that the cleaner also removes part of the chrome, so use it sparingly.

---

3   Vinyl trim - maintenance

Don't clean vinyl trim with detergents, caustic soap or petroleum-based cleaners. Plain soap and water works just fine, with a soft brush to clean dirt that may be ingrained. Wash the vinyl as frequently as the rest of the vehicle.

After cleaning, application of a high quality rubber and vinyl protectant will help prevent oxidation and cracks. The protectant can also be applied to weather-stripping, vacuum lines and rubber hoses (which often fail as a result of chemical degradation) and to the tires.

---

4   Upholstery and carpets - maintenance

1   Every three months remove the carpets or mats and clean the interior of the vehicle (more frequently if necessary). Vacuum the upholstery and carpets to remove loose dirt and dust.

2   Leather upholstery requires special care. Stains should be removed with warm water and a very mild soap solution. Use a clean, damp cloth to remove the soap, then wipe again with a dry cloth. Never use alcohol, gasoline, nail polish remover or thinner to clean leather upholstery.

3   After cleaning, regularly treat leather upholstery with a leather wax. Never use car wax on leather upholstery.

4   In areas where the interior of the vehicle is subject to bright sunlight, cover leather seats with a sheet if the vehicle is to be left out for any length of time.

---

5   Body repair - minor damage

See photo sequence

Repair of minor scratches

1   If the scratch is superficial and does not penetrate to the metal of the body, repair is very simple. Lightly rub the scratched area with a fine rubbing compound to remove loose paint and built-up wax. Rinse the area with clean water.

2   Apply touch-up paint to the scratch, using a small brush. Continue to apply thin layers of paint until the surface of the paint in the scratch is level with the surrounding paint. Allow the new paint at least two weeks to harden, then blend it into the surrounding paint by rubbing with a very fine rubbing compound. Finally, apply a coat of wax to the scratch area.

3   If the scratch has penetrated the paint and exposed the metal of the body, causing the metal to rust, a different repair technique is required. Remove all loose rust from the bottom of the scratch with a pocket knife, then apply rust inhibiting paint to prevent the formation of rust in the future. Using a rubber or nylon applicator, coat the scratched area with glaze-type filler. If required, the filler can be mixed with thinner to provide a very thin paste, which is ideal for filling narrow scratches. Before the glaze filler in the scratch hardens, wrap a piece of smooth cotton cloth around the tip of a finger. Dip the cloth in thinner and then quickly wipe it along the surface of the scratch. This will ensure that the surface of the filler is slightly hollow. The scratch can now be painted over as described earlier in this Section.

Repair of dents

4   When repairing dents, the first job is to pull the dent out until the affected area is as close as possible to its original shape. There is no point in trying to restore the original shape completely as the metal in the damaged area will have stretched on impact and cannot be restored to its original contours. It is better to bring the level of the dent up to a point which is about 1/8-inch below the level of the surrounding metal. In cases where the dent is very shallow, it is not worth trying to pull it out at all.

5   If the back side of the dent is accessible, it can be hammered out gently from behind using a soft-face hammer. While doing this, hold a block of wood firmly against the opposite side of the metal to absorb the hammer blows and prevent the metal from being stretched.

6   If the dent is in a section of the body which has double layers, or some other factor makes it inaccessible from behind, a different technique is required. Drill several small holes through the metal inside the damaged area, particularly in the deeper sections. Screw long, self-tapping screws into the holes

just enough for them to get a good grip in the metal. Now the dent can be pulled out by pulling on the protruding heads of the screws with locking pliers.

7   The next stage of repair is the removal of paint from the damaged area and from an inch or so of the surrounding metal. This is done with a wire brush or sanding disk in a drill motor, although it can be done just as effectively by hand with sandpaper. To complete the preparation for filling, score the surface of the bare metal with a screwdriver or the tang of a file, or drill small holes in the affected area. This will provide a good grip for the filler material. To complete the repair, see the subsection on filling and painting later in this Section.

Repair of rust holes or gashes

8   Remove all paint from the affected area and from an inch or so of the surrounding metal using a sanding disk or wire brush mounted in a drill motor. If these are not available, a few sheets of sandpaper will do the job just as effectively.

9   With the paint removed, you will be able to determine the severity of the corrosion and decide whether to replace the whole panel, if possible, or repair the affected area. New body panels are not as expensive as most people think and it is often quicker to install a new panel than to repair large areas of rust.

10   Remove all trim pieces from the affected area except those which will act as a guide to the original shape of the damaged body, such as headlight shells, etc. Using metal snips or a hacksaw blade, remove all loose metal and any other metal that is badly affected by rust. Hammer the edges of the hole in to create a slight depression for the filler material.

11   Wire brush the affected area to remove the powdery rust from the surface of the metal. If the back of the rusted area is accessible, treat it with rust inhibiting paint.

12   Before filling is done, block the hole in some way. This can be done with sheet metal riveted or screwed into place, or by stuffing the hole with wire mesh.

13   Once the hole is blocked off, the affected area can be filled and painted. See the following subsection on filling and painting.

Filling and painting

14   Many types of body fillers are available, but generally speaking, body repair kits which contain filler paste and a tube of resin hardener are best for this type of repair work. A wide, flexible plastic or nylon applicator will be necessary for imparting a smooth and contoured finish to the surface of the filler material. Mix up a small amount of filler on a clean piece of wood or cardboard (use the hardener sparingly). Follow the manufacturer's instructions on the package, otherwise the filler will set incorrectly.

15   Using the applicator, apply the filler paste to the prepared area. Draw the applicator across the surface of the filler to achieve

the desired contour and to level the filler surface. As soon as a contour that approximates the original one is achieved, stop working the paste. If you continue, the paste will begin to stick to the applicator. Continue to add thin layers of paste at 20-minute intervals until the level of the filler is just above the surrounding metal.

16   Once the filler has hardened, the excess can be removed with a body file. From then on, progressively finer grades of sandpaper should be used, starting with a 180-grit paper and finishing with 600-grit wet-or-dry paper. Always wrap the sandpaper around a flat rubber or wooden block, otherwise the surface of the filler will not be completely flat. During the sanding of the filler surface, the wet-or-dry paper should be periodically rinsed in water. This will ensure that a very smooth finish is produced in the final stage.

17   At this point, the repair area should be surrounded by a ring of bare metal, which in turn should be encircled by the finely feathered edge of good paint. Rinse the repair area with clean water until all of the dust produced by the sanding operation is gone.

18   Spray the entire area with a light coat of primer. This will reveal any imperfections in the surface of the filler. Repair the imperfections with fresh filler paste or glaze filler and once more smooth the surface with sandpaper. Repeat this spray-and-repair procedure until you are satisfied that the surface of the filler and the feathered edge of the paint are perfect. Rinse the area with clean water and allow it to dry completely.

19   The repair area is now ready for painting. Spray painting must be carried out in a warm, dry, windless and dust free atmosphere. These conditions can be created if you have access to a large indoor work area, but if you are forced to work in the open, you will have to pick the day very carefully. If you are working indoors, dousing the floor in the work area with water will help settle the dust which would otherwise be in the air. If the repair area is confined to one body panel, mask off the surrounding panels. This will help minimize the effects of a slight mismatch in paint color. Trim pieces such as chrome strips, door handles, etc., will also need to be masked off or removed. Use masking tape and several thickness of newspaper for the masking operations.

20   Before spraying, shake the paint can thoroughly, then spray a test area until the spray painting technique is mastered. Cover the repair area with a thick coat of primer. The thickness should be built up using several thin layers of primer rather than one thick one. Using 600-grit wet-or-dry sandpaper, rub down the surface of the primer until it is very smooth. While doing this, the work area should be thoroughly rinsed with water and the wet-or-dry sandpaper periodically rinsed as well. Allow the primer to dry before spraying additional coats.

21   Spray on the top coat, again building up the thickness by using several thin layers of paint. Begin spraying in the center of the

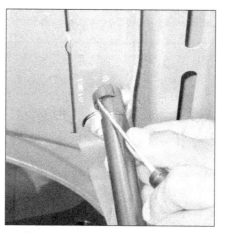

9.2 Use a small screwdriver to pry the clips out, then detach the ends of the strut from the locating studs

repair area and then, using a circular motion, work out until the whole repair area and about two inches of the surrounding original paint is covered. Remove all masking material 10 to 15 minutes after spraying on the final coat of paint. Allow the new paint at least two weeks to harden, then use a very fine rubbing compound to blend the edges of the new paint into the existing paint. Finally, apply a coat of wax.

6   Body repair - major damage

1   Major damage must be repaired by an auto body shop specifically equipped to perform these repairs. Most shops have the specialized equipment required to do the job properly.

2   If the damage is extensive, the body must be checked for proper alignment or the vehicle's handling characteristics may be adversely affected and other components may wear at an accelerated rate.

3   Due to the fact that all of the major body components (hood, fenders, etc.) are separate and replaceable units, any seriously damaged components should be replaced rather than repaired. Sometimes the components can be found in a wrecking yard that specializes in used vehicle components, often at considerable savings over the cost of new parts.

7   Hinges and locks - maintenance

Once every 3000 miles, or every three months, the hinges and latch assemblies on the doors, hood and trunk should be given a few drops of light oil or lock lubricant. The door latch strikers should also be lubricated with a thin coat of grease to reduce wear and ensure free movement. Lubricate the door and trunk locks with spray-on graphite lubricant.

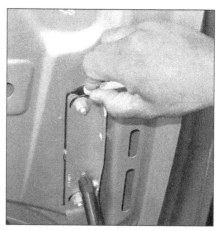

10.2 Before removing the hood, draw a mark around the hinge plate

8   Windshield and fixed glass - replacement

Replacement of the windshield and fixed glass requires the use of special fast-setting adhesive/caulk materials and some specialized tools. It is recommended that these operations be left to a dealer or a shop specializing in glass work.

9   Hood support struts - removal and installation

Refer to illustration 9.2

1   Open the hood and support it securely.

2   Using a small screwdriver, release the retaining clips at both ends of the support strut. Then pry or pull sharply to detach it from the vehicle **(see illustration).**

3   Installation is the reverse of removal.

10   Hood - removal, installation and adjustment

Note: *The hood is heavy and somewhat awkward to remove and install - at least two people should perform this procedure.*

Removal and installation

Refer to illustrations 10.2 and 10.4

1   Use blankets or pads to cover the cowl area of the body and fenders. This will protect the body and paint as the hood is lifted off.

2   Make marks or scribe a line around the hood hinges to ensure proper alignment during installation **(see illustration).**

3   Disconnect any cables or wires that will interfere with removal.

4   Have an assistant support one side of the hood while you support the other. Simultaneously remove the hinge-to-hood

These photos illustrate a method of repairing simple dents. They are intended to supplement *Body repair - minor damage* in this Chapter and should not be used as the sole instructions for body repair on these vehicles.

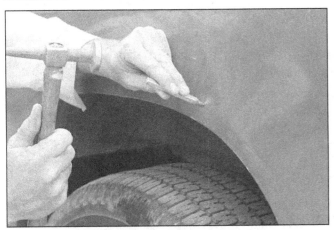

1 If you can't access the backside of the body panel to hammer out the dent, pull it out with a slide-hammer-type dent puller. In the deepest portion of the dent or along the crease line, drill or punch hole(s) at least one inch apart . . .

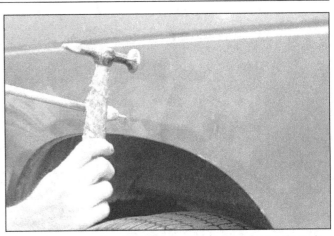

2 . . . then screw the slide-hammer into the hole and operate it. Tap with a hammer near the edge of the dent to help 'pop' the metal back to its original shape. When you're finished, the dent area should be close to its original contour and about 1/8-inch below the surface of the surrounding metal

3 Using coarse-grit sandpaper, remove the paint down to the bare metal. Hand sanding works fine, but the disc sander shown here makes the job faster. Use finer (about 320-grit) sandpaper to feather-edge the paint at least one inch around the dent area

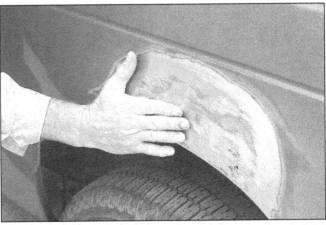

4 When the paint is removed, touch will probably be more helpful than sight for telling if the metal is straight. Hammer down the high spots or raise the low spots as necessary. Clean the repair area with wax/silicone remover

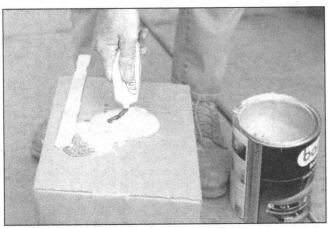

5 Following label instructions, mix up a batch of plastic filler and hardener. The ratio of filler to hardener is critical, and, if you mix it incorrectly, it will either not cure properly or cure too quickly (you won't have time to file and sand it into shape)

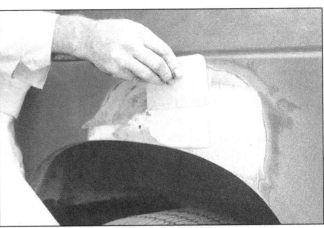

6 Working quickly so the filler doesn't harden, use a plastic applicator to press the body filler firmly into the metal, assuring it bonds completely. Work the filler until it matches the original contour and is slightly above the surrounding metal

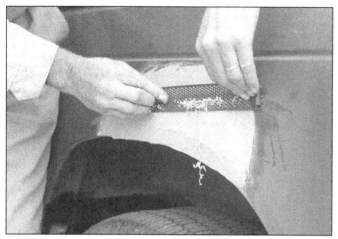

7 Let the filler harden until you can just dent it with your fingernail. Use a body file or Surform tool (shown here) to rough-shape the filler

8 Use coarse-grit sandpaper and a sanding board or block to work the filler down until it's smooth and even. Work down to finer grits of sandpaper - always using a board or block - ending up with 360 or 400 grit

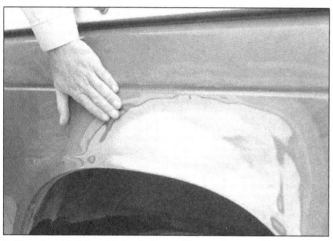

9 You shouldn't be able to feel any ridge at the transition from the filler to the bare metal or from the bare metal to the old paint. As soon as the repair is flat and uniform, remove the dust and mask off the adjacent panels or trim pieces

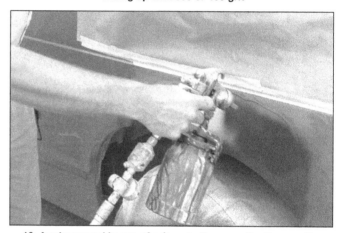

10 Apply several layers of primer to the area. Don't spray the primer on too heavy, so it sags or runs, and make sure each coat is dry before you spray on the next one. A professional-type spray gun is being used here, but aerosol spray primer is available inexpensively from auto parts stores

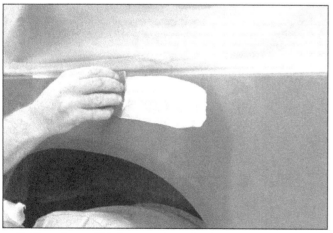

11 The primer will help reveal imperfections or scratches. Fill these with glazing compound. Follow the label instructions and sand it with 360 or 400-grit sandpaper until it's smooth. Repeat the glazing, sanding and respraying until the primer reveals a perfectly smooth surface

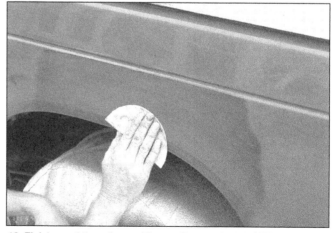

12 Finish sand the primer with very fine sandpaper (400 or 600-grit) to remove the primer overspray. Clean the area with water and allow it to dry. Use a tack rag to remove any dust, then apply the finish coat. Don't attempt to rub out or wax the repair area until the paint has dried completely (at least two weeks)

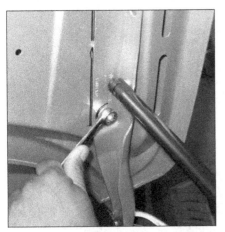

10.4 Remove the hinge-to-hood retaining nuts and lift off the hood with the help of an assistant

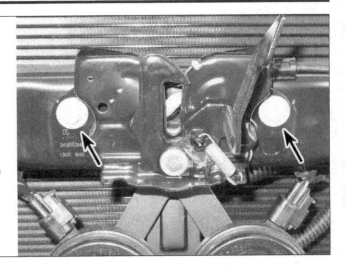

10.10 Make a mark around the latch to use as a reference point. To adjust the hood latch, loosen the retaining bolts, move the latch and retighten bolts, then close the hood to check the fit

nuts **(see illustration).**
5   Lift off the hood.
6   Installation is the reverse of removal.

Adjustment

Refer to illustrations 10.10 and 10.11

7   Fore-and-aft and side-to-side adjustment of the hood is done by moving the hinge plate slot after loosening the bolts or nuts.
8   Mark around the each hinge plate so you can determine the amount of movement **(see illustration 10.2).**
9   Loosen the bolts or nuts and move the hood into correct alignment. Move it only a little at a time. Tighten the hinge bolts and carefully lower the hood to check the position.
10   If necessary after installation, the hood latch can be adjusted up-and-down as well as from side-to-side on the radiator support so the hood closes securely and flush with the fenders. To make the adjustment, scribe a line or mark around the hood latch mounting bolts to provide a reference point, then loosen them and reposition the latch, as necessary **(see illustration).** Following adjustment, retighten the mounting bolts.

11   Finally, adjust the hood bumpers on the radiator support so the hood, when closed, is flush with the fenders **(see illustration).**
12   The hood latch assembly, as well as the hinges, should be periodically lubricated with white, lithium-base grease to prevent binding and wear.

11   Hood latch and release cable - removal and installation

Latch

1   Scribe a line around the latch to aid alignment when reinstalling the latch assembly.
2   Remove the latch retaining bolts securing the latch to the radiator support **(see illustration 10.10)** and remove the latch.
3   Disconnect the hood release cable by disengaging the cable from the back of the latch assembly.
4   Installation is the reverse of the removal procedure. **Note:** *Adjust the latch so the hood engages securely when closed and the hood bumpers are slightly compressed.*

Cable

5   Remove the hood latch as described earlier in this Section, then detach the cable from the latch.
6   Working in the engine compartment, detach the cable from all of its retaining clips. It may be necessary to cut some of the clips to free the cable.
7   Working under the instrument panel, remove the screws and detach the hood release handle. Dislodge the grommet and pull the cable through the firewall and into the cab to remove it.
8   Pull the new cable through the firewall and into the engine compartment. Seat the grommet in the firewall.
9   The remainder of installation is the reverse of removal.

12   Radiator grille - removal and installation

Refer to illustrations 12.2 and 12.3

1   Open the hood.
2   Remove the upper radiator panel. It's secured by push-pin retainers and plastic rivets **(see illustration).**

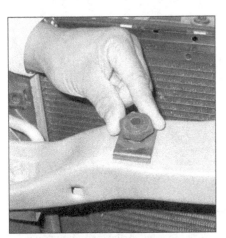

10.11 Adjust the hood closing height by turning the hood bumpers in or out

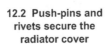
12.2 Push-pins and rivets secure the radiator cover

12.3 These four screws retain the top of the grille - it's held by clips at the bottom

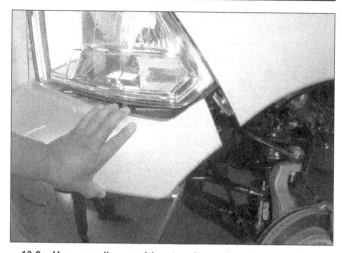

13.6a Use a small screwdriver to release the tabs and pull the ends of the bumper fascia free . . .

3   Remove the four top grille mounting bolts **(see illustration)**.
4   Reach behind the grille and use a small tool to release the clip from each lower inside corner of the grille, then remove the grille.
5   Installation is the reverse of removal.

13   Bumpers - removal and installation

Front bumper

Refer to illustrations 13.6a, 13.6b and 13.6c

1   Front bumpers on all models are composed of a plastic or chrome steel fascia, or exterior skin, and a structural beam.
2   Remove the two push-pins from each end of the bumper fascia to separate it from the inner wheel well. Move the inner wheel well out of the way slightly.
3   Remove the screws from each end of the bumper fascia.
4   On sport models, remove the fasteners

from the bottom of the bumper fascia.
5   Remove the grille (see Section 12).
6   Use a small screwdriver to detach the top clips from the bumper fascia, then remove it **(see illustrations)**.
7   Mark the position of the bumper brackets to the frame rail, or the bumper-to-bracket nuts. Unplug the electrical connectors to the fog lights, if equipped.
8   With an assistant supporting the bumper, remove the nuts/bolts retaining the bumper to the frame rails, or the nuts retaining the bumper to the brackets.
9   Detach the bumper from the frame rails or brackets.
10   Installation is the reverse of removal.

Rear bumper

Refer to illustrations 13.12 and 13.14

11   Raise the vehicle and support it securely on jackstands.
12   Remove the license plate and the bumper mounting bolts behind it **(see illustration)**.

13.6b . . . then release the center of the fascia and lift it off

13   Unplug any electrical connectors that interfere with bumper removal.
14   With an assistant or a secure jack supporting the bumper, remove the bolts retaining

13.6c Remove the plastic panel for access to the bumper bolts

13.12 Remove these two bolts securing the center of the bumper

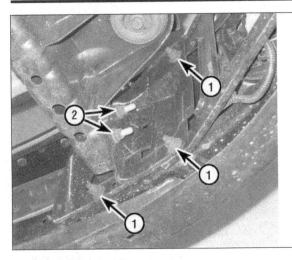

13.14 The bumper can either be detached from the mounting brackets, or the mounting brackets can be removed along with the bumper by detaching the brackets from the frame

1   Bumper-to-bracket nuts (fourth nut not visible in this photo)
2   Bumper bracket-to-frame nuts/bolts

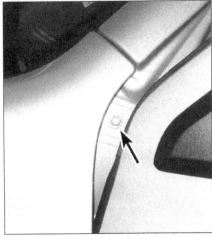

14.4 The top rear fender bolt is accessible with the door open

the bumper brackets to the frame rails **(see illustration)**.
15   Installation is the reverse of removal.

14.7 Remove the pin retainers from the inner fender splash shield (A) and the lower rear section of the fender (B)

14   Fender - removal and installation

Refer to illustrations 14.4, 14.7, 14.9 and 14.11

1   Raise the vehicle, support it securely on jackstands and remove the front wheel.
2   Remove the antenna if required (see Chapter 12).
3   Remove the headlight (see Chapter 12).
4   With the door open, remove the top rear-most fender bolt **(see illustration)**.
5   Remove the front bumper fascia (see Section 13).
6   Remove the front wheel opening molding if one is installed.
7   Remove the inner fender splash shield **(see illustration)**.
8   Remove the two bolts from the lower rear part of the fender **(see illustration 14.7)**.
9   Working inside the wheel opening, remove the bolts from the center rear part of the fender **(see illustration)** and the front mounting bracket.

10   On models with fender-mounted batteries, remove the battery and the battery tray (see Chapter 5).
11   Remove the four bolts from the top inner edge of the fender **(see illustration)**.
12   Detach the fender. It's a good idea to have an assistant support the fender while it's being moved away from the vehicle to prevent damage to the surrounding body panels.
13   Installation is the reverse of removal.

15   Door trim panels - removal and installation

Refer to illustrations 15.1, 15.2a, 15.2b, 15.3, 15.4, 15.5a, 15.5b and 15.7

1   On manual window models, remove the window crank **(see illustration)**.
2   On power window models, use a plastic

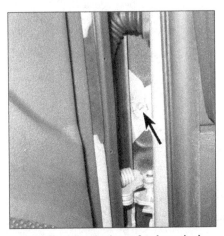

14.9 Remove the inner fender splash shield to reach this fender bolt and the two at the front of the fender

14.11 There are four bolts along the top edge of the fender - don't forget the one at the extreme rear

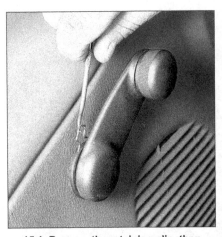

15.1 Remove the retaining clip, then detach the window crank handle. A hooked tool like this can be used, but special window crank clip removal tools are available at auto parts stores, and make this step much easier

15.2a Use a soft plastic tool like this to pry up the rear of the power window switch assembly . . .

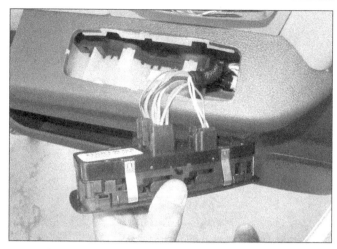

15.2b . . . then disconnect the wiring and remove it

trim tool or a screwdriver wrapped with tape to pry up the switch from the top of the armrest, if so equipped **(see illustration)**. Disconnect the switch **(see illustration)**.

3   On front doors, use the tool to remove the panel from the inside of the mirror **(see illustration)**. On rear doors, use it to remove the trim panel from the upper rear section of the door.

4   Remove the screw and the inside door handle trim plate **(see illustration)**.

5   Remove the plastic screws/push-pins from the outer door trim panel **(see illustrations)**.

6   Pull upward to release the door panel hooks from the door. The door panel must also clear the top of the door lock button.

7   Once all of the hooks are disengaged, raise the trim panel up and off the door **(see illustration)**. Disconnect any wiring harness connectors and remove the trim panel from the vehicle.

8   Connect the wiring harness connectors and place the panel in position on the door.

15.3 Carefully pry up the inner mirror trim panel to avoid scratches

Press the trim panel straight against the door until the clips on the trim panel align with all

15.4 There is a concealed screw under the inside door handle

the holes in the door, then push down on the panel until the clips are seated.

9   The remainder of installation is the reverse of removal.

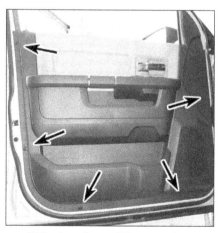

15.5a Locations of the outer door panel fasteners

15.5b The outer door panel fasteners can be removed with a screwdriver, then pushed back into place later

15.7 Lift the door trim panel straight upwards to release the clips - don't try to pry it off

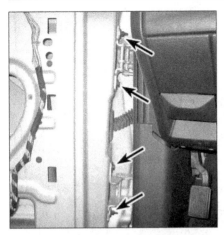

16.3 Mark their locations, then remove the door retaining bolt and nut at each hinge

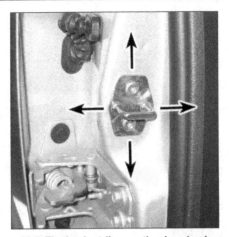

16.5 The latch striker on the door jamb can be adjusted slightly up/down or in/out

17.2 Rear door access panel on crew cab models - use it to reach the glass clips

16 Door - removal, installation and adjustment

Refer to illustrations 16.3 and 16.5

1   Remove the door trim panel (see Section 15). Disconnect any electrical connectors and push them through the door opening so they won't interfere with door removal. Leave the wiring harness boot attached to the body but disconnected from the door.
2   Place a jack under the door or have an assistant on hand to support it when the hinge bolts are removed. **Note:** *If a jack is used, place rags between it and the door to protect the door's painted surfaces.*
3   Scribe around the mounting bolt/nut heads and hinges with a marking pen, remove the fasteners and carefully lift off the door **(see illustration).**
4   Installation is the reverse of removal. Align the hinge with the marks made during removal before tightening the bolts.
5   Following installation of the door, check the alignment and adjust the hinges, if neces-sary. Adjust the door latch striker, centering it in the door latch **(see illustration).**
6   Rear doors are removed and installed as described above for front doors, except that the door is attached to the body B pillar.

17 Door carrier plate, latch, handles and lock cylinder - removal and installation

Carrier plate

Refer to illustrations 17.2, 17.4, 17.6, 17.10, 17.12, 17.13a and 17.13b

1   Remove the door trim panel (see Section 15).
2   On crew cab models, remove the speaker (see Chapter 12) and the access panel at the rear of the door **(see illustration).** On club cab models, remove the two small access plugs in the upper center of the door.
3   Detach the clips at the bottom edge of the glass from the window regulator lift plates (see Section 18). **Note:** *You'll have to tempo-*rarily install the window crank or the power window switch so you can move the glass to make the clips accessible through the holes.
4   Raise the glass to the top and secure it in place using tape strips over the top of the door **(see illustration).**
5   Disconnect the main wiring connector inside the door.
6   Pull the rubber wiring grommet off of the door jamb, then gently move it downwards to expose the lock tabs of the inside wiring connector **(see illustration).**
7   Push the lock tabs in until they're flush using a small screwdriver.
8   Gently manipulate the connector out of the hole in the metal. **Caution:** *Use care to avoid damaging the wiring, the connector or the sheet metal.*
9   Remove the access plate, then disconnect the control rods from the outside handle **(see illustration 17.20).**
10   Remove the two screws and detach the wire harness from the front edge of the door **(see illustration).**
11   On standard cab models, remove the screw from the lower belt channel.

17.4 Tape will support the glass out of the way while you're working inside the door

17.6 Carefully pull this grommet loose to reach the wiring connector

17.10 Two screws secure this wiring connector to the door assembly

17.12 Door latch mounting screws

17.13a Lift off the door carrier . . .

latch. Detach the latch from the bracket to remove it **(see illustration)**.
17   Installation is the reverse of removal.

Outside handle

Refer to illustrations 17.20, 17.21a and 17.21b

18   Disconnect the cable(s) from the negative battery terminal(s) (see Chapter 5).
19   Remove the carrier plate (see Steps 1 through 13).
20   Disconnect the two control rods from the handle **(see illustration)**.
21   Remove the door handle bolts from the inside of the door, then remove the handle **(see illustrations)**.
22   Installation is the reverse of removal.

Inside handle

Refer to illustration 17.25

23   Remove the door trim panel (see Section 15).
24   Disconnect the control rod from the handle, then pull it out of its retainer if necessary.
25   Remove the mounting screw from the top of the handle hinge, then remove the han-

12   Remove the door latch screws from the edge of the door **(see illustration)**.
13   Remove the bolts from the perimeter of the carrier plate, then remove it **(see illustrations)**.
14   Installation is the reverse of removal.

Latch

Refer to illustration 17.16

15   Remove the carrier plate (see Steps 1 through 13).
16   Disconnect the wiring from the door

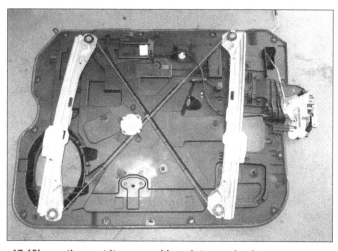

17.13b . . . then set it on a workbench to service its components

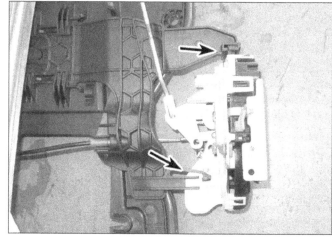

17.16 The latch assembly is attached to the door carrier plate by these clips

17.20 Detach these two control rods from the inside of the door handle

17.21a Exterior door handle mounting bolts

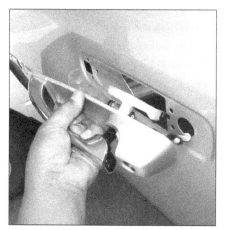

17.21b Pull the handle straight off to remove it

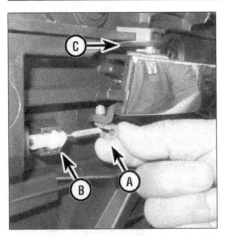

17.25 Detach the rod from the handle (A), pull the rod from the clip (B), then remove the screw (C) to remove the inside door handle

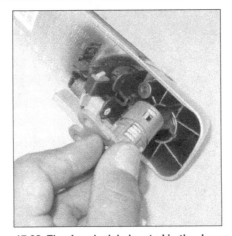

17.28 The door lock is located in the door handle assembly and can be removed without tools

18.3 Pull up on the inner weather seal and remove it from the door glass opening

dle **(see illustration)**.
26   Installation is the reverse of removal.

Lock cylinder

Refer to illustration 17.28

27   Remove the outside door handle (see Steps 18 through 21).
28   Pull the lock cylinder from the inside of the door handle **(see illustration)**.
29   Installation is the reverse of removal.

---

18   Door window glass - removal and installation

Front

Refer to illustrations 18.3, 18.4, 18.5a and 18.5b

1   Remove the door trim panel (see Section 15).
2   Remove the speaker (see Chapter 12).
3   Pry the inner weather seal out of the door glass opening **(see illustration)**.
4   Remove the access hole cover, then

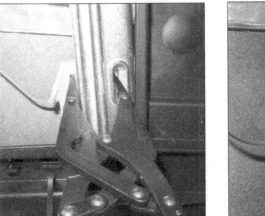

18.5a Release the glass clips . . .

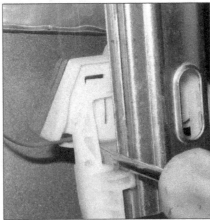

18.4 The glass clips can be reached through the speaker opening and the rear access hole

temporarily install the window crank or the power window switch and move the window so you can reach the glass clips through the speaker opening and the access hole **(see illustration)**.
5   Unclip the glass from the window regulator plates **(see illustrations)**.

6   Remove the window by tilting it forward to release it from the front and rear channels, then lift it out of the door.
7   If necessary, use a vacuum cleaner hose inserted through the holes to remove all traces of any broken glass. **Note:** *Glass debris left in the door will create noise.*
8   To install, lower the glass into the door, slide it into position and secure the clips.
9   The remainder of installation is the reverse of removal.

Rear

10   Remove the carrier plate (see Section 17).
11   Remove the glass through the lower section of the door.
12   Installation is the reverse of removal.

---

19   Door window glass regulator - removal and installation

   The window regulator is an integral part of the carrier plate **(see illustration 17.13b)**. Refer to Section 17 for the carrier plate removal and installation procedures.

18.5b . . . then move the glass out of the retainer

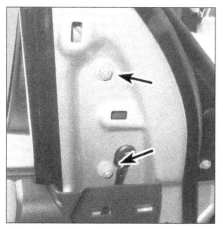

20.3 Remove the nuts and detach the mirror from the door

20.10 The interior mirror is secured with this set screw

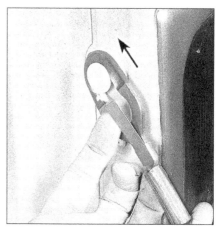

21.1 Lift the spring retainer up and slide the cable end off the pin

21.3 Align the flat on the right side hinge pin with the slot in the hinge pocket and lift the tailgate off the vehicle

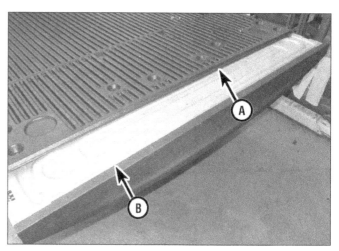

22.1 Remove the plastic liner (A) to reach the tailgate access panel (B)

20 Mirrors - removal and installation

Outside mirrors
Mirror assembly

Refer to illustration 20.3

1   Remove the door trim panel (see Section 15).
2   On power mirrors, unplug the electrical connector.
3   Remove the nuts and detach the mirror from the door **(see illustration).**
4   Installation is the reverse of removal.

Mirror glass

5   Tilt the mirror fully downward.
6   Carefully pull the top edge of the mirror out to release the upper clips.
7   Detach the lower clips and lift off the mirror.
8   Disconnect any wiring.
9   Installation is the reverse of removal.

Inside mirror

Refer to illustration 20.10

10   Remove the setscrew, then slide the mirror up off the support base on the windshield **(see illustration).** On models with optional automatic day/night mirror, disconnect the electrical connector first.
11   Installation is the reverse of removal.
12   If the support base for the mirror has come off the windshield, it can be reattached with a special mirror adhesive kit available at auto parts stores. Clean the glass and support base thoroughly and follow the directions on the adhesive package. **Note:** *Don't try to glue the bracket on in cold weather.*

21 Tailgate - removal and installation

Refer to illustrations 21.1 and 21.3

1   Open the tailgate and detach the retaining cables **(see illustration).**

2   Disconnect any wiring harnesses.
3   Angle the tailgate until the flat on the right side hinge pin aligns with the slot in the hinge pocket. Lift the tailgate out of the pocket **(see illustration).** With the help of an assistant to support the weight, withdraw the left hinge pin from the body and remove the tailgate from the vehicle.
4   Installation is the reverse of removal.

22 Tailgate latch and handle - removal and installation

Refer to illustration 22.1

1   Lower the tailgate and remove the plastic tailgate liner and the access cover **(see illustration).**

Latch

Refer to illustrations 22.2 and 22.3

2   Remove the latch mounting fasteners

22.2 Remove the tailgate latch retaining screws; they're tight, so you may have to use the type of impact driver that is struck with a hammer

22.3 Reach into the tailgate to disconnect the control rod from the latch

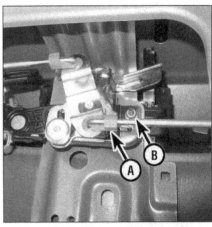

22.6 Disengage the handle-to-latch rods (A) and remove the handle retaining nuts (B)

(see illustration). It may be necessary to use an impact-driver to loosen them.
3   Disconnect the control rod from the latch (see illustration).

22.8 Remove these screws to separate the tailgate lock cylinder from the handle assembly

4   Remove the latch from the tailgate.
5   Installation is the reverse of removal.

Handle

Refer to illustrations 22.6 and 22.8
6   Disconnect the control rods from the handle (see illustration).
7   Disconnect any wiring harnesses.
8   Remove the retaining nuts and remove the handle assembly from the tailgate. The lock cylinder can now be removed from the handle assembly (see illustration).
9   Installation is the reverse of removal.

---

23   Dashboard trim panels - removal and installation

Warning: *The models covered by this manual are equipped with Supplemental Restraint Systems (SRS), more commonly known as airbags. Always disable the airbag system before working in the vicinity of any airbag system component to avoid the possibility of*

accidental deployment of the airbags, which could cause personal injury (see Chapter 12).
1   Disconnect the cable(s) from the negative battery terminal(s) (see Chapter 5).

Instrument cluster bezel

Refer to illustration 23.2
2   Remove the three screws from the top of the bezel (see illustration).
3   Use a plastic trim tool or a screwdriver wrapped with tape to carefully release the clips around the edges of the bezel, then remove it.
4   Installation is the reverse of removal.

Center instrument panel bezel

Refer to illustrations 23.5 and 23.9
5   Use a plastic trim tool or a screwdriver wrapped with tape to carefully pry up the liners from the center trays to expose the screws under them (see illustration). Remove the two screws under each liner. **Note:** *Some models don't have a lower tray.*
6   On column shift models with floor consoles, remove the rubber mats from the con-

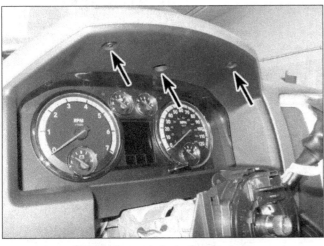

23.2 Instrument cluster bezel screws

23.5 Peel up the cover to expose the screws under it

23.9 On vehicles without a center console, carefully pry off this lower trim panel

23.13 Use a plastic tool to remove the instrument panel end cap

23.14 Carefully pry upward to remove the door threshold trim strips

sole to expose console top trim panel mounting screws and remove the screws. Raise the armrest then carefully separate the console top trim panel using a trim panel tool and remove the panel.

7   On early models, release the clips around the trim ring of the right-side power outlet and remove it. Then remove the screw under the power outlet trim ring, if equipped. **Note:** *Some models have a power outlet to the left of the center trim bezel. These models don't have a screw under them.*

8   On later models, remove the retaining screw above the outlet.

9   Using the trim tool, pry up the floor console trim ring and the shifter bezel, if so equipped. Remove the lower trim panel on vehicles without a console **(see illustration)**. Remove the screws under the trim panel. **Note:** *There are five screws that secure the center bezel on most models - two at the top under the tray liner, two at the bottom, and one at the center right.*

10   Release the clips that secure the center instrument panel bezel and remove it using a non-scratching plastic trim tool.

11   Disconnect the wiring harnesses.

12   Installation is the reverse of removal.

Left lower instrument trim panel

Refer to illustrations 23.13, 23.14, 23.15a and 23.15b

13   Use a plastic trim tool or a screwdriver wrapped with tape to pry off the cover from the left side of the instrument panel **(see illustration)**.

14   Remove the left threshold trim panel **(see illustration)**.

15   Remove the screws that secure the bezel to the main section of the instrument panel **(see illustrations)**.

16   Use the plastic tool to release the clips, then tilt the top of the bezel down and disconnect any wiring harnesses. Remove the bezel by detaching it from the hooks.

17   Installation is the reverse of removal.

Headlight switch bezel

Refer to illustration 23.18

18   Use a plastic trim stick to carefully pry around the edge of the headlight switch bezel **(see illustration)**. **Note:** *There are two clips on each side.*

23.15a Two screws secure the lower left trim panel; one screw not shown can be reached with the door open

19   Pull the panel out far enough to disconnect all electrical connectors, then remove the bezel.

20   Installation is the reverse of removal.

Glove box

Refer to illustration 23.23

21   Open the glove box door.

23.15b There is another screw under the parking brake release handle on some models

23.18 The headlight switch is retained by spring clips on each side

23.23 The glove box can be removed from its hinges by fully opening it

23.26 This screw secures the storage bin

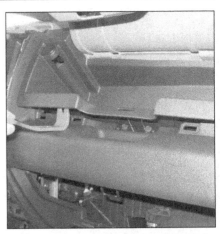

23.27 After removing the screw through the glove box opening, pry out the storage bin with a plastic tool

22   Push the sides of the glove box inward so the stops clear the instrument panel, then lower the door completely.

23   Pull the glove box door away from the instrument panel to detach the hinge from the instrument panel **(see illustration)**. Remove the glove box.

24   Installation is the reverse of removal.

Storage bin

Refer to illustrations 23.26 and 23.27

25   Remove the glove box (see Steps 20 through 22).

26   Working through the glove box opening, remove the screw from the rear of the storage bin **(see illustration)**.

27   Open the bin door and release the clips that retain the bin **(see illustration)**.

28   Pull the bin out far enough to disconnect the wiring harness, then remove it.

29   Installation is the reverse of removal.

Windshield pillar trim panels

Refer to illustrations 23.30 and 23.31

30   Use a small screwdriver to open the covers in the trim panel **(see illustration)**. **Note:** *There's a notch at the bottom to insert the tool.*

31   Remove the bolts under the covers **(see illustration)**, then pull the cover free.

32   Installation is the reverse of removal.

Cup holder

Refer to illustrations 23.35 and 23.36

33   Remove the center instrument panel bezel (see Steps 5 through 11).

34   Pull off the lower trim panel **(see illustration 23.8)**.

35   Remove the screws from the side trim panels and remove the panels **(see illustration)**.

36   Remove the cup holder **(see illustration)**.

37   Installation is the reverse of removal.

Driver's knee trim panel

Refer to illustrations 23.37 and 23.38

38   Carefully pry the hood release handle off of the trim panel **(see illustration)**. Push in the tabs on the data link connector and push it through the hole in the rear of the cover.

39   Remove the two lower mounting screws

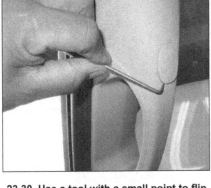

23.30 Use a tool with a small point to flip open the windshield trim panel bolt covers . . .

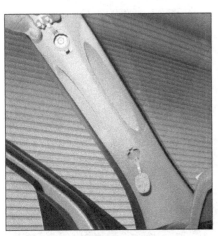

23.31 . . . then remove the bolts

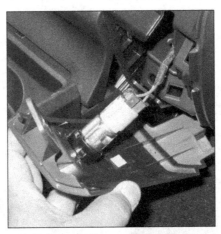

23.35 Remove the screws from the top and bottom of each side trim panel and disconnect the wiring

23.36 Remove the cup holder

23.38 Use a small screwdriver to release the clip that retains the hood release handle

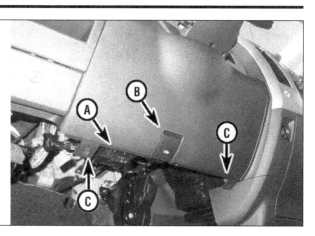

23.39 Data link connector (A) and hood release handle (B); the lower cover can be pried off when the screws (C) are removed

(see illustration). Use a plastic trim tool or a screwdriver wrapped with tape to carefully pry the cover off, then remove the trim panel.
40   Installation is the reverse of removal.

Instrument panel cover

Refer to illustrations 23.46, 23.47, 23.54 and 23.55

Note: *This is a difficult and time-consuming procedure. If you decide to attempt it, make notes or take pictures as you go to keep track of the locations of all components. Lay all parts out in order as you remove them and label any complex assemblies so they can be installed correctly.*
41   Remove the steering column covers (see Section 26).
42   Remove the steering column if necessary (see Chapter 10).
43   Remove the four retainers from the wireless ignition node, disconnect the wiring and remove it.
44   Remove the floor console from vehicles so equipped (see Section 24).
45   Use a plastic trim tool or a screwdriver wrapped with tape to pry off the lower center

trim bezel **(see illustration 23.9)**.
46   Remove the windshield pillar trim panels (see Steps 29 and 30) and the defrost grilles **(see illustration)**.
47   Remove the covers from the bolts along the windshield line, then remove the bolts **(see illustration)**.
48   Remove the center instrument panel bezel (see Steps 5 through 11).
49   Remove the radio (see Chapter 12).
50   Remove both power outlets and their bezels.
51   Remove the instrument cluster bezel (see Steps 2 and 3), then remove the cluster (see Chapter 12).
52   Remove the bezel from the parking brake release handle.
53   Remove the headlight switch and the air register above it by prying both out with a plastic trim tool (see Chapter 12).
54   Remove the instrument panel speakers and the light sensor **(see illustration)**.
55   Use a plastic trim tool or a screwdriver wrapped with tape to remove the right air register **(see illustration)**.
56   Remove the glove box (see Steps 21 through 23).
57   Remove the storage bin (see Steps 25 through 28).
58   Remove the passenger's airbag (see

Chapter 12).
59   Remove all of the 26 instrument panel cover screws, pull the cover rearward, then disconnect the passenger's airbag wiring harness.
Note: *Count the screws to make sure they're all accounted for. It also helps to make notes of where they belong so you don't end up with leftover screws and a noisy instrument panel. Alternatively, a small dot of bright paint at each screw location will help with installation.*
60   Remove the instrument panel cover.
61   Installation is the reverse of removal.

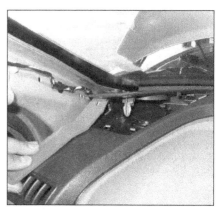

23.46 Pry up the defrost registers

23.47 There are several concealed screws along the base of the windshield

23.54 Use a plastic trim removal tool or a screwdriver wrapped with tape to remove the light sensor as well as the speaker grilles

23.55 Remove the right air register by prying it out with a plastic tool

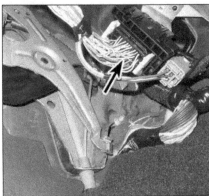

25.5 Wiring harness connector in the left footwell

25.9 Instrument panel left-side mounting bolts

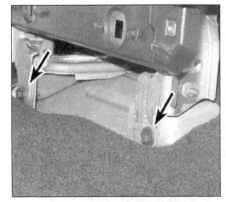

25.11 Instrument panel center mounting bolts

24 Console - removal and installation

Warning: *The models covered by this manual are equipped with Supplemental Restraint Systems (SRS), more commonly known as airbags. Always disable the airbag system before working in the vicinity of any airbag system component to avoid the possibility of accidental deployment of the airbags, which could cause personal injury (see Chapter 12).*

All except 65RFE 6-speed models.

1   Use a plastic trim tool or a screwdriver wrapped with tape to pry the covers from the sides of the console.
2   Pry up and remove the shifter trim ring and the cover. Remove the shifter bezel.
3   Detach the shift cable from the shift lever and the console.
4   Remove the console mounting screws and lift the console.
5   Disconnect all wiring harnesses, then remove the console.
6   Installation is the reverse of removal.

65RFE 6-speed models

7   Use a plastic trim tool or a screwdriver wrapped with tape to pry the trim ring from the

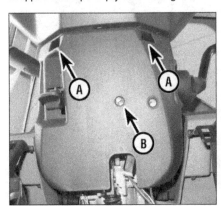

26.1 Two screws (A) attach the column covers together - another screw (B) secures the lower cover to the column

sides of the console bezel. Remove the bezel.
8   Remove the rubber insert from the console bezel.
9   Remove the two front mounting screws from the bezel.
10   Using a plastic trim tool or a screwdriver wrapped with tape to pry the console bezel, lift up and remove the console bezel.
11   Remove the center storage tray's rubber liner, remove the fasteners and remove the storage tray.
12   Installation is the reverse of removal.

25 Instrument panel - removal and installation

Refer to illustrations 25.5, 25.9 and 25.11

Warning: *The models covered by this manual are equipped with Supplemental Restraint Systems (SRS), more commonly known as airbags. Always disable the airbag system before working in the vicinity of any airbag system component to avoid the possibility of accidental deployment of the airbags, which could cause personal injury (see Chapter 12).*
Note: *This is a difficult and time-consuming procedure. If you decide to attempt it, make notes or take pictures as you go to keep track of the locations of all components. Lay all parts out in order as you remove them and label any complex assemblies so they can be installed correctly.*
1   Disconnect the cable(s) from the negative battery terminal(s) (see Chapter 5). Turn the front wheels to the straight-ahead position and lock the steering column.
2   Remove the front seat(s) (see Section 27)
3   Remove both door threshold/kick panel trim panels by pulling them upward.
4   Remove the lower steering column cover (see Section 26).
5   Working under the instrument panel near the brake pedal, disconnect the wiring harness from the firewall connector (see illustration).
6   Remove the steering column (see Chapter 10). Remove the parking brake release handle and the hood release handle.
7   Remove both windshield pillar trim pan-

els and the glove box (see Section 23).
8   Use a plastic trim tool or a screwdriver wrapped with tape to pry both end caps from the instrument panel (see illustration 23.12). **Note:** *There's a notch at the bottom to insert the tool.*
9   Remove the three bolts from the left end of the instrument panel (see illustration).
10   Remove the console (see Section 24). Remove the brake sled mounting bolts.
11   Remove the two instrument panel mounting bolts from the bottom center near the front of the console (see illustration).
12   Working under the right windshield pillar trim panel, disconnect the wiring harness.
13   Remove the instrument panel mounting screw in the back of the glove box opening.
14   Disconnect the five wiring harnesses and the antenna cable at the right kick panel area.
15   Remove the three bolts from the bracket at the right end of the instrument panel.
16   Remove the screw covers along the windshield, then remove the screws under them (see illustration 23.47).
17   Slide the carpet to the rear enough to reach the chassis ground screw of the two electronic modules. Disconnect the wire and remove the wiring harness.
18   On models so equipped, remove the two bolts securing the instrument panel to the floor pan.
19   Have an assistant help you to raise the instrument panel off of its mounts, then remove it from the vehicle.
20   Installation is the reverse of removal.

26 Steering column covers - removal and installation

Refer to illustrations 26.1 and 26.3

1   Remove the two screws that connect the upper and lower parts of the covers (see illustration).
2   Remove the lower cover mounting screw.
3   Pinch the upper column cover sides together to detach it from the lower cover, then remove both covers (see illustration).
4   Installation is the reverse of removal.

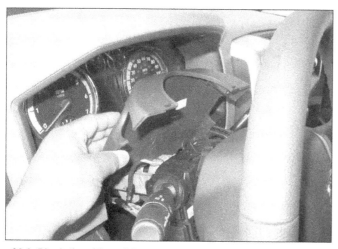

26.3 Pinch the sides of the upper cover near where it meets the lower cover to detach them, then lift the upper cover off

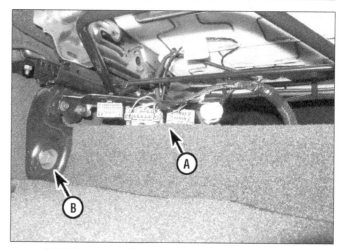

27.1 Disconnect the wiring (A), then remove the seat mounting bolts (B)

27 Seats - removal and installation

Warning: *The models covered by this manual are equipped with Supplemental Restraint Systems (SRS), more commonly known as airbags. Always disable the airbag system before working in the vicinity of any airbag system component to avoid the possibility of accidental deployment of the airbags, which could cause personal injury (see Chapter 12).*

Front seats

Driver's seat

Refer to illustration 27.1

1   Detach the plastic bolt trim covers, then disconnect the wiring harnesses **(see illustration)**.
2   Remove the front seat mounting bolts.
3   Remove the rear seat bolts and remove the seat.
4   Installation is the reverse of removal. Install all of the mounting bolts finger tight until they're all in place. Tighten the mounting bolts to the torque listed in this Chapter Specifications in the following sequence: front inner bolt, front outer bolt, rear inner bolt, rear outer bolt.

Passenger's seat

5   Detach the front clip of the jack cover that's at the base of the seat. Pull the front of the cover free.
6   Unhook the rear of the cover and remove it.
7   Disconnect any wiring harnesses.
8   Remove the forward seat mounting bolts.
9   If the vehicle has a center seat, remove its mounting bolts.
10   Remove the passenger's seat rear mounting bolts.
11   Detach the under-seat air duct by moving the part that's connected to the seat to the rear.
12   Remove the seat assembly from the vehicle.

27.19 Rear seat mounting bolts

13   Installation is the reverse of removal. Tighten the mounting bolts to the torque listed in this Chapter Specifications in the following sequence: front inner bolt, center seat front bolt, center seat left bolt, right outer bolt, center seat right bolt, right outer bolt.

Front center seat

Note: *This seat is not installed on all models.*
14   Remove the front passenger's seat assembly (see Steps 5 through 12).
15   Remove the screws that secure the center seat and remove it.
16   Installation is the reverse of removal. Tighten the mounting bolts to the torque listed in this Chapter Specifications in the correct sequence (see Step 13).

Rear seat

Refer to illustration 27.19

17   Disconnect any wiring harnesses attached to the seat.
18   Remove the bolt from the retractor-end of the center shoulder belt, then move the belt out of the way.
19   Remove the bolts from the seat mounts

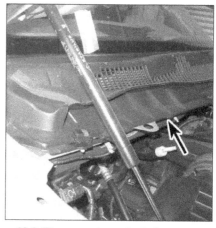

28.2 There are six push-pin fasteners along the front edge of the cowl - remove them, then pull the cowl straight forward to disconnect the washer hose

(see illustration).
20   Lift the seat assembly so that the tabs on the top of the seat back disengage from the body.
21   Remove the seat.
22   Installation is the reverse of removal. Tighten the mounting bolts to the torque listed in this Chapter Specifications.

28 Cowl cover - removal and installation

Refer to illustration 28.2

1   Remove the windshield wiper arms (see Chapter 12).
2   Remove the push-pins from the front edge of the cowl **(see illustration)**.
3   Slide the cowl forward to detach the rear section from the base of the windshield.
4   Disconnect the windshield washer hose and remove the cowl.
5   Installation is the reverse of removal.

Notes

Chapter 12
Chassis electrical system

Contents

1 General information

The electrical system is a 12-volt, negative ground type. A lead/acid-type battery that is charged by the alternator supplies power for the lights and all electrical accessories.

This Chapter covers repair and service procedures for the various electrical components not associated with the engine. Information on the battery, alternator, distributor and starter motor can be found in Chapter 5. **Warning:** *When working on the electrical system, disconnect the cable(s) from the negative battery terminal(s) to prevent electrical shorts and/or fires (see Chapter 5).*

2 Electrical troubleshooting - general information

Refer to illustrations 2.5a, 2.5b, 2.6, 2.9 and 2.15

A typical electrical circuit consists of an electrical component, any switches, relays, motors, fuses, fusible links or circuit breakers related to that component and the wiring and connectors that link the component to both the battery and the chassis. To help you pinpoint an electrical circuit problem, wiring diagrams are included at the end of this Chapter.

Before tackling any troublesome electrical circuit, first study the appropriate wiring diagrams to get a complete understanding of what makes up that individual circuit. You can often narrow down trouble spots, for instance, by noting whether other components related to the circuit are operating correctly. If several components or circuits fail at one time, chances are that the problem is in a fuse or ground connection, because several circuits are often routed through the same fuse and ground connections.

Electrical problems usually stem from simple causes, such as loose or corroded connections, a blown fuse, a melted fusible link or a failed relay. Visually inspect the condition of all fuses, wires and connections in a problem circuit before troubleshooting the circuit.

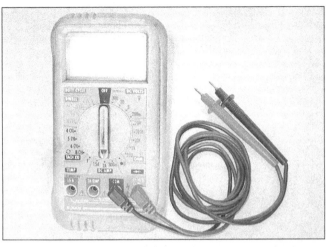

2.5a The most useful tool for electrical troubleshooting is a digital multimeter that can check volts, amps, and test continuity

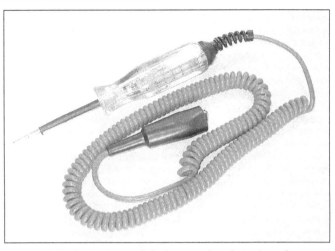

2.5b A simple test light is a very handy tool used for testing voltage

If test equipment and instruments are going to be utilized, use the diagrams to plan ahead of time where you will make the necessary connections in order to accurately pinpoint the trouble spot.

For electrical troubleshooting you'll need a circuit tester or voltmeter, a continuity tester, which includes a bulb, battery and set of test leads, and a jumper wire, preferably with a circuit breaker incorporated, which can be used to bypass electrical components **(see illustrations)**. Before attempting to locate a problem with test instruments, use the wiring diagram(s) to decide where to make the connections.

Voltage checks

Voltage checks should be performed if a circuit is not functioning properly. Connect one lead of a circuit tester to either the negative battery terminal or a known good ground.

Connect the other lead to a connector in the circuit being tested, preferably nearest to the battery or fuse **(see illustration)**. If the bulb of the tester lights, voltage is present, which means that the part of the circuit between the connector and the battery is problem free. Continue checking the rest of the circuit in the same fashion. When you reach a point at which no voltage is present, the problem lies between that point and the last test point with voltage. Most of the time the problem can be traced to a loose connection. **Note:** *Keep in mind that some circuits receive voltage only when the ignition key is in the ACC or RUN position.*

Finding a short

One method of finding shorts in a circuit is to remove the fuse and connect a test light or voltmeter to the fuse terminals. There should be no voltage present in the circuit

when it is turned off. Move the wiring harness from side-to-side while watching the test light. If the bulb goes on, there is a short to ground somewhere in that area, probably where the insulation has rubbed through. The same test can be performed on each component in the circuit, even a switch.

Ground check

Perform a ground test to check whether a component is properly grounded. Disconnect the battery and connect one lead of a continuity tester or multimeter (set to the ohm scale), to a known good ground. Connect the other lead to the wire or ground connection being tested. If the resistance is low (less than 5 ohms), the ground is good. If the bulb on a self-powered test light does not go on, the ground is not good.

Continuity check

A continuity check determines whether there are any breaks in a circuit (whether it's conducting electricity correctly). With the circuit off (no power in the circuit), use a self-powered continuity tester or multimeter to check the circuit. Connect the test leads to both ends of the circuit (or to the power end and a good ground). If the test light comes on, the circuit is conducting current correctly **(see illustration)**. If the resistance is low (less than 5 ohms), there is continuity; if the reading is 10,000 ohms or higher, there is a break somewhere in the circuit. The same procedure can be used to test a switch, by connecting the continuity tester to the switch terminals. With the switch turned on, the test light should come on (or low resistance should be indicated on a meter).

Finding an open circuit

When diagnosing for possible open circuits, it is often difficult to locate them by sight because the connectors hide oxidation or terminal misalignment. Merely wiggling a

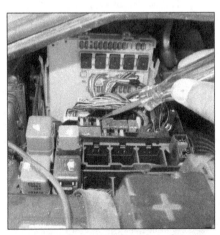

2.6 In use, a basic test light's lead is clipped to a known good ground, then the pointed probe can test connectors, wires or electrical sockets - if the bulb lights, battery voltage is present at the test point

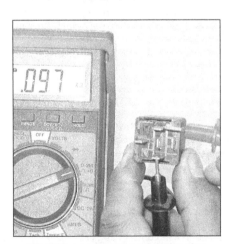

2.9 With a multimeter set to the ohm scale, resistance can be checked across two terminals - when checking for continuity, a low reading indicates continuity, a very high or infinite reading indicates lack of continuity

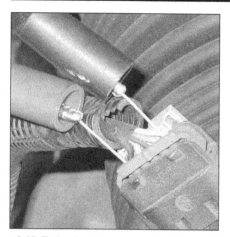

2.15 To backprobe a connector, insert a small, sharp probe (such as a straight-pin) into the back of the connector alongside the desired wire until it contacts the metal terminal inside; connect your meter leads to the probes - this allows you to test a functioning circuit

3.1 The fuse and relay box is located at the left front corner of the engine compartment. To locate a fuse or relay, refer to the fuse and relay guide imprinted on the underside of the lid

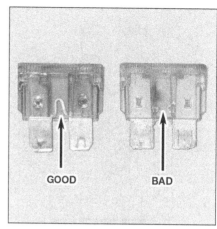

3.3 When a fuse blows, the element between the terminals melts

connector on a sensor or in the wiring harness may correct the open circuit condition. Remember this when an open circuit is indicated when troubleshooting a circuit. Intermittent problems may also be caused by oxidized or loose connections.

Electrical troubleshooting is simple if you keep in mind that all electrical circuits are basically electricity running from the battery, through the wires, switches, relays, fuses and fusible links to each electrical component (light bulb, motor, etc.) and to ground, from which it is passed back to the battery. Any electrical problem is an interruption in the flow of electricity to and from the battery.

Connectors

Most electrical connections on these vehicles are made with multi-wire plastic connectors. The mating halves of many connectors are secured with locking clips molded into the plastic connector shells. The mating halves of large connectors, such as some of those under the instrument panel, are held together by a bolt through the center of the connector.

To separate a connector with locking clips, use a small screwdriver to pry the clips apart carefully, then separate the connector halves. Pull only on the shell, never pull on the wiring harness as you may damage the individual wires and terminals inside the connectors. Look at the connector closely before trying to separate the halves. Often the locking clips are engaged in a way that is not immediately clear. Additionally, many connectors have more than one set of clips.

Each pair of connector terminals has a male half and a female half. When you look at the end view of a connector in a diagram, be sure to understand whether the view shows the harness side or the component side of the connector. Connector halves are mirror images of each other, and a terminal shown on the right side end-view of one half will be on the left side end view of the other half.

It is often necessary to take circuit voltage measurements with a connector connected. Whenever possible, carefully insert a small straight pin (not your meter probe) into the rear of the connector shell to contact the terminal inside, then clip your meter lead to the pin. This kind of connection is called backprobing **(see illustration)**. When inserting a test probe into a male terminal, be careful not to distort the terminal opening. Doing so can lead to a poor connection and corrosion at that terminal later. Using the small straight pin instead of a meter probe results in less chance of deforming the terminal connector.

3   Fuses - general information

Refer to illustrations 3.1 and 3.3

The electrical circuits of the vehicle are protected by a combination of fuses and circuit breakers. The fuse and relay box is located on the left side of the engine compartment **(see illustration)**. At a dealer parts department you might hear the phrase "Power Distribution Center" or "Integrated Power Module." These are just manufacturer terms for the fuse and relay box. In this manual, we simply use the term "fuse and relay box."

Each of the fuses is designed to protect a specific circuit, and the various circuits are identified on the fuse panel itself.

Different sizes of fuses are employed in the fuse blocks. There are mini and maxi sizes, with the larger located in the fuse and relay box. The maxi fuses can be removed with your fingers, but the mini fuses require the use of pliers or the small plastic fuse-puller tool found in most fuse boxes. If an electrical component fails, always check the fuse first. The best way to check the fuses is with a test light. Check for power at the exposed terminal tips of each fuse. If power is present at one side of the fuse but not the other, the fuse is blown. A blown fuse can also be identified by visually inspecting it **(see illustration)**.

Replace blown fuses with the correct type. Fuses of different ratings are physically interchangeable, but only fuses of the proper rating should be used. Replacing a fuse with one of a higher or lower value than specified is not recommended. Each electrical circuit needs a specific amount of protection. The amperage rating of each fuse is molded into the fuse body.

If the replacement fuse immediately fails, don't replace it again until the cause of the problem is isolated and corrected. In most cases, the cause will be a short circuit in the wiring caused by a broken or deteriorated wire.

4   Circuit breakers - general information

Circuit breakers protect certain heavy-load circuits. Depending on the vehicle's accessories, there may be one to three circuit breakers located in the main fuse panel and also in the fuse and relay box.

Because the circuit breakers reset automatically, an electrical overload in a circuit-breaker-protected system will cause the circuit to fail momentarily, then come back on. If the circuit does not come back on, check it immediately.

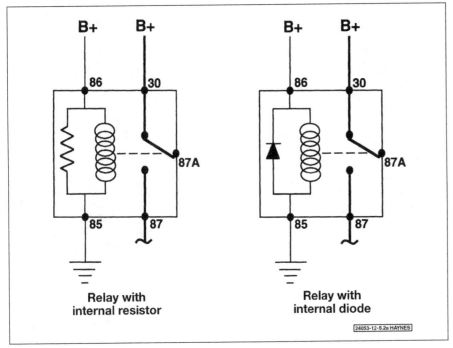

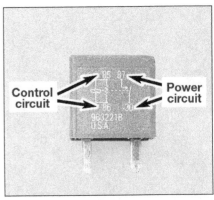

5.5b Most relays are marked on the outside to easily identify the control circuit and power circuits - this one is of the four-terminal type

5.5a Typical ISO relay designs, terminal numbering and circuit connections

For a basic check, pull the circuit breaker up out of its socket on the fuse panel, but just far enough to probe with a voltmeter. The breaker should still contact the sockets.

With the voltmeter negative lead on a good chassis ground, touch each end prong of the circuit breaker with the positive meter probe. There should be battery voltage at each end. If there is battery voltage only at one end, the circuit breaker must be replaced.

5   Relays - general information and testing

General information

1   Many electrical accessories in the vehicle utilize relays to transmit current to the component. If the relay is defective, the component won't operate properly.
2   Most relays are located in the engine compartment fuse and relay box **(see illustration 3.1)**.
3   Some relays are located in other parts of the vehicle, primarily in various wiring harnesses underneath the instrument panel.
4   If a faulty relay is suspected, it can be removed and tested using the procedure below or by a dealer service department or a repair shop. Defective relays must be replaced as a unit.

Testing

Refer to illustrations 5.5a and 5.5b

5   Most of the relays used in these vehicles are of a type often called ISO relays, which refers to the International Standards Organiza-

tion. The terminals of ISO relays are numbered to indicate their usual circuit connections and functions. There are two basic layouts of terminals on the relays used in the vehicles covered by this manual **(see illustrations)**.
6   Refer to the wiring diagram for the circuit to determine the proper connections for the relay you're testing. If you can't determine the correct connection from the wiring diagrams, however, you may be able to determine the test connections from the information that follows.
7   Two of the terminals are the relay control circuit and connect to the relay coil. The other relay terminals are the power circuit. When the relay is energized, the coil creates a magnetic field that closes the larger contacts of the power circuit to provide power to the circuit loads.
8   Terminals 85 and 86 are normally the control circuit. If the relay contains a diode, terminal 86 must be connected to battery positive (B+) voltage and terminal 85 to ground. If the relay contains a resistor, terminals 85 and 86 can be connected in either direction with respect to B+ and ground.
9   Terminal 30 is normally connected to the battery voltage (B+) source for the circuit loads. Terminal 87 is connected to the ground side of the circuit, either directly or through a load. If the relay has several alternate terminals for load or ground connections, they usually are numbered 87A, 87B, 87C, and so on.
10   Use an ohmmeter to check continuity through the relay control coil.

   a) *Connect the meter according to the polarity shown in the illustration for one check; then reverse the ohmmeter leads and check continuity in the other direction.*

   b) *If the relay contains a resistor, resistance will be indicated on the meter, and should be the same value with the ohmmeter in either direction.*
   c) *If the relay contains a diode, resistance should be higher with the ohmmeter in the forward polarity direction than with the meter leads reversed.*
   d) *If the ohmmeter shows infinite resistance in both directions, replace the relay.*

11   Remove the relay from the vehicle and use the ohmmeter to check for continuity between the relay power circuit terminals. There should be no continuity between terminal 30 and 87 with the relay de-energized.
12   Connect a fused jumper wire to terminal 86 and the positive battery terminal. Connect another jumper wire between terminal 85 and ground. When the connections are made, the relay should click.
13   With the jumper wires connected, check for continuity between the power circuit terminals. Now there should be continuity between terminals 30 and 87.
14   If the relay fails any of the above tests, replace it.

6   Multi-function switch and switch pod – replacement

Refer to illustrations 6.2 and 6.3
Warning: *The models covered by this manual are equipped with a Supplemental Restraint System (SRS), more commonly known as airbags. Always disarm the airbag system before working in the vicinity of any airbag system component to avoid the possibility of accidental deployment of the airbag, which could cause personal injury (see Section 25). Do not use a memory-saving device to preserve the PCM's memory when working on or near airbag system components.*

Multi-function switch

1   Remove the upper and lower steering column covers (see Chapter 11).

6.2 Disconnect the wiring harness from the rear of the multi-function switch

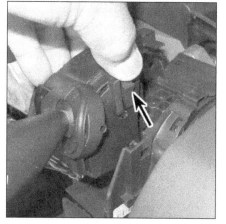

6.3 Press this tab to release the switch, then pull it up

8.2 Pry the headlight switch from the instrument panel with a plastic tool to avoid scratching the plastic panels

2   Unplug the wiring harness from the rear of the switch **(see illustration)**.

3   Use a small screwdriver or your thumb to release the catch in the small access port, then pull the switch upward to remove it **(see illustration)**.

4   Installation is the reverse of removal.

Switch pod

Note: *The switch pod housing has multiple configurations; all configurations will have the hazard warning switch but may include trailer brake control switches, air suspension switch, Electronic Stability Control (ESC) Off switch, tow/haul mode switch, heated seats (and on later models, cooled seats), Start/ Stop switch (2013 and later), heated steering wheel, exhaust brake On switch (diesel) or park assist Off switch. The individual switches in the instrument panel switch pod cannot be repaired and are not serviced individually. If any component within the switch pod is defective, the entire switch pod must be replaced.*

5   Disconnect the cable(s) from the negative battery terminal(s) (see Chapter 5).

6   Remove the center instrument trim bezel (see Chapter 11).

7   Disconnect the electrical connectors to the pod.

8   Remove the pod four mounting screws and separate the pod from the bezel.

9   Installation is the reverse of removal.

7   Key lock cylinder and ignition switch - replacement

Warning: *The models covered by this manual are equipped with a Supplemental Restraint System (SRS), more commonly known as airbags. Always disarm the airbag system before working in the vicinity of any airbag system component to avoid the possibility of accidental deployment of the airbag, which could cause personal injury (see Section 25). Do not use a memory-saving device to preserve the PCM's memory when working on or*

near airbag system components.

1   The models covered by this manual are not equipped with conventional ignition switches. Instead, they have a Wireless Ignition Node (WIN) that senses when the ignition key is near.

2   The WIN module can be easily replaced, but diagnosis requires scan tools and complex procedures. Also, a replacement WIN module must be programmed to the vehicle and to the keys. These procedures are beyond the scope of this manual.

8   Headlight switch - replacement

Refer to illustration 8.2

1   Disconnect the cable(s) from the negative battery terminal(s) (see Chapter 5).

2   Use a plastic trim tool or a screwdriver wrapped with tape to carefully pry the headlight switch loose from the dash **(see illustration)**.

3   Disconnect the electrical connector from the headlight switch, then remove it.

4   Installation is the reverse of removal.

9   Instrument cluster - removal and installation

Refer to illustrations 9.3a and 9.3b

1   Disconnect the cable(s) from the negative battery terminal(s) (see Chapter 5).

2   Remove the instrument cluster bezel (see Chapter 11).

3   Remove the two top instrument cluster mounting screws, pull out the cluster and disconnect the electrical connectors from the rear **(see illustrations)**.

4   Installation is the reverse of removal.

10   Windshield wiper motor - replacement

Refer to illustrations 10.4, 10.8, 10.9 and 10.10

1   Disconnect the cable(s) from the negative battery terminal(s) (see Chapter 5).

2   Make sure that the wipers are in the Park position.

9.3a After removing the two screws, pull the cluster out from the top . . .

9.3b . . . then disconnect the wiring from the rear

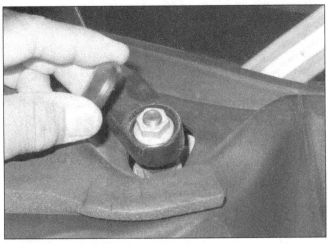

10.4 Lift the plastic caps to expose the wiper arm mounting nuts

10.8 Disconnect the wiring from the wiper motor

10.9 Remove this nut to detach the motor from the linkage, mark the position of the arm to the shaft, then remove the motor mounting screws

3   Mark the positions of the wiper blades on the windshield using tape or a marker.

4   Pivot each wiper arm to lock it in its Up position, then remove the plastic nut cap **(see illustration)**.

5   Remove the arm retaining nut.

6   Rock the wiper arm up and down to break it free from the shaft, then lift off the arm.

7   Remove the cowl cover (see Chapter 11).

8   Disconnect the electrical connector from the windshield wiper motor **(see illustration)**.

9   Remove the nut and separate the motor's arm from its output shaft **(see illustration)**. **Note:** *Leave the arm and the linkage mechanism in the cowl.*

10   Remove the windshield wiper motor mounting screws **(see illustration)**, then maneuver it from under the linkage bracket and out of the cowl.

11   Installation is the reverse of removal.

11   Radio and speakers - removal and installation

Radio

Refer to illustrations 11.3a, 11.3b, 11.3c, 11.4 and 11.5

1   On REQ radios, turn the ignition system On, then push the Scan and Set buttons at the same time. The display should read "Transportation."

2   Disconnect the cable(s) from the negative battery terminal(s) (see Chapter 5).

3   Remove the center instrument panel bezel **(see illustrations)**.

4   Remove the radio mounting screws **(see illustration)** and pull the radio out of the dash.

5   Disconnect the electrical connectors **(see illustration)**. Pull out the antenna cable lock and disconnect the antenna cable, then

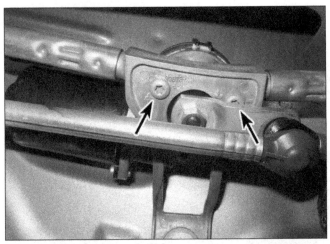

10.10 Wiper motor mounting screws (lower one hidden)

11.3a After removing the two screws under the upper storage tray, use a plastic trim tool to pull the center bezel rearward

11.3b Carefully pull the bezel to the rear . . .

11.3c . . . then disconnect the wiring

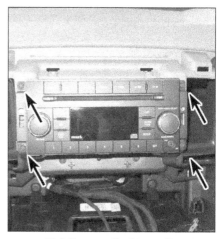

11.4 Remove the four radio mounting screws

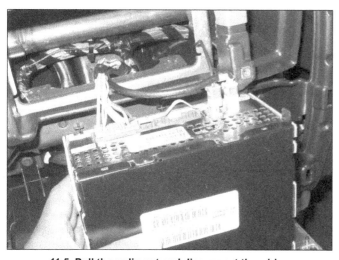

11.5 Pull the radio out and disconnect the wiring

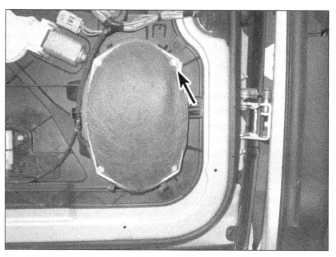

11.9 Remove the door trim panel for access to the speaker mounting screws

remove the radio.

Caution: *Pulling the antenna cable straight out of the radio without pulling on the antenna locking connector could damage the cable or radio.*

6   Installation is the reverse of removal.

Speakers

7   Disconnect the cable(s) from the negative battery terminal(s) (see Chapter 5).

Front door speakers

Refer to illustrations 11.9 and 11.10

8   Disconnect the door trim panel (see Chapter 11).

9   Remove the four speaker mounting screws **(see illustration)**.

10   Pull out the speaker and disconnect the electrical connector **(see illustration)**.

11   Installation is the reverse of removal.

Instrument panel center speaker and end speakers

Refer to illustrations 11.12 and 11.14

12   Remove the covers from the instrument panel near the windshield **(see illustration)**.

11.10 Pull the speaker out, then disconnect the wiring

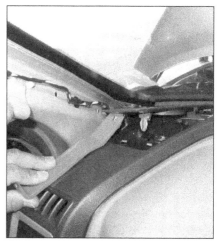

11.12 Carefully pry up the end speaker grilles - the center cover is removed in the same way

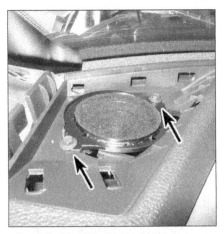

11.14 Typical speaker mounting screws

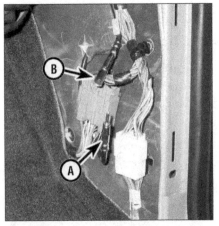

12.2 Working in the right kick panel area, disconnect the antenna side of the cable (A) from the radio side of the cable (B)

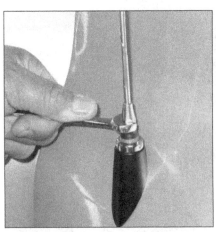

12.5 Using a small wrench of the appropriate size, unscrew the antenna from its mounting base

13   Remove the screws under the covers, then remove the top cover.
14   Remove the speaker mounting screws **(see illustration)**.
15   Pull out the speaker and disconnect the electrical connector.
16   Installation is the reverse of removal.

Rear cab side speakers or rear door speakers

17   Disconnect the B-pillar lower trim or the rear door trim panel (see Chapter 11).
18   Remove the three speaker mounting screws.
19   Pull out the speaker and disconnect the electrical connector **(see illustration 11.10)**.
20   Installation is the reverse of removal.

Subwoofer

21   On quad cab models, put the right rear seat into the Up position.
22   Disconnect the wiring from the subwoofer.
23   Remove the mounting bolts and remove the subwoofer.
24   Installation is the reverse of removal.

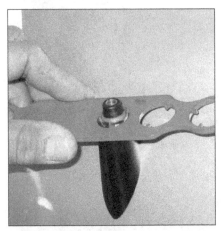

12.6 Using an antenna wrench (available at auto parts stores) unscrew the nut from the antenna mounting base, then remove the base by pulling it straight up

Headliner speaker (quad cab models)

25   Use a plastic trim tool or a screwdriver wrapped with tape to pry off the speaker cover.
26   Remove the two clips that secure the speaker.
27   Disconnect the wiring harness from the speaker and remove it.
28   Installation is the reverse of removal.

---

12   Antenna and cable - replacement

1   Disconnect the cable(s) from the negative battery terminal(s) (see Chapter 5).

Radio antenna

Refer to illustrations 12.2, 12.5 and 12.6

2   Remove the right-side kick panel/threshold trim panel **(see illustration)**.
3   Disconnect the antenna cables.
4   Tie a piece of string to the antenna end of the cable.
5   Unscrew the antenna mast from its base **(see illustration)**.
6   Remove the base retaining nut and the antenna base **(see illustration)**.
7   Open the passenger's door, then pull the antenna downward and out. Don't allow the pull cord to come out.
8   If you're installing a new antenna, tie the cord to the end of its cable, then use it to pull the cable through the body panels. Ensure that the rubber grommet is fully seated.
9   The remainder of installation is the reverse of removal.

Satellite radio antenna

10   The rear of the headliner must be pulled down for access to this antenna. Because this requires the removal of several interior trim panels, we recommend that you have this procedure done by a shop familiar with this type of work.

Navigation system antenna

11   Remove the radio (see Section 11).
12   Disconnect the antenna from its bracket and remove it.
13   Installation is the reverse of removal.

---

13   Rear window defogger - check and repair

---

1   The rear window defogger consists of a number of horizontal heating elements baked onto the inside surface of the glass. Power is supplied through a relay and fuse from the interior fuse/relay box. A defogger switch on the instrument panel controls the defogger grid.
2   Small breaks in the element can be repaired without removing the rear window.

Check

Refer to illustrations 13.5, 13.6 and 13.8

3   Turn the ignition and defogger switches to the ON position.
4   Using a voltmeter, place the positive probe against the defogger grid positive side and the negative probe against the ground side. If battery voltage is not indicated, check that the ignition switch is On and that the feed and ground wires are properly connected. Check the two fuses, defogger switch, defogger relay and related wiring. The dealer can scan the body control module if necessary. If voltage is indicated, but all or part of the defogger doesn't heat, proceed with the following tests.
5   When measuring voltage during the next two tests, wrap a piece of aluminum foil around the tip of the voltmeter positive probe and press the foil against the heating element with your finger **(see illustration)**. Place the negative probe on the defogger grid ground terminal.
6   Check the voltage at the center of each heating element **(see illustration)**. If the voltage is 5 to 6 volts, the element is okay (there

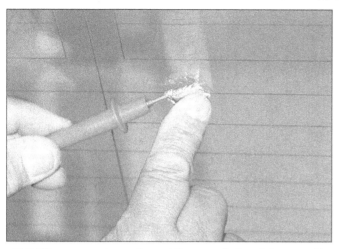

13.5 When measuring voltage at the rear window defogger grid, wrap a piece of aluminum foil around the positive probe of the voltmeter and press the foil against the wire with your finger

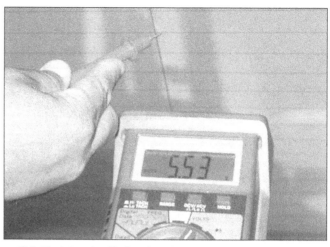

13.6 To determine if a heating element has broken, check the voltage at the center of each element - if the voltage is 6-volts, the element is unbroken

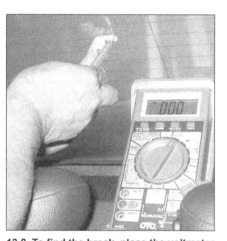

13.8 To find the break, place the voltmeter negative lead against the defogger ground terminal, place the voltmeter positive lead with the foil strip against the heat wire at the positive terminal end and slide it toward the negative terminal end. The point at which the voltmeter deflects from several volts to zero volts is the point at which the wire is broken

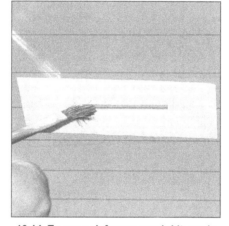

13.14 To use a defogger repair kit, apply masking to the inside of the window at the damaged area, then brush on the special conductive coating

14.1 Disconnect the wiring from the headlight bulb by first releasing the orange lock tab

is no break). If the voltage is 0 volts, the element is broken between the center of the element and the positive end. If the voltage is 10 to 12 volts, the element is broken between the center of the element and the ground side. Check each heating element.

7   If none of the elements are broken, connect the negative probe to a good chassis ground. The voltage reading should stay the same, if it doesn't the ground connection is bad.

8   To find the break, place the voltmeter negative probe against the defogger ground terminal. Place the voltmeter positive probe with the foil strip against the heating element at the positive side and slide it toward the negative side. The point at which the voltmeter deflects from several volts to zero is the

point where the heating element is broken **(see illustration)**.

Repair

Refer to illustration 13.14

9   Repair the break in the element using a repair kit specifically for this purpose, such as Dupont paste No. 4817 (or equivalent). The kit includes conductive plastic epoxy.

10   Before repairing a break, turn off the system and allow it to cool for a few minutes.

11   Lightly buff the element area with fine steel wool; then clean it thoroughly with rubbing alcohol.

12   Use masking tape to mask off the area being repaired.

13   Thoroughly mix the epoxy, following the kit instructions.

14   Apply the epoxy material to the slit in the masking tape, overlapping the undamaged area about 3/4-inch on either end **(see illustration)**.

15   Allow the repair to cure for 24 hours before removing the tape and using the system.

14 Headlight bulb - replacement

Refer to illustrations 14.1 and 14.2

Warning: *Halogen bulbs are gas-filled and under pressure and they can shatter if the surface is scratched or the bulb is dropped. Wear eye protection and handle the bulbs carefully, grasping only the base whenever possible. Don't touch the surface of the bulb with your fingers because the oil from your skin could cause it to overheat and fail prematurely. If you do touch the bulb surface, clean it with rubbing alcohol.*

1   Remove the entire headlight housing (see Section 15). Disconnect the wiring harness **(see illustration)**.

14.2 Twist the bulb housing counterclockwise to remove it

15.2 Remove the radiator top cover for access to the upper screw (A), then use a magnetic socket with a long extension to remove the vertical screw (B) - other components removed here for clarity

2   On models with two headlights, rotate the bulb counterclockwise to remove it **(see illustration)**.

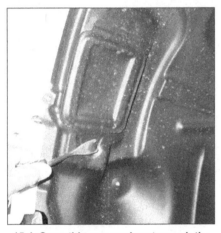

15.4 Open this access door to reach the headlight slide lock

3   On models with four headlights, twist the bulb access cover counterclockwise and remove it. Rotate the bulb counterclockwise to remove it.

4   Installation is the reverse of removal.

15   Headlight housing - removal and installation

Refer to illustrations 15.2, 15.4, 15.5, 15.6 and 15.7

1   Remove the six push-pin clips, then remove the plastic shield from the top of the radiator.

2   Remove the top headlight mounting screw **(see illustration)**.

3   Remove the lower headlight mounting screw by reaching a socket and extension vertically between the grille and the radiator core support structure **(see illustration 15.2)**. **Note:** *The screw must not be dropped after it's unscrewed*

so you'll have to use a magnetic socket. If you don't have one, put some sticky substance such as butyl windshield adhesive or silicone adhesive into the socket to hold the screw.

4   Release the bottom of the tab at the lower section of the access panel to the wheel well splash shield **(see illustration)**.

5   Working through the opening in the wheel well splash shield, move the slide lock upward to disengage it from the rear of the headlight housing **(see illustration)**.

6   Pull the headlight toward the front of the vehicle to pop it loose from the plastic retainer **(see illustration)**.

7   Disconnect the wiring harnesses from the headlight, then remove it **(see illustration)**.

8   Installation is the reverse of removal.

16   Headlights - adjustment

Warning: *The headlights must be aimed correctly. If adjusted incorrectly, they could*

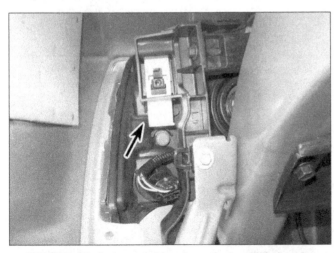

15.5 Slide this lock upward to release the headlight housing

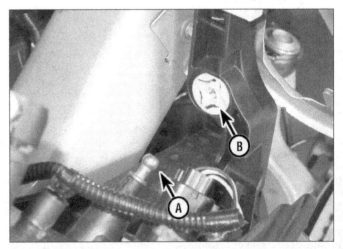

15.6 The ball (A) will pop out of socket (B) when the headlight is pulled forward

15.7 Disconnect the wiring harnesses from the rear of the headlight housing (arrow indicates the headlight adjustment screw)

16.1 Insert a socket through these holes to engage the headlight adjustment screws

temporarily blind the driver of an oncoming vehicle and cause an accident or seriously reduce your ability to see the road. The head- lights should be checked for proper aim every 12 months and any time a new headlight is installed or front-end bodywork is performed.

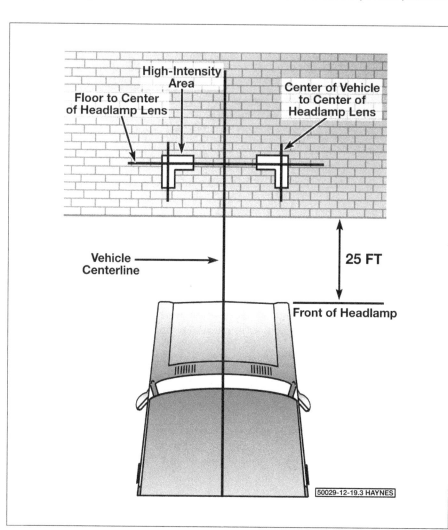

16.2 Headlight adjustment screen details

The following procedure is only intended to provide temporary adjustment until you can have the headlights professionally adjusted by a dealer service department.

Headlights

Refer to illustrations 16.1 and 16.2

1   Each headlight has an adjusting screw for vertical adjustments only (**see accompanying illustration and illustration 15.7**). The screw access holes are in the top of the fenders and are indicated by arrows.

2   There are several methods of adjusting the headlights. The simplest method requires an open area with a blank wall and a level floor (**see illustration**).

3   Position masking tape vertical on the wall in reference to the vehicle centerline and the centerlines of both headlights.

4   Position a horizontal tape line in reference to the centerline of the headlights. **Note:** *It might be easier to position the tape on the wall with the vehicle parked only a few inches away.*

5   Adjustment should be made with the vehicle parked 25 feet from the wall, sitting level, the gas tank full and no unusually heavy load in the vehicle.

6   The high intensity zone should be vertically centered with the exact center, about three inches below the horizontal line.

7   Have the headlights adjusted by a qualified technician at the earliest opportunity.

Fog lights

8   Fog lights are optional on these vehicles.

9   Park the vehicle 25 feet from the wall.

10   Tape a horizontal line on the wall that represents the height of the fog lights and tape another line four inches below that line.

11   Using the adjusting wheel on the rear of each fog light, adjust the pattern on the wall so that the top of the fog light beam meets the lower line on the wall.

Bulb removal

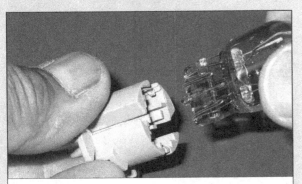

To remove many modern exterior bulbs from their holders, simply pull them out

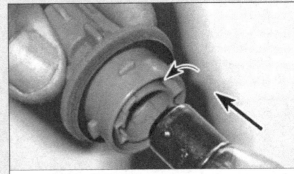

On bulbs with a cylindrical base ("bayonet" bulbs), the socket is spring-loaded; a pair of small posts on the side of the base hold the bulb in place against spring pressure. To remove this type of bulb, push it into the holder, rotate it 1/4-turn counterclockwise, then pull it out

If a bayonet bulb has dual filaments, the posts are staggered, so the bulb can only be installed one way

To remove most overhead interior light bulbs, simply unclip them

17 Bulb replacement

Exterior lights
Front turn signal/parking light/ sidemarker light bulbs

Refer to illustrations 17.1 and 17.2

1   Remove the headlight housing (see Section 15) and remove the bulb holder for the front turn signal/parking light/sidemarker light bulb **(see illustration)**.
2   Remove the turn signal/parking light/sidemarker light bulb from its bulb holder by pulling it straight out of the holder **(see illustration)**.
3   Install the new bulb by pushing it straight into the bulb holder until it's fully seated.
4   Installation is otherwise the reverse of removal.

High-mounted brake light and cargo light bulbs

Refer to illustrations 17.5, 17.6 and 17.7

5   Remove the screws that attach the high-mount brake light housing **(see illustration)**.
6   Disconnect the electrical connectors

from the center high-mounted brake light housing **(see illustration)**.
7   To remove a bulb holder from the center high-mounted brake light housing, rotate the bulb holder counterclockwise, then remove it from the housing **(see illustration)**.

8   To remove the old bulb, pull it straight out of the bulb holder.
9   To install a new bulb, push it straight into the bulb holder. Make sure that the bulb is fully seated in the holder.
10   Installation is the reverse of removal.

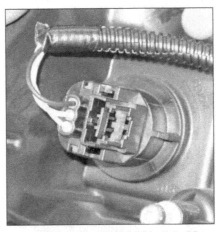

17.1 Turn the parking light bulb holder counterclockwise to remove it . . .

17.2 . . . then pull the bulb straight out

17.5 Remove these screws to detach the light housing

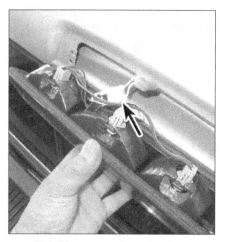

17.6 Disconnect the main wiring connector to lift the housing free

17.7 Twist each bulb holder and remove it from the housing, then pull out the defective bulb

Fog lamp bulbs

11   Reach behind the bumper and disconnect the wiring from the fog lamp.
12   Turn the bulb counterclockwise, then remove it.
13   Installation is the reverse of removal.

License plate light bulbs

Refer to illustration 17.14

14   To remove a license plate light bulb holder, rotate it counterclockwise **(see illustration)**.
15   To remove a bulb from its holder, pull it straight out of the bulb holder.
16   To install a new bulb in its holder, push it straight into the holder until it's fully seated.
17   Installation is the reverse of removal.

Mirror-mounted lights

18   These are LED lights, and do not use bulbs. If the light unit is faulty, it must be replaced with a new LED assembly.

Rear fender lights

Note: *These clearance lights are used only on 3500 series trucks with dual rear wheels.*
19   Gently push the back of the light lens toward the front of the vehicle, then pull it out to remove it.
20   Turn the bulb holder counterclockwise and pull it out.
21   Pull the bulb straight out of the socket.
22   Installation is the reverse of removal.

Tailgate lights

Note: *These lights are used only on 3500 series vehicles with dual rear wheels.*
23   Remove the screws from the light lens, then pull it out to access the bulb holders.
24   Turn the bulb holder counterclockwise and pull it out.
25   Pull the bulb straight out of the socket.
26   Installation is the reverse of removal.

Roof lights

Note: *These lights are only used on heavy-duty models with clearance lights.*
27   Remove the screws from the roof light lens, then remove it.
28   Turn the bulb holder counterclockwise and pull it out.
29   Pull the bulb straight out of the socket.
30   Installation is the reverse of removal.

Rear turn signal/brake light/back-up light bulbs

Refer to illustrations 17.31, 17.32, 17.33 and 17.34

31   Open the tailgate, then remove the taillight housing mounting screws from the jamb **(see illustration)**
32   Pull rearward on the light assembly to snap it out of the retainers at the outside edge **(see illustration)**.
33   Disconnect the electrical connectors

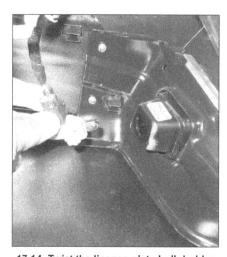

17.14 Twist the license plate bulb holder to remove it, then pull the bulb from the holder

17.31 Remove these screws to remove the taillight assembly

17.32 Pull straight to the rear to release the taillight from these retainers

17.33 Disconnect the wiring harnesses to remove the taillight

17.34 The bulb holders are removed by twisting them counterclockwise - the bulbs can then be removed

17.42 Pry down on the left side of the dome light lens with a small screwdriver

from the taillight assembly **(see illustration)** and remove the taillight housing.

34   Remove the bulb holder from the taillight housing by turning it counterclockwise, then pulling it out **(see illustration)**.

35   Pull the bulb that you want to replace from its socket.

36   To install a new bulb into the socket, insert it into the socket and push it in until it snaps into place.

37   Installation is otherwise the reverse of removal.

Bed storage bin light bulbs

38   Pull up on the lens tab and pull the bottom of the lens off. Slide it down to release it from the top.

39   Turn the bulb holder counterclockwise and pull it out.

40   Pull the bulb straight out of the socket.

41   Installation is the reverse of removal.

Interior lights

Warning: *The models covered by this manual are equipped with a Supplemental Restraint System (SRS), more commonly known as airbags. Always disarm the airbag system before working in the vicinity of any airbag system component to avoid the possibility of accidental deployment of the airbag, which could cause personal injury (see Section 25). Do not use a memory-saving device to preserve the PCM's memory when working on or near airbag system components.*

Dome light bulb

Refer to illustrations 17.42 and 17.43

42   Using a small flat-bladed screwdriver, carefully pry down the side of the dome light lens that has the small notch in it **(see illustration)**. Swing down the lens and allow it to hang.

43   Remove the dome light bulb from the dome light housing **(see illustration)**.

44   Insert a new dome light bulb into its socket in the dome light housing until it's fully seated.

45   Installation is otherwise the reverse of removal.

Glove box light bulb

Refer to illustration 17.46

46   To remove the glove box light bulb, pull it straight out toward the front of the vehicle **(see illustration)**.

47   To install a new glove box light bulb, push it straight into its socket until it's fully seated.

Reading light bulbs

Refer to illustration 17.48

48   Use a small screwdriver to pry the sides near the front of the lens to pop it loose from the pivots **(see illustration)**.

49   Slide the lens forward to disengage it from the rear, then remove it.

50   Pull the bulb from its terminals.

51   Installation is the reverse of removal.

Vanity light bulbs

52   Use a small screwdriver to carefully pry the lens from the light assembly

53   With small pliers, gently pull the bulb from its socket without excessive squeezing.

54   Installation is the reverse of removal.

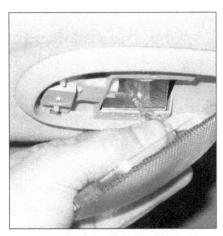

17.43 Pull the bulb straight out

17.46 Pull the glove box light bulb straight out to remove it

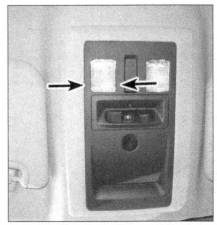

17.48 Pry the front of the reading light lens down using these slots, then slide it forward to remove it

Console storage bin light bulb

55   This storage bin is at the front of the console. Use a small screwdriver to press down on the lens tab.
56   Pull the light assembly down and to the right to remove it from its hole.
57   Disconnect the wiring from the light.
58   Push down on the release tab, then slide the cover to expose the bulb.
59   Pull the bulb straight out of the socket.
60   Installation is the reverse of removal.

18   Horn - replacement

Refer to illustration 18.2

1   Remove the plastic push-pins, then lift off the radiator top cover.
2   Disconnect the electrical connectors from the horns **(see illustration)**.
3   Remove the horn mounting bracket bolt and remove the horns and mounting bracket as a single assembly.
4   Remove the retaining nut for the horn that you're replacing and remove that horn from the mounting bracket.
5   Installation is the reverse of removal.

19   Electric side-view mirrors - general information

1   The electric side-view mirrors can be adjusted up-and-down and left-to-right by a driver's side switch located on the left door trim panel. On models with factory-installed dual power mirrors, each mirror is also equipped with a heater grid behind the mirror glass to clear the mirror surface of fog, ice or snow. On these models, the mirror heater grid is an integral component of each mirror. If a heater grid fails, replace the mirror (see Chapter 11). The heater grid switches and the heated mirror system indicator light are integral components of the heater/air conditioning control panel on the dash. If one of these components fails, replace the heater/air conditioning control assembly (see Chapter 3). The heated mirror relay is located in the engine compartment fuse and relay box (see Section 3).
2   The mirror control switch has a LEFT-RIGHT selector switch that allows you to send voltage to the side-view mirror that you want to adjust. With the ignition switch in the ACC position, roll down the windows and operate the mirror control switch through all functions (left-right and up-down) for both the left and right side-view mirrors.
3   Listen carefully for the sound of the electric motors running in the mirrors.
4   If you can hear the motors but the mirror glass doesn't move, the problem is probably a defective drive mechanism inside the mirror, which will necessitate replacement of the mirror.
5   If the mirrors don't operate and no sound comes from the mirrors, check the fuse in the

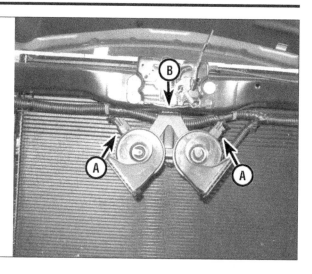

18.2 Disconnect the horn wiring connectors (A), then unbolt the mounting bracket (B) and remove the horns along with the bracket (grille removed here for clarity)

engine compartment fuse and relay box (see Section 3).
6   If the fuse is OK, remove the door panel for access to the back of the mirror control switch, without disconnecting the wires attached to it (see Chapter 11). Turn the ignition ON and check for voltage at the switch. There should be voltage at one terminal. If there's no voltage at the switch, check for an open in the wiring between the fuse panel and the switch.
7   If there's voltage at the switch, disconnect it. Check the switch for continuity in all its operating positions. If the switch does not have continuity, replace it.
8   Reconnect the switch. Locate the wire going from the switch to ground. Leaving the switch connected, connect a jumper wire between this wire and ground. If the mirror works normally with this wire in place, repair the faulty ground connection.
9   If the mirror still doesn't work, remove the mirror and check the wires at the mirror for voltage. Check with the ignition key turned to ON and the mirror selector switch on the appropriate side. Operate the mirror switch in all its positions. There should be voltage at one of the switch-to-mirror wires in each switch position, except the neutral (off) position.
10   If voltage is not present in each switch position, check the wiring between the mirror and control switch for opens and shorts.
11   If there's voltage, remove the mirror and test it off the vehicle with jumper wires. Replace the mirror if it fails this test.

20   Cruise control system - general information

1   The cruise control system maintains vehicle speed with the Powertrain Control Module (PCM), throttle actuator control motor, brake switch, control switches and associated wiring. There is no mechanical connection, such as a vacuum servo or cable. Some features of the system require special testers and diagnostic procedures that are beyond

the scope of the home mechanic. Listed below are some general procedures that may be used to locate common problems.
2   Check the fuses (see Section 3).
3   The Brake Pedal Position (BPP) switch (or brake light switch) deactivates the cruise control system. Have an assistant press the brake pedal while you check the brake light operation.
4   If the brake lights do not operate properly, correct the problem and retest the cruise control.
5   Check the wiring between the PCM and throttle actuator motor for opens or shorts and repair as necessary.
6   The cruise control system uses information from the PCM, including the Vehicle Speed Sensor (VSS). Refer to Chapter 6 for more information on the VSS, and to check for the presence of any stored Diagnostic Trouble Codes (DTCs).
7   If no obvious problems are found, take it to a dealer service department or other qualified repair shop for further diagnosis.

21   Power window system - general information

1   The power window system controls the electric motors, mounted inside the doors, that lower and raise the windows. The power window system consists of the control switches, the fuse, the circuit breaker, the motors, the window regulators (the scissors-like mechanisms that raise and lower the window glass) and the wiring connecting the switches to the motors. When the ignition switch is turned to ON, current flows through the power window fuse in the engine compartment fuse and relay box to a circuit breaker located in the instrument panel wiring harness (located near the parking brake pedal). From there, current flows to the power window switches.
2   The power windows are wired so that they can be lowered and raised from the master control switch by the driver or by passengers using remote switches located at each passenger window. Each window has a sepa-

rate motor that is reversible. The position of the control switch determines the polarity and therefore the direction of operation.

3   The power window system will only operate when the ignition switch is turned to ON. In addition, a window lockout switch at the master control switch can, when activated, disable the power window switches on the other doors. Always check these items before troubleshooting a window problem.

4   These procedures are general in nature, so if you can't find the problem using them, take the vehicle to a dealer service department.

5   If the power windows don't work at all, check the fuse or circuit breaker.

6   If only the rear windows are inoperative, or if the windows only operate from the master control switch, check the window lockout switch for continuity in the unlocked position. If it doesn't have continuity, replace it.

7   Check the wiring between the switches and the fuse for continuity. Repair the wiring, if necessary.

8   If only one window is inoperative from the master control switch, try the control switch at the window that doesn't work. **Note:** *This doesn't apply to the driver's door window.*

9   If the same window works from one switch, but not the other, check the switch for continuity.

10   If the switch tests OK, check for a short or open in the wiring between the affected switch and the window motor.

11   If one window is inoperative from both switches, remove the trim panel from the affected door (see Chapter 11), then check for voltage at the switch and at the motor while operating the switch. First check for voltage at the electrical connectors for the circuit. With the ignition key turned to ON and the connectors all connected, backprobe at the designated wire (see the wiring diagrams at the end of this Chapter) with a grounded test light. Pushing the driver's window switch to the DOWN position, there should be voltage at one terminal. Pushing the same switch to the UP position, there should be voltage at another terminal. If these voltage checks are OK, disconnect the electrical connector at the driver's motor, and check it for voltage when the switch is operated.

12   If voltage is reaching the motor and the switch is OK, disconnect the door glass from its regulator (see Chapter 11). Move the window up and down by hand while checking for binding and damage. Also check for binding and damage to the regulator. If the regulator is not damaged and the window moves up and down smoothly, replace the motor. If there's binding or damage, lubricate, repair or replace parts, as necessary.

13   If voltage isn't reaching the motor, check the wiring in the circuit for continuity between the switches and motors (see the wiring diagram at the end of this Chapter).

14   If you have to replace the main power window switch, pry it out of the door trim panel, then disconnect the electrical connector(s) from the switch.

15   When you're done, test the windows to confirm that the window system is functioning correctly.

22 Power door lock system - general information

1   The power door lock system operates the power door motors, which are integral components of the door latch units in each door. The system consists of a fuse (in the engine compartment fuse and relay box), the instrument cluster, the control switches (in each of the front doors), the power door motors and the electrical wiring harnesses connecting all of these components.

2   The lock mechanisms in the door latch units are actuated by a reversible electric motor in each door. When you push the door lock switch to LOCK, the motor operates one way and locks the latch mechanism. When you push the door lock switch the other way, to the UNLOCK position, the motor operates in the other direction, unlocking the latch mechanism. Because the motors and lock mechanisms are an integral part of the door latch units, they cannot be repaired. If a door lock motor or lock mechanism fails, replace the door latch unit (see Chapter 11).

3   Even if you don't manually lock the doors or press the door lock switch to the LOCK position before driving, the instrument cluster automatically locks the doors when the vehicle speed exceeds 15 mph, as long as all the doors are closed and the accelerator pedal is depressed. You can turn off this feature if you don't want the doors to lock automatically. Refer to your owner's manual.

4   Some vehicles have an optional Remote Keyless Entry (RKE) system that allows you to lock and unlock the doors from outside the vehicle. The RKE system consists of the transmitter (the electronic push-button key) and a receiver located on the instrument cluster. The RKE receiver, which operates all the time, is protected by a fuse in the engine compartment fuse and relay box. Vehicles are shipped from the factory with two RKE transmitters but, if you want to purchase extra units, the RKE receiver can actually handle up to four vehicle access codes.

5   Some features of the door lock system on these vehicles rely on resources that they share with other electronic modules through the Programmable Communications Interface (PCI) data bus network. Proper diagnosis of these modules and the PCI data bus network requires the use of a DRB III (proprietary factory) scan tool and factory diagnostic information. At-home repairs are therefore limited to inspecting the wiring for bad connections and for minor faults that can be easily repaired. If you are unable to locate the trouble using the following general steps, consult your dealer service department.

6   Always check the circuit fuses (in the engine compartment fuse and relay box) first.

7   When depressed, each power door lock switch locks or unlocks *all* of the doors. The easiest way to verify that each door lock switch is operating correctly is to watch the door lock button in each door as you operate the switch. The door lock buttons should all go down when you push the door lock switch to the LOCK position, and go up when you push the door lock switch to the UNLOCK position. Also, with the engine turned off so that you can hear better, operate the door lock switches in both directions and listen for the faint click of the motors locking and unlocking the latch mechanisms.

8   If there's no click, check for voltage at the switches. If no voltage is present, check the wiring between the fuse and the switches for shorts and opens (see the wiring diagrams at the end of this chapter).

9   If voltage is present, but no clicking sound is apparent, remove the switch from the door trim panel (see Chapter 11) and test it for continuity. If there is no continuity in either direction, replace the switch.

10   If the switch has continuity but the latch mechanism doesn't click, check the wiring between the switch and the motor in the latch mechanism for continuity. If the circuit is open between the switch and the motor, repair the wiring.

11   If all but one motors is operating, remove the trim panel from the affected door (see Chapter 11) and check for voltage at the motor while operating the lock switch. One of the wires should have voltage in the LOCK position; the other should have voltage in the UNLOCK position.

12   If the inoperative motor is receiving voltage, replace the latch mechanism.

13   If the inoperative motor isn't receiving voltage, check for an open or short in the circuit between the switch and the motor. **Note:** *It's common for wires to break in the harness between the body and the door because repeatedly opening and closing the door fatigues and eventually breaks the wires.*

23 Power seats - general information

Warning: *The models covered by this manual are equipped with a Supplemental Restraint System (SRS), more commonly known as airbags. Additionally, some models are equipped with seat belt pre-tensioners, which are explosive devices. Always disarm the airbag/restraint system before working in the vicinity of any airbag/restraint system component to avoid the possibility of accidental deployment of the airbag/seat belt pre-tensioners, which could cause personal injury (see Section 25). Do not use a memory-saving device to preserve the PCM's memory when working on or near airbag system components.*

1   Some models feature an optional eight-way power seat system that allows the driver and passenger to adjust the front seats up, down, front up, front down, rear up, rear down, forward and rearward. The system consists of

the driver's power seat switch, the passenger's power seat switch, the driver's power seat track, the passenger's power seat track and, on some models, the optional power lumbar adjusters.

2   The power seat switches are located on the outboard side of the seat cushions, on the seat cushion side panels. If the vehicle is equipped with the optional power lumbar adjusters, the lumbar switches are located on the power seat switch assemblies. Each switch assembly is attached to the seat side panel by two Torx screws. Refer to your owner's manual for instructions regarding switch functions. Individual switches in the power seat switch assemblies cannot be repaired or replaced separately. If one of the switches in a power seat switch assembly fails, replace the entire switch assembly.

3   The seats are powered by three reversible motors that are attached to the upper half of the power seat track assembly. These motors are controlled by the power seat switches on the sides of the seats. Each switch changes the direction of seat travel by reversing polarity to the drive motor. The motors are an integral part of the power seat track assembly and cannot be repaired or replaced separately. If a motor fails, replace the power seat track assembly.

4   The optional power lumbar adjuster and motor are located on the back of the seat, under the seat trim cover and padding, where they're attached to a molded plastic back panel and to the seat back frame. The power lumbar adjuster and motor cannot be repaired or replaced separately from the seat back frame. If either the adjuster or the motor fails, replace the entire seat back frame unit.

5   Diagnosis is usually a simple matter, using the following procedures.

6   Look under the seat for any object which may be preventing the seat from moving.

7   If the seat won't work at all, check the fuse, which is located in the engine compartment fuse and relay box.

8   With the engine off to reduce the noise level, operate the seat controls in all directions and listen for sound coming from the seat motors.

9   If the motor doesn't work or make noise, check for voltage at the motor while an assistant operates the switch.

10   If the motor is getting voltage but doesn't run, test it off the vehicle with jumper wires. If it still doesn't work, replace it. The individual components are not available separately. The whole power seat track must be purchased as an assembly.

11   If the motor isn't getting voltage, remove the seat side panel to access the switch and check for voltage. If there's no voltage at the switch, check the wiring between the fuse and the switch. If there's voltage at the switch, check for a short or open in the wiring between the switch and the motor. If that circuit is okay, replace the switch. No further testing is recommended. If the power seat system is still malfunctioning at this point, have the system checked out by a dealer service department.

24   Daytime Running Lights (DRL) - general information

Canadian models are equipped with Daytime Running Lights (DRL). The DRL system illuminates the headlights whenever the engine is running and the parking brake is disengaged. The DRL system provides reduced power to the headlights so that they won't be too bright for daytime use and it prolongs the headlight bulbs' service life. It does this by modulating the pulse-width of the power to the headlights. The duration and interval of these power pulses is programmed into the Front Control Module (FCM), which is located on the instrument cluster. If you want to alter the pulse-width, you must have it done by a dealer service department.

25   Airbag system - general information

These models are equipped with a Supplemental Restraint System (SRS), more commonly called an airbag system. There are at least two airbags, one for the driver and one for the front seat passenger, on all models. The SRS system is designed to protect the driver and passenger from serious injury in the event of a head-on or frontal collision. The airbag control module is located on the transmission tunnel, below the center of the instrument panel. Some models are also equipped with optional side curtain airbags. Vehicles with this option can be identified by the "SRS - AIRBAG" logo printed on the headliner above the B-pillar.

Airbag modules

The airbag module houses the airbag and the inflator unit. The inflator unit is mounted on the back of the housing over a hole through which gas is expelled, inflating the bag almost instantaneously when an electrical signal is received from the airbag control module. On the driver's airbag, the specially wound wire that carries this signal to the module is called a clockspring. The clockspring is a flat, ribbon-like electrically conductive tape that winds and unwinds as the steering wheel is turned so it can transmit an electrical signal regardless of wheel position. The procedure for removing the driver's airbag is part of *Steering wheel - removal and installation* in Chapter 10.

The passenger airbag is located in the top of the dashboard, above the glove box. There's also a passenger airbag ON/OFF switch located at the lower right corner of the center instrument panel bezel. This switch allows you to deactivate the passenger airbag if you're transporting an infant or a young child in a child safety seat. We don't recommend removing the passenger airbag because there is no reason to do so unless it has been activated during an accident and needs to be replaced afterward. Although the electrical

connector for the passenger airbag must be disconnected when removing the instrument panel (see Chapter 11), the airbag module itself does not need to be removed.

Optional side-curtain airbags, if equipped, are located on each roof side rail, above the headliner, and they extend from the A-pillar to the C-pillar on quad cab models. Again, we don't recommend trying to remove the side-curtain airbags because there is no reason to do so unless they've been deployed in an accident and must be replaced.

Airbag Control Module (ACM) and Side Impact Airbag Control Modules (SIACMs)

The Airbag Control Module (ACM) is the microprocessor that monitors and operates the airbag system. The ACM checks the system every time the vehicle is started. When you start the car, an AIRBAG indicator light comes on for about six seconds, then goes off, if the system is operating properly. If there is a fault in the system, the ACM stores a Diagnostic Trouble Code (DTC) and illuminates the AIRBAG indicator light, which remains on until the problem is repaired and the ACM memory is cleared of any DTCs. If the AIRBAG indicator light comes on at any time other than the bulb test and remains on, or doesn't come on at all, there's a problem in the system. A DRBIII scan tool is the only means by which the system can be diagnosed. Take the vehicle to your dealer immediately and have the system professionally diagnosed and repaired.

The ACM controls the operation of the standard driver and passenger airbags. Vehicles with optional side-curtain airbags are also equipped with Side Impact Airbag Control Modules (SIACMs). There are two SIACMs, one for each side-curtain airbag. The SIACMs are located behind the B-pillar trim, above the outboard front seat belt retractor inside each B-pillar.

Servicing components near the SRS system

There are times when you need to remove the steering wheel, the instrument cluster, the radio, the heater/air conditioning control assembly or other components on or near the dashboard. At these times you'll be working around components and wire harnesses for the SRS system. Do not use electrical test equipment on airbag system wires; it could cause the airbag(s) to deploy. **ALWAYS DISABLE THE SRS SYSTEM BEFORE WORKING NEAR THE SRS SYSTEM COMPONENTS OR RELATED WIRING.**

Disabling the system

Whenever working in the vicinity of the steering wheel, steering column, floor console or near other components of the airbag system, the system should be disarmed. To do this, perform the following steps:

a) *Turn the ignition switch to the OFF position.*

b) *Disconnect the cable(s) from the negative battery terminal(s) (see Chapter 5).*
c) *WAIT FOR AT LEAST TWO MINUTES before beginning work (during this two-minute interval the capacitor that provides emergency back-up power to the system loses its charge).*

Enabling the system

To enable the airbag system, perform the following steps:

a) *Turn the ignition switch to the OFF position.*
b) *Connect the cable(s) to the negative battery terminal(s).*

c) *Without putting your body in front of either airbag, turn the ignition switch to the ON position. Note whether the airbag indicator light glows for six seconds, then goes out. If it does, this indicates that the system is functioning properly.*

26   Wiring diagrams - general information

Since it isn't possible to include all wiring diagrams for every year covered by this manual, the following diagrams are those that are typical and most commonly needed.

Prior to troubleshooting any circuits, check the fuse and circuit breakers (if equipped) to make sure they are in good condition. Make sure the battery is properly charged and has clean, tight cable connections (see Chapter 1).

When checking the wiring system, make sure that all electrical connectors are clean, with no broken or loose pins. When disconnecting an electrical connector, do not pull on the wires, only on the connector housings.

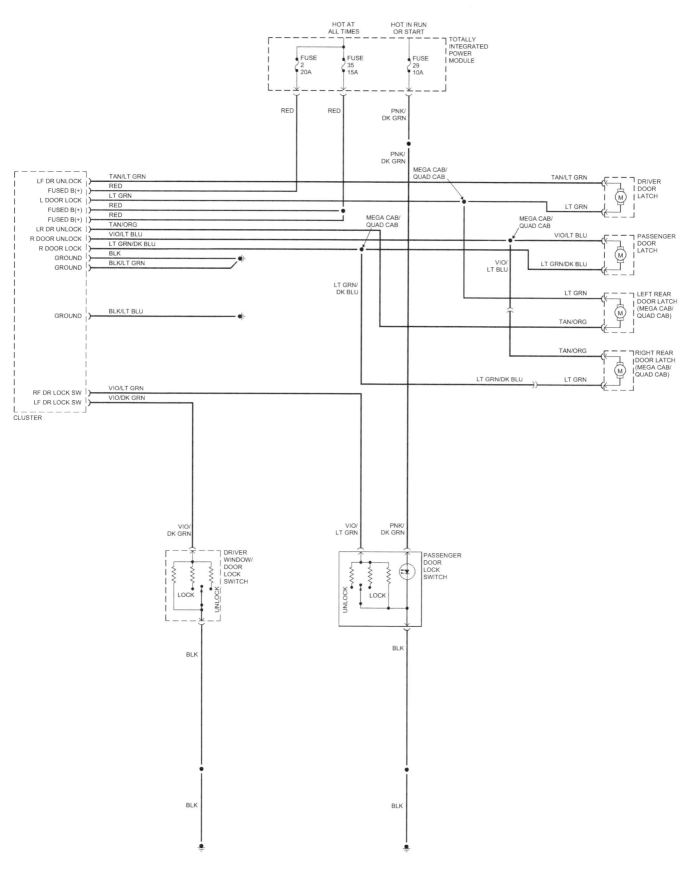

Power door locks - 2009 2500/3500 models

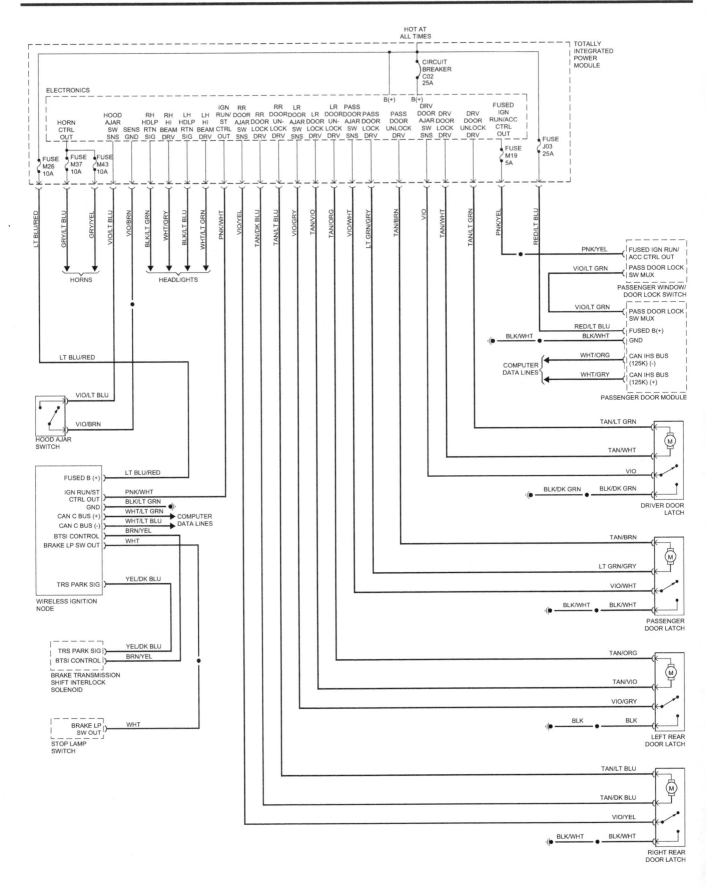

Power door locks - 2009 1500 models

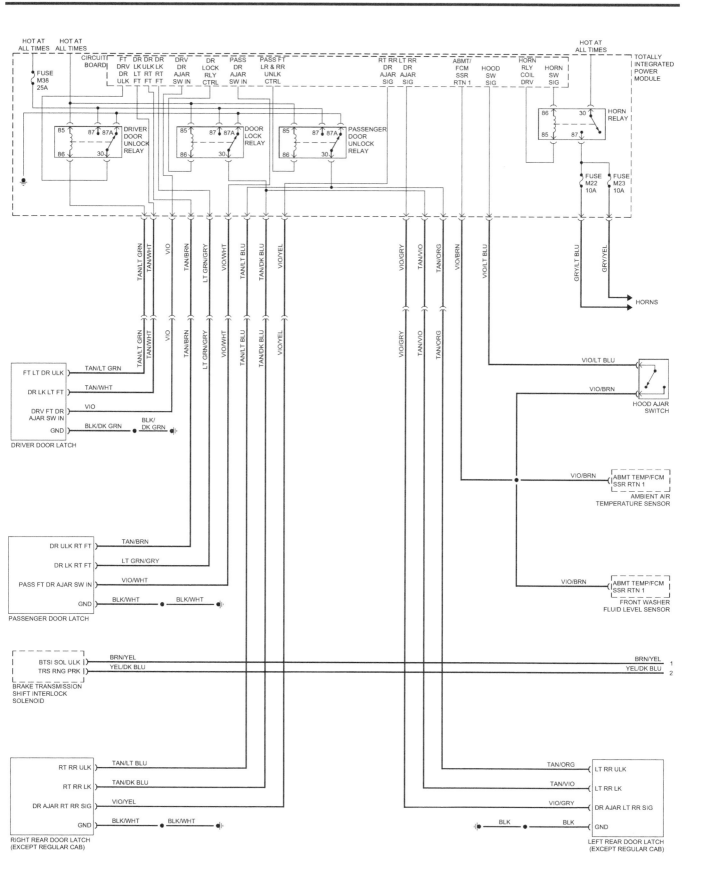

Power door locks - 2010 and later models (1 of 2)

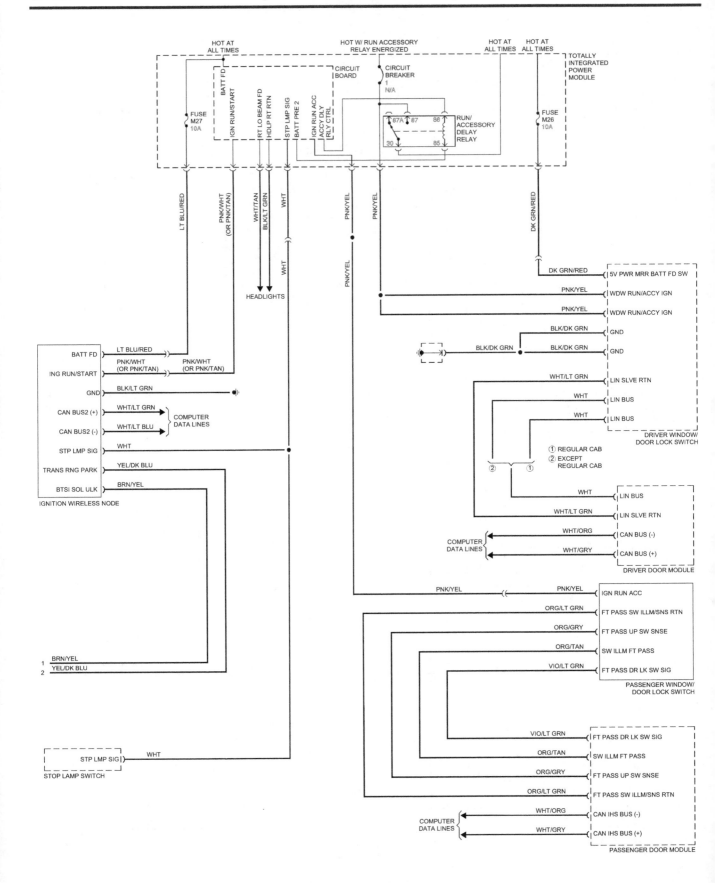

Power door locks - 2010 and later models (2 of 2)

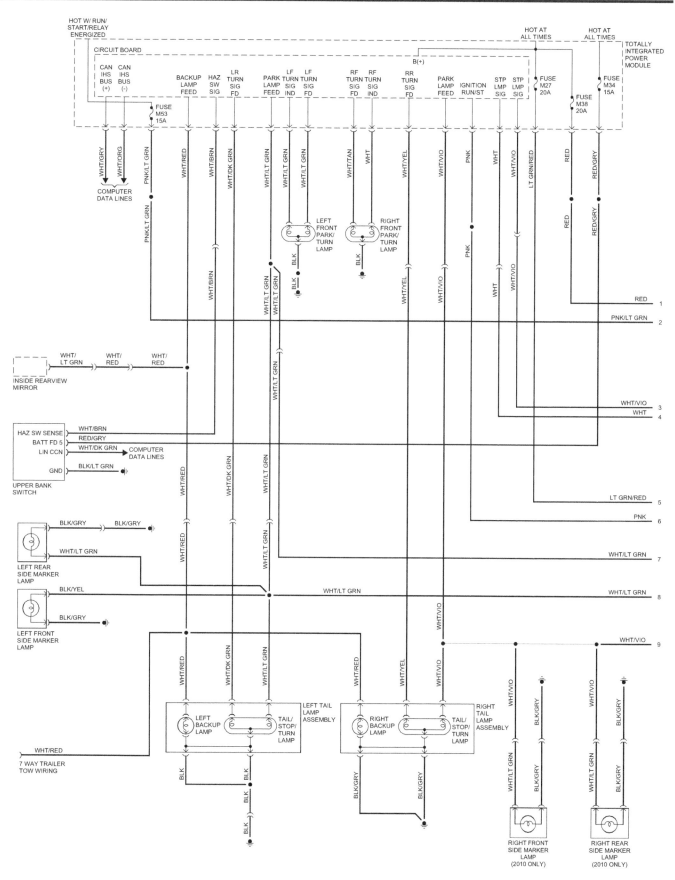

Exterior lighting system - 2009 and 2010 models (1 of 2)

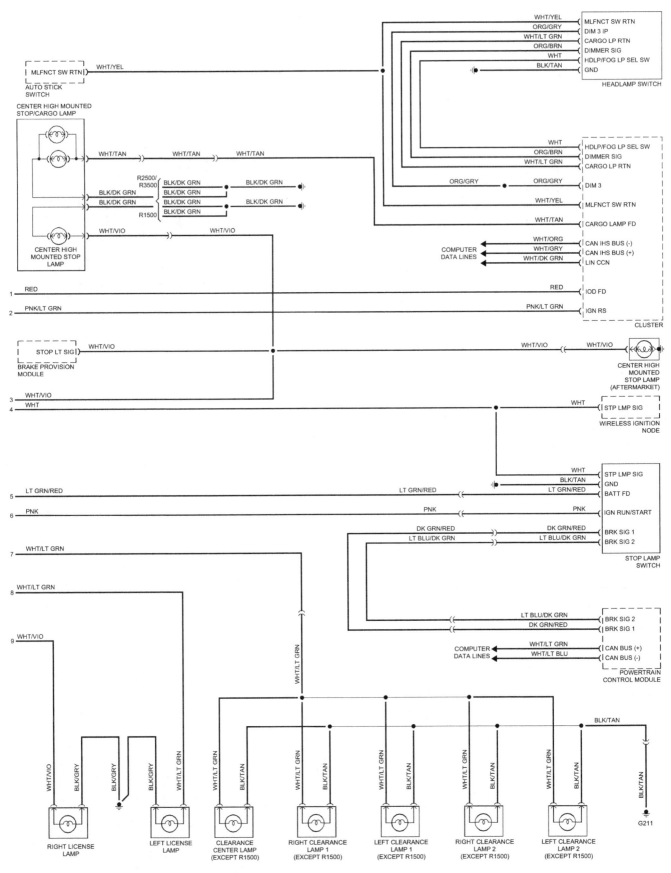

Exterior lighting system - 2009 and 2010 models (2 of 2)

…

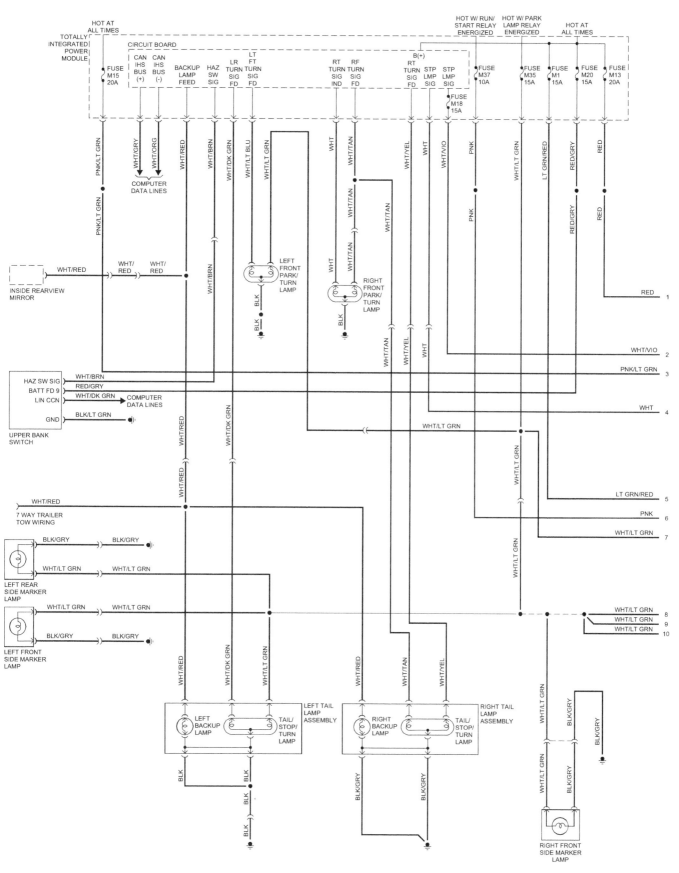

Exterior lighting system - 2011 and later models (1 of 2)

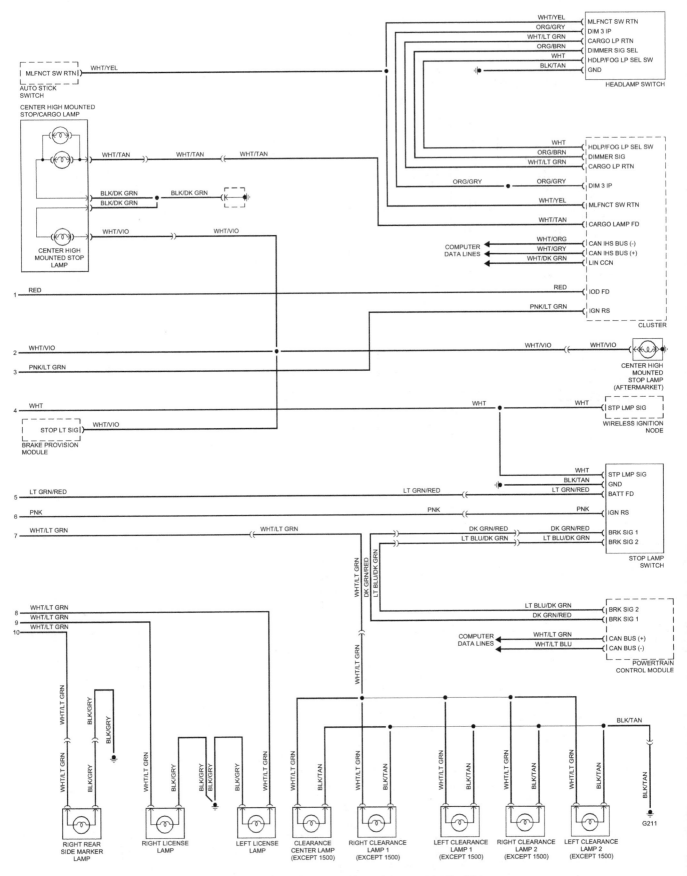

Exterior lighting system - 2011 and later models (2 of 2)

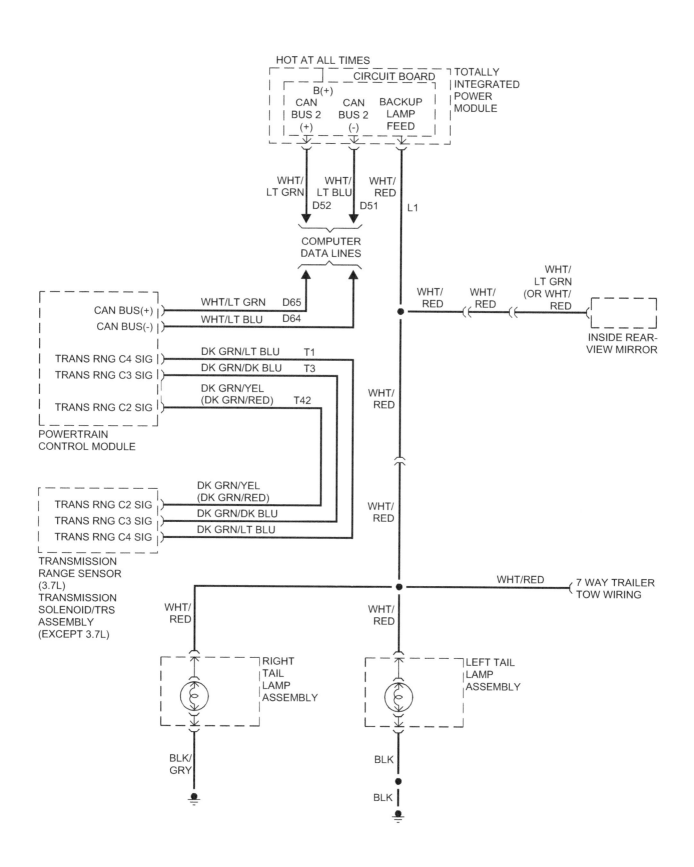

Back-up light system

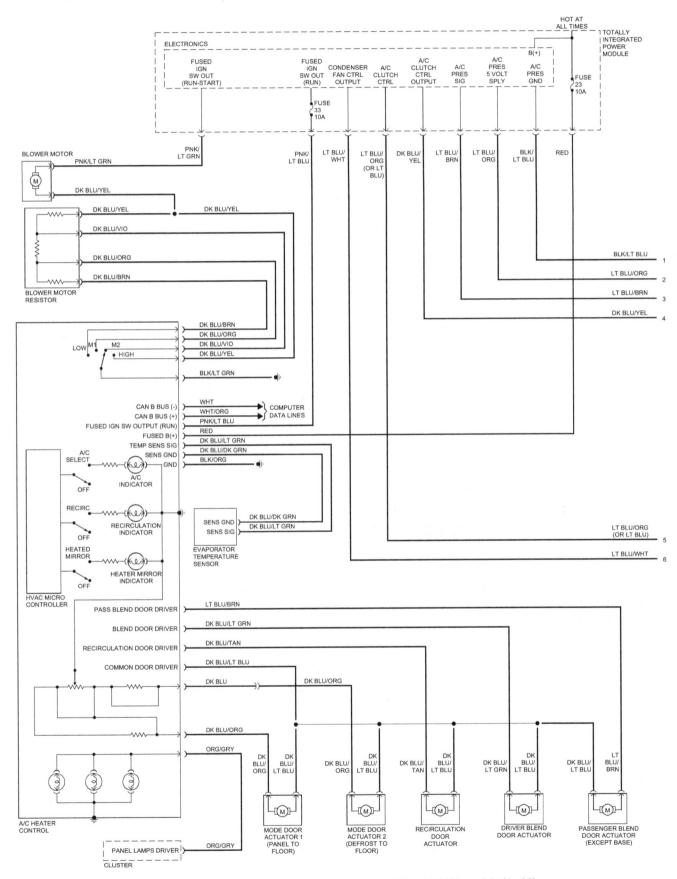

Heating and air conditioning system (manual) - 2009 2500/3500 models (1 of 2)

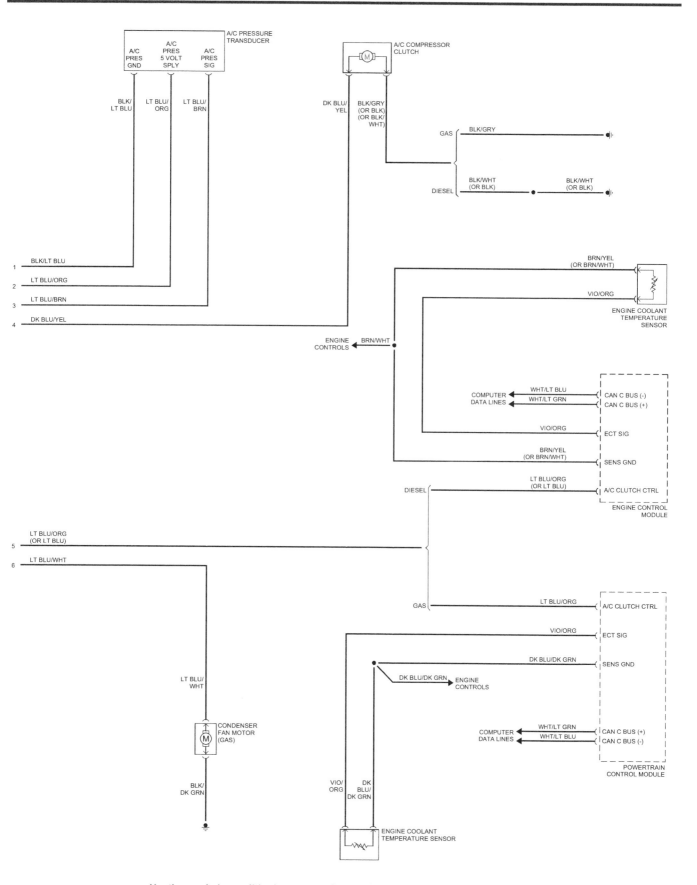

Heating and air conditioning system (manual) - 2009 2500/3500 models (2 of 2)

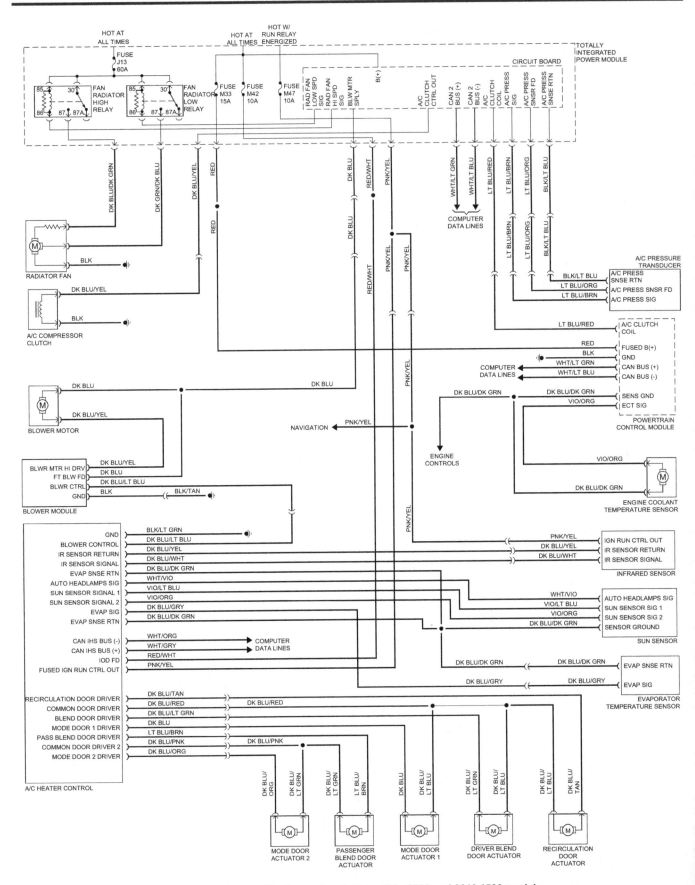

Heating and air conditioning system (automatic) - 2009 and 2010 1500 models

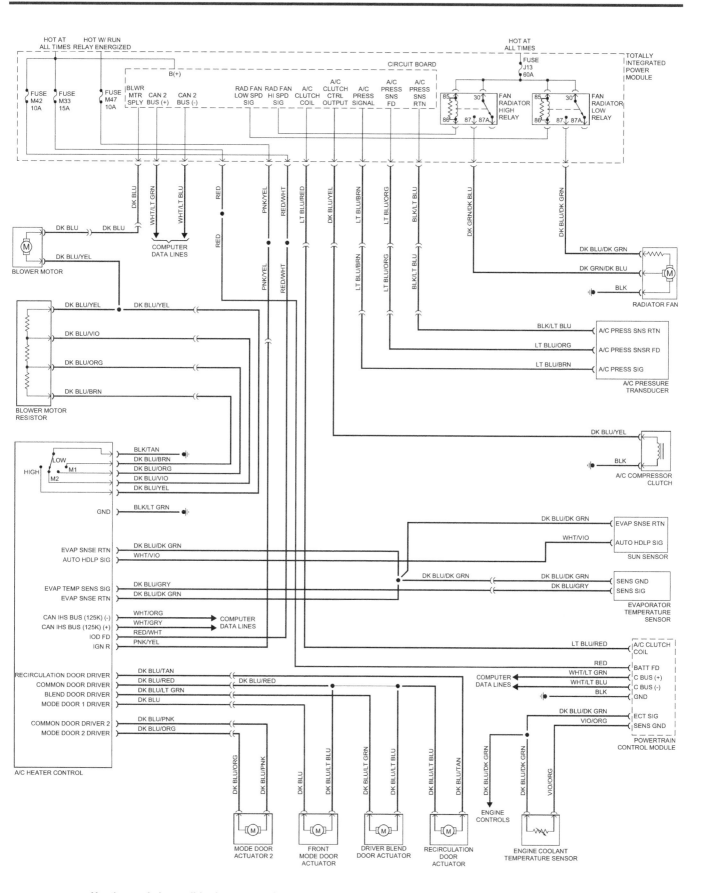

Heating and air conditioning system (manual) - 2009 and 2010 1500 models, 2010 2500/3500 models

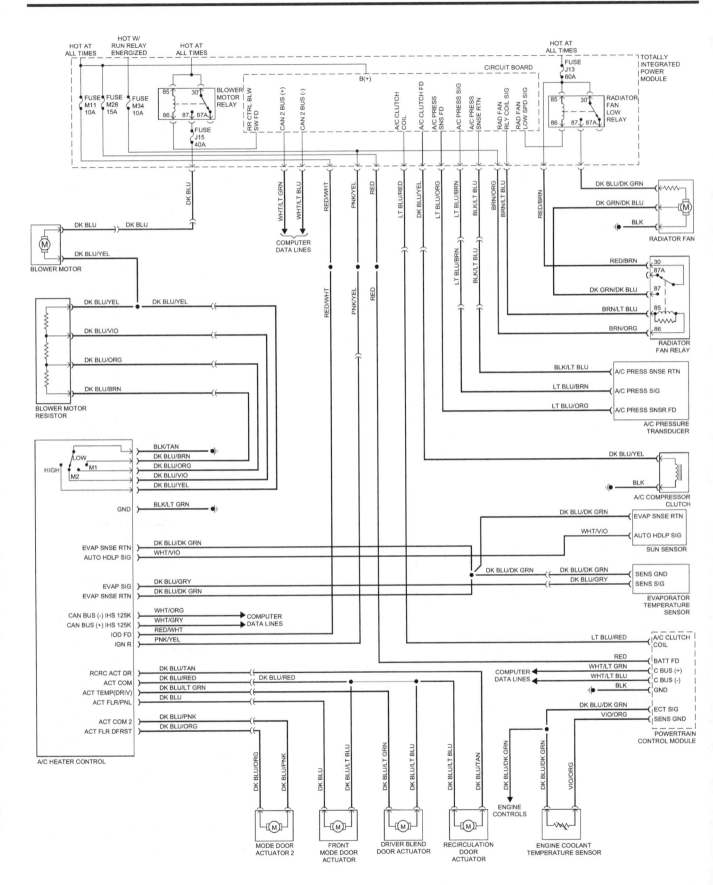

Heating and air conditioning system (manual) - 2011 and later models

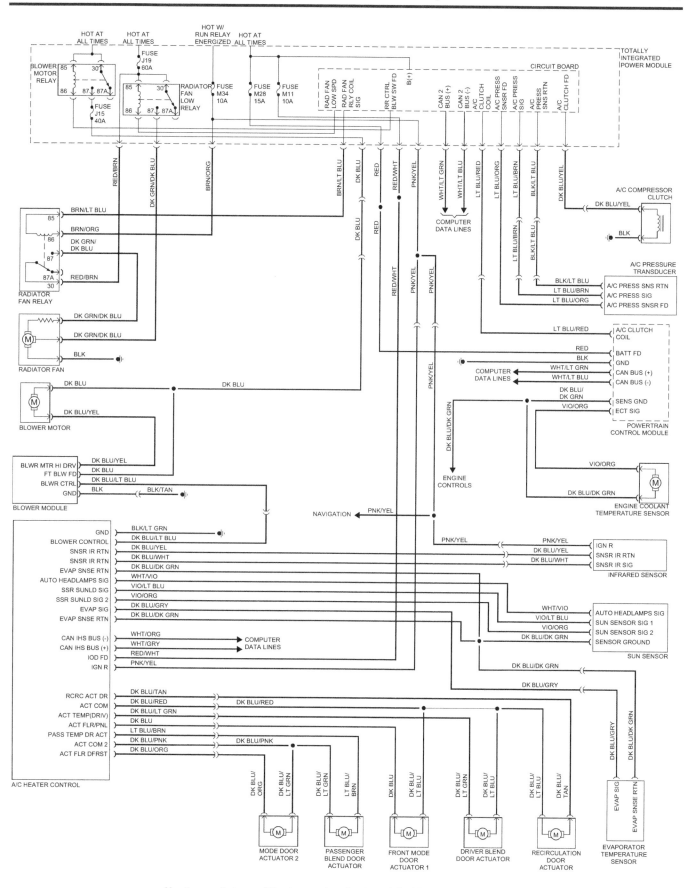

Heating and air conditioning system (automatic) - 2011 and later models

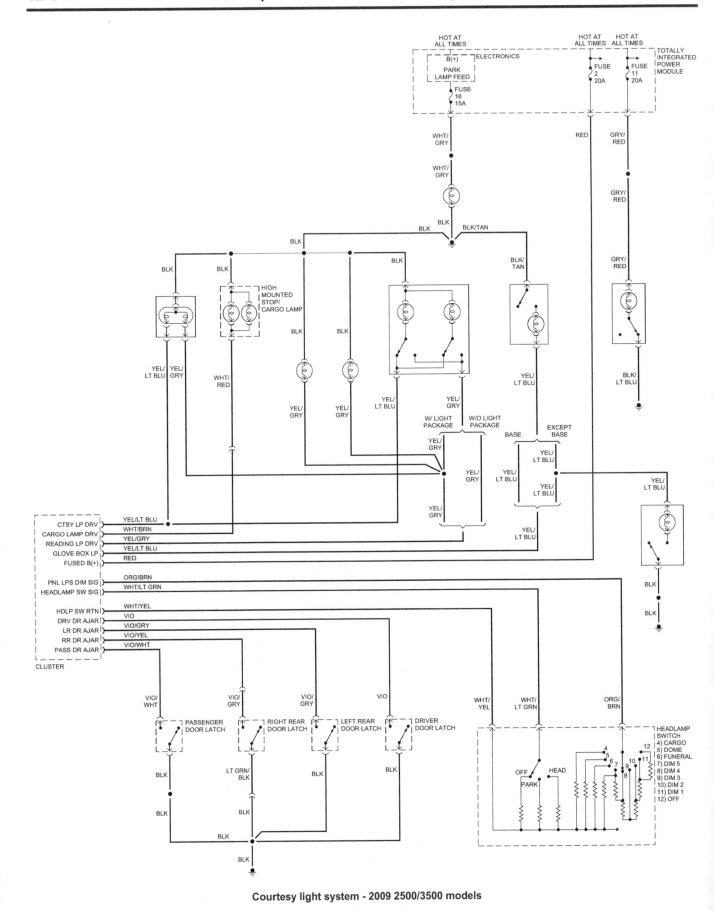

Courtesy light system - 2009 2500/3500 models

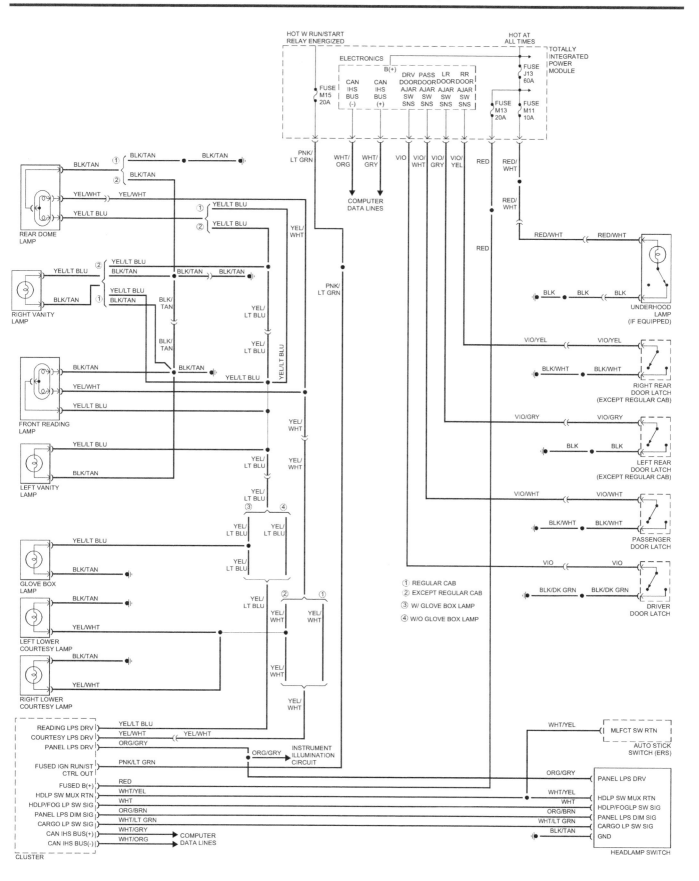

Courtesy light system - all 1500 models, 2010 and later 2500/3500 models

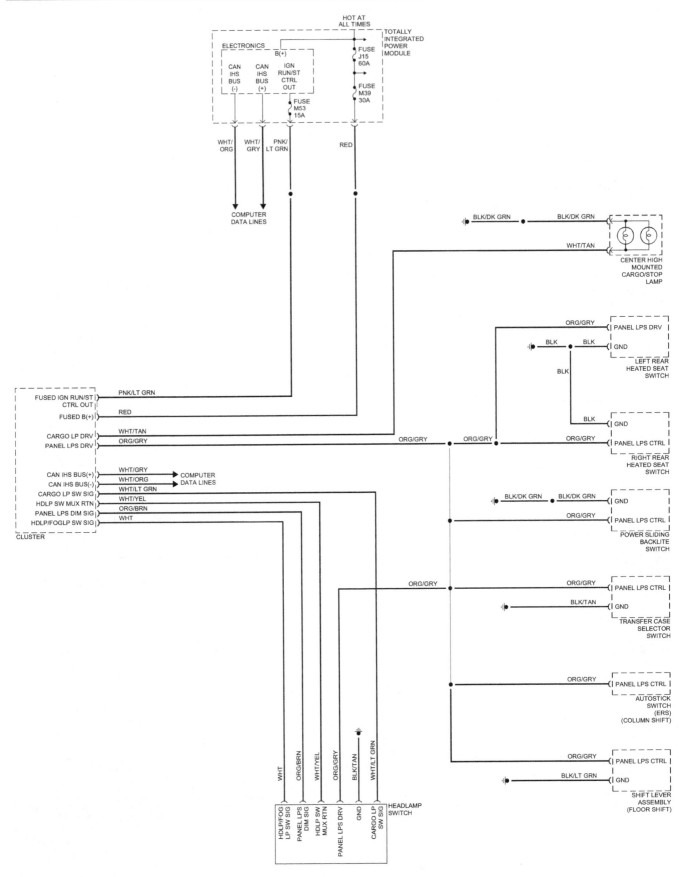

Switch illumination - 2009 2500/3500 models

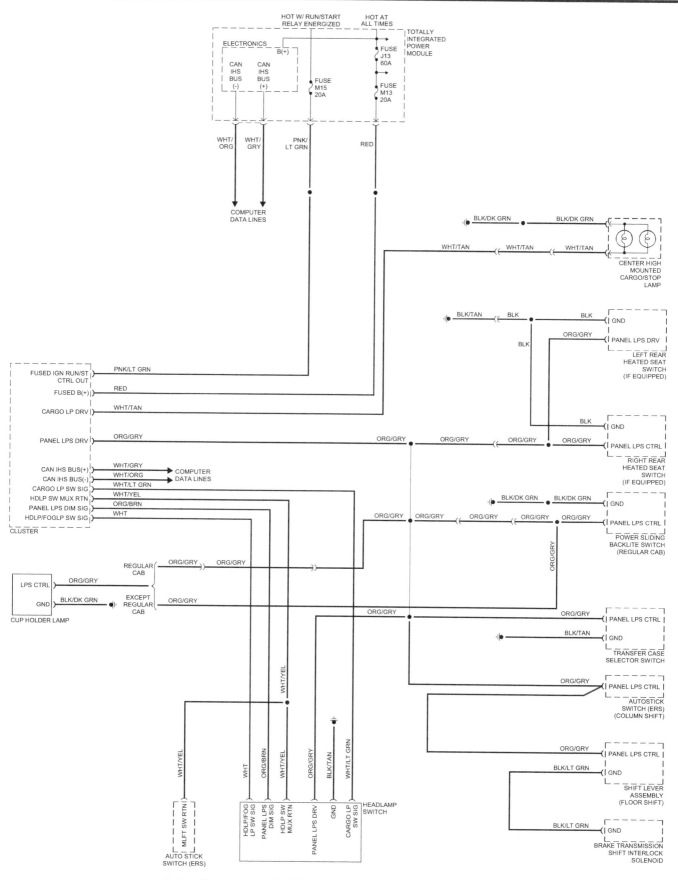

Switch illumination - all 1500 models, 2010 and later 2500/3500 models

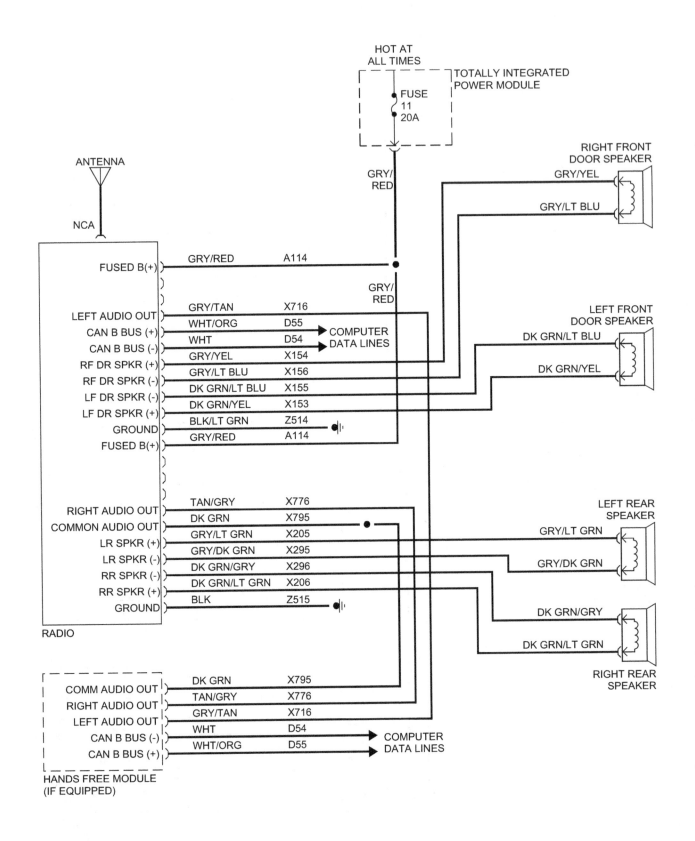

Audio system - 2009 2500/3500 models

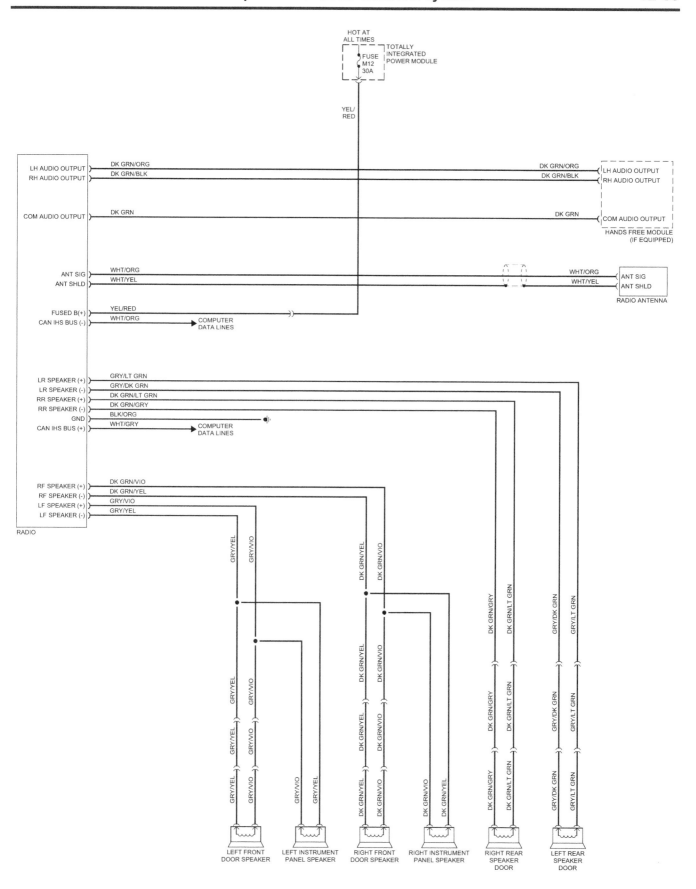

Audio system - all 1500 models, 2010 and later 2500/3500 models

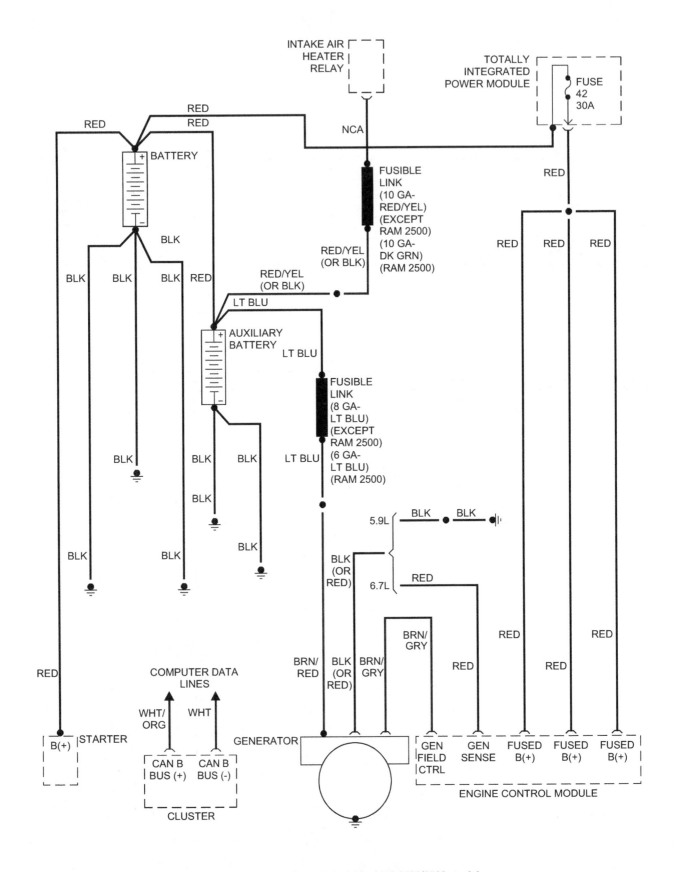

Charging system (diesel engine) - 2009 2500/3500 models

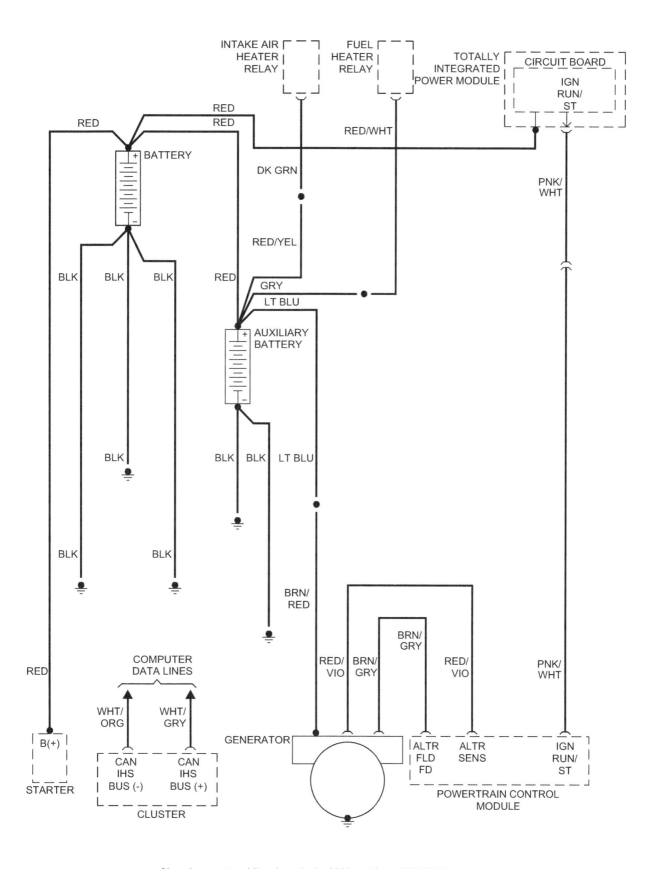

Charging system (diesel engine) - 2010 and later 2500/3500 models

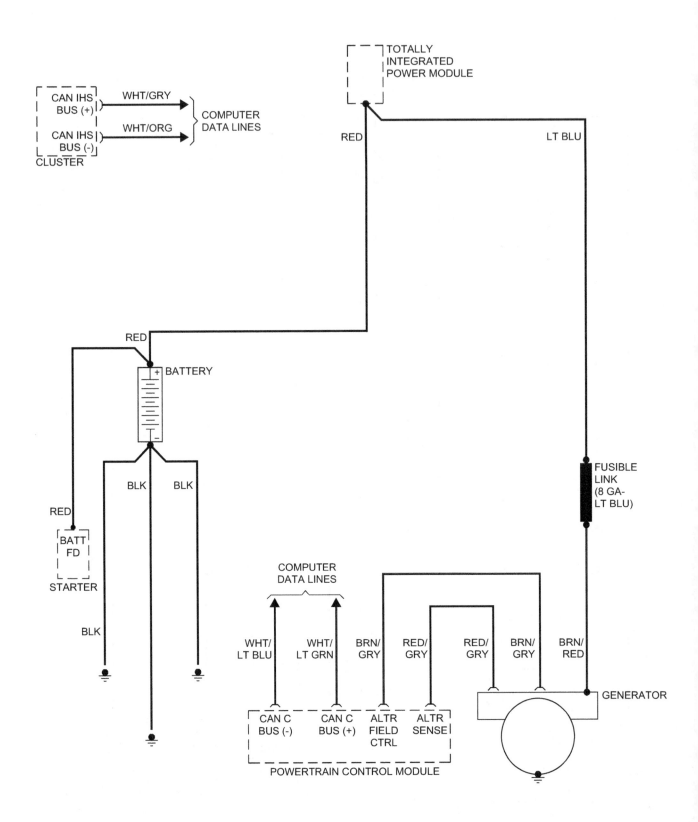

Charging system (gasoline engines) - all 1500 models, 2010 and later 2500/3500 models

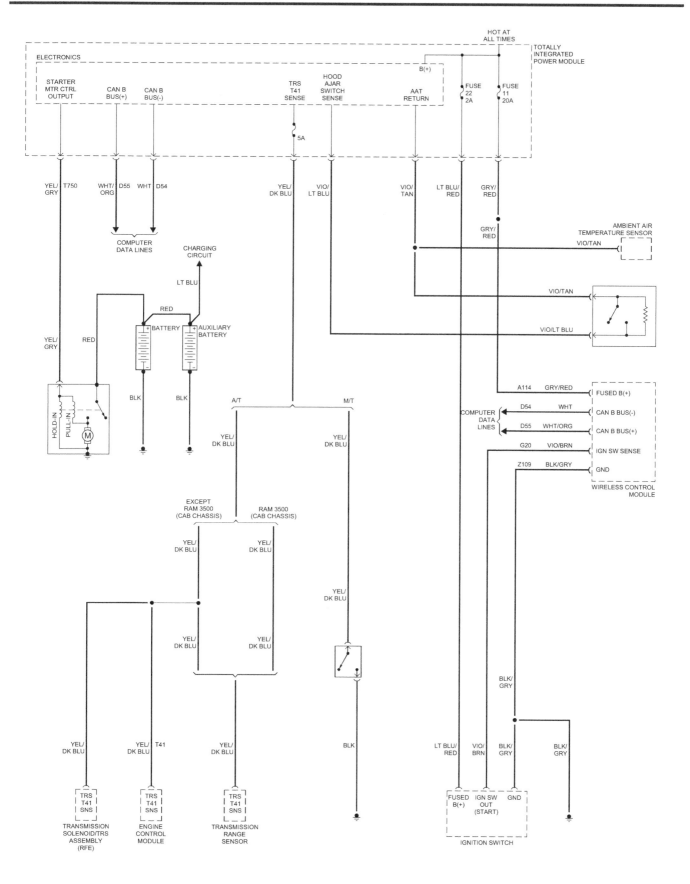

Starting system (diesel engine) - 2009 2500/3500 models

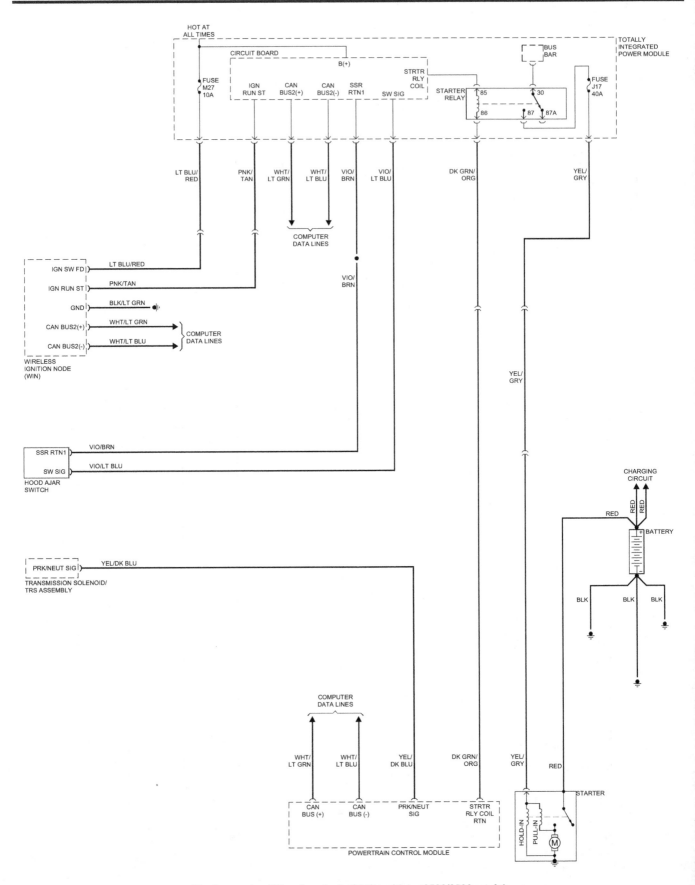

Starting system (diesel engine) - 2010 and later 2500/3500 models

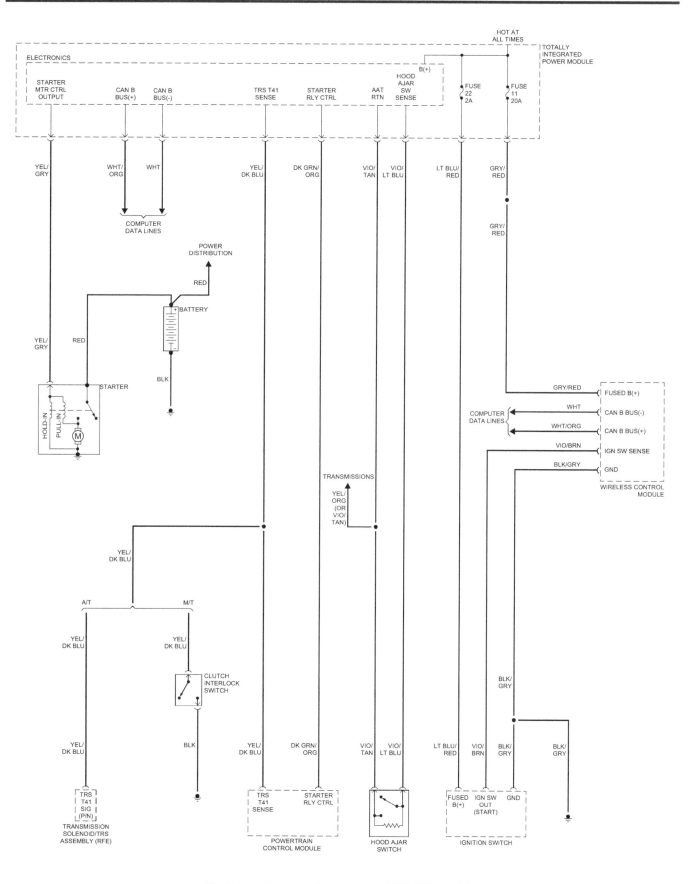

Starting system (gasoline engines) - 2009 1500 models

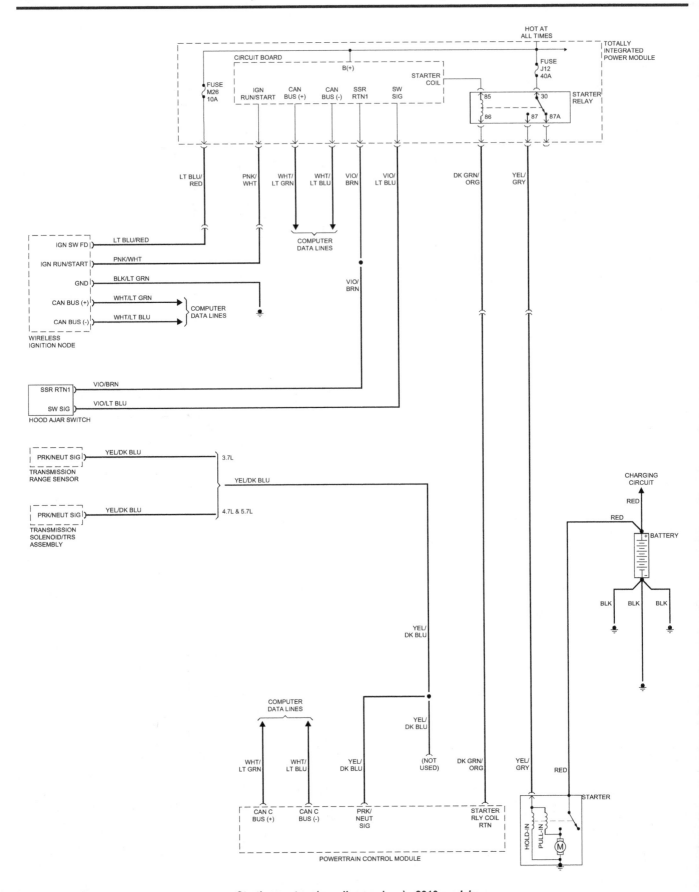

Starting system (gasoline engines) - 2010 models

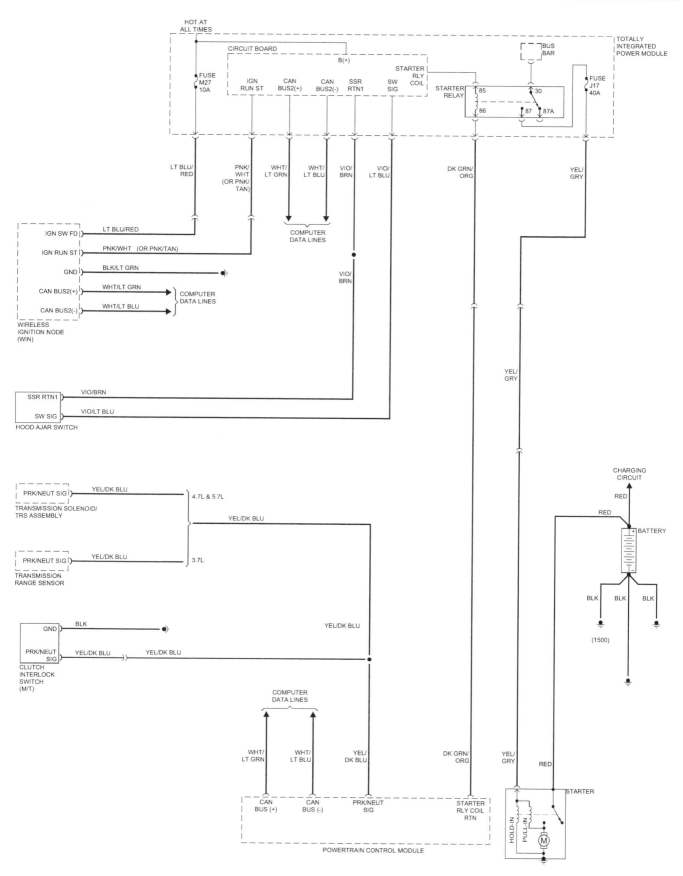

Starting system (gasoline engines) - 2011 and later models

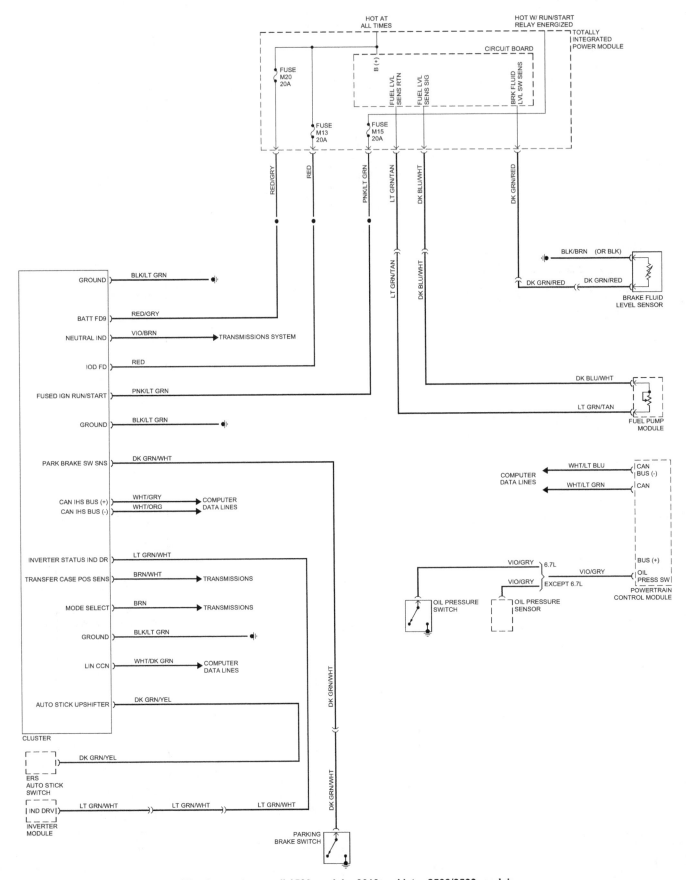

Warning systems - all 1500 models, 2010 and later 2500/3500 models

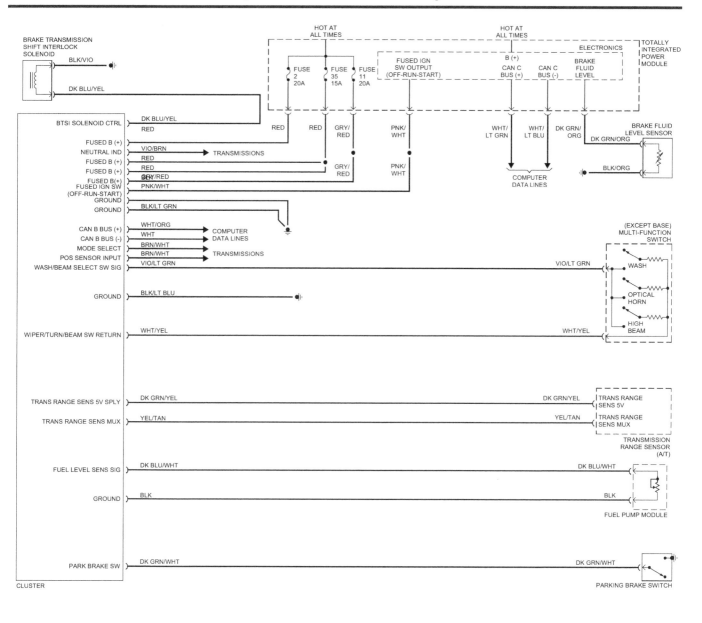

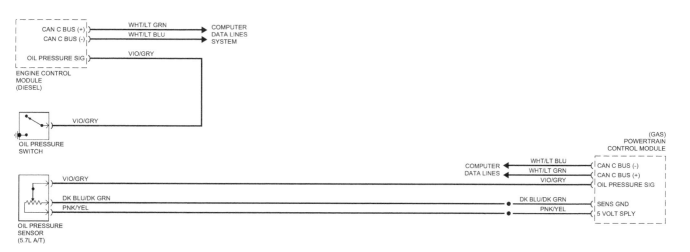

Warning systems - 2009 2500/3500 models

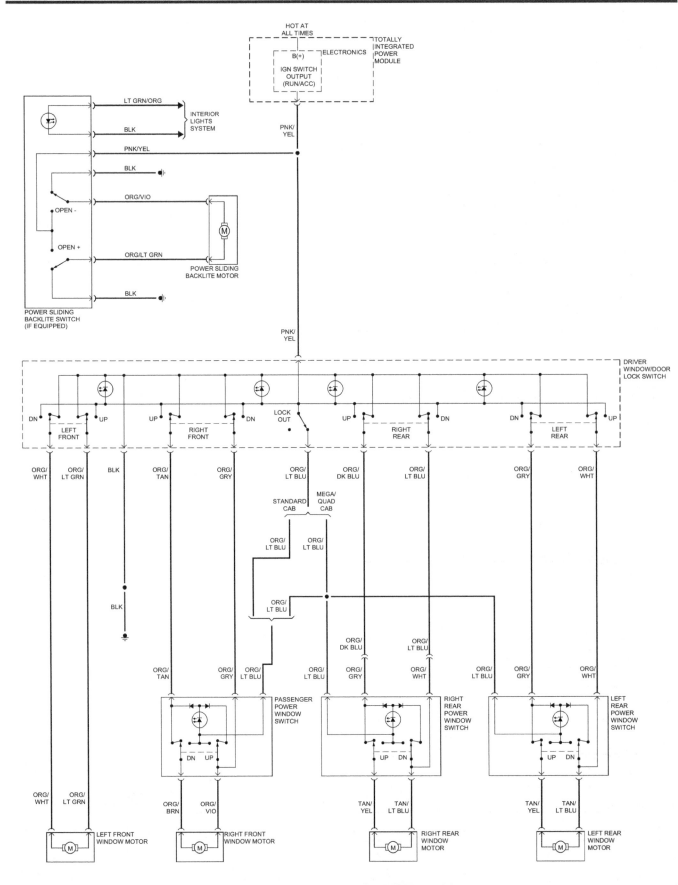

Power window system - 2009 2500/3500 models

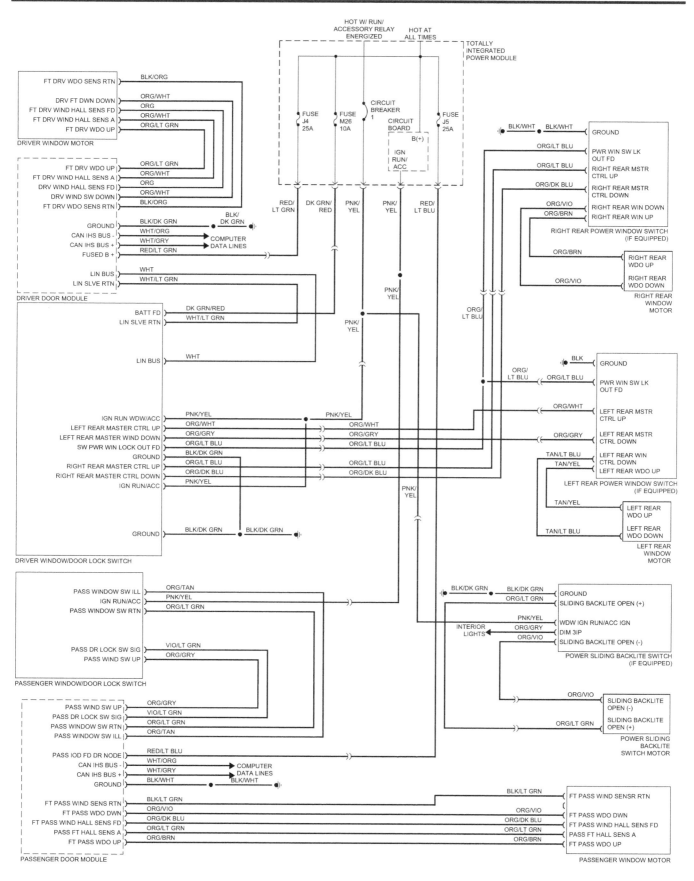

Power window system - all 1500 models, 2010 and later 2500/3500 models

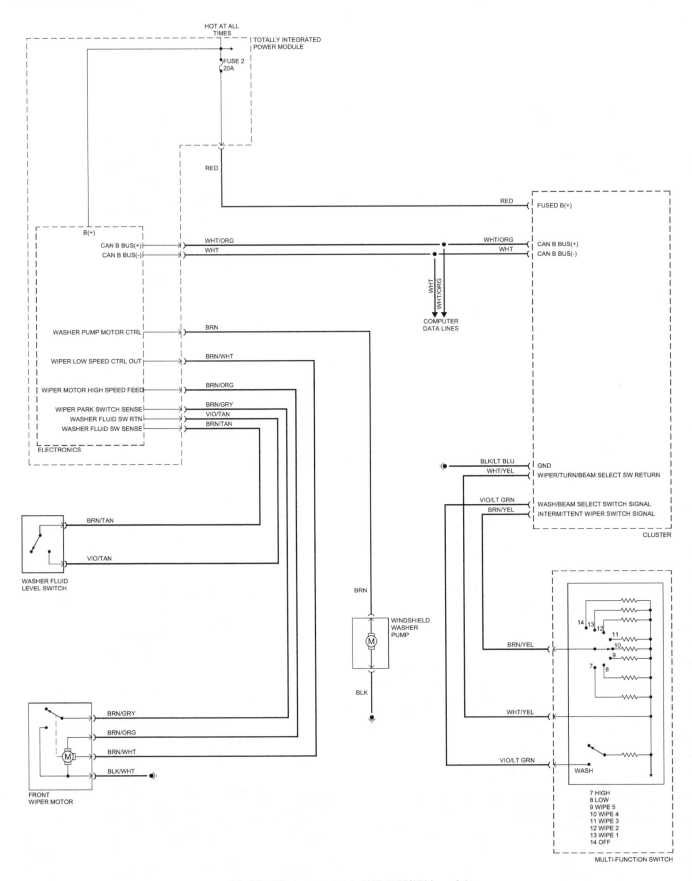

Windshield wiper system - 2009 2500/3500 models

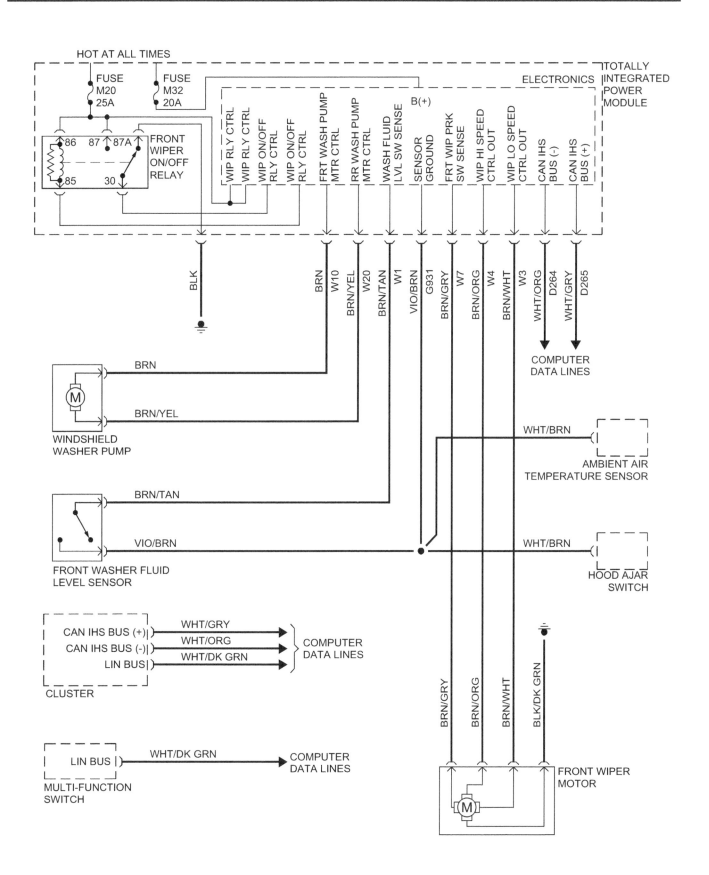

Windshield wiper system - 2009 1500 models

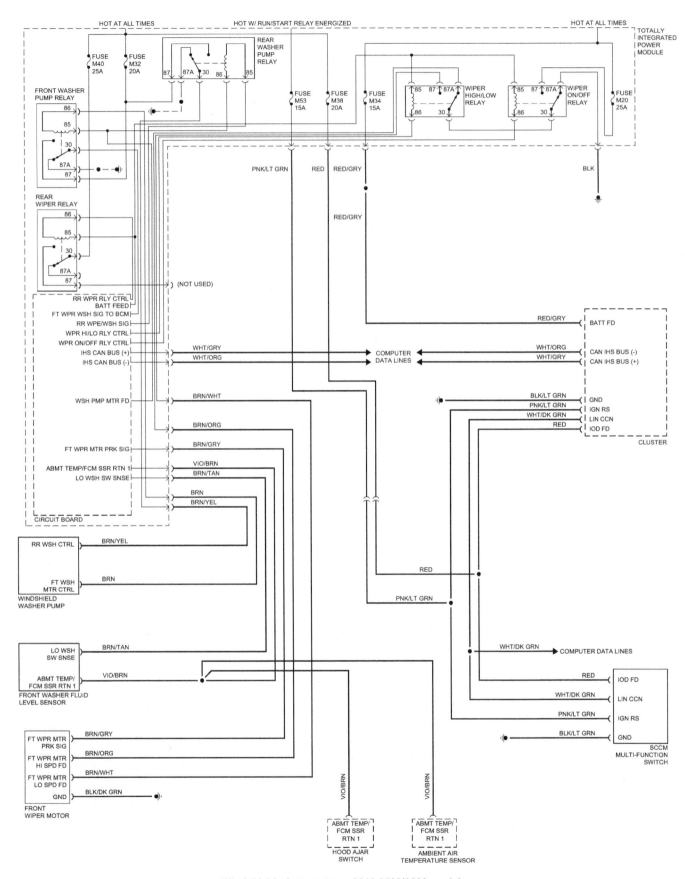

Windshield wiper system - 2010 2500/3500 models

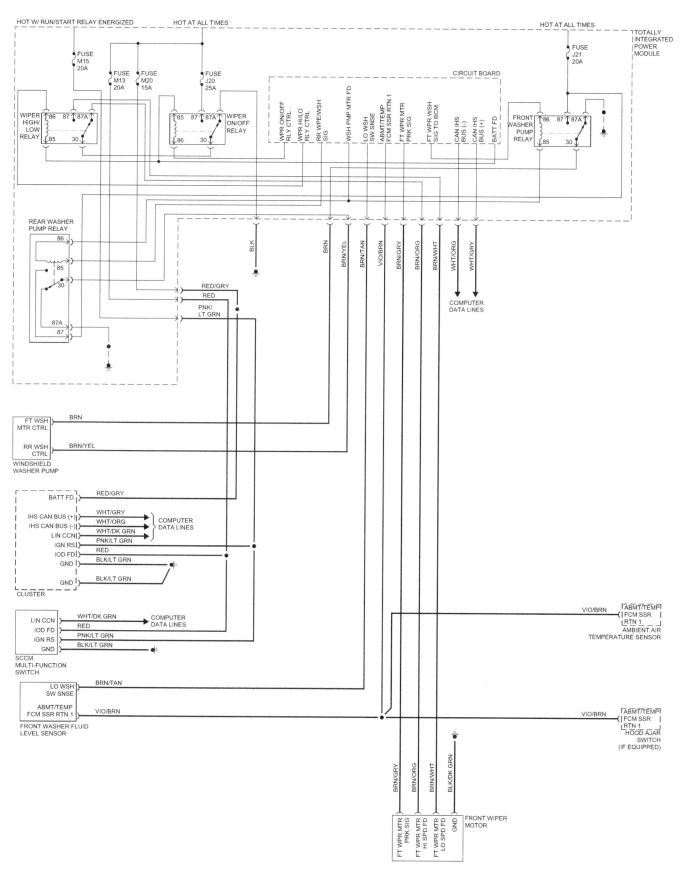

Windshield wiper system - 2010 1500 models, 2011 and later 2500/3500 models

Notes

Index

Haynes Automotive Manuals

*NOTE: If you do not see a listing for your vehicle, please visit **haynes.com** for the latest product information and check out our **Online Manuals!***

ACURA
12020 Integra '86 thru '89 & Legend '86 thru '90
12021 Integra '90 thru '93 & Legend '91 thru '95
Integra '94 thru '00 - *see HONDA Civic (42025)*
MDX '01 thru '07 - *see HONDA Pilot (42037)*
12050 Acura TL all models '99 thru '08

AMC
14020 Mid-size models '70 thru '83
14025 (Renault) Alliance & Encore '83 thru '87

AUDI
15020 4000 all models '80 thru '87
15025 5000 all models '77 thru '83
15026 5000 all models '84 thru '88
Audi A4 '96 thru '01 - *see VW Passat (96023)*
15030 Audi A4 '02 thru '08

AUSTIN-HEALEY
Sprite - *see MG Midget (66015)*

BMW
18020 3/5 Series '82 thru '92
18021 3-Series incl. Z3 models '92 thru '98
18022 3-Series incl. Z4 models '99 thru '05
18023 3-Series '06 thru '14
18025 320i all 4-cylinder models '75 thru '83
18050 1500 thru 2002 except Turbo '59 thru '77

BUICK
19010 Buick Century '97 thru '05
Century (front-wheel drive) - *see GM (38005)*
19020 Buick, Oldsmobile & Pontiac Full-size (Front-wheel drive) '85 thru '05 Buick Electra, LeSabre and Park Avenue; Oldsmobile Delta 88 Royale, Ninety Eight and Regency; Pontiac Bonneville
19025 Buick, Oldsmobile & Pontiac Full-size (Rear wheel drive) '70 thru '90 Buick Estate, Electra, LeSabre, Limited, Oldsmobile Custom Cruiser, Delta 88, Ninety-eight, Pontiac Bonneville, Catalina, Grandville, Parisienne
19027 Buick LaCrosse '05 thru '13
Enclave - *see GENERAL MOTORS (38001)*
Rainier - *see CHEVROLET (24072)*
Regal - *see GENERAL MOTORS (38010)*
Riviera - *see GENERAL MOTORS (38030, 38031)*
Roadmaster - *see CHEVROLET (24046)*
Skyhawk - *see GENERAL MOTORS (38015)*
Skylark - *see GENERAL MOTORS (38020, 38025)*
Somerset - *see GENERAL MOTORS (38025)*

CADILLAC
21015 CTS & CTS-V '03 thru '14
21030 Cadillac Rear Wheel Drive '70 thru '93
Cimarron - *see GENERAL MOTORS (38015)*
DeVille - *see GENERAL MOTORS (38031 & 38032)*
Eldorado - *see GENERAL MOTORS (38030)*
Fleetwood - *see GENERAL MOTORS (38031)*
Seville - *see GM (38030, 38031 & 38032)*

CHEVROLET
10305 Chevrolet Engine Overhaul Manual
24010 Astro & GMC Safari Mini-vans '85 thru '05
24013 Aveo '04 thru '11
24015 Camaro V8 all models '70 thru '81
24016 Camaro all models '82 thru '92
24017 Camaro & Firebird '93 thru '02
Cavalier - *see GENERAL MOTORS (38016)*
Celebrity - *see GENERAL MOTORS (38005)*
24018 Camaro '10 thru '15
24020 Chevelle, Malibu & El Camino '69 thru '87
Cobalt - *see GENERAL MOTORS (38017)*
24024 Chevette & Pontiac T1000 '76 thru '87
Citation - *see GENERAL MOTORS (38020)*
24027 Colorado & GMC Canyon '04 thru '12
24032 Corsica & Beretta all models '87 thru '96
24040 Corvette all V8 models '68 thru '82
24041 Corvette all models '84 thru '96
24042 Corvette all models '97 thru '13
24044 Cruze '11 thru '19
24045 Full-size Sedans Caprice, Impala, Biscayne, Bel Air & Wagons '69 thru '90
24046 Impala SS & Caprice and Buick Roadmaster '91 thru '96
Impala '00 thru '05 - *see LUMINA (24048)*
24047 Impala & Monte Carlo all models '06 thru '11
Lumina '90 thru '94 - *see GM (38010)*
24048 Lumina & Monte Carlo '95 thru '05
Lumina APV - *see GM (38035)*
24050 Luv Pick-up all 2WD & 4WD '72 thru '82
24051 Malibu '13 thru '19
24055 Monte Carlo all models '70 thru '88
Monte Carlo '95 thru '01 - *see LUMINA (24048)*
24059 Nova all V8 models '69 thru '79
24060 Nova and Geo Prizm '85 thru '92
24064 Pick-ups '67 thru '87 - Chevrolet & GMC
24065 Pick-ups '88 thru '98 - Chevrolet & GMC
24066 Pick-ups '99 thru '06 - Chevrolet & GMC
24067 Chevrolet Silverado & GMC Sierra '07 thru '14
24068 Chevrolet Silverado & GMC Sierra '14 thru '19
24070 S-10 & S-15 Pick-ups '82 thru '93, Blazer & Jimmy '83 thru '94,
24071 S-10 & Sonoma Pick-ups '94 thru '04, including Blazer, Jimmy & Hombre
24072 Chevrolet TrailBlazer, GMC Envoy & Oldsmobile Bravada '02 thru '09
24075 Sprint '85 thru '88 & Geo Metro '89 thru '01
24080 Vans - Chevrolet & GMC '68 thru '96
24081 Chevrolet Express & GMC Savana Full-size Vans '96 thru '19

CHRYSLER
10310 Chrysler Engine Overhaul Manual
25015 Chrysler Cirrus, Dodge Stratus, Plymouth Breeze '95 thru '00
25020 Full-size Front-Wheel Drive '88 thru '93
K-Cars - *see DODGE Aries (30008)*
Laser - *see DODGE Daytona (30030)*
25025 Chrysler LHS, Concorde, New Yorker, Dodge Intrepid, Eagle Vision, '93 thru '97
25026 Chrysler LHS, Concorde, 300M, Dodge Intrepid, '98 thru '04
25027 Chrysler 300 '05 thru '18, Dodge Charger '06 thru '18, Magnum '05 thru '08 & Challenger '08 thru '18
25030 Chrysler & Plymouth Mid-size front wheel drive '82 thru '95
Rear-wheel Drive - *see Dodge (30050)*
25035 PT Cruiser all models '01 thru '10
25040 Chrysler Sebring '95 thru '06, Dodge Stratus '01 thru '06 & Dodge Avenger '95 thru '00
25041 Chrysler Sebring '07 thru '10, 200 '11 thru '17 Dodge Avenger '08 thru '14

DATSUN
28005 200SX all models '80 thru '83
28012 240Z, 260Z & 280Z Coupe '70 thru '78
28014 280ZX Coupe & 2+2 '79 thru '83
300ZX - *see NISSAN (72010)*
28018 510 & PL521 Pick-up '68 thru '73
28020 510 all models '78 thru '81
28022 620 Series Pick-up all models '73 thru '79
720 Series Pick-up - *see NISSAN (72030)*

DODGE
400 & 600 - *see CHRYSLER (25030)*
30008 Aries & Plymouth Reliant '81 thru '89
30010 Caravan & Plymouth Voyager '84 thru '95
30011 Caravan & Plymouth Voyager '96 thru '02
30012 Challenger & Plymouth Sapporro '78 thru '83
30013 Caravan, Chrysler Voyager & Town & Country '03 thru '07
30014 Grand Caravan & Chrysler Town & Country '08 thru '18
30016 Colt & Plymouth Champ '78 thru '87
30020 Dakota Pick-ups all models '87 thru '96
30021 Durango '98 & '99 & Dakota '97 thru '99
30022 Durango '00 thru '03 & Dakota '00 thru '04
30023 Durango '04 thru '09 & Dakota '05 thru '11
30025 Dart, Demon, Plymouth Barracuda, Duster & Valiant 6-cylinder models '67 thru '76
30030 Daytona & Chrysler Laser '84 thru '89
Intrepid - *see CHRYSLER (25025, 25026)*
30034 Neon all models '95 thru '99
30035 Omni & Plymouth Horizon '78 thru '90
30036 Dodge & Plymouth Neon '00 thru '05
30040 Pick-ups full-size models '74 thru '93
30042 Pick-ups full-size models '94 thru '08
30043 Pick-ups full-size models '09 thru '18
30045 Ram 50/D50 Pick-ups & Raider and Plymouth Arrow Pick-ups '79 thru '93
30050 Dodge/Plymouth/Chrysler RWD '71 thru '89
30055 Shadow & Plymouth Sundance '87 thru '94
30060 Spirit & Plymouth Acclaim '89 thru '95
30065 Vans - Dodge & Plymouth '71 thru '03

EAGLE
Talon - *see MITSUBISHI (68030, 68031)*
Vision - *see CHRYSLER (25025)*

FIAT
34010 124 Sport Coupe & Spider '68 thru '78
34025 X1/9 all models '74 thru '80

FORD
10320 Ford Engine Overhaul Manual
10355 Ford Automatic Transmission Overhaul
11500 Mustang '64-1/2 thru '70 Restoration Guide
36004 Aerostar Mini-vans all models '86 thru '97
36006 Contour & Mercury Mystique '95 thru '00
36008 Courier Pick-up all models '72 thru '82
36012 Crown Victoria & Mercury Grand Marquis '88 thru '11
36014 Edge '07 thru '19 & Lincoln MKX '07 thru '18
36016 Escort & Mercury Lynx all models '81 thru '90
36020 Escort & Mercury Tracer '91 thru '02
36022 Escape '01 thru '17, Mazda Tribute '01 thru '11, & Mercury Mariner '05 thru '11
36024 Explorer & Mazda Navajo '91 thru '01
36025 Explorer & Mercury Mountaineer '02 thru '10
36026 Explorer '11 thru '17
36028 Fairmont & Mercury Zephyr '78 thru '83
36030 Festiva & Aspire '88 thru '97
36032 Fiesta all models '77 thru '80
36034 Focus all models '00 thru '11
36035 Focus '12 thru '14
36045 Fusion '06 thru '14 & Mercury Milan '06 thru '11
36048 Mustang V8 all models '64-1/2 thru '73
36049 Mustang II 4-cylinder, V6 & V8 models '74 thru '78
36050 Mustang & Mercury Capri '79 thru '93
36051 Mustang all models '94 thru '04
36052 Mustang '05 thru '14
36054 Pick-ups & Bronco '73 thru '79
36058 Pick-ups & Bronco '80 thru '96
36059 F-150 '97 thru '03, Expedition '97 thru '17, F-250 '97 thru '99, F-150 Heritage '04 & Lincoln Navigator '98 thru '17
36060 Super Duty Pick-ups & Excursion '99 thru '10
36061 F-150 full-size '04 thru '14
36062 Pinto & Mercury Bobcat '75 thru '80
36063 F-150 full-size '15 thru '17
36064 Super Duty Pick-ups '11 thru '16
36066 Probe all models '89 thru '92
Probe '93 thru '97 - *see MAZDA 626 (61042)*
36070 Ranger & Bronco II gas models '83 thru '92
36071 Ranger '93 thru '11 & Mazda Pick-ups '94 thru '09
36074 Taurus & Mercury Sable '86 thru '95
36075 Taurus & Mercury Sable '96 thru '07
36076 Taurus '08 thru '14, Five Hundred '05 thru '07, Mercury Montego '05 thru '07 & Sable '08 thru '09
36078 Tempo & Mercury Topaz '84 thru '94
36082 Thunderbird & Mercury Cougar '83 thru '88
36086 Thunderbird & Mercury Cougar '89 thru '97
36090 Vans all V8 Econoline models '69 thru '91
36094 Vans full size '92 thru '14
36097 Windstar '95 thru '03, Freestar & Mercury Monterey Mini-van '04 thru '07

GENERAL MOTORS
10360 GM Automatic Transmission Overhaul
38001 GMC Acadia '07 thru '16, Buick Enclave '08 thru '17, Saturn Outlook '07 thru '10 & Chevrolet Traverse '09 thru '17
38005 Buick Century, Chevrolet Celebrity, Oldsmobile Cutlass Ciera & Pontiac 6000 all models '82 thru '96
38010 Buick Regal '88 thru '04, Chevrolet Lumina '88 thru '04, Oldsmobile Cutlass Supreme '88 thru '97 & Pontiac Grand Prix '88 thru '07
38015 Buick Skyhawk, Cadillac Cimarron, Chevrolet Cavalier, Oldsmobile Firenza, Pontiac J-2000 & Sunbird '82 thru '94
38016 Chevrolet Cavalier & Pontiac Sunfire '95 thru '05
38017 Chevrolet Cobalt '05 thru '10, HHR '06 thru '11, Pontiac G5 '07 thru '09, Pursuit '05 thru '06 & Saturn ION '03 thru '07
38020 Buick Skylark, Chevrolet Citation, Oldsmobile Omega, Pontiac Phoenix '80 thru '85
38025 Buick Skylark '86 thru '98, Somerset '85 thru '87, Oldsmobile Achieva '92 thru '98, Calais '85 thru '91, & Pontiac Grand Am all models '85 thru '98
38026 Chevrolet Malibu '97 thru '03, Classic '04 thru '05, Oldsmobile Alero '99 thru '03, Cutlass '97 thru '00, & Pontiac Grand Am '99 thru '03
38027 Chevrolet Malibu '04 thru '12, Pontiac G6 '05 thru '10 & Saturn Aura '07 thru '10
38030 Cadillac Eldorado, Seville, Oldsmobile Toronado & Buick Riviera '71 thru '85
38031 Cadillac Eldorado, Seville, DeVille, Fleetwood, Oldsmobile Toronado & Buick Riviera '86 thru '93
38032 Cadillac DeVille '94 thru '05, Seville '92 thru '04 & Cadillac DTS '06 thru '10
38035 Chevrolet Lumina APV, Oldsmobile Silhouette & Pontiac Trans Sport all models '90 thru '96
38036 Chevrolet Venture '97 thru '05, Oldsmobile Silhouette '97 thru '04, Pontiac Trans Sport '97 thru '98 & Montana '99 thru '05
38040 Chevrolet Equinox '05 thru '17, GMC Terrain '10 thru '17 & Pontiac Torrent '06 thru '09

GEO
Metro - *see CHEVROLET Sprint (24075)*
Prizm - *'85 thru '92 see CHEVY (24060), '93 thru '02 see TOYOTA Corolla (92036)*
40030 Storm all models '90 thru '93
Tracker - *see SUZUKI Samurai (90010)*

(Continued on other side)

Haynes Automotive Manuals (continued)

NOTE: If you do not see a listing for your vehicle, please visit **haynes.com** for the latest product information and check out our **Online Manuals!**

GMC
Acadia - see GENERAL MOTORS (38001)
Pick-ups - see CHEVROLET (24027, 24068)
Vans - see CHEVROLET (24081)

HONDA
42010 **Accord CVCC** all models '76 thru '83
42011 **Accord** all models '84 thru '89
42012 **Accord** all models '90 thru '93
42013 **Accord** all models '94 thru '97
42014 **Accord** all models '98 thru '02
42015 **Accord** '03 thru '12 & **Crosstour** '10 thru '14
42016 **Accord** '13 thru '17
42020 **Civic 1200** all models '73 thru '79
42021 **Civic 1300 & 1500 CVCC** '80 thru '83
42022 **Civic 1500 CVCC** all models '75 thru '79
42023 **Civic** all models '84 thru '91
42024 **Civic & del Sol** '92 thru '95
42025 **Civic** '96 thru '00, **CR-V** '97 thru '01 & **Acura Integra** '94 thru '00
42026 **Civic** '01 thru '11 & **CR-V** '02 thru '11
42027 **Civic** '12 thru '15 & **CR-V** '12 thru '16
42030 **Fit** '07 thru '13
42035 **Odyssey** all models '99 thru '10
Passport - see ISUZU Rodeo (47017)
42037 **Honda Pilot** '03 thru '08, **Ridgeline** '06 thru '14 & **Acura MDX** '01 thru '07
42040 **Prelude CVCC** all models '79 thru '89

HYUNDAI
43010 **Elantra** all models '96 thru '19
43015 **Excel & Accent** all models '86 thru '13
43050 **Santa Fe** all models '01 thru '12
43055 **Sonata** all models '99 thru '14

INFINITI
G35 '03 thru '08 - see NISSAN 350Z (72011)

ISUZU
Hombre - see CHEVROLET S-10 (24071)
47017 **Rodeo** '91 thru '02, **Amigo** '89 thru '94 & '98 thru '02 & **Honda Passport** '95 thru '02
47020 **Trooper** '84 thru '91 & **Pick-up** '81 thru '93

JAGUAR
49010 **XJ6** all 6-cylinder models '68 thru '86
49011 **XJ6** all models '88 thru '94
49015 **XJ12 & XJS** all 12-cylinder models '72 thru '85

JEEP
50010 **Cherokee, Comanche & Wagoneer Limited** all models '84 thru '01
50011 **Cherokee** '14 thru '19
50020 **CJ** all models '49 thru '86
50025 **Grand Cherokee** all models '93 thru '04
50026 **Grand Cherokee** '05 thru '19 & **Dodge Durango** '11 thru '19
50029 **Grand Wagoneer & Pick-up** '72 thru '91 Grand Wagoneer '84 thru '91, Cherokee & Wagoneer '72 thru '83, Pick-up '72 thru '88
50030 **Wrangler** all models '87 thru '17
50035 **Liberty** '02 thru '12 & **Dodge Nitro** '07 thru '11
50050 **Patriot & Compass** '07 thru '17

KIA
54050 **Optima** '01 thru '10
54060 **Sedona** '02 thru '14
54070 **Sephia** '94 thru '01, **Spectra** '00 thru '09, **Sportage** '05 thru '20
54077 **Sorento** '03 thru '13

LEXUS
ES 300/330 - see TOYOTA Camry (92007, 92008)
ES 350 - see TOYOTA Camry (92009)
RX 300/330/350 - see TOYOTA Highlander (92095)

LINCOLN
MKX - see FORD (36014)
Navigator - see FORD Pick-up (36059)
59010 **Rear-Wheel Drive Continental** '70 thru '87, **Mark Series** '70 thru '92 & **Town Car** '81 thru '10

MAZDA
61010 **GLC** (rear-wheel drive) '77 thru '83
61011 **GLC** (front-wheel drive) '81 thru '85
61012 **Mazda3** '04 thru '11
61015 **323 & Protegé** '90 thru '03
61016 **MX-5 Miata** '90 thru '14
61020 **MPV** all models '89 thru '98
Navajo - see Ford Explorer (36024)
61030 **Pick-ups** '72 thru '93
Pick-ups '94 thru '09 - see Ford Ranger (36071)
61035 **RX-7** all models '79 thru '85
61036 **RX-7** all models '86 thru '91
61040 **626** (rear-wheel drive) all models '79 thru '82
61041 **626 & MX-6** (front-wheel drive) '83 thru '92
61042 **626** '93 thru '01 & **MX-6/Ford Probe** '93 thru '02
61043 **Mazda6** '03 thru '13

MERCEDES-BENZ
63012 **123 Series Diesel** '76 thru '85
63015 **190 Series** 4-cylinder gas models '84 thru '88
63020 **230/250/280** 6-cylinder SOHC models '68 thru '72
63025 **280 123 Series** gas models '77 thru '81
63030 **350 & 450** all models '71 thru '80
63040 **C-Class:** C230/C240/C280/C320/C350 '01 thru '07

MERCURY
64200 **Villager & Nissan Quest** '93 thru '01
All other titles, see FORD Listing.

MG
66010 **MGB** Roadster & GT Coupe '62 thru '80
66015 **MG Midget, Austin Healey Sprite** '58 thru '80

MINI
67020 **Mini** '02 thru '13

MITSUBISHI
68020 **Cordia, Tredia, Galant, Precis & Mirage** '83 thru '93
68030 **Eclipse, Eagle Talon & Plymouth Laser** '90 thru '94
68031 **Eclipse** '95 thru '05 & **Eagle Talon** '95 thru '98
68035 **Galant** '94 thru '12
68040 **Pick-up** '83 thru '96 & **Montero** '83 thru '93

NISSAN
72010 **300ZX** all models including Turbo '84 thru '89
72011 **350Z & Infiniti G35** all models '03 thru '08
72015 **Altima** all models '93 thru '06
72016 **Altima** '07 thru '12
72020 **Maxima** all models '85 thru '92
72021 **Maxima** all models '93 thru '08
72025 **Murano** '03 thru '14
72030 **Pick-ups** '80 thru '97 & **Pathfinder** '87 thru '95
72031 **Frontier** '98 thru '04, **Xterra** '00 thru '04, & **Pathfinder** '96 thru '04
72032 **Frontier & Xterra** '05 thru '14
72037 **Pathfinder** '05 thru '14
72040 **Pulsar** all models '83 thru '86
72042 **Roque** all models '08 thru '20
72050 **Sentra** all models '82 thru '94
72051 **Sentra & 200SX** all models '95 thru '06
72060 **Stanza** all models '82 thru '90
72070 **Titan pick-ups** '04 thru '10, **Armada** '05 thru '10 & **Pathfinder Armada** '04
72080 **Versa** all models '07 thru '19

OLDSMOBILE
73015 **Cutlass** V6 & V8 gas models '74 thru '88
For other OLDSMOBILE titles, see BUICK, CHEVROLET or GENERAL MOTORS listings.

PLYMOUTH
For PLYMOUTH titles, see DODGE listing.

PONTIAC
79008 **Fiero** all models '84 thru '88
79018 **Firebird** V8 models except Turbo '70 thru '81
79019 **Firebird** all models '82 thru '92
79025 **G6** all models '05 thru '09
79040 **Mid-size Rear-wheel Drive** '70 thru '87
Vibe '03 thru '10 - see TOYOTA Corolla (92037)
For other PONTIAC titles, see BUICK, CHEVROLET or GENERAL MOTORS listings.

PORSCHE
80020 **911** Coupe & Targa models '65 thru '89
80025 **914** all 4-cylinder models '69 thru '76
80030 **924** all models including Turbo '76 thru '82
80035 **944** all models including Turbo '83 thru '89

RENAULT
Alliance & Encore - see AMC (14025)

SAAB
84010 **900** all models including Turbo '79 thru '88

SATURN
87010 **Saturn** all S-series models '91 thru '02
Saturn Ion '03 thru '07- see GM (38017)
Saturn Outlook - see GM (38001)
87020 **Saturn L-series** all models '00 thru '04
87040 **Saturn VUE** '02 thru '09

SUBARU
89002 **1100, 1300, 1400 & 1600** '71 thru '79
89003 **1600 & 1800** 2WD & 4WD '80 thru '94
89080 **Impreza** '02 thru '11, **WRX** '02 thru '14, & **WRX STI** '04 thru '14
89100 **Legacy** all models '90 thru '99
89101 **Legacy & Forester** '00 thru '09
89102 **Legacy** '10 thru '16 & **Forester** '12 thru '16

SUZUKI
90010 **Samurai/Sidekick & Geo Tracker** '86 thru '01

TOYOTA
92005 **Camry** all models '83 thru '91
92006 **Camry** '92 thru '96 & **Avalon** '95 thru '96
92007 **Camry, Avalon, Solara, Lexus ES 300** '97 thru '01

92008 **Camry, Avalon, Lexus ES 300/330** '02 thru '06 & **Solara** '02 thru '08
92009 **Camry, Avalon & Lexus ES 350** '07 thru '17
92015 **Celica Rear-wheel Drive** '71 thru '85
92020 **Celica Front-wheel Drive** '86 thru '99
92025 **Celica Supra** all models '79 thru '92
92030 **Corolla** all models '75 thru '79
92032 **Corolla** all rear-wheel drive models '80 thru '87
92035 **Corolla** all front-wheel drive models '84 thru '92
92036 **Corolla & Geo/Chevrolet Prizm** '93 thru '02
92037 **Corolla** '03 thru '19, **Matrix** '03 thru '14, & **Pontiac Vibe** '03 thru '10
92040 **Corolla Tercel** all models '80 thru '82
92045 **Corona** all models '74 thru '82
92050 **Cressida** all models '78 thru '82
92055 **Land Cruiser** FJ40, 43, 45, 55 '68 thru '82
92056 **Land Cruiser** FJ60, 62, 80, FZJ80 '80 thru '96
92060 **Matrix** '03 thru '11 & **Pontiac Vibe** '03 thru '10
92065 **MR2** all models '85 thru '87
92070 **Pick-up** all models '69 thru '78
92075 **Pick-up** all models '79 thru '95
92076 **Tacoma** '95 thru '04, **4Runner** '96 thru '02 & **T100** '93 thru '08
92077 **Tacoma** all models '05 thru '18
92078 **Tundra** '00 thru '06 & **Sequoia** '01 thru '07
92079 **4Runner** all models '03 thru '09
92080 **Previa** all models '91 thru '95
92081 **Prius** all models '01 thru '12
92082 **RAV4** all models '96 thru '12
92085 **Tercel** all models '87 thru '94
92090 **Sienna** all models '98 thru '10
92095 **Highlander** '01 thru '19 & **Lexus RX330/330/350** '99 thru '19
92179 **Tundra** '07 thru '19 & **Sequoia** '08 thru '19

TRIUMPH
94007 **Spitfire** all models '62 thru '81
94010 **TR7** all models '75 thru '81

VW
96008 **Beetle & Karmann Ghia** '54 thru '79
96009 **New Beetle** '98 thru '10
96016 **Rabbit, Jetta, Scirocco & Pick-up** gas models '75 thru '92 & Convertible '80 thru '92
96017 **Golf, GTI & Jetta** '93 thru '98, **Cabrio** '95 thru '02
96018 **Golf, GTI, Jetta** '99 thru '05
96019 **Jetta, Rabbit, GLI, GTI & Golf** '05 thru '11
96020 **Rabbit, Jetta & Pick-up** diesel '77 thru '84
96021 **Jetta** '11 thru '18 & **Golf** '15 thru '19
96023 **Passat** '98 thru '05 & **Audi A4** '96 thru '01
96030 **Transporter 1600** all models '68 thru '79
96035 **Transporter 1700, 1800 & 2000** '72 thru '79
96040 **Type 3 1500 & 1600** all models '63 thru '73
96045 **Vanagon Air-Cooled** all models '80 thru '83

VOLVO
97010 **120, 130 Series & 1800 Sports** '61 thru '73
97015 **140 Series** all models '66 thru '74
97020 **240 Series** all models '76 thru '93
97040 **740 & 760 Series** all models '82 thru '88
97050 **850 Series** all models '93 thru '97

TECHBOOK MANUALS
10205 **Automotive Computer Codes**
10206 **OBD-II & Electronic Engine Management**
10210 **Automotive Emissions Control Manual**
10215 **Fuel Injection Manual** '78 thru '85
10225 **Holley Carburetor Manual**
10230 **Rochester Carburetor Manual**
10305 **Chevrolet Engine Overhaul Manual**
10320 **Ford Engine Overhaul Manual**
10330 **GM and Ford Diesel Engine Repair Manual**
10331 **Duramax Diesel Engines** '01 thru '19
10332 **Cummins Diesel Engine Performance Manual**
10333 **GM, Ford & Chrysler Engine Performance Manual**
10334 **GM Engine Performance Manual**
10340 **Small Engine Repair Manual**, 5 HP & Less
10341 **Small Engine Repair Manual**, 5.5 thru 20 HP
10345 **Suspension, Steering & Driveline Manual**
10355 **Ford Automatic Transmission Overhaul**
10360 **GM Automatic Transmission Overhaul**
10405 **Automotive Body Repair & Painting**
10410 **Automotive Brake Manual**
10411 **Automotive Anti-lock Brake (ABS) Systems**
10420 **Automotive Electrical Manual**
10425 **Automotive Heating & Air Conditioning**
10435 **Automotive Tools Manual**
10445 **Welding Manual**
10450 **ATV Basics**

Over a 100 Haynes motorcycle manuals also available

10/22